I0032884

The Case of the Sexual Cosmos

Everything You Know About Nature is Wrong

The Case of the Sexual Cosmos:
Everything You Know About Nature is Wrong

©2025 Howard Bloom. All Rights Reserved. No part of this publication may be reproduced, stored in a retrieval system or transmitted in any form by any means, electronic, mechanical, or photocopying, recording, or otherwise without the permission of the author.

For more information, please contact:

World Philosophy and Religion Press & Dandy Lion Publishing Group 4401 Friedrich Lane #302 Austin, TX 78744

Library of Congress Control Number: 2023923020
CPSIA Code: PRV0124A

ISBN-13: 979-8-9928719-4-4
Printed in the United States

The Case of the Sexual Cosmos

Everything You Know About Nature is Wrong

We Have Radically Misinterpreted Nature

And That Misinterpretation Is Killing Our Dreams
and Poisoning Our Possibilities

But a Radically New and Scientifically More
Accurate View of Nature

Can Give Us Powers Beyond Belief

HOWARD BLOOM

ENDORSEMENTS

"A massive achievement! A gigantic achievement. WOW!"
— *Richard Foreman, MacArthur Genius Award winner, officer of the French Academy of Arts and Letter, the reigning god of avant-garde theater*

"The Case of the Sexual Cosmos is a triumph. I particularly enjoyed the way Bloom's writing mirrored the content: full of surprises, unexpected connections, complexifying outcomes. I look forward to reading it again. And again!"
— *James Burke, creator of seven BBC-TV series including the classic Connections*

"Howard Bloom makes a forceful and interesting argument that our understanding of nature is wrong. He argues that we are not savaging the earth as some would have it, but instead are growing the cosmos. A fascinating read."
— *Ellen Langer, Professor of Psychology, Harvard University, author of Mindfulness*

"I believe The Case of the Sexual Cosmos is a masterwork. It captures wonderfully the elements of cosmic self-organization. At last a popular book that brings you a magisterial command of the disciplines, yet is optimistic about the human future."
— *Gregory Matloff, Fellow of the British Interplanetary Society, Advisor to Yuri Milner's Breakthrough Initiative Project Starshot, author of Biosphere Extensions*

"Bloom's writing is like getting tickled by a goddess. Invigorating and all consuming. A timely salve for all the doomers on X, an inspiration and a handbook. I look forward to savoring parts every day."
— *Nando Pelusi, co-founder of the Applied Evolutionary Psychology Society (AEPS), board of advisers of the National Association of Cognitive Behavioral Therapists*

"There is no one who writes about science in a more engaging and entertaining way than Howard Bloom. His new book, The Case of the Sexual Cosmos, is a wild ride with an optimistic message for humanity and a call to action."
— *Bobby Azarian, cognitive neuroscientist, author of The Romance of Reality, contributor to The Atlantic, The New York Times, and the Scientific American*

"As a tree-hugging dirt worshipper, I don't take kindly to whizbang. Yet by the time I reached Bloom's defense of space solar power, as an option for getting to net zero, I trusted him enough to consider it—to view it as part of a continuum of climate stabilization measures (starting with furs and fire) rather than a radical departure from what "nature" has ordained.

I applaud Bloom's efforts to pull us humans—and our propensity for fucking up stunningly—out of Greta Thunberg's (and others') joy-free self-hate machine. We can't prove that we're awful; nor can we prove that we're awesome. But how many more miracles might we open our eyes to if we chose—as Bloom does—to tell the latter story?

The true tragedy, according to The Case of the Sexual Cosmos, is not war, or climate catastrophe. It's rejecting the exuberant flamboyance of exactly who we are."
— *Helen Zuman, author of Mating in Captivity*

"Engaging, enjoyable, and compelling."
— *Conrad Labandeira, Smithsonian Institution*

"Howard Bloom's mega-dose of genius and high-protein prose is a mind-expanding brew that will rocket you into the darkly creative sexual heart of nature, from biology to cosmology.'
— *Nova Spivack, entrepreneur, next-technology venture capitalist*

TABLE OF CONTENTS

INTRODUCING HOWARD BLOOM:

SCIENTIFIC GENIUS, CULTURAL VISIONARY, STONE~COLD ATHEIST, OR MATERIALIST MYSTIC

BY DR. MARC GAFNI

I am delighted in this brief preface to share something of the unusual mystery of Howard Bloom—as a person, a scientist, a seminal philosopher of science, and a cultural visionary.

In 2016, Howard and I each received an ecstatically urgent call from our mutual friend, Barbara Marx Hubbard, the world's leading proponent of conscious evolution. With her trademark perspicacity and enthusiasm, she told us that we *must* meet each other and that she was convening a bi-weekly dialogue series to ensure that this happened.

Barbara, with her wondrously energized and sometimes imperious finality, declared that the dialogues would be called "Evolutionary Geniuses Co-Creating."

"But who will participate?" we each queried independently.

"Just the two of you," said Barbara, "and I will both join you and moderate."

Naturally, we each protested, claiming—with some level of relatively cheerful martyrdom—that we were among the busiest people on the planet, doing vital source-code-changing work, and *could not possibly* devote that kind of regular time to dialogues.

But saying *no* to Barbara was a losing proposition.

And, as was so often the case, the genius of Barbara's creative intuition was profound.

Today, we are intellectual interlocutors, creative partners, dear friends, and co-authors of two forthcoming volumes, both of which are in preparation.

Indeed, from that moment back in 2017, the three of us met twice a month for three years until Barbara passed at the age of 89. We have continued the same pace of conversations, and at some point in the not-too-distant future, a volume of those dialogues will appear.

During these past years, Howard has been thinking, creating, writing, and teaching nonstop—both at the Howard Bloom Institute and on 545 radio stations across the country. I have been doing the same at the think tank that I founded with my close friend Ken

Wilber, and we now co-lead with Dr. Zachary Stein. Howard has also been a senior scholar at our think tank, as I have been at the Howard Bloom Institute.

Ken remarked to me with delight when I first mentioned that I was talking to Howard, "Reading Howard is like reading postcards from the future." What he meant was that Howard, on so many crucial issues of science and culture, has been ahead of his time, opening the door for many to follow.

Several examples: First, evolutionary biologist David Sloan Wilson credits Howard's early formulations of group selection with being ahead of the academic pack, and with making a formerly disreputable idea respectable.

A second example: Yuri Ozhigov of Moscow University invited Howard to deliver a paper at an international conference on quantum physics and informatics in Moscow in 2006. The paper was entitled "Why Everything You Know About Quantum Physics Is Wrong." The conceptual structure that Howard introduced later was central to Ozhigov's book proposing a new direction for quantum physics.

The third example is in an article in the physics magazine *PhysicaPlus*, where Howard showed the same evolutionary patterns at work in everything from the Big Bang and the formation of stars and galaxies to bacteria.

Then, in a paper in the journal *Biosystems*, Howard demonstrated that the core pattern of stretching out to explore, then rushing together again to compare notes appears all the way up and down the chain of life, from bacteria and chimpanzees to human beings. In that same *Biosystems* article, Bloom showed democracy versus authoritarianism at work from microorganisms to societies of today.

And in his book *The God Problem: How a Godless Cosmos Creates,* Howard called these patterns that show up on one evolutionary level after another "Ur patterns." What's more he used these Ur patterns to explain why science's math and metaphors work.

Current science has yet to catch up with these insights.

This volume, *The Case of the Sexual Cosmos: Everything You Know About Nature Is Wrong,* describes key dimensions of what we call together the new Universe Story, which Howard and I understand in an overlapping manner, even as we have arrived at our conclusions through very different routes.

Indeed, when you read each of our works, you will notice that we deploy different languages—which at first blush might seem to contradict each other—as we articulate our respective Universe stories.

And this brings us to the crux of what I would like to share in this introduction—something about the essential paradox that is Howard. Bloom participates in a paradox that is part of a larger puzzle, a puzzle that we *must inhabit* to create what I would like to refer to as a *shared Story of Value.*

We live in a time between stories and a time between worlds. We need a shared story of value that overcomes the polarization of our era. *The Case of the Sexual Cosmos: Everything You Know About Nature Is Wrong* is a crucial part of that story.

As television news script writer Michael Shore says, *The Case of the Sexual Cosmos: Everything You Know About Nature Is Wrong* is:

> Challenging and reassuring, confounding and delightful, infuriating and enlightening, mysterious and insightful, contradictory and confirmatory... yeah, just like the sexy perverse beast of Life Itself that is its subject. I return to it often—though not often enough—to be challenged, enlightened, confounded, and delighted when I need my spirit refreshed.

And, as Scotland's Elaine Henderson adds:

> Bloom has consumed a vast library of information in arriving at his conclusions, and every single assertion can be followed up. This kind of research and resource is quite astonishing. Bloom's style—fizzing phrases, drum beat delivery, repetition of key ideas—relentlessly draws one forward.

Here is the Bloomian Paradox: Howard describes himself in two different ways, ways that might rightly appear to be opposites. And indeed they are—but they do not contradict. Rather, they exist in the wondrous dialectical tension of paradox.

Or, in Howard's words, "Opposites are joined at the hip."

Howard himself, incarnating the really real, is exactly that set of opposites joined at the hip—in ways that are significant to all of us.

They are significant to science, significant to the culture, and they are significant to you and me.

First, Howard often describes himself as "a stone-cold atheist."

Second, Howard just as often uses a phrase that one reader long ago pinned on him. He describes himself as "a materialist mystic."

Whether these two formally contradict or not, they certainly seem to point in quite different directions.

Howard will regularly—for example, in his book *The Lucifer Principle*—talk about reality from a decidedly postmodern perspective. In that perspective, rightness or value is ultimately a social construct, a fiction, or a figment of our imagination. As Howard puts it, "Reality is a shared hallucination."

And yet, in phrases like *The God Problem*, it's very clear to Howard that value is real. That is an axiom of his thought, which he formally denies in one breath and embodies, embraces, and assumes in another breath. In fact, one of his key mottoes is "the truth at any price, including the price of your life."

Howard and I come from radically different sources and have undergone radically different interior processes. But despite that, we've actually traveled much of the same path, both internally and in terms of the way we think, write, and teach about the world.

Much of our discussion over the years has been about the nature of the Universe, the nature of the human being in the Universe, and the great question of normativity, what we are called to do at this moment in time.

Howard talks about reality as a *Conversational Cosmos*. I described reality as a *Storied Cosmos*. What we are both saying is that story and its organic subset "conversation" are not merely reductive materialist social constructions, but rather Ontologies of Cosmos, which fabric Reality at every level of the evolutionary chain, from matter to live to the depth of the self-reflective human mind.

In conversing about the Storied Cosmos or the Conversational Cosmos, we are describing Cosmos as defined by what I call *intimate communion*, or what Howard calls *sociality*.

Howard calls this a communicative Cosmos, I call it an Intimate Universe or Amorous Cosmos. Howard also calls it a sexual Cosmos. But whatever our differences of intention in these overlapping terms, what we both agree is that we live in a Cosmos in which meaning, or what we might also call value or information, is constantly being exchanged from the very first nanoseconds after the Big Bang until this very moment.

Emergent from our distinctive vectors of research over many decades, coupled with our years of conversation, we have come to a definitive conclusion that not only is value real, but is also woven into the fabric of the Cosmos.

All of Reality is coded with meaning, with value, with information— but not in the sense that Claude Shannon deployed the term in his information theory. As Howard elegantly formulates it, "Shannon got the math right but the metaphor wrong." Shannon inappropriately uses the term information to refer only to bits and bytes.

Howard incisively invokes Shannon's co-author, Warren Weaver, at Bell Labs, who points out that information is meaning, the exchanges of meaning between communicants. Citing from a shared soon-to-be-published essay, "Meaning or value or information implies a unique quality that incarnates a dimension of cosmic quality or rightness—a value of Cosmos which is at once distinct and at the same time indistinct from the deeper Field of Value in which everything arises."

Value, as we suggest, "is a continuum which moves from mathematical value to musical value—music is after all sung mathematics—to molecular value to metabolic value to moral value and more. There is a core continuity of meaning in all these deployments of value even as each is individually distinguished and distinct."

Or, the way we say it in same unpublished article, first drafted in 2021, there is "meaning all the way down and all the way up the evolutionary chain."

But however we got here, we are together taking a panoramic view of Reality, in which science is core to the new revelation. But the sciences, to deserve their name, must be radically empirical. Hence, they must describe not only exterior mechanics but also what Nietzsche referred to as the music of Cosmos.

Or, as Howard says, "Science must attempt to understand the emotions we experience when we are caught up in what one founding father of sociology, Emil Durkheim, calls 'collective effervescence.' Science must come to understand the sort of ecstatic passions we feel when we are on our feet shouting at the top of our lungs in a crowd at a hockey game, a rock concert, or a political rally. Science must come to understand the experiences that some would call 'spiritual'."

Notice that I am deploying the word value, not values. We are not affirming a particular social system's vision of values. Rather, we are talking about a Field of Value or meaning that lives beneath and beyond all specific values or meanings.

Howard talks about this in terms of what he calls "Omnology" and "Ur patterns." I talk about this same phenomenon as a "Universal Grammar of Value, grounded in evolving First Principles and First Values, that come together to articulate a shared Story of Value as a context for our diversity."

Howard and I have spent the last decade trying to articulate the value structure of Cosmos. Again, based on different sets of sources and different internal processes—which ultimately overlap and converge—we both share the realization that Cosmos has inherent telos, that Cosmos is clearly going somewhere, and that Cosmos is not only moved by the past but called by the future. Reality, and we, as participants in the real, are called by memories of the future. And part of what Cosmos remembers is its inherent allurement to ever deeper contact and greater wholeness—or what Howard loves to call "Super-sized surprises."

We both share a scientific realization that what some ancient sages referred to as the *eternal Tao* is also the *evolving Tao*. We might easily embrace a slightly reformulated version of Blake, who talks about *eternity being in love with the productions of time.*

We both engage in intense readings of science. Howard's, however, are more learned and generally just better than mine. He knows a lot more science than I do. He is generous when he calls me a brilliant reader of science. But really, I am a simple reader who does not how to do the math that underlies so much of the sciences. But I have tried to train my readings from a first simplicity to a second simplicity that seeks to understand the phenomenology that underlies the often obfuscating jargon and to point toward the implications of a close reading of the sciences: a Cosmos coded with value.

I have, for the last ten years, worked with my students in classical readings of molecular chemistry and biology, demanding that we take them out of their mechanical metaphors and actually begin to hear their music, the primary currents of attraction and repulsion, allurement and autonomy, moving and animating them.

Howard had, shockingly, come to a similar read of science, but as already noted, being an exponentially more brilliant scientist than myself, he was able to validate my intuitions in multiple ways that were radically illuminating and would have been profoundly unavailable to me.

Gradually, the following became clear to us.

Howard has immersed himself, since his early boy-genius days, in the canons of the entire gamut of the classical sciences.

Since my early days, I have immersed myself in canons of what I refer to as the interior sciences, the original lineage sources of the great wisdom traditions in multiple vectors.

I have spent my life challenging the orthodoxies of these lineages while remaining immersed in their realizations. Together with my close colleagues Dr. Zachary Stein and Ken Wilber, I have codified them into a new Story of Value that we call CosmoErotic Humanism.

As I shared these sources with Howard, we began to realize that these sources were underneath their veneer, non-dogmatic, radically empirical, and paradoxically aligned and deeply fructifying to Howard's own classically scientific intuitions.

Howard is a pure man of science. He is a radical empiricist who is willing to challenge the dogmas of scientism to formulate a more accurate science.

Howard, at some point, did rebel against the dogmas of science. However, he remained, and remains, a scientist—the very core of his being—and yet, in some sense, a maverick who is pushing science to evolve to its next stage. Or, as he puts it, "I'm trying to kick down the door to the next paradigm." He is grounded in the science of the living Universe, but in a radically original way that capacitates coherence and confidence, even as it inspires the covenant between generations.

At the core of all of our conversations has been a radical kind of love—a love of truth, a love of joy, a love of fullness, and a horror in response to suffering, pain, and the agonies that so many in the world experience. The mutual fructification that has taken place between us is indescribable, becoming a source of enormous joy and ever-growing clarity for both of us.

It is my utter delight, pleasure, and great honor to present this book by my dear friend Howard Bloom, who is, by all accurate accounts, a world treasure.

Dr. Marc Gafni
Co-President, Center for World Philosophy and Religion

[Note: A more complete version of this prefatory essay, with a fuller exploration of Howard's intellectual work and the context of our shared story, appears in an essay on Substack, entitled "Introducing Howard Bloom," https://worldphilosophyreligion. substack.com/preface-by-marc-gafni-for-case-of-sexual-cosmos.]

THE CASE OF THE SEXUAL COSMOS: EVERYTHING YOU KNOW ABOUT NATURE IS WRONG

It's the most complicated thing in the cosmos. And the most expensive. It draws us with visions of paradise, but drags us through agonies of hell. And it has inflicted these aches and ecstasies on plants and animals for roughly two billion years.[1] It's sex.

And sex is the ultimate disproof of two scientific concepts that have thrown our views of nature violently off track—the Principle of Least Action and the Second Law of Thermodynamics. What's more, sex is the ultimate disproof of our current view of the planet and our relationship to it. And sex is a crucial cue to what nature really demands of us.

In this book, I will show you that four common sins are not the transgressions of humans, but are tricks of nature:

- materialism,
- consumerism,
- waste, and
- vain display.

In fact, I will show you how nature has reveled in these sins for billions of years. Why? Because they are tools with which nature creates. And sex is nature's most stunningly disruptive creation. So far.

Oh, and one more thing. I will give you a new way to reach net-zero. A new way to end our emission of greenhouse gases. A new way to achieve the Green New Deal.

* * *

Why is a new view of nature necessary? Because we are told that we humans are about to crash and burn our planet. And in many ways that may be true. But the accusation comes from a false understanding. An understanding that can be toxic to our civilization. And poisonous to our happiness.

Behind this venomous perception of nature are two of the most incorrect ideas ever to gain control over your mind and mine. They began in roughly 1230 AD when Robert Grosseteste, an English philosopher and Bishop of Lincoln, distilled this phrase from the works of Aristotle—"Nature operates in the shortest way possible."[2] Nature always takes the shortest path. That idea was inserted into modern science in 1744 when Pierre Louis de Maupertuis, the Director of the French Academy of Sciences and the first President of the Prussian Academy of Science, announced his Principle of Least Action in a memo to the French Academy.[3] His principle? "Nature always uses the simplest means to accomplish its effects."[4] "Nature is thrifty in all its actions."[5] An idea de Maupertuis conceived to prove that the universe was designed by God. An idea that encourages us to believe that we live in a chaste and constrained cosmos. A thrifty, penny-pinching cosmos. A cosmos of "the simplest means."

The fact is, we live in just the opposite. We live in a flamboyant cosmos, a daring cosmos, an innovative cosmos, a spendthrift cosmos, a cosmos with moxie, a cosmos that counts on us to reinvent her. A cosmos that counts on us to help her penetrate new realms of possibility.

A cosmos that prizes both finding the shortest path between two points and creating the longest.

In 1852, De Maupertuis' mistaken notion was augmented by two men—Germany's Hermann von Helmholtz and Scotland's William Thomson. You probably know Thomson better by the title he acquired forty years later, in 1892: Lord Kelvin.[6] Thomson and Helmholtz brainstormed via snail mail, traveled to each other's countries for personal visits,[7] and answered each other's articles with articles of their own.[8] And they came up with the idea that the universe will die in the random whizzle of what Helmholtz[9] called heat death.[10] What's heat death? Put a sugar cube in a glass of water, stir for fifteen minutes, and you will see heat death at work. Your water glass will eventually look like just plain old water. With nothing in it. Your sugar cube will have disappeared, dissolved into an invisible, random whizzle. That, said the pair, is what will happen to the universe.

But the universe begs to disagree.

Here's this book's view of nature in a nutshell. Nature is not what we think she is. Her most astonishing creation on this planet is life.

And life does not live in harmony with nature. Far from it. Life does not take nature lying down. Life is obstreperous. Life is uppity. Life is impertinent. Life is not a mere survivor. Life is a doomrider and a catastrophe tamer. Life takes nature apart and puts her back together in whole new ways.

Which makes life nature's servant in something she prizes above all else: the invention game.

***

Life is an invasive species. And nature has a special place in her heart for invasive species. Look at the finches and the iguanas of the Galapagos Islands, the islands visited by Charles Darwin in 1835 during his five-year Voyage of the Beagle. Today we regard the finches and the iguanas of the Galapagos as the symbols of nature at her most pristine. But the iguanas were an invasive species who first reached the Galapagos Islands roughly 10.5 million years ago.[11] And the finches[12] were an invasive species that first reached the Galapagos 2.3 million years ago.[13] From as far as 1,705 miles away.[14]

These species changed the Galapagos' ecology.[15] And they in turn were challenged by ecological change as the earth bounced from deep freeze to thaw and back[16] and as new invasive species of plants and insects arrived.[17]

Invasive species are nature incarnate. They don't take no for an answer. Invasive species spread life. Invasive species remake nature. Yes, invasive species reinvent the nature from which they sprang. And that is nature's mandate. Reinvent me. Upgrade me.

Give me new powers. Haul me from the present and the past into the landscape of the impossible. Pole jump me into the future.

As you and I will soon see, nature is a search engine exploring her potential, probing her future possibilities. Nature is an invention engine. Using finches, iguanas, and you and me as feelers into the dark realm of the extraordinary.

What does this mean to you and me? Be bodacious. Explore the borders of the everyday and go beyond them. When you do, you push the envelope of nature's possibilities. You expand nature's sway.

Understanding this is vital to understanding what we've achieved... and to seeing our future opportunities.

***

The notion that we have raped Mother Nature is false. So is the idea that we have reached the limits of our resources and that we are banging our heads against the prison bars known as the limits of growth. Nature is the opposite of what we imagine her to be. Nature demands that we conserve her past. But nature also demands that we expand her future. Nature demands that we enlarge her bounds. Nature demands that we amplify her exultations and her ingenuities.

Getting nature wrong has consequences. Big ones. The errors in our vision of nature are blinding us. They are crippling us with a claustrophobic picture of our world. Far worse, the errors in our vision of nature are killing our dreams and poisoning our possibilities. They have been convincing young girls like the fifteen-year-old Greta Thunberg in 2018 that we live in the age of nightmare and apocalypse. Thunberg's warnings were useful. But, in fact, the opposite is also true. We live on the cusp of the greatest triumphs nature has ever seen.

Why the disconnect? Because the real cosmos is far, far different than you and I have been led to believe. This is not a pinch-penny, miserly cosmos conserving energy and teetering on the brink of disaster. It is a flamboyant, profligate, creative cosmos marching toward higher degrees of intricacy.

And nature does not want to sleep forever in some sort of perfect state. Nature abhors what science calls an equilibrium. Nature is the mother of invention—from the creation of stars and stones to the evolution of DNA.

Nature rebels against her own rules, shatters her own status quo, and goes for breakthroughs. Which means that nature is not violated by technology. Technology is nature's invention.

Change your scientific lens and a very different nature reveals herself. Lay out the 13.8 billion year[18] history of the cosmos on a timeline. Insert the data from every science—theoretical physics, cosmology, chemistry, geology, paleontology, evolutionary biology, environmental science, and even anthropology, sociology, political science, and economics. Toss in the evidence of human history and of current events right down to Joe Biden versus Donald Trump. The result is a timeline that tells a story—the tale of the evolution of the universe from the Big Bang to what's going on in your brain as you read this page.

The science in that chronology reveals a very different big picture from the one being preached to you and me. Nature's reality is challenging, brutal, appalling, bracing, and exhilarating. And we must wake up to nature's reality. We must add to nature's creativity. And we must rebel against nature's brutality.

Nature is the ultimate materialist, consumerist generator of waste. Nature is innovative, restless, rebellious, outrageous, cruel, and extravagant. In fact, nature uses trash and litter to innovate. She uses garbage to carve out startling new possibilities.

A truer view of nature can help us increase the justice and compassion in this cosmos and dial down the violence with which nature is obsessed.

Oh, and one more thing. A big one. Nature adores those who oppose her most. Yes, nature loves those who break her laws. And if you learn nature's secrets, you can be one of those law-breakers, a law-breaker who gives nature vast new realms in which to create. That is the vision of nature at the heart of The Case of the Sexual Cosmos.

CHAPTER 2

NATURE GOES FOR BROKE: THE MOST COMPLEX TRICK IN HISTORY, SEX

Pierre Louis de Maupertuis declared in 1744 that "nature is thrifty in all its actions."[19] Nature always takes the shortest path between two points.[20] And de Maupertuis' "law of least effort"[21] has been at modern science's core ever since.[22]

But the experts disagree. Says William Shakespeare, in Sonnet 129,

> The expense of spirit in a waste of shame
> Is lust in action

Adds a quote attributed to Lord Chesterfield, "the pleasure is momentary, the position ridiculous, and the expense damnable."[23]

Then there's late 19th century[24] poet A.E. Housman:

> When I was one-and-twenty
>   I heard a wise man say,
> "Give crowns and pounds and guineas
>   But not your heart away;

Give pearls away and rubies
    But keep your fancy free."
But I was one-and-twenty,
    No use to talk to me.

When I was one-and-twenty
    I heard him say again,
"The heart out of the bosom
    Was never given in vain;
'Tis paid with sighs a plenty
    And sold for endless rue"
And I am two-and-twenty,
    And oh, 'tis true, 'tis true.

What are Shakespeare, Chesterfield, and Housman moaning about? Sex. In reality, sex can be a lot more joyful than Shakespeare, Chesterfield, and Housman make it sound. But the three of them put their fingers on a key aspect of sexuality, an aspect we often disregard: its cost, its astonishing "expense." Sex is not thrifty. It is not the shortest path between two points.

Here's how Swiss/British essayist, TV personality, and Fellow of the Royal Society of Literature Alain de Botton sums up your sex life and mine. "We are universally deviant," he says. We all want very odd things to bring us to sexual climax. We want our partner to call us baby or daddy. We want to make love like dogs or cats. We want to pretend to be a patient and his nurse. Or we want our partner to wear her highest heels to bed. Sometimes we are more interested in the heels than in our partner.

Meanwhile, we are obsessed with pushing "a hand inside an unfamiliar skirt or pair of trousers," to quote de Botton. But the opportunities to do that pushing are few and far between. Explains de Botton, "the majority of people we encounter will be not merely uninterested in having sex with us but positively revolted by the idea." And what does all this expense of your energy and mine, this daily preoccupation that seldom leaves our minds, sex, lead to? Says de Botton, "To feel only intermittent affection for a spouse, to have mediocre sex six times a year, [and] to keep a marriage going for the wellbeing of the children."[25] In other words, sex is so expensive that you pay for it with your life.

If this is a thrifty cosmos, a cosmos that pinches her pennies and nurses her resources, a cosmos that always takes the shortest path, how in the world did the "expense" of sex come to be? To find the answer, let's look at a bit of history.

Over 3.5 billion years ago, life shattered the natural order. Yes, life destroyed this planet's harmony and status quo. Life imprisoned water in membranes and in that captured water created macro-molecules. Massive, unaccountably complex macromolecules. Macromolecules of roughly 7,709,842[26] atoms each. A wee bit less than 8 million atoms. These macromolecules were not just big and intricate. They were capable of doing something radically new in this cosmos. Something acquisitive and materialistic. Something that showed no respect for the environment. They were able to grab dead atoms and enslave them, turning them into links in the chains of yet more new macromolecules. Turning dead atoms into raw ingredients, ingredients for precise copies of the macromol-ecules themselves. How absurdly selfish. How ridiculously vain.

How utterly narcissistic. Making copies of themselves. Manically mass producing those copies. Yes, life sinned against the existing order over 3.5 billion years ago when it invented cells, RNA, and DNA.[27]

Then life bashed the status quo again when it saturated the ocean and turned the waters from a virginal and pure, life-free chemical soup into the muddy, polluted puddle of a biosphere.

3.22 billion years ago,[28] life became outrageous once again. It defied gravity and lifted itself to the shelves of rock above the sea. This new stuff, life, was not thrifty. It did not live in harmony with its environment. It was uppity and pertinacious. It did not worship the status quo. It did the very opposite: it showed no respect. It remanufactured nature.

And life committed another absolute outrage 2.4 billion years ago.[29] It dared to reengineer the very atmosphere, poisoning the air with oxygen. But more about the killing spree of The Great Oxygen Catastrophe later.

Remaking the nature of the sea and probing the cracks of the land were just the beginning. 1.63 billion years ago,[30] congregations of cells invented a whole new way to work together in teams. For two billion years, cells of bacteria had worked by the billions[31] within colonies. Communicating like kids on iPhones. Using a chemical language. Gossiping and integrating. Turning their communities into massive collective exploration and learning machines. But inside those colonies and on their margins, cells had been free to swim around on their own. Well, not exactly on their own. It turns out that even on the outskirts of their colonies, restless, rebellious cells preferred to join groups of ten thousand travelers, migrating armies.[32] Armies of scouts, discoverers, and pioneers. Armies swimming into the wilderness to probe their environment and to hunt for new riches and new homes. But, still, individual bacterial cells possessed a relative independence.

Then 1.63 billion years ago,[33] a mother cell gave birth to a whole new kind of multitude. She commanded her daughter cells to stick together. And she ordered them to specialize. The new, tightly bound, totalitarian congregations of cells that resulted were the first multicellular beings, the first plants and animals. And as far back as seven hundred million years ago,[34] some of these cheeky new totalitarian sheets and towers of cells, these new multicellulars, followed the path of the bacteria who had gone beyond the seas 3.22 billion years ago. The new multicellulars, like the bacteria who had first probed the land, defied gravity. They sinned against nature's old water-bound ways and, like astronauts setting up colonies on another planet, they left the comfort of the ocean to join the bacterial pioneers in moving on up and raping the rock of the land. This was bravery, heroism, and audacity. It was not the shortest path.

When these first multicellulars—plants—arrived on the rock face, they fed off the heaps of garbage and toxic waste that bacterial land-pioneers had left behind. Garbage whose miracles we will dig into in a bit. But the multicellulars did more. These outrageous plants despoiled the purity of the landscape. They dug their roots into the microscopic cracks that temperature changes had opened in the stone, and they split the virgin rock.[35] They desecrated nature. Or did this desecration create nature?

Was this rape? Was this shattering of nature's status quo a sin? Two hundred million years later, evolution produced the earthworm.[36] If Charles Darwin is right, every fruitful field now covered with soil is the product of a massive landscaping effort left to us by millions of generations of earthworms who sinned against nature by doing plastic surgery on her face, then sinned again by littering a virginal wasteland of stone with their droppings, their sewage, their shit. Were the earthworms rapists?

In 1881, Darwin dedicated an entire book to the earthworms' depredations: The Formation of Vegetable Mould through the Action of Worms. Said Darwin,[37] the earthworms turned jagged outcrops and crevasses into the vast, flat meadows we cherish today. Darwin added that, "When we behold a wide, turf-covered expanse, we should remember that its smoothness, on which so much of its beauty depends, is mainly due to all the inequalities having been slowly levelled by worms."[38]

"It is a marvellous reflection," Darwin concluded, "that the whole of the superficial mould over any such expanse has passed, and will again pass, every few years through the bodies of worms." Again, were these earthworms desecrators?

Kevin Laland and John Odling-Smee sum up the modern view:

> Across the globe, earthworms have dramatically changed the structure and chemistry of soil by burrowing, dragging plant material into the soil, mixing it up with inorganic material such as sand, and mulching the lot by ingesting and excreting it as worm casts. The scale of these earthworks is vast.[39]

In other words, the worms left the trail of their toxic waste, their feces, wherever they went. They despoiled nature. But we use the worms' violation of Mother Nature... the worms' sewage... to grow our harvests of food and our backyard flowers.[40] What's more, we worship the worms' pileup of throwaways, their garbage. Their vast litter, their materialist, consumerist legacy includes soil, rainforests, and greenery. Did earthworms "rape" the landscape? Or did they make it?

Earthworms did not honor nature. They changed her. They coated her with their shit. Once upon a time, that shit was radically unnatural. But we worship that shit today as, guess what? Nature. Earth.

Nature favors those who oppose her most. Nature insists on audacity. Which is where sex comes in.

CHAPTER 3

FEASTING ON CATASTROPHE: THE TALE OF THE FIRST TEASPOON OF LIFE

The earth beneath your feet and mine is the opposite of what you and I have been told. This planet is not a warm and nurturing place.

Over 3.5 billion years ago in the cold black of space, there was a poison pill of stone, a hostile planet, a planet scalded by liquids, gassed and poisoned by its atmosphere, lashed by floods of photons, and bashed by cosmic rays. That planet was the mother of all climate change. The mother of all climate catastrophe. In those murderous, bad old days, there was less than half a teaspoon of life. Half a teaspoon threatened with extinction by a toxic planet. How did that half a teaspoon survive?

By exulting in materialism, consumerism, and, yes, waste. By seething with ambition, competition, collaboration, sociality, in-novation, and creativity. By itching to turn things on their head. By transforming poisons into pleasures, disasters into delights, and wastelands into seas of waving grain.

Life survived by opposing nature. Life survived by violating nature's ways. Oxygen and phosphorus, for example, were deadly. They were poisons. So life invented ways to use them as pistons in the machinery of life. Life tapped the chemical tricks of phosphorus to produce and distribute energy, turning the burning, white substance from a toxin to a daily necessity. Life inserted these murderously flammable white phosphorus atoms into a molecular merry-go-round called the Krebs Cycle and turned these toxic phosphorous atoms into energy storage and energy transport devices. Yes, life turned biohazards into batteries. Into spokes and gears of her machinery. These were sins against nature's status quo. They were unnatural acts. But they were not alone.

Nature confronted life with vents at the bottom of the sea so scalding that they sometimes hit temperatures nearly 500 degrees above the boiling point. The ocean surface was no more welcoming. It was whipped by turbulence and bombarded by radiation. All of these were mortal threats. But life turned those dangers to its advantage, too. Life turned the sea vents into energy sources and a cozy home. At the ocean surface, life harvested the incoming bullets of radiation—photons—and used them to break down CO_2 to oxygen and carbon, thus providing energy and the building blocks of photosynthesis, the building blocks from which life could make more life. Life thrived by upsetting the existing order. Life thrived by turning the "nature" of nature upside down.

What's more, life thrived by eating its environment. Life thrived by remanufacturing its surroundings. Life thrived by turning the toxic spaces of an alien planet green.

Was this a blissful harmony with the status quo? Far from it. Our foremothers, the first lifeforms, bacteria, desecrated, and "raped" their environment. They reengineered it. Bacteria three billion years ago changed the chemical composition of the very atmo-sphere.[41] They rejiggered the planet's weather.[42] They altered the chemistry of the sea. They despoiled their virginal surroundings. And in the process, they transformed a raging heavenly body into a nurturing mother—an earth. They made the nature we worship today. Again, they did not do it through a false worship of a past that never existed. They did not do it by holding tight to the way things used to be. They did it by raging against the way things were. They did it by grabbing the future. They did it by making change.

But that was just the beginning. In the next 3.5 billion years, Mother Nature tortured the planet with 142 mass extinctions.[43] How did life survive? By explosive growth, by explosive multiplication. By explosive real estate development. And by explosive invention. Invention of new technologies, new strategies, new techniques, new engineering schemes, new ways to turn poisons into pistons, new ways to turn dangerous spaces into niches and parks, new ways to remake, rejigger, rearrange, evolve, transform, revolutionize, and reinvent the natural order. New ways to upend, upset, and transmogrify Mother Nature's ways.

And here's the clincher. At the heart of life's consumerism was waste—garbage, and excess. At the heart of life's consumerism was the throwaway. But who really invented the habit of building material things at extravagant cost, then carelessly tossing them away? Who really invented consumerism? Nature. She did it when she invented death. She did it less than a billion years[44] after the Big Bang, when she tossed her most complex and expensive contrivances—stars—away. Yes, nature killed stars off. And she did it wastefully. Flamboyantly. She did it by exploding stars in the gaudy and unbelievably huge blasts of supernovas.[45] She did it by using her most intricate creations as luminescent garbage, as fireworks displays.

Ten billion years later, nature did it again when she decreed that every organism ever born shall die. And nature went farther. She invented cruelty. She invented torture. She invented pain. But from mounds of trash and from indulgence in consumerism, materialism, waste, and vain display, nature made amazements. How and why? And what do her strategies mean for you and me today? What demands are hidden in the natural laws that eco-catastrophists fail to see?

The Case of the Sexual Cosmos will show you where our concepts of materialism, consumerism, waste, and vain display go wrong. *The Case of the Sexual Cosmos* will show you why theories of limits to growth, notions of shrinking resources, and the anti-growth movement[46] make big mistakes. It will show you how these ecological concepts have produced massive benefits, but have gone overboard and are now dumbing us down. It will show you how a sneaky collaboration between plants, animals, humanity, and technology has dramatically upped the GAS, the Gross Amount of Sentience, the Gross Amount of living Spirit, on this planet. And *The Case of the*

Sexual Cosmos will show you how, without knowing it, we humans have carried out the mandate of the first teaspoon of life over 3.5 billion years ago. *The Case of the Sexual Cosmos* will show you how we have carried out nature's most basic command in the most surprising way: by turning nature upside down.

What's more, *The Case of the Sexual Cosmos* will show you how we've been misled about consumerism and materialism. We've been told that our civilization, Western civilization, is plundering the planet and producing a population explosion that will usher in apocalypse. We've been told that the agricultural revolution of twelve thousand years ago,[47] the Industrial Revolution of 250 years ago, the rise of the West and the West's offspring, capitalism, consumerism and materialism, are forms of rape.

Yet the evidence indicates that consumerism, materialism, waste, vain display, and the Western System are not enemies of nature and of humankind. They are the very opposite. They have produced new empowerments:

- Consumerism, materialism, waste, vain display, and the Western System have more than doubled the human lifespan in 200 years.[48]
- Consumerism, materialism, waste, vain display, and the Western System have upped the peace among humans since 1650 by a factor of ten. Yes, thanks to the Western system your odds of dying a violent death at the hands of a fellow human being are nearly one-tenth what they would have been 350 years ago. And one tenth what they would have been in one of those lovely indigenous cultures that we are

told live in peace with their neighbors and in harmony with the earth.[49]

- Consumerism, materialism, waste, vain display, and the Western System have allowed a woman earning minimum wage in a shoe factory in Massachusetts in 2023 to earn what nearly seventeen women working in a shoe factory earned in 1850.[50]

- Consumerism, materialism, waste, vain display, and the Western System have boosted the IQ of the average child to a level that would have measured close to genius on an IQ test from 1916.[51] Which means that today's average dummy is 31 IQ points[52] smarter than the average teenager a little over 100 years ago.

- And materialism, consumerism, waste, and vain display have increased the spread of trees on this planet in the 35 years from 1983 to 2018 by an area over twice the size of France. That's an increase of 864,868 square miles[53] of greenery. In other words, materialism, consumerism, waste and vain display have increased the dominion of what we call nature. Yes, we humans have increased nature's sway.

How the hell have we pulled this off?

***

But that's just the beginning.

- Consumerism, materialism, waste, vain display, and the Western System have increased the amount of conscious life on this planet by a factor of 8,000[54] since the early agricultural revolution 12,000 years ago.[55] More on that to come.

- Consumerism, materialism, waste, vain display, and the Western System have expanded your senses and your transport powers in ways that would have taken biological evolution half a billion years to achieve.

- Consumerism, materialism, waste, vain display, and the Western System have created a privilege that scarcely existed before 1750—personal freedom—a wide range of daily personal choices.

- Consumerism, materialism, waste, vain display, and the Western System have expanded something that's become a cliché—freedom to access knowledge. Not to mention freedom of speech.

- Consumerism, materialism, waste, vain display, and the Western System have increased the amount of information going on tap per year from two zettabytes in 2010 to nearly 120 zettabytes in 2023. That's 60 times more information at your fingertips in just 13 years.[56]

- Consumerism, materialism, waste, vain display, and the Western System have allowed a mere middle class person to buy medicines and machines that the richest English tech-lover of 1850, Prince Albert, husband of Queen Victoria, could not afford...or even conjure in his wildest dreams. Among other things, Albert died at the age of 42. Of a bout of typhoid fever[57] that could probably have been cured by an antibiotic.[58] Yes, Prince Albert died at 42.[59] You are likely to live to 80.

- Thanks to consumerism, materialism, waste, vain display, and the Western System you are likely to own a car, something that for all his riches and power, tech-loving Prince Albert could never have achieved.

- And, to repeat, consumerism, materialism, waste, vain display, and the Western System have vastly expanded the activity of something else that's vital but hard to quantify—the human spirit. Yes, consumerism, materialism, waste, vain display, and the Western System have upped the amount of what legendary paleontologist, evolutionary thinker, and Catholic priest Teilhard de Chardin called the "noosphere." The realm of thought and feeling with which the planet has become aware of herself. The realm of thought and feeling with which the cosmos has come to regard herself in a mirror. Yes, the realm of thought and feeling with which the very universe has come to understand herself through you and me.

One more thing. Consumerism, waste, vain display, and the Western system have given us new concepts: ecology, the goal of saving endangered species, and the concept of saving the planet. Not to mention anti-slavery movements, anti-imperialist movements, anti-colonialist movements, anti-racist movements, peace movements, social justice movements, and the notion of human rights. No other civilization, not the Chinese, the Muslim, the Marxist, the Olmec, or the Inca, has given its citizens these concepts.

Did Western Civilization accomplish this by plundering nature? Far from it. We did it by adding to nature's powers. We did it by increasing nature's sway. How?

How in the world did a culture decried for its materialism, consumerism, waste, and vain display pull off a string of material miracles? How did that culture's materialism, consumerism, and waste actually add to the biosphere? And how did surplus and vain display play a vital role?

For the answer, read on.

***

But first, there's a moral to this story. A powerful one.

If our great, great grandparents could give us an extra 40 years of life, surely we owe an extra 40 to our great, great grandkids. If our great, great grandparents could multiply the incomes of the poorest paid workers among us seventeen times, surely we owe another seventeen to our great, great grandkids. If our great, great grandparents could up the average IQ by 31 points, surely we owe another 31 IQ points to our great, great grandkids. And if our great, great grandparents could increase the peace in the world by a factor of ten, surely we owe our great, great grandkids ten times more. Yes, ten times more peace.

But to carry out this obligation to our great, great descendants, we need to see and value what we've achieved. And we need to defend the system that has given us these gifts. We need to realize that the Western System is not the worst system in the history of humankind, it is the best. And we need to defend its values—human rights, freedom of speech, tolerance, pluralism, and democracy. Topping all that, we need to defend the Western System's greatest hidden secret—a perpetual balancing act between three elements—government, private enterprise, and the protest industry.

But that's another book. My book, *The Genius of the Beast: A Radical Re-Vision of Capitalism.*[60] So let's get back to the mother of all surprise, nature.

CHAPTER 4

YOU ARE A STONE`S WAY OF IMAGINING NEW STONES[61]

The path that led to *The Case of the Sexual Cosmos* began in 2006 when I was asked to speak on sustainability at Yale. Creating technological and agricultural processes that do not damage the biosphere is a great goal. But there is a side to sustainability that's like the frigid touch of liquid nitrogen to the soul. It's the insistence on sacrifice, self-denial, and, most of all, shame. The insistence that we stop our exuberant creativity, despise our technology, and crawl back to a new stone age. So my Yale speech was not the praise for sustainability that the occasion called for. When I phoned the conference organizer to tell him what I wanted to speak about, he nearly fell off his chair laughing: "Fuck Sustainability." We toned the title down to "Screw Sustainability."

Then I was asked to repeat the key points of the Yale speech at a conference of rocket scientists and space advocates—the 2006 Space Frontier Foundation conference in Las Vegas.

The creator of The Foundation for the Future was in that audience and organized a symposium in Seattle based on the same key points, then had his organization fly me out to pontificate. And I was asked to repeat those points again in a keynote lecture at an international conference on governance sponsored by the United Nations Department of Economic and Social Affairs in Seoul, Korea—the World Civic Forum 2009.

The path to *The Case of the Sexual Cosmos* was pushed a step farther in 2010 when three of Pakistan's leading industrialists gathered on a conference call to meet me. One owned a four-billion-dollar business, another owned a $750 million company, and the third had headed Pakistan's software association during the years in which Pakistan's software industry had doubled. Before the conversation could begin, the three wanted to apologize for something. They wanted to apologize for Dubai. Why? Because of Dubai's materialism, consumerism, waste, and gaudy display.

That apology made something obvious. There is a huge subculture in the 57-nation[62] Muslim world that's modernist, pluralist, and passionate about free speech. But out of fear of having its throat slit by militants, it dares not raise its voice. One modernist, secular Pakistani governor was shot 28 times[63] by his chief security guard for an act of tolerance, visiting a Christian woman imprisoned for blasphemy.[64] In a country where blasphemy brings a death sentence. Within an hour of the killing, the security guard was elevated to a culture hero.

Dubai speaks up for the Muslim world's tolerant, secular subculture. Outrageously. Courageously. Not with words, but with architecture. There's the 163-story high Burj Khalifa, which "has been the world's tallest building since 2009."[65] And there's the fact that in 2023, Dubai had more quarter-of-a-mile high skyscrapers than any other city on earth.[66] Which means that there is something more to materialism, consumerism, waste, and flamboyant display than at first there seems. Something related to the way that the human spirit and its aspirations reshape reality.

What do the craze for sustainability and the shame of the Pakistanis have in common? Both are tied to one of the most basic questions of our time. Have we raped the planet? Are we running out of resources? And the answer is surprising. No, we are not running out of resources. We are running out of imagination.

For every ounce of living biomass on this planet there are three and a half billion ounces of dead stuff,[67] raw material waiting to be kidnapped, seduced, and recruited into the grand enterprise of life.

Yes, we are using less than one three-and-a-half-billionth of the resources of this planet. There are 1.097 sextillion cubic meters[68] of rock, magma, and iron beneath our feet. That's over 137 billion cubic meters per person. Over 137 billion! Just waiting to be kidnapped, seduced, and recruited into the megaproject of life.

Wait a minute. Is the hulking quantity of raw stone and magma beneath our feet really raw material for living things? Isn't that just more of the rape-and-pillage philosophy with which we have perverted nature and polluted the planet? More of the evil, extractive way of thinking? You be the judge.

At this very minute twelve miles beneath your feet and mine, bacteria are turning granite and basalt into food and fuel. Bacteria—chemolithoautotrophs[69]—are turning raw rock into resources.[70] They are turning barren stone into the stuff of life.[71] In fact, all of earth's underground that we've explored contains bacteria, no matter how deep we've gone.[72] According to University of Tennessee microbiologist Karen Lloyd, there are "10,000 times more" bacteria "in Earth's crust than there are number of stars in the universe."[73] All turning raw stone into the stuff of life.[74] Are bacteria profit-maddened capitalists? Are they rapists and plunderers? Or are they nature?

If it's true that we are smarter than bacteria, then why don't we see the potential of the rock on which we stand? Why are bacteria doing more creative R&D? Why are bacteria doing more to expand life's sway, the realm of cells and DNA? Why are bacteria doing more to green and garden the place? It's simple. A false concept of nature is leading us astray.

One more thing. We have invented environmentalism and are using it to save other species, something no species before us and no civilization before us has ever achieved.

Yes, environmentalism is a product of the civilization of the West. Environmentalism is a gift for which we give Western civilization no credit. That is an enormous mistake.

So what's going on here? Not only is our view of nature way off base, but so is our perception of the civilization in which we live. And our perception of our civilization's potential.

Get our concepts of nature and of our civilization right and you might just save the consciousness of nature that nature has used us to achieve. And you just may save our ability to add new powers to nature's range of capabilities.

You just may save the life of nature herself.

CHAPTER 5

WHAT`S AN OMNISCOPE? THE TALE OF THE TIMELINE

In the spring of 2021, the British magazine *The New Scientist* announced that Michio Kaku, a charismatic physicist from the City University of New York who had appeared on TV steadily over the previous 40 years, would be giving a virtual lecture. Kaku was going to explain grand unified theories of everything.

First, he was going to explain how Isaac Newton had unified gravity with astronomy and had come up with the laws of motion that explain the solar system. Then he was going to show how two hundred years later, James Clerk Maxwell had unified electricity, magnetism, and light with four equations that took Newton one step farther and described what electromagnetism was.[75] Finally, Kaku was going to explain how physicists today are trying to unify electromagnetism, the weak force, the strong force, and gravity[76] to come up with an even more inclusive understanding. Says Kaku, physicists want an equation that you can put on a T-shirt. That to them will be "the Grand Unified Theory of Everything." The GUT.

In other words, physicists are looking for what Kaku's 2021 book title calls "The God Equation."[77]

But this Grand Unified Theory of Everything isn't really going to be what it pretends to be. There are vast levels of reality that this grand equation, this God Equation, will never help us understand. The equation won't give us insight into how life pulled itself together in a poisonous sea on a poisonous planet over 3.5 billion years ago.[78] It won't explain how an egg hatches into a chick. It won't explain how we humans come together in clans, bands, tribes, nations, and civilizations. It won't help us understand the ecstasies we can experience when we make love with each other. Or even the ecstasy at a rock concert. Not to mention how to stop war.

The mass raptures that the equation will not capture utterly defy belief. When crowds have ecstasies, those ecstasies can be forces of history.

Look at Adolf Hitler... he was an artist of group ecstasies. In the 1930s, Hitler gave Germans the feeling of being a part of something far, far bigger than themselves. When he gathered the German people in crowds, he was able to evoke collective raptures. And those ecstatic experiences fueled the Germans to start a war of conquest—World War II.

Ecstasies, things of the spirit, can power new realities. New realities in the material world.

Which means that we don't just need a science that uses the atom smasher at CERN to look at things so small that even a microscope can't see them. We don't just need a grand unified equation that physicists can claim sums up everything. That God Equation, when it arrives, will be important. But it will sum up almost nothing!

***

Someone has to put together the opposite of the atom smasher. Someone has to assemble the opposite of the instrument that lets us break things down to their tiniest bits in order to understand everything. The simple tool that grabbed hold of me early in my life and has never let me go reaches for the biggest picture possible, not the smallest. That tool has been the timeline of the cosmos. The history of the whole thing, of everything that ever was, so that you can see how the burst of the Big Bang and the birth of the first galaxies relates to the strange flickers of the human spirit in you and me.

The aim has been a Grand Unified Theory that puts physics, chemistry, biology, neuroscience, genetics, the evolution of human societies, poetry, the arts, and the strange emotions of the human mind into a single big picture, a grand panoramic vision, a real Grand Unified Theory of Everything. A theory that filmmaker David Van Taylor, the producer of eighteen films,[79] would someday take a look at and call The Grand Unified Theory of Everything in the Universe Including Sex, Violence, and The Human Soul. But that theory will have to wait for my next book.

Meanwhile, you could say that instead of using a microscope or a telescope, the new Grand Unified Theory I've been piecing together has used an omniscope.

For the story of the omniscope, let me put you in my shoes.

In a sense, this book. *The Case of the Sexual Cosmos* really begins in 1955 when you are twelve years old in Buffalo, New York. You've been reading two books a day since you were ten. You've read one book under your school desk instead of paying attention in class, so you are not popular with your teachers. Then you've gone home and read another book in the quiet of your beige bedroom overlooking the trees of Delaware Park.

You are being bullied by your public school teacher. So your parents are kind enough to try to find a school that will fit your oddness. When you are twelve, they send you to an interview with the headmaster of the Park School of Buffalo, a private school founded in 1912[80] with a lot of personal input from famed philosopher John Dewey, the founder of "progressive education." John Dewey stressed letting children follow their curiosities. An approach whose appeal we will see more of later.

The Park School's headmaster, E. Barton Chapin, is a graduate of Harvard. He is a big man who augments his authority by cultivating a Sphinx-like image. He sits behind his desk, wears a tweed jacket, smokes a pipe, leans back, and hides his face in the shadows to give an impression of power and inscrutability. But you don't humbly implore this imposing figure to let you into his school. Instead, you tell him, "I will only come to your school on the following conditions." Yes, you make a set of demands. Demands that are just plain absurd. Remember, you are twelve years old. A pip. A squeak. A person guaranteed ignorant by sheer virtue of your age. And thanks to your two-books-a-day reading habit, the grades you've received in your previous school are abysmal. So what do you insist on?

Demand number one is that the headmaster teach you Russian, the language of a superpower you suspect will soon surprise us, the rapidly rising Soviet Union. Apparently, you are onto something. Two years later the USSR will startle the world and launch the first satellite in human history, Sputnik.

Demand number two is that the poor headmaster reorganize his science program for you. Science in those days is taught in the following order: biology first, chemistry next, and physics last. You demand a reversal. Why?

Because physics, you tell this poor, patient, red-faced man, is the beginning of the story of the universe. It is the tale of the Big Bang, a theory that is new and still fighting for survival in 1955.[81] Physics, you say, is the saga of the birth of elementary particles and the origin of atoms. Then you demand chemistry as course number two. Why? Because chemistry is the tale of how atoms got together in molecules and the saga of what molecules did when they mixed and mingled. Next you want biology because biology is the story of how molecules assembled life. After biology, you demand something no high school teaches—anthropology. You want to be told how human societies evolved. You want to be taught how small-scale human societies have done their thing and complexified. And finally, at the end of all the rest, you want history. In other words, you want all the sciences and humanities laid out in a timeline. In a single story.

What a cheeky little brat you are.

Lord knows what the headmaster thinks of this scientific churlishness. But he is kind enough to tolerate it. The proof? He lets you in to his school. And he reverses the order of your science courses for you. However, he feels that you are so deeply marinated in science that you've lost touch with the daily rituals of humanity. So he invents a special one-student tutorial, a tutorial in which he, personally, becomes your teacher. To make up for your social deficiencies, he has you read Victorian novels of manners. Apparently, he hopes that the manners will wear off on you. They don't.

Seven years later, you drop out of Reed College in Portland, Oregon, ride the rails, and hitchhike up and down the coast of California. You accidentally help start the hippie movement. But that's an odyssey for another book, *How I Accidentally Started the Sixties*, a book available from Amazon. Then you live in a Marxist agricultural commune, a kibbutz in Israel, for a year to see if a change in social structure changes human nature. It doesn't. Finally, you go back to college. This time at New York University.

And you discover something. College plunges you into a state of utter confusion. You are studying seventeenth century English poetry, physics, biology, psychology, probability theory, art history, and normal history. They seem like a total jumble. You can't keep them straight. And without being able to make sense of them, you can't remember them. When you are walking down the streets of Manhattan's Lower East Side near NYU, you feel like an Alzheimer's patient...always in a cross between a swirl and a kafuffle. And when you glue yourself and your books to a long wooden table in the NYU library, you have a hard time staying awake. So you try something.

You Scotch-Tape six pieces of notebook paper side by side to make a continuous, horizontal, accordion-folding panel. You don't do the arithmetic, so you don't realize that the result is over four feet wide. All you know is that it weighs nearly nothing, it fits into your three-ring binder, and you can slip it into your knapsack with your class notes. Another fact escapes you. Your accordion-folded horizontal diorama of paper is an instrument in tune with the demands that you made eight years earlier on your poor, badgered headmaster. It's a timeline.

Whenever you get a fact, you pin down the date.

- Hammurabi's Code, 1772 BC.
- China's Han Dynasty, 220 BC.
- Michelangelo's David, 1504.
- Henry VIII escapes the grip of the Catholic Church and establishes the Church of England, 1534.
- Establishment of the first English global trading company, the East India Company, 1600.
- King Charles I grants a charter to a guild of clockmakers in London and clocks become common in the households of England, 1631.[82]
- Andrew Marvell's poem "To His Coy Mistress," 1649.
- The first book on Probability Theory, 1657.
- The founding of Britain's Royal Academy of Science, The Royal Society of London for Improving Natural Knowledge, 1660.[83]

Then you put each event on the timeline. And an amazing thing happens. Relationships pop out. The timeline tells a story.

For example, Andrew Marvell[84] writes "To His Coy Mistress" 115 years after England escapes the headlock of the Papacy and almost fifty years after the establishment of one of the first global trading companies. He pens his lines a mere eight years before the first book on probability, and eleven years before the founding of The Royal Society of London for Improving Natural Knowledge.

Some sort of common zeitgeist is knitting itself together. And you can X-ray its mind by reading "To His Coy Mistress." Marvell's poem exults in living a physical life in a physical body. It exudes the sort of exhilaration in the physical that Michelangelo put on the map back in 1504 with his sculpture of a tall, muscular, naked David dangling a slingshot over his shoulder. What's more, Marvell focuses on something he can measure with the new household clocks of 1649—time. And what does Marvell use the concept of time to accomplish? To seduce a girl. Or at least to give it one of the world's most inventive tries.

The poet implores the object of his desire to sleep with him before the bloom of her youth—her gorgeousness—is gone. But he does it in British East India Company terms: the terms of geography, measurement, commerce, and global trade. Terms reducible to numbers. And terms with a historical scope.

> Had we but world enough, and time,

Writes Marvell,

> This coyness, lady, were no crime.
> We would sit down and think which way
> To walk, and pass our long love's day;
> Thou by the Indian Ganges' side
> Shouldst rubies find; I by the tide
> Of Humber would complain. I would
> Love you ten years before the Flood;

And you should, if you please, refuse
Till the conversion of the Jews.
My vegetable love should grow
Vaster than empires, and more slow.
An hundred years should go to praise
Thine eyes, and on thy forehead gaze;
Two hundred to adore each breast,
But thirty thousand to the rest;
An age at least to every part,
And the last age should show your heart.
For, lady, you deserve this state,
Nor would I love at lower rate.

"Rate" is a vital word in trade and taxation in 1649.[85] It's the price of something. For example, the price per ton of freight a ship owner might charge the British East India Company for a 20-month voyage to Asia and back.[86] Or the price of something new—shares,[87] shares of the profits from the voyage of one of the East India Company's ships.[88] Shares Marvell could have seen traded at a London coffee house when he visited the city.[89]

Then there's the Ganges, a river in India, where the East India Company is making a fortune on treasures like rubies.[90] And there are numbers: a hundred years to praise his inamorata's eyes, two hundred to each breast, and 30,000 to the rest.

Marvell urges the woman he is flirting with, a woman who is in all probability a virgin, to "tear" her "pleasures with rough strife thorough the iron gates of life." He urges her to keep her eye on the deadlines of aging and death. "The grave's a fine and private place," he says, "But none I think do there embrace."

However, there's something strange about Marvell's timescape. Despite a biblical reference to the flood, it is intensely secular. It has no heaven, no hell, and no afterlife. Despite the infinite wait for "the conversion of the Jews," it has no God, no Jesus, and no salvation. It has no sin and no demand for self-denial. Instead, it is riddled with something very non-religious—sexual pleasure. Try this for sensuality:

> Now therefore, while the youthful hue
> Sits on thy skin like morning dew,
> And while thy willing soul transpires
> At every pore with instant fires,
> Now let us sport us while we may;
> And now, like am'rous birds of prey,
> Rather at once our time devour,
> Than languish in his slow-chapp'd power.

There's something more than just sexual delectation here. Marvell exudes a glorious sense of control, a sense that you can seize joy and you can triumph over time despite the fact that someday you will die. A sense that you can have a sensual salvation down here on earth, a secular salvation born of your own powers.

Just look again at his conclusion:

> Let us roll all our strength, and all
> Our sweetness, up into one ball;
> And tear our pleasures with rough strife

Thorough the iron gates of life.
Thus, though we cannot make our sun
Stand still, yet we will make him run.

We can't stop time. We can't dodge death. But we can give time and death such a run for their money that we can beat them at their own game.

"To His Coy Mistress" is a poem about beating the odds. Remember, Marvell's focus on numbers and his sense of racing with time has something akin to the obsession with numbers and risk that will appear in the first book on probability theory eight years later in 1657. It has something akin to the sense that there are earthly forces without supernatural components, earthly forces that can be grasped with the senses and numbered with the merchants' math. And this sense of forces that you can master with your senses, with numbers, and with an exuberant sense of control has everything in the world to do with the invention of a new thing that will someday be called "science," a new thing that is then called "natural philosophy," a new thing that will be given validity and heft in 1660 by the establishment of the Royal Society—the organization more formally known as The Royal Society of London for the Improving of Natural Knowledge.[91] That beacon of science would be founded just eleven years after Marvell's poem.

Marvell does what you will do someday. He puts commerce, the knowledge of distant lands, their resources, their "rubies," and mathematics, into a timeline. And with a timeline, your timeline, you can see the currents of history that give birth to Marvell's poem and to your main anchoring point in life—science. You can take a chaos and see its hidden story. And your story-revealing-device, your timeline, does two more things: it keeps you awake at your wooden library table. Awake and fascinated by every new detail. And it turns a senseless jumble into an amazing tale. A tale that enthralls you. A tale you can remember.

As new findings from physics and cosmology enter the picture in the coming years, you will add six more pages to the timeline, stretch it to eight feet long, and flesh out the continuum that you outlined to your headmaster when you were thirteen. You will put in the 13.8 billion years of cosmic evolution that started with the Big Bang.[92] You will add in the birth of the first elementary particles, the first atoms, the first wisps of gas, the first galaxies, the first stars, and the first complex molecules. You will insert the results of new scientific studies as they pour in, whether they are research findings in astronomy, particle physics, geology, paleontology, anthropology, neurobiology, mathematics, or psychology. To top it off, you will add in history, literature, art, geopolitics, and current events. You will put in all the things that you'd begged your high school headmaster to teach you.

The result? You, a person with less memory than a decorticated frog, will graduate from NYU four years later magna cum laude and Phi Beta Kappa.

But after your graduation, you will continue to add to your timeline. And your timeline will make something obvious. Glaringly obvious. Hitting-you-over-the-head and stopping-you-dead-in-your-tracks obvious. This universe is on a constant climb toward higher degrees of flamboyance. Higher degrees of order and complexity. It is not doing what the believers in one of science's most sacred principles, the Second Law Of Thermodynamics, insist. It is not tumbling downward toward heat death. It is climbing. It is step-by-stepping on a constant staircase up. And that upward staircase is unlikely to end anytime in the next twenty, thirty, or forty million years.

The Second Law Of Thermodynamics, that all things tend toward entropy, that all things tend toward a random whizzle, that all things fall apart, will not fit the cosmos emerging from your timeline. In fact, your timeline will make the Second Law seem ridiculous.

We will poke into the story of how that Second Law of Thermodynamics got a headlock on modern science and how its biggest prediction failed later.

But the Second Law is not the only scientific assumption that your timeline will question. From the 1950s onward there will be another apocalyptic drumbeat, the drumbeat of ecological extremism. A lot of ecological thinking is right. But some of it is wrong. Not just wrong, but upside down, backwards, and perversely counterfactual. And you will be exposed to these topsy-turvy extremes of ecological thinking earlier than most.

CHAPTER 6

WHALES

Let's go back to high school. The high school that the headmaster you tortured was kind enough to let you into—The Park School of Buffalo.

You are no longer a brash twelve-year-old. You are now sixteen. The other kids do not like you. You have worked at the world's largest cancer research facility, the Roswell Park Memorial Cancer Research Institute. You have come up with a theory of the beginning, middle, and end of the universe.[93] A theory that makes a prediction that will be found true 38 years later—dark energy.[94] You are a science nerd, a misfit from another planet. The other kids will not vote you president, vice president, or treasurer of the class. Those are popularity positions, and you are not popular.

But when it comes to positions that demand actual work, well, they do not like to get their hands dirty. And you don't seem to mind slaving away. So they vote you the head of the Program Committee at Park School for two years in a row, from 1959 to 1961. Which means that you program two eight a.m. school assemblies a week and emcee five. Not a big deal. This is a high school in one of America's least distinguished cities. But it is more of an opportunity than it might seem. More of a periscope position—an opportunity to peer into strange corners where unusual things are going on.

One day in roughly 1959, you book a speaker on the plight of whales. This is twenty years before saving whales will become a symbol of the eco-vanguard. The speaker arrives and is strangely off-putting. He is big, dour, austere, and angry. He is not interested in socializing. Warmth and simple human gestures like hello, small talk, and goodbye are not his thing. When he takes the podium and shows his slides, his pictures of whales being butchered are ghastly. Profoundly disturbing. But the man is as memorable as his slides. He may be your first encounter with a genuine puritan.

But that is just your first run-in with the new field of environmentalism, your first encounter with ecological thinking. From the age of twelve, you have been trying to use the tools of science to understand mass human behavior and mass human emotions. You have been attempting to use the modes of thought of the first sciences you plunged into, the ones that consumed you at the age of ten— theoretical physics and microbiology. You've been using those ways of thought to examine the mass passions that fuel the forces of history.

So in 1968, nine years after your encounter with the whale activist, you graduate NYU and are offered four grad school fellowships in physiological psychology—a field that will someday be renamed neuroscience. Those graduate fellowships are thanks, in part, to your timeline. But you bypass the fellowships, jump ship from academia, and go into a field you know nothing about, popular culture. Why? You want to dissect human mass emotions at work in the real world. You want to see the forces of history *in vivo*. You hope that a scientific expedition into pop culture will take you to the dark underbelly where new myths and new movements are made.

One of those movements will be environmentalism. In 1969, a friend of yours is involved in the first Earth Day. It is a great vision, making sure this planet will stay green. And the way of thinking that it promotes—environmental thinking—cleans up a problem.

When you first came to New York City to attend NYU in 1964, you tried wearing white pants and white shirts in the early fall. It didn't work. By noon there were lines of black in the horizontal creases of your shirt and pants. And by 4pm your white outfits were black, white, and gray.

Why? There were four huge Consolidated Edison smokestacks, smokestacks for a power plant generating 660 megawatts of electricity, on the East River at Fourteenth Street. Those fifty-story[95] tall coal-smoke belchers were just east of NYU and seven blocks from the slum apartment on Seventh Street between Avenue A and Avenue B where you were living. But twenty years later, in 1984, the smoke was gone. Again, why? The eco-movement's pressures had resulted in the creation of an Environmental Protection Agency, and the EPA had turned the smoke coming from the Con Ed towers from black to white. The environmental movement had also cleaned up the air in cities all over North America, England, and Europe.

Yet there was something not quite right in the underlying attitude of a few eco-enthusiasts. Something you'd glimpsed in the harshness of the whale activist. Some eco-extremists hated the Agricultural Revolution and the Industrial Revolution. They opposed modernity. They thought that technology was the problem and that technology should be muzzled, leashed, and rolled back. But you had a sense that technology was the answer. In fact, you had a sense that without technology you would not be who you are. And you had the impression that if the environmental extremists had their way, you would be stripped and helpless. So would at least a few billion others.

Was there any evidence that you might be right?

THE ERROR OF PAUL EHRLICH`S BUTTERFLIES

Remember, you have been a science nerd since the age of ten, when you started in theoretical physics and microbiology. Among other things, at the age of twelve you've spent an hour discussing Big Bang versus steady state theory of the universe and the interpretation of the Doppler Shift with the head of the graduate school of physics at the University of Buffalo.[96] In the very year when Fred Hoyle, leader of the steady state movement, was certain he was about to consign Big Bang theory to the ash heap.[97] And you've been tutored in one-on-one sessions on outside-of-the-box science by the head of Research and Development for the Moog Valve Corporation, the company that made the valves for the first airplane to break the sound barrier and for the first airplane to reach the edge of space.

But in 1976, you start a public relations firm in a field that, until five years earlier, you knew nothing about—rock music. As you know, you are seeking a deep dive into the dark underbelly of society, the power pits where new myths and movements are made. And that dive into rock proves to be what you hoped it would—a tiny window into the forces of history.

You help champion subcultures—group identities—that use music to assert their right to exist—from a Southern community trying to break out of the ghetto of the Bible Belt with crossover country music; a Texas culture trying to express its right to respect through the music of ZZ Top; the gay community seeking to liberate itself through its chosen music, disco; to the white, middle-class rebellions of heavy metal and punk, and to two musical forms that the black community is using to assert its identity: rap and the sort of black crossover manifested in two of your clients, Prince and Michael Jackson.

What's more, you are asked to help expand the visibility of Amnesty International in North America. You are asked to kick off Farm Aid. You work with the Black United Fund, the United Negro College Fund, and the NAACP. You are named Ambassador of Texas Culture to the World by the mayor of Houston and the unofficial musical spokesman of the nationwide gay community in the same month, despite the fact that you are neither Texan nor gay.

Insiders let you into their communities. They let you see firsthand how their subcultures have distinct worldviews, distinct ways of interpreting the world. And they let you see how they become true believers, and how they sometimes pay no attention when the belief systems of their subcultures make predictions that prove to be wrong. Wrong as wrong can be. As you and I are about to see.

***

One of the musicians you work with is a founding father of New Age Music, Paul Winter. And New Age Music is the pop expression of ecological thinking. Paul Winter is marvelous. He is a saxophonist who duets with wolves and whales. And he is deeply committed to bettering the world. He is deeply committed to saving the planet. Your commitment is to help him get his beliefs across. But there is a problem.

In 1978, Paul tells you point blank that all the whales in the sea will be gone within two years. Off the planet. Out of the oceans. Disappeared. Extinct. Again, that is in 1978. If Paul's prediction is on target, there will be no whales left outside of Sea World by 1980. And Paul's information comes from some of the greatest ecological activists on the planet.

Yet there will actually be an estimated 1.5 million whales still swimming the oceans in 2023.[98] In part thanks to environmental activists. But only in part. Something is not right.

Paul speaks on behalf of an eco-extremist community. A community dedicated to predictions of apocalypse. And there is apparently something askew in that community's claims.

***

In 2013, you have friends who promote modern updates of Paul Winter's point of view: Dorion Sagan, son of Carl Sagan and of National-Medal-of-Science-winning biologist Lynn Margulis. Dorion is a terrific science writer with at least 20 books to his credit. Then there's Don Davis, one of the leading space artists in the world, the man behind some of NASA's most iconic space paintings. And Daniel Pinchbeck, author of *Breaking Open the Head* and *2012: The Return of Quetzalcoatl*, not to mention founder of an impressive psychedelic and ecological community.

All of these friends are convinced of a somber truth with dark implications—the earth is vastly overpopulated and we are on the path to the end. We have doomed the planet. And we have doomed ourselves.

This is a very strange point of view. Why? Because it has a track record. And that track record is one of failure.

HOW PAUL EHRLICH AND HIS BUTTERFLIES GOT IT WRONG

Paul Winter probably did not realize it, but his whale-conservation activist friends—folks on a crusade that made a positive difference in the real world—were being led astray. By butterflies.

In 1968, Stanford University population biologist Paul Ehrlich published a book, *The Population Bomb*, and went on a crusade to save humanity. Here's a glimpse of that crusade as described by Yale's Paul Sabin in his book *The Bet*:

> In June 1974, Paul Ehrlich and his close friend John Holdren flew from California to Washington D.C., to testify before Congress. [...] "A new era in the world" was coming, Ehrlich said, taking people from an "age of abundance to an age of scarcity." Within a decade, food and water would become difficult to obtain and expensive, resulting in a billion or more people starving to death. [...]

> Holdren insisted that ecosystems could not be maintained by artificial means. No "technological equivalent" existed for nat-

ural ecosystems. Humans could not run Earth like an "Apollo capsule," he said, referring to the space program. He saw "unwarranted technological optimism" as the "most dangerous tendency" facing society.

Ehrlich and Holdren spoke freely and easily of a "no-growth" society and "de-development," the coming period of austerity. [...] The Ehrlichs considered the American model simply doomed. "We are facing, within the next three decades, the disintegration of an unstable world of nation-states infected with growthmania,"[99] the Ehrlichs [Paul and his wife Anne] wrote. "The game of unlimited growth is ending, like it or not. We are approaching the limits."

The Ehrlichs argued, in their typical rhetorical style, that society should make voluntary changes today to avoid having future changes imposed by catastrophe.[100]

A billion people would die by 1984. That was Ehrlich's prediction. But did it come true? Did the human population of the planet decrease by a billion in the 1980s? No, it increased. The human population jumped by more than four billion between 1974 and 2021.[101] It more than doubled.

Yet Ehrlich's views way back in the 1970s fed the popular notion that humans are a cancer on the planet. A cancer? Isn't that the rhetoric of irrationality and the language of fear, not the language of discourse and debate? Yes. But is it merited?

There was a problem with Ehrlich and Holdren's dire pronouncements. Both men were scientists. Holdren, in fact, went on to become Assistant to the President for Science and Technology, Director of the White House Office of Science and Technology Policy, and Co-Chair of the President's Council of Advisors on Science and Technology under president Barack Obama.[102] Which means Holdren was not just any scientist. He was a scientist at the top of the heap.

Above all other people, scientists know the nature of nature. Don't they? When Holdren talks about "fundamental ecological limits," he is not just talking off the top of his hat. He is talking about an inescapable reality. He is talking about conclusions derived from data. Mountains of data. Data acquired painstakingly over more than five hundred years of modern scientific pursuit. He is talking about conclusions drawn not from opinion, but from facts. Drawn with detachment. By men and women whose lack of bias, whose objectivity, is basic to their disciplines. Right?

Wrong.

In science, you test your hypotheses. You make predictions. And you see if those predictions come true. Einstein and his followers, for example, derived eight predictions from his theory of relativity in the 20th century. If those predictions had proven false, his theory would have gone out the window. In science, if your predictions fail, you abandon your hypotheses and look for new ones. Or you adjust your hypotheses until they can accurately predict.

Ehrlich and Holdren's predictions failed. Big time. The pair told the Senate in 1974 that we would run out of resources, hit the wall, and plunge into mass starvation by 1984. A billion people would die. And the "catastrophe" of starvation and resource drought would only begin in the 1980s. It would get worse from there. But guess what? The catastrophe never arrived. People in the 1980s and 1990s were better fed,[103] richer, longer-lived,[104] and were subject to less poverty[105] than they had been in the 1960s and in the 1970s when Holdren and Ehrlich had made their predictions. But Holdren and Ehrlich never gave up their hypothesis.

In fact, they ceased regarding "the population bomb"[106] as an educated guess, a hypothesis, and came to see it as a fundamental reality. That's not science. It's faith. It's religion.

How did two intelligent men with impeccable scientific credentials go so wrong?

The nature Holdren and Ehrlich talked about does not exist. Why? They ignored the big picture. They ignored the real evidence on nature, evidence of the kind that began to emerge from 1792 to 1803, when Charles Darwin's grandfather, Erasmus Darwin, put together a timeline of the universe and wrote three evolutionary accounts of the cosmos and of life.[107] Holdren and Ehrlich ignored the timeline.

And they failed to observe one of the primary rules of science: look at things right under your nose as if you've never seen them before, then proceed from there. Look at the evidence.

Ehrlich looked at too little of the evidence. He looked at only one thing, the creatures he'd been obsessed with since he was a kid in Maplewood, New Jersey. He looked at butterflies. Butterflies go through cycles of boom and crash. As Paul Sabin put it in *The Bet*,

Butterflies existed in tenuous balance with available resources and external threats from predators and disease. No gentle "balance of nature" stabilized butterfly populations. Rather, booming and crashing population cycles characterized all animal species. Populations that grew beyond a certain threshold were brought down by resource shortages, disease, and other population-dependent factors. "The shape of the population-growth curve is one familiar to the biologist," Ehrlich wrote in a 1969 essay on human overpopulation entitled "Eco-Catastrophe." "It is the outbreak part of an outbreak-crash sequence. A population grows rapidly in the presence of abundant resources, finally runs out of food or some other necessity, and crashes to a low level or extinction."[108]

Extending the example of the butterfly to human history was a big mistake. Why? It ignored something right under Ehrlich's nose. Again, the timeline. The history of the cosmos that had been pieced together by over 500 years of science. The history that began with a Big Bang.

And Ehrlich ignored a lesson from that history. The ultimate resource is invention. And invention turns poisons into pistons and wastelands into fields of waving grain.

Invention creates new resources. Yes, invention creates new resources. Invention turns nightmares and nothings into abundance.

Ehrlich imagined that lifeforms reach the "carrying capacity"[109] of their environments, then crash. Crash and disappear. But his butterflies have been around for one hundred and fifty million years[110] and have not disappeared. Some have proven so astonishingly inventive that they've figured out a way to feast on milkweed in Canada, then to outfox the climate changes of summer, winter, fall, and spring by wintering in Mexico. That's a vast navigational feat. And it is a butterfly solution to the problem of climate change, a solution to the problem of winter that goes way, way outside the box. Nearly five thousand[111] miles outside the box.

Monarch butterflies have pulled this off without the use of satellite photos to check out the landscape, remote sensors to measure the resources of Ottawa and Mexico, and GPS to plot a route.

The monarch butterfly's migration is an invention beyond belief. Invention by creatures with brains the size of pinheads, brains whose central location is, of all places, in their chests.[112] And invention has existed since the universe began. But more on the monarchs and invention later.

Isaac Newton portrayed God as the great contriver, the great maker of machines. But evolution is more. She is the great breakthrough maker. The great creator of explosive inventions. Inventions that change the nature of the game. Inventions that change the very nature of nature. As you and I will soon see.

Like butterflies, we are now among evolution's tools. Tools for what? Tinkering, inventing, giving new powers to the grand enterprise of life.

Ehrlich and Holdren overlooked the timeline. And what shows up when you look at the timeline? Breakthroughs. The inventive capacities of a self-upgrading universe. The confabulating capabilities of a cosmos that constantly reinvents herself. The inventive compulsions of a universe that rejiggers the very nature of her nature.

CHAPTER 8

THE STRANGE STORY OF CLIMATE APOCALYPSE

Let's recap. In 1969, butterfly scientist Paul Ehrlich predicted apocalypse. By the 1980s, he said, we humans would be blasted by a population bomb. Because of our population explosion, the number of people would grow faster than the amount of food. A billion people, he said, would die. They would die of starvation, disease, and a lack of wiggle room. That mass death would arrive in the late 1970s and early 1980s.[113]

The 1980s came and went, but Ehrlich's prediction did not come true.

Despite that failure, Ehrlich's claims grew in popularity. In other words, Ehrlich's ideas were touted as science. But they were not. In science, you throw away a hypothesis if its predictions fail.

There is, however, a realm in which theories that produce false predictions can grow explosively. As you know, it's called religion. For example, two thousand years ago Jesus predicted that the Kingdom of God would arrive when he went to Jerusalem for a Passover dinner. Instead of witnessing the arrival of the Kingdom of God, Jesus was crucified. Jesus' prediction proved to be false.

Then his followers predicted that the Kingdom of God would arrive no later than tomorrow, then no later than next week. Then next year. Then the next millennium.[114] But those predictions, too, proved wrong.

Yet Christianity has survived as the leading religion on the planet.[115] Why? Christianity preaches the threats of hell and judgment day. It preaches that after death, you may be broiled, sizzled, and sautéed for eternity. Or you may be flattened by the terror of the end times, Armageddon, and the rapture. Christianity preaches fear. Fear can be a bonding mechanism. Visions of apocalypse can send us running into each other's arms.

Christianity fails in its predictions, but it taps something deep in human nature. It taps our addiction to visions of the end of the world. And it acts as social glue. It gives us something we deeply need. Recognition and affection from our fellow believers.

Paul Ehrlich's end-of-the-world thinking resembles the religion of an upstate New York Christian farmer named William Miller who predicted the end of the world in 1843.[116] Miller gathered a following that he estimated to be as high as 100,000[117] around his prediction. Many sold all their earthly possessions, then, on October 21, 1843,[118] donned white robes,[119] climbed to the top of Cobb's Hill in Rochester, New York,[120] and waited in groups for the appointed hour. When the world did not end, Miller secluded himself, reworked his mathematical calculations to eliminate their errors, and came up with a new end-of-the-world date: 1844.[121] On the new date, the end of the world stubbornly refused to arrive. Again. That's two big blunders. Based on math. Math and the Bible.

You'd think that would have been the end. But it wasn't. Today, William Miller's belief system has over 18.7 million followers.[122] They're called Seventh Day Adventists and Advent Christians.

Which means, to repeat, that we humans sometimes love predictions of apocalypse. In fact, we cling to visions of Armageddon even when they fail to come true. Why? Again, because fear bonds us in tight social units.[123] And bonding to our fellow humans is the very stuff of life.

But there's more. There's one more lure of apocalyptic thinking. In the great conflagration, everyone will die, right? Not quite. Us believers will survive.

Remember, there's a big difference between science and religion. In science, perpetually wrong predictions discredit a theory. Yet the apocalypse-obsessed point of view that Paul Ehrlich promoted in his 1968 book *The Population Bomb* and in his 1969 paper on "Eco-Catastrophe" was repeated two years later by the Club of Rome in its 1972 book *The Limits to Growth*.[124]

The Club of Rome used an MIT computer model[125] to prove conclusively and "scientifically" that Paul Ehrlich was right. The Club of Rome warned that if we don't solve our problems with "the arms race, environmental deterioration, the population explosion, and economic stagnation," these crises "will have reached such staggering proportions that they will be beyond our capacity to control." More specifically, we will run out of copper by 1991, new farmland for food production by 2000, and aluminum by 2001.[126] And we will also run out of food and water. "Food prices," said *The*

Limits to Growth, "will rise so high that some people will starve." When will we reach this "crisis point"? By 1979.[127]

But that mathematically-based, computer-verified, absolutely scientific prediction about running out of key materials like copper, aluminum, and farmland did not come true. In fact, it failed as dramatically as saxophonist Paul Winter's prediction that all the world's whales would be extinct by 1980. Yet in the 2020s, millions of people still believed the *Limits to Growth's* predictions of apocalypse. Passionately.

OK, Adventism is religion. But in science, when a belief is proven wrong, it's supposed to be revised until it can make accurate predictions. Or it's shelved. It is not exalted as the holy creed of a new orthodoxy. Right?

Alas, that's not always the case. For example, the idea that the sun revolves around the earth was retooled and rejiggered from 150 BC[128] until 1630 AD. That's a long time—1,780 years. It wasn't until the 1600s that a new form of thinking was able to elbow earth-centric beliefs aside. The new way of thinking was called "natural philosophy." And natural philosophy finally put the idea that the sun revolves around the earth into the trash bin of history. Then, in the late 1800s, natural philosophy underwent a name change. As you know, today it's called science. But more about that peculiar evolution of science later.

Remember, in religion, an idea that's proven false can grow stronger. Especially if it offers a vision of the end of the world. But not in science. Right?

Yet science-based friends like the ones I mentioned a minute ago—NASA space artist Don Davis, *Breaking Open the Head* author Daniel Pinchbeck, and Carl Sagan's oldest son, science author Dorion Sagan—believe in Paul Ehrlich's apocalyptic visions to this day. Powerfully. Passionately.

Meanwhile the leading science museum in the United States—The Smithsonian[129]—objects to this obsessive pessimism. In 2017, it started an initiative called Earth Optimism.[130] Explained Smithsonian Secretary Lonnie Bunch, Earth Optimism "shows us how to find hope in the face of odds that might seem overwhelming. It reminds us that change happens when we focus on what works—when we collaborate to find solutions and celebrate our successes." This is a bit more in tune with nature's real nature. As we are about to see.

***

Ehrlich and the Club of Rome's apocalyptic belief system has been proven wrong over and over again. Yet, as you know, the apocalyptic view claims to be science. The only science. And like religion, it will not tolerate opposition. It hunts down heretics and punishes them. It punishes them with excommunication from the scientific community. It calls them "climate deniers." And being labeled a climate denier can end your career. But what if you are not a climate denier? What if you are a climate amplifier? What if you try to point out the dangers of the next Ice Age or of the next global warming caused by nature, not just by human beings? And what if you point out climate-change solutions?

I ran into the sort of pressure that makes dissidents keep their mouths shut when I wrote an opinion piece in 2009 in a publication often disapproved of by the environmental community... and by me, *The Wall Street Journal*.[131] My essay pointed out that climate change is not just a human creation, it is a perpetual twitch of nature. We have a sun that's increased 30% in warmth since this planet was born.[132] We've had 80 sudden global warmings[133] and ten ice ages, ten glaciations[134] in just the 120,000 years during which we etched the first symbols into an aurochs bone,[135] invented ways to use sea shells to decorate ourselves,[136] and climbed from stone-tool wielding *Homo sapiens* to *Homo urbanis*, man the inventor of the city.[137] All of these climate catastrophes happened long before the invention of tailpipes, smokestacks, and capitalism.

Now, frankly, I disagreed with the single-minded conservatism of the *Wall Street Journal*. And I disagreed with Rupert Murdoch—the *Wall Street Journal*'s owner at the time—and his pigheaded insistence on destroying climate change arguments no matter what their validity. However in 2012, three years after my *Wall Street Journal* article appeared, I wrote to a cosmologist whose work I admire, Lee Smolin, author of books like *The Trouble With Physics* and *Einstein's Unfinished Revolution*. Smolin wrote back with cordiality, but had this to say:

> btw I saw your piece in the WSJ on climate change and disagree very strongly. I think it is very irresponsible of you to ignore all the evidence that the climate has warmed the last 50 years due to the CO_2 and other greenhouse gases put into the atmosphere by we humans. If we are to have a chance of avoiding the

consequences, which range from challenging to calamitous, we have to understand the causes as this determines what we can do about it.

But my piece had not been an attack on the concept of human-produced global warming. Not at all. It had been a warning about nature's addiction to her own forms of climate change. Not to mention our need to prepare for them. Which is why this book presents solutions that will help us transcend two different forms of climate change: climate change caused by humans, and climate change inflicted by the planet itself.

But back to our story. Lee Smolin is an astronomer and theoretical physicist. His books and ideas are magnificent. Yet he implied that debate is a form of treason to the scientific community. At least on the topic of climate change. Yet I thought debate was the very thing that makes science science.

Yes, life depends on interlaced systems. Yes, we can easily poison our air and our water. And yes, at the end of this book I will show you a technology that can end the use of fossil fuels for transportation and energy production. For good. A technology that can bring us to net-zero. A technology that can bring us the Green New Deal.

But no, we are not hitting the carrying capacity of our environment. We are not running out of resources. We are not destroying a garden of Eden. And, no, reversing the Agricultural Revolution and undoing the Industrial Revolution will not return us to a perfect state of green, to a Garden of Eden, to a global petting zoo like something out of Bambi.

Most important, taking greenhouse gases out of the air will not return us to a state in which climate change ceases to be. Why? Because nature doesn't work that way. Climate change is not just a result of disturbing nature. It is nature's status quo. Climate change and species extinction are not inventions of human beings. They are Mother Nature's way of filing her nails.

It's our vision of paradise and of an unchanging equilibrium that is unnatural. So is our vision of a loving Mother Nature. Extremely unnatural. But guess what? Nature favors the unnatural. Nature favors those who oppose her most. We are capable of conceiving things that have never been and turning them into new realities. Like compassion, peace, and justice. And that ability is part of nature's creative process. We humans are tools nature uses to create, tools that nature uses to reinvent herself.

***

But why is nature not the nurturing mother we imagine? The answer is in the timeline. The answer is in the story of the cosmos.

CHAPTER 9

IS MATERIALISM A SIN?

Remember my conference call with three major figures in Pakistan's economy? They apologized for Dubai's materialism, waste, and vain display. And I suddenly realized that Dubai's materialism, consumerism, waste, and vain display were not a waste. Dubai's flamboyance spoke without words on behalf of the Muslim world's tolerant, secular, modernists. Dubai's skyscrapers were a silent voice for those who dared not speak. They were rousers of the human spirit. And they were lifters of the human sense of possibility.

Starting with that phone call, something stirred in the back of my mind. A heresy. A revelation arising from the omniscope, from the timeline. What was that insight? What was that heresy? Materialism, consumerism, waste, and vain display are not inventions of humankind. They are built into the very operating system of the universe. They are part of Mother Nature's DNA.

They began long before this universe gave birth to life. In fact, materialism, consumerism, waste, and vain display are clues to Nature's way. They are clues to the secrets of her least-explored and most important gift—her creativity. Her urge to invent.

***

Let's start with materialism. Materialism is not a product of capitalism. It is not a product of the Western way of life. Materialism is a product of nature. Materialism goes back over 13 billion years. Materialism goes all the way back to a mere 380,000 years after the Big Bang.

380,000 years after the Big Bang, electrons and protons mated and formed the first atoms.[138] And materialism began. The new atoms discovered that they were being whispered to, recruited, and seduced by a force so weak that it had been hidden until then. That force of summons and seduction was gravity.

Gravity, it would soon turn out, was the greatest materialist this cosmos would ever see. It was a social gatherer. It was the cosmos' ultimate attractor. And it was the generator of attraction's opposite, competition. Yes, the birth of the first atoms set off a competition to the death. Gravity's appearance kicked off the era of something utterly unimaginable until then, The Era of the Great Gravity Crusades. Teams of atoms competed. Yes, atom teams competed. To win the prize of a snatch-and-grab sociality. To kidnap, seduce, and recruit yet more atoms. To kidnap, seduce, and recruit more and more material things.

Masses of atoms pulled together in gravity balls. Gravity balls with greed. The gravity balls that grew the fastest sucked in and swallowed the gravity balls that grew more slowly. The big ate the small in what astrophysicists and astronomers call "cannibalism."[139] The biggest winners became galaxies. The smaller winners became stars. The winners in the number three slot became planets. And the runners up became moons.

In other words, a force of nature, gravity, led to materialism. Materialism led to competition. Competition led to cooperation. And cooperation led, in its own strange way, to the primal ancestor of pecking orders. Yes, pecking orders. Dominance hierarchies. With galaxies on top, stars next in line, planets at number three, and moons at the bottom.

Competition and dominance arose in the first hundred million years of the cosmos. All thanks to materialism. Not human materialism. Cosmic materialism. Cosmic greed. All without a capitalist, a patriarch, or a smokestack in sight. Which hints that materialism, competition, and hierarchy may be stamped into the source code of this cosmos. Materialism, competition, and hierarchy may be stamped into Mother Nature's DNA.

How can we tell? The materialism, competition, and hierarchy of the Great Gravity Crusades continue today. As you read this sentence, our sun—a winner in the Great Gravity Crusades—is demonstrating the ultimate consumerism, burning through 4.29 billion kilograms of hydrogen per second, and spilling a toxic trash of 385 trillion terawatts per second of radiation.[140] That's materialism, consumerism, and waste on a massive scale.

And at this very moment, galaxies are gathering together in clusters. The biggest clusters are eating their smaller brethren. They are bulking up and becoming superclusters.[141] Why? Because galaxy clusters are flocks in which galaxies gather like sheep. Shepherded together by gravity. And these galactic flocks compete. They compete to suck in matter. They compete to become more materialistic than their peers. They compete to gather as much matter as they can. To gather enough to literally make a dent in space.

It's true: the more massive you are, the more you can bend the space around you.[142] And the bigger your dent in space, the more you can attract even more material stuff. The more humongous you become, the more pull you have. And that's not anthropomorphism. That's physics.

Hovering above this mega-materialism is something that's not material at all—the unique personality of each galaxy. The galaxy's group identity. The deliciousness of its spiral shape. But more about group identities in a minute.

That's not the end of nature's materialism. Nature is a consumerist. She does more than merely gather gargantuan masses of material stuff to make galaxies and stars. She manically mass produces stars and galaxies. She creates galaxies and stars by the billions of trillions.[143]

Then nature shows her delight in waste. Her delight in manic mass destruction. A mere 232 million years[144] into the existence of the cosmos, nature casually kills stars off. She lets them shut down and die. Or she turns them into the star explosions called supernovas. She turns them into careless, extravagant throwaways.[145] Yes, nature sometimes celebrates death with fireworks displays.

One way or the other, nature turns discarded stars into waste. Nature is not just a materialist, she is a consumerist. And she loves to litter. She loves to trash the place.

So this is a materialist cosmos. This is a consumerist cosmos. This is a cosmos that loves one-use throwaways.

But nature treats every form of waste as an opportunity. She turned the death agonies of stars, those showy wastes of material things, into amazements. She used the incredible crushing power of star death 13.6 billion years ago to mash together 26 new forms of social groups, 26 new forms of atoms, in the dying stars' hearts.[146] She crunched together unwilling protons and neutrons to invent 26 new kinds of teams with 26 new forms of group identity.[147] Atoms with a core of six protons were carbon. Atoms with a core of 8 protons were oxygen. And atoms with a core of 26 protons were iron. All elements far stranger than just the protons and neutrons at their hearts. And all elements created by brand new forms of social schmooze. All elements that emerged from the group identities of whole new kinds of teams. Whole new kinds of organizational personalities. Whole new atomic nuclei. Nuclei for brand new elements. Nuclei mashed together in the hearts of dying stars.

Yes, nature is the ultimate materialist. But there is something spooky in her way of doing things. With every ornate display of throwaways, Mother Nature invents. She summons forth things far, far stranger than the sum of their parts. Things unexplainable by the laws of logic, unexplainable by the laws of one plus one equals two. Unexplainable by the laws of garbage in, garbage out. Unexplainable by the rule of the shortest path between two points. She brings forth things far beyond the sum of the mere matter they contain.

Something arises from each social grouping. A hard and fast new reality brought to life by the team. Yes, to repeat, nature invents new group identities. Identities that hover like ghosts over the particles that make them up. Nature generates the group identities we know as hydrogen, helium, lithium, and beryllium.[148] Not to mention carbon, oxygen, and iron.

Which leads to a question. Is materialism really all about something that transcends the material—that umphish thing that I've been calling group identity? You experience a group identity every minute of every day. You consist of a hundred trillion cells[149] each of which has all it takes to be independent. Not a single one of those cells has a clue that it is helping generate the identity that is you. In fact, more than 330 billion[150] of those cells die and are replaced every day, and yet your name, your childhood memories, and your sense of your self remain the same. Yes, somehow your identity arises over and above your hundred trillion individual cells. And that identity is real as real can be. Like the you that you call your "self," an overarching identity also springs from the teamwork of protons and electrons.

Like you, hydrogen, helium, and lithium are ethereal and ghost-like. They are the higher selves of societies. What in the world do I mean? Hydrogen is the intimate wedding of a single electron and a single proton. Aristotle says that if you break things down to their smallest parts and understand the laws of those parts, you understand everything you need to know. But even if you understand all there is to know about protons and electrons, that knowledge will never allow you to predict or explain the magic of the group identity that arises from a single proton mated to a single electron. Knowing everything about protons and electrons will never allow you to predict the utter weirdness that together they generate: the weirdness called hydrogen. Just as knowing the laws of the cells of which you are made will never allow even the greatest expert to predict your personality. Why? Because your personality is a group identity. So are the properties of hydrogen, helium, and lithium. They are group identities. Identities that spring into being from protons and electrons gathered in teams. Identities that arise like ghosts, like spirits, like the flickers of flames. Yet despite this ethereal nature, hydrogen, helium, and lithium are elements. They are the "materials" of materialism.

What are group identities? And what is so ghost-like about them? Put eleven guys from Boston and eleven guys from Atlanta together on the pitch of a football stadium in Foxborough, Massachusetts. Then put 60,000 random bystanders in the stadium's seats. What will happen? Nothing. You will have 22 accountants, lawyers, and truck drivers milling around on the turf, in the bullseye of attention, without a purpose. That's not a thrilling spectacle for the befuddled folks in the seats. What's missing? Group identities. Organizational personalities. The higher identities that groups generate when they're organized as teams. When you give them a purpose. And when you let them compete.

You are missing the New England Patriots and the Atlanta Falcons. Not to mention the group identities of football and the NFL.

The unique characteristics of the New England Patriots do not emerge from just the quarterback or from one of the defensemen. The group identity of the Patriots emerges like a ghost from the interaction of all of the members of the team. And from that team's competition with the other teams in its league.

Not to mention from the vision of its owners and its coaches, people who *do* see that higher identity and use it to sell tickets and TV rights. And to motivate the team.

The group identity of the New England Patriots emerges like the flame that hovers over a charcoal briquette. Take one charcoal briquette out of a bag. Put it on your driveway. What happens? Nothing. Now saturate it with lighter fluid. Put a match to it. What happens? A flame. But is that flame real? It dances around, never occupying exactly the same space twice. You could run a spatula through it and it will offer no resistance. It's immaterial. So it isn't real, is it? But boogie and flick as it may, it has its own unique and ever-shifting outline, its own flickering, disco-dancing self. Put your hand into it for ten seconds. It burns you. It is very real indeed.

Group identities are like the flame. They are not material things. In that sense, they are not real. But when one football team or one army goes up against another, group identities are very real indeed. In fact, in many ways they are more real than the people who make them up.

The New England Patriots was founded in 1959.[151] Its first members have long since retired and been replaced. But the team called the Patriots still remains. That's a group identity. An identity rising like a flame above the sum of its parts. And persisting for over sixty years. In fact, the identity of the New England Patriots is likely to persist long after its founding members are dead.

In the world of atoms, group identities like these are as hard and fast as the protons of which they are made. And as radically surprising. Hydrogen is one group identity. It's a higher quirkiness that arises from the team of one proton and one electron. Helium is a very different envelope of social quirks that arises from the team of two protons and two electrons.[152] And 99.9 percent[153] of the material stuff in this cosmos is made up of these two very different teams with two very different ways of taking on their surroundings, two very different selves, two very different group identities, two very different organizational personalities. 99.9 percent of the cosmos is made up of these two group identities, hydrogen and helium.

What's the place of group identities in your life and mine? At 6:05 pm on April 4th, 1968, a charismatic man stood on a balcony at the Lorraine Motel in Memphis, Tennessee, talking to his aides.[154] His preachings had been the voice of a group identity, the identity of the black community and of its white sympathizers. But that man on the balcony was also the product of a far more personal group identity. Like you and me, he was a hundred trillion cells, each cell totally unaware of the larger self it was powering. Then came a sniper's bullet and struck him in the neck. When that bullet hit, it only damaged a tiny percentage of his 100 trillion cells. The rest were still capable of carrying on their functions. As a gathering of matter, the man was still almost entirely the same. But something bigger had died: the group's overarching identity. And with that organizational identity died the voice of more than just one man. With it died the voice of the group identity of millions of human beings. That group identity of millions of humans was called the civil rights movement. The group identity of the hundred trillion cells was known as Martin Luther King Jr.[155]

The flame that was extinguished in Martin Luther King was immaterial. From Aristotle's reductionist point of view, it did not exist. It was not explainable by the rules of the atoms or the cells that were its "elements." But it was a human with flaws and magnificence. An individual who added to the rushing river of group identities we call history.

You, too, are a group identity. I am a group identity. In a sense, we are not real. We are made of constantly changing matter. We lose and replace nearly 35 million cells a minute.[156] And yet we remain us. I can touch your skin. But I can't touch that immaterial flicker called you. Yet you are real as real can be.

When protons and electrons come together in teams, they too give birth to group identities. For that's what hydrogen, helium, and lithium really are. The group identities of teams.

And even these flickers that arise from mere groupings are the fruits of one of nature's most important lusts, her acquisitiveness, her greed, her love of gathering material things. Her love of kindling new group identities. Not to mention one of her deepest addictions of all: invention.

CHAPTER 10

THE HIDDEN STING IN GLOBAL WARMING: THE DISASTER IS NOT WHERE YOU THINK

This is not a book about anthropogenic climate change. It is not a book about man-made greenhouse gases and their effect on earth's atmosphere. Thousands of other scientific thinkers and activists have applied themselves to those topics. They are being well researched and well publicized. Well taken care of.

This is a book about what is being ignored.

There's a scene in Joseph Heller's *Catch 22* in which a World War II aircraft, a B-25 Mitchell bomber, is hit by flack over Avignon. A radio-gunner in the transparent tail-gun bubble[157] is wounded, and the central character—Yossarian—is sent to the damaged rear section of the plane to save the injured man's life. The radio-gunner has lost his leg. Yossarian sees a thigh wound, a massive one, pouring blood. So Yossarian works frantically to stop the bleeding and to sanitize what's left of the thigh. Meanwhile, he tries to comfort the radio-gunner as the wounded man weakens and begs for help. The radio-gunner whimpers "I'm cold, I'm cold, I'm cold, I'm cold,"[158] a sign that he is bleeding to death.

So Yossarian tends even more to the massive wound where a thigh used to be. He tends to it with panic and urgency. He needs to stop the flow of blood before the gunner can die of blood loss. Finally Yossarian gives "a long sigh of relief." Why? He sees the unmistakable sign that the gunner is "not in danger of dying." How does he know? The blood is coagulating in the wound and it's "simply a matter of bandaging him up."

But the gunner dies. Despite all of Yossarian's work. Why? There's something Yossarian failed to see. Hidden in the radio-gunner's "quilted, armor-plate flack suit," the man had not just one wound but two. Shrapnel had pierced his stomach wall. His intestines were falling out. The real blood loss—the real threat—was not just from the wound in his thigh.[159]

Mistaken perceptions, mistaken visions of what the problem is, can lead us astray.

Our obsession with man-made global warming may well be like Yossarian's mistake. Anthropogenic climate change is important. But it's not the really big source of climate instability. It's not the 800-pound gorilla standing on the living room couch. It's not the elephant eating the living room rug. So what is?

This is a planet of climate change. In fact, it's a planet of climate catastrophe. It's a planet of ice ages and of hot-house eras, a planet of periods when even the icy Arctic[160] and Antarctic[161] have turned to tropical swamps. It's a planet that alternates between a sauna and an ice ball. Or, to put it in anthropomorphic terms, it's a planet whose nature is not nice. A planet whose nature is not mothering and kind.

CHAPTER 11

BIRTHING THE PLANET OF CATASTROPHE

To see nature's real nature, let's go back to a time over 3.5 billion years ago when there was less than a teaspoonful of life. A teaspoon of life on a planet of punishment. 3.5 billion years ago the earth was still at the tail end of a billion-year-long merger and acquisition binge. This earth was, guess what? Materialistic. For 200 million years,[162] it had grabbed as many material things as possible and held on to them with all its might. Why? In reality, the early earth wasn't what you and I would call a planet. It was a greedy scrapheap competing in a non-stop tournament of gravitational showdowns. Competing in the Great Gravity Crusades. Two gravity piles would come within each other's reach and the bigger would win. Its prize? To swallow the other whole.[163] As you know, astronomers have a name for this rapacious violence. It's cannibalism.[164]

Over 3.5 billion years ago[165] when life first dared to yank itself together, the proto-earth had gone through over ten billion[166] of these showdowns and had won every time. And this battling rock pile wasn't finished. It was still running into high speed chunks of stone, massive asteroids, snagging them with its gravity, reeling them into its surface, and smashing them into its pile of winnings, into its great, galumphing, growing gravity ball.[167] It was still pulling chunks of stone at full speed into its face. Which means that the surface on which life was trying to get a start was not a warm and nurturing place. It was like the surface of a pudding smacked by the back of a tablespoon. It was splattered and earthquaked.

What's worse, this bald, barren, shuddering ball of gravel on which life first dared arise was slapped around by a motley rogues' gallery of additional cataclysms. Take the simple six-hour cycle of torture that came from one of the new planet's many hyperkinetic tics, its rotation. Our growing gravity ball spun at five times the speed of sound—close to 4,000 miles an hour. Which means that this planet wannabe swiveled around its center, its axis, like a top every six hours.[168] Four times as fast as today. And the six-hour swivel was a non-stop climate disaster. It prevented any sort of equilibrium. For one thing, the spin was an instant climate-changer, a creator of non-stop weather shock. It refrigerated the temperature, ratcheting it down roughly 100 degrees every three hours, then sizzled the temperature back up by 100 degrees again.[169] Every three hours it took the surface from unbearable heat to shivering freeze and back to heat again.

But this roller-coaster of temperature extremes was just one of the insults delivered by the planet's spin. There was yet another way in which the environment whiplashed as abruptly as a Tesla Roadster rocketing forward at full throttle, then shifting into reverse. Every three hours the surface was flooded, lashed, and whipped by toxic radiation. Then the surface was plunged into something equally intolerable—three hours of utter darkness. These climate disasters were day and night.

And there were other cycles of climate apocalypse. There was the manic-depressive climate gyration caused by the earth's tilt on its axis. We call this cycle of freezes and fries summer, winter, fall, and spring. Then there were the Milankovitch Cycles,[170] cycles generated by a tilt and a wobble of the earth, and by a periodic shift in the orbit of our infant planet around the sun from circular to egg-shaped and back.[171] These peculiarities smacked the planet upside the head every 23,000, 41,000, 100,000, and 413,000 years,[172] ripping apart what few dependable patterns of climate there might have been. The Milankovitch Cycles are climate shifters that still go on today. And there was yet another nasty trick of the new planet, a twitch that wrung, racked, and tortured it once every 26 million years.[173] A twitch that would produce mass extinctions that remain unexplained[174] to this day.

Frankly, even the sun around which this cataclysmic embryo of a planet rotated could not be counted on to anchor things. That sun turned up its intensity by roughly 10% every billion years.[175] Yes, it kept getting hotter. And it shimmied through a tinier tremble of unreliability every eleven years, its sunspot cycle, a cycle in which the sun's magnetic pole reversed utterly,[176] a cycle that added yet one more punch to the planet's temperature slaps.[177] Topping it all off, the sun was seized by spasms called coronal mass ejections, convulsions whose electromagnetic shockwaves wreaked havoc on everything that dared orbit within its vicinity.

Yes, this loving planet of ours was the mother of climate change and the creator of climate catastrophe.

But these were not the only reasons that this third gravel pile from the sun was so unsafe and so riven by climate flips. Among other things, this infant earth was not tucked into some cozy corner of the cosmos where it could be sheltered and coddled. It was not nested, swaddled, and walled off. Far from it. The newly clenching fist of stone we call terra firma was a voyager yoked to its very young sun. It was an unwilling traveler trekking through dangers that would have frozen the blood of Frodo the Hobbit. Like a duckling following its mother, the infant planet and its seven siblings were dragged by their youthful sun on a giant orbit around the core of the galaxy, a circular trip that took 240 million years.[178] On their way, the sun and its baby planets traveled through spiral arms[179] that showered the new gravity balls with comets and cosmic rays, catastrophes that spurred the formation of continents and racked the infant earth with salvos of savage high-energy particles, particles that sand-blasted our planet's atmosphere and whacked its climate into appalling change.[180]

In addition, the eight infant gravity balls—the eight planets—and their mother sun snowplowed through dust clouds of "galactic fluff."[181] Which means that in a normal year, our planet, the planet infected by life's first teaspoonful of cells and DNA, accumulated 15,000 tons of space dust in its outer atmosphere.[182] That was a normal year. But once every 100,000 years, things went abnormal and our newborn planet whiffled through a cloud of interplanetary powder that shot that amount up one hundredfold.[183]

Cosmic rays from the spiral arms of the galaxy and the pileups of space dust—each of these travel mishaps changed the newborn planet's weather dramatically. Each brought a tsunami of climate change. And those recurring travel disasters still thwack this planet today.

But there's more. Environmental scientist Vaclav Smil calculates that supernova explosions have machine-gunned this solar system with high-energy radioactive particles every two million years. Smil says that ten times in the last five hundred million years alone supernovas have strafed our rock pile with almost enough radioactivity to kill off every animal with a backbone that has ever crawled, walked, or flown.[184]

And putting the cherry on the climate-upheaval cake, at least twice[185] this planet has frozen completely and has been encrusted in three thousand feet of ice—yes, over half a mile of ice—at its warmest point, the equator.[186] This third stone from the sun has turned into an ice ball earth not just once but twice. And possibly three times.[187]

This is not harmony. This is not equilibrium. This is not the sort of loving, balanced nature some call on to nourish the human soul. This is bakeoff, burn, broil, and freeze. This is non-stop climate catastrophe. On this harsh pebble circling a middling yellow star, about the only thing you can count on is climate change. Dramatic, atom-shuddering climate change.

Yes, nature herself has given us climate change's crazy pinball.[188] Our loving Mother Nature. To repeat, we've had 80 sudden global warmings[189] and ten ice ages, ten glaciations[190] in just the 120,000 years during which we etched the first symbols into aurochs bone and advertised ourselves with the first jewelry.

But there's more. 1.4 billion years ago, carbon dioxide in the atmosphere shot up to between 20 and 200 times what it is today.[191] And until 10,000 years ago, the Gulf Stream's center see-sawed back and forth between the North and South poles every 1,500 years,[192] shoving the weather into devastatingly alien patterns, making cold places warm, warm places cold, wet places dry, and dry places into marsh and flood. Then there were the collisions of tectonic plates, leading to a pileup of earthquakes that shoved seabeds onto mountaintops.[193] And there were volcanos whose sunlight-blocking spew of black smoke plunged the planet into a dark ice box. Into winters that lasted years. Not to mention the occasional chunk of rock from space, the occasional meteor striking with the power of more than 4.5 billion atomic bombs.[194] Like the meteor that killed off the dinosaurs 65 million years ago. The Chicxulub impactor.[195] But more on that murderous meteor later.

No, our little rock heap, our little orbiting clump of gravel, was never a warm and fuzzy place. It was never loving and kind. In fact, its insistent upheavals have produced roughly 142 mass extinctions.[196] All without tailpipes and smokestacks. So if you'd been real estate shopping for a nice place to settle down 3.5 billion years ago and you'd had any brains, this poison pill of a planet would have been the last place you would have picked to land. And if you'd been betting on this gravity ball's odds of harboring life, you'd have been forced to bet against it thirty billion to one.[197] Yet we would someday call this smashing, searing, freezing ball of stone, this hell hole of climate change, our earth. Our precious mother. Our nurturing bosom. We would someday call it our home. How in hell would this turnaround take place? How would an entire planet go through a plastic surgery so radical? How would it go through such a total identity change? Such a total change of organizational personality?

Or, to put it differently, how would the fragile first teaspoon of life manage to thread this planet's murderous obstacle course? How would life turn its torturer into a nurturer? How would life so radically resculpt its environment that we would someday misinterpret the result as a harmony that had been here all along? How would life twist the arm of its punisher into becoming what we'd someday call nature? And a mothering nature at that?

But before we answer that question, we have to watch out for the Yossarian mistake. We have to prevent a climate Catch 22. We have to recognize that climate change is inevitable. Taking man-made greenhouse gases out of the atmosphere is a promising move. But it won't stop the massive climate shifts that come with this gravity ball that we call home.

***

If removing man-made greenhouse gases is a step toward climate stabilization technologies, that's great. We humans have been working on climate stabilization technologies ever since we tamed fire one and a half million years ago.[198] Ever since we sliced the first fur coat off of a bear 300,000 years ago[199] and wore it ourselves. Ever since we built the first tent of mammoth ribs and mammoth tusks supporting mammoth hide walls and roofs 25,000 years ago.[200] Now it's time to take climate stabilization technologies to a higher plane if we're going to avoid the Yossarian mistake. If we're going to make it through this planet's next totally natural climate change. If we're going to make it through the next great global warming or the next ice age.

But before we get to climate stabilization technologies, let's dig into a mystery. How did life thrive on a planet of disaster? And what role, if any, did an exuberant exercise of materialism, consumerism, waste, and vain display play? Not to mention invention?

CHAPTER 12

RAPING GLASS AT THE BOTTOM OF THE SEA

"We have lived our lives by the assumption that what was good for us would be good for the world. We have been wrong. We must change our lives so that it will be possible to live by the contrary assumption, that what is good for the world will be good for us. And that requires that we make the effort to know the world and learn what is good for it." Wendell Berry, 1969

"The flourishing of non-human life requires...a substantial decrease of the human population." Arne Naess and George Sessions, *The Deep Ecology Platform,*[201] 1984

"Discovery is always rape of the natural world. Always." Michael Crichton, *Jurassic Park,* 1990

"The planet cannot sustain continued global economic growth." *Nature Magazine,* 2010[202]

How did life survive on this planet of disaster?

Bear with me while I repeat something crucial. Nature is not a benevolent earth goddess. She will not give us peace and a leafy paradise if we remove the greenhouse gasses with which we are disturbing her balance. Nature herself has sometimes driven the level of greenhouse gas to over a thousand[203] times what it is today.

Our job is to adapt to the challenge that nature has thrown us ever since we invented the first stone tool 3.3 million years ago.[204] What is that challenge? Invent ways to triumph over nature's bread and butter—climate change. Invent ways to be doom riders and catastrophe tamers.

To understand why nature sometimes calls on us to do the opposite of what we are told, let's go back 3.42 billion years and spy on the strange doings at the bottom of the seas. Let's spy on the years when life had just gotten its start.

***

Nature has a nasty habit of destroying herself.

Today, the rocky outer crust of the earth's surface is dominated by eight huge sheets of cooled, solidified rock millions of square miles in size, floating on an oozy sea of hot, melted, liquid stone. Eight very distinct group identities. Eight tectonic plates. These plates are accompanied by a flock of smaller floating stone saucers—volcanoes and chains of islands like Hawaii. The planet-girdling rafts of rock and their smaller outliers float on the viscoelastic rind of mantle called the asthenosphere.[205] They float on a liquid of molten rock.[206] And, tectonic plates don't play nice in the sandbox. Ever so slowly, they grind up against each other and battle for dominance. It's called "subduction."[207] When they meet, one plate tries to climb on the other's back.[208] Just like puppies in their playing do. How animal-like. But there is no conscious struggle for status among these rocky sheets. There is no conscious anything. These slabs of rock have no mind. They simply obey nature's laws. The laws that we are told lead to harmony.

Needless to say, harmony is seldom the name of nature's game. As you would suspect, at a tender billion years old, this adolescent planet is still a troublemaker, a violator of nature, and a wrecker of its environment. It is even a violator of its environment in outer space. The newborn planet still bullies and threatens the calm of a floating sea of stones[209] beyond our skies[210]—space stones, asteroids, planetesimals, and meteors.

On the young planet's surface, things are just as turbulent. Take the Pacific Plate, 39.76 million square miles of slowly moving rock, a tectonic slab like a parking lot ten times the size of China. Where the Pacific Plate drifts over a deep, stationary hot spot, a "melting anomaly" in the asthenosphere,[211] it throws a fit, producing volcanoes like Mauna Loa in Hawaii, a volcano that can rip 376 million cubic meters[212] of red-hot stone from deep beneath the earth and fling it high above the surface in a single eruption.[213] That's in recent times, in the last 700,000 years. That's just to give you an idea of how the competition between group identities called plates works in plate tectonics today.

Let's lunge back to 3.47 billion years ago.[214] It's only a little more than a billion years after this third rock from the sun has pulled itself together. Life has taken hold and is churning forward, seeping into as many new territories as it can eat. But life thrives because it is a doom rider and a catastrophe tamer. Life thrives because it is invasive, because it is a weed. Life thrives because it innovates. Life thrives because it invents new resources.

There's a massive supercontinent called Vaalbara,[215] a supercontinent that in the distant future will rip itself in two, a supercontinent that will someday split into Africa and Australia. Violently.[216] The battles of tectonic platelets around Vaalbara are fierce. Geological wounds and gore emerge from these struggles. They emerge in the form of volcanoes. And the volcanoes do something common to nature. They spew so much soot into the atmosphere that they change the climate for a year or three. Yet from trauma, volcanoes produce astonishments.

One wonder that emerges from these molten stone gashes, sores, and blisters, these volcanoes of the adolescent earth, is a form of stone scab. It's accidental jewelry. It's black and shiny or quartz-gray and translucent.[217] Its surfaces are surprising—slick in a way that's alien to a planet of gravel. The facets of these shards are as reflective as curved mirrors, flashing and glinting in amazing ways. They are volcanic glasses.[218] They are obsidian. And they are not like anything the earth has known before.

But nature has very little respect for her creations, no matter how brilliantly they shine. She assaults them with desecrators, spoilers, looters, plunderers, and opportunists. She opens her treasures to those who would dare scar them for selfish, material gain. She throws her treasures open to what some of us might call "rapists." And nature's most creative rapists 3.47 billion years ago are the descendants of the first teaspoon of life. They are bacteria.[219]

Bacteria are our foremothers. They are at the base of humanity's family tree. They are microscopic creatures who function in armies of billions.[220] Invisibly thin armies with tight communal cohesion. Organized hordes that specialize in migrating, scavenging, and settling down. Hordes that specialize in growth, expansion, and colonialism.[221] Hordes whose individuals gather knowledge, communicate it in a chemical language, and add it to their society's data base.[222] A database that constantly updates the community on new ways to digest its environment. New ways to plunder the innocent. New ways to feast and feed. And when it comes to the manner of that feasting, bacteria invent.

These hordes obey life's 3.5-billion-year-old imperative: kidnap, seduce, and recruit as many dead atoms as you can into the project of life. Be fruitful and multiply. Show no mercy for the status quo. Violate nature. Invade every crease and cranny. Make cracks where there was perfection. Make rough footholds where there was beauty and smoothness. Gorge on your environment. Do it fast. Do it with speed and greed. Do it with all the ingenuity this cosmos has planted within you. Why?

Because you are up against a deadline. Before the next great catastrophe arrives—the next meteor, the next wave of cosmic rays from a spiral arm of your galaxy, the next climate fry or climate freeze—you have to place your bets on so many possibilities in so many places with so many metabolic innovations that a few of you will survive. But how?

It's not easy. It takes radical disrespect for the current status quo. It takes rapaciousness. It takes inventiveness. It takes making new niches where there were none. It takes violating the environment's "virginity". It takes creating new appetites. It takes inventing new recipes. It takes eating your environment in radically new ways. It takes what Oxford University's John Odling-Smee[223] calls "niche construction."[224]

CHAPTER 13

NATURE SLASHES HERSELF

Remember, once upon a time, roughly 3.5 billion years ago, there was only a teaspoon of life on this planet. And that teaspoon was up against a deadline. In its future were 142 mass extinctions. Mass extinctions caused by guess what? Nature. Yes, Mother Nature herself.

The first generations of bacterial ocean-water-and-stone invaders were rough riders, they figured out how to live off the land. How? By inventing mind-boggling ways to eat that land. Technically they were called autotrophs.[225] They did not need other lifeforms to lay a path and make things easy for them. They themselves were the creators of new paths. They were the pioneers, scouts, and thugs that specialized in kidnapping, seducing, and recruiting truly dead atoms and knitting them into the lofty enterprise of life. And they had no sense of reverence, limitations, or deference.

They were called primary producers.[226] They corroded raw glass and rock. They snatched dissolved carbon dioxide from the water of the seas[227] and pocketed the carbon molecules that they needed to knit protein, cell walls, and DNA. They kidnapped, seduced, and recruited molecules of sulfur dioxide from the seawater and used them as electron acceptors.[228] And they pried atoms of iron and manganese out of nature's jewelry, volcanic glass, cemented those atoms and molecules into their own bodies, and used them as electron donors.

Meanwhile their cousins grabbed virgin methane and hydrogen sulphide emerging from sea floor seeps and used them for food and fuel, thus advancing the common project of life.[229]

Some desecrated the virginal glasses, the mirror-like gems churned out by volcanoes. They despoiled obsidian. Raping its jewel-like gorgeousness. But the plunderers were not impressive. In fact, they were appallingly mundane. They were a mere two microns in size. Two millionths of a meter. A forty thousandth of an inch. Despoiling a natural wonder.

Some bacteria looked deceptively ordinary. They were tiny, Frisbee-like discs. Some were a bit more distinguished. Their surfaces were wrinkled. And others were "bacillar"—they were shaped like sausages.[230] But these harmless-looking little things were more disruptive than they looked. They were capable of clawing nature and slashing their surroundings. They were "endolithic"—they invented new ways to live inside of rock. And they were "chasmoliths"—they perfected the art of living in cracks, fissures, bubbles, and the stone micro-pockets called "vesicles."[231]

Which means that these professional looters showed no respect. They sought out weaknesses—hairline fractures. Then they invaded.[232] They penetrated. And they made the cracks and fractures bigger.[233] To put it in environmentalist terms, they ravaged, pillaged, and raped. They were masters of extraction, colonialism, and manifest destiny, virtuosos of the development of real estate. They built homes. Not individual homes. They built architectural structures that housed entire colonies. They built the equivalent of domes enclosing entire nations.

Some microbial armies constructed tubes of titanium oxide to house their megalopolises. Some built capsules and sheaths.[234] Others laid down massive platforms, "bedding planes"[235] five to ten microns thick on which an entire colony of billions[236] could settle down, platforms embedded with primitive data networks, chemical data webs.[237] And these bacteria became masters of materials engineering.

Like concrete craftsmen impregnating their slurry[238] with gravel, the platform-makers strengthened their building material by mixing in aragonite, calcite, gypsum, quartz, and halite.[239] Minerals "stolen" from their "innocent" environment. Some platform-makers went farther. They manufactured a construction material with the sort of molecular structure we use in plastic—polymers.[240]

We make our plastics by linking identical hydrocarbon[241] molecules in long chains. But bacteria cranked out their polymers, their plastics, by linking long chains of identical sugar molecules.[242] By the trillions.[243] This was materialism taken to the nth degree.

What technological hubris! Tubes of titanium oxide? Polymer sheets? Bedding planes? Surely these were Franken-products. Things that had never existed in nature. Things that should never have been. But there's more.

The bacterial manufacturers of these unnatural chemical constructions at the bottom of the sea ravaged the untouched nakedness of the volcanic glasses and the rest of nature's pure rock face. Some microbial societies used their polymers to turn tiny tunnels five to ten microns long into communal homes.[244] Other bacterial infrastructure builders produced what we call biofilms, thin film coatings that housed entire colonies.[245] Or films that functioned like bacterial rainforests, housing colonies of multitudinous species, a plethora of colonies of different kinds of microcreatures that worked together to do their thing.[246] These biofilms were so flexible that they could graffiti every wrinkle of the underside of a complex "irregular pumice-encrusted surface,"[247] clinging to it as skin-tight as spray paint.

The colony-building was often ugly. Some invaders manufactured an ooze, a gel, within which the entire community of colony members could hive.[248] Others mass produced another kind of semi-liquid protective material in which to establish their housing tracts—slime.[249] And some oozed a noxious sludge that required a complex chemical trick, a materials engineering high-wire act—a metabolic strategy, an industrial process—called fractionation.[250]

Fractionation is the process that we would someday use in oil refineries. It churns out foul-smelling wastes. The end product for bacterial fractionators was a black, sometimes solid, sometimes liquid hydrocarbon called bitumen.[251] A pollutant. The stuff from which we get oil and natural gas. In other words, the first lifeforms used the sort of industrial processes that we are taught to loathe. The sort that we call environmental nightmares.

But the lifeforms that showed up to eat what volcanic and tectonic conflict had produced were not satisfied creating environmental outrages in the cracks of volcanic glass. They also invaded the volcanic sands and dusts in between these cracks. They infested individual sand particles. They invaded and colonized the space between the particles.[252] And they went beyond the sands, overrunning and plundering the tiny bubbles left behind when lava cooled to rock.

The volcanic glass looters, methane eaters, and hydrogen sulphide swallowers were not alone. They had trillions of trillions of distant relatives. The descendants of the first teaspoon of life turned themselves into a vast family of lifeforms that insinuated themselves wherever they could invade. The extended family of plunderers specialized in twisting nature's arm and distorting her beyond belief.

Audacious bacteria even worked where "mantle rock has been thrust up through the seafloor, exposing it to seawater and serpentinization."[253] They exploited and desecrated the "chemical rich water [that] oozes from the seafloor... around the edges of continents."[254] They even invaded rock over 900 feet beneath the bottom of the sea and drew "their energy from chemical compounds, hydrogen and sulfates—produced by the slow decay of radioactive elements in the rocks."[255]

These extended clans specialized in upsetting "nature's balance" by turning nature's tortures into treats. Bacteria feasted in the "abyssal plains that cover vast stretches of the deep."[256] They feasted over 2.5 miles beneath the surface of the sea where the temperature was just above freezing. They feasted in the inconceivable depths where it can be "extremely cold, utterly dark, often low in oxygen, and smothered by a pressure 1,000 times greater than at the surface—a pressure so immense it alters biochemistry."[257] Yes, bacteria ate their fill in pressures so immense that the crush alters biochemistry itself.

And these ravenous hordes feasted in deep water plumes where the thermal weather was like a deep-fat-fryer engulfed in flames, plumes where the temperatures were from 250 degrees to 700 degrees Fahrenheit[258]—nearly 500 degrees above the boiling point. Hells where only the mash of pressure kept the water from turning to steam. The gangs of bacterial bandits also feasted near eighteen-story-high undersea chimneys of heat and chemicals called "black smokers."

Meanwhile, the clan also sated its appetites around heat vents with a very different chemistry, a chemistry whose dominant colors were not black, but white and gray,[259] underwater spires and cliffs where the hordes of microorganisms tunneled into "the porous channels and crevices of the carbonate towers," where they attached themselves in "polymer-encased biofilms" to the outsides of the chimneys, where they established footholds in the chimneys' dark, oxygen-less insides,[260] and where they formed disgustingly mucus-like "dense strands of filamentous bacteria...in the warm fluids issuing from the summit."[261]

But these invaders didn't just adapt to the way things were, they adapted to the way that things refuse to stay the same. They adapted to nature's very backbone—change. Says the University of Washington team that explored the Lost City Hydrothermal Field twelve miles west of the mid-Atlantic Ridge, "Results from next-generation pyrosequencing show that the archaeal and bacterial biofilm communities underwent dramatic changes as environmental conditions in the chimneys changed over a 1,000-year period."[262]

English translation? Bacteria like those in the Lost City Field surfed the waves of insult, threat, and change. How did communities of mere microorganisms ride the tides of environmental pitch, plunge, and sway? They reveled in ripping up the "natural order" and putting it back together again in whole new ways. They reveled in transforming every catastrophe into an opportunity.

Bacteria like those of the Lost City Hydrothermal Field even used the chemical energy[263] of rock emerging from the staggering pressures within the earth and decompressing, the "serpentization reaction within the Atlantis Massif,"[264] to pull off "abiotic hydrocarbon production."[265]

And voracious bacterial colonies seized dead carbon and hydrogen atoms and compacted them into a comfort food for specialized eaters called methanotrophs.[266] The meat and potatoes for these gourmet microbes with an offbeat taste was methane, one carbon and four hydrogen atoms.[267] Methane is the core ingredient of the group identity we call natural gas.

The clans of microbial plunderers paid no attention to the fact that nature rebuffed them in every conceivable way. Remember, they were disaster-riders and catastrophe-tamers. They worked in waters almost as acidic as the caustic stuff that you pour down your drain to unclog it, Liquid-Plumr.[268]

You read that right. Life feasted in seas of acid. As you've seen, some bacterial colonies even invented techniques with which to harvest the fruits of another deadly poison, radioactivity.[269] Others invaded "seamounts, the peaks of undersea volcanic mountains,"[270] where up to half the species that showed up to plunder the place rejiggered their genes and invented utterly unique adaptations, utterly unique ways to turn their toxic environment into Franken-snacks and grotesque treats. These peak-conquerors also invented utterly unique new metabolisms, new species, that showed up no place else on earth.[271] The inhabitants of white smokers at the bottom of the sea are still inventing new forms and new ways to eat their environment today.

The result? In some extreme environments "there is at least 100 times greater species diversity" among bacteria "than had been expected."[272] One hundred times more bacterial techniques, tricks, and tools for exploitation than we thought. One hundred times more forms of opportunism and invention whose instruction manuals are written into strings of genes. One hundred times more ways of violating the status quo with bio-technologies. One hundred times more forms of materialism, consumerism, and waste. One hundred times more bacterial ways to "despoil" the virginal and the pristine.

How in the world did the first teaspoon of life take hold on a planet of climate catastrophe? Life's secret will not come as a surprise. Life did not triumph by living in harmony with its environment. Far from it. Life survived by lashing out against the "natural order." Life survived by sinning against nature and inventing a freakfarm of Franken-forms. Franken-forms able to eat the roiling disaster of their environment and turn it into bio-stuff.

Life survived by radically reinventing the hell of stone that nature had made. In fact, life survived by remanufacturing the planet. Remaking it right down to its very chemistry.

As you and I are about to see.

CHAPTER 14

HOW PILLAGE AND PLUNDER PLOWED A PLANET

Wait, what's this claim about exploitation, violation, and despoilation? Not to mention colonialism and growth? Isn't this just anthropomorphic fantasy? No, it's a demonstration of how nature used materialism, waste, and vain display to bioform the place. To green and garden a poison pill of stone, a hostile planet. To green and garden a planet of climate catastrophe. But there's more. See what you think.

The first wave of volcanic glass desecrators were damned good at what they did. And what, pray tell, was that? Acquiring material things. And turning them into gadgetry.

They snabbed, grabbed, corralled, hoarded, chewed, and used the stuff that a materialist covets and accumulates. They clawed pieces of the existing order and used them as Lego blocks to assemble outrages, alien engines, materials, and processes that should never have been. They pulled together matter with an inventiveness that makes Leonardo da Vinci, Steve Jobs, and Elon Musk look like midgets of ingenuity.

For example, they turned the crushing pressure of the deep sea into an energy source.[273] Small as they were, the bacteria turned hundreds of billions of tons of dead matter into bio-stuff. Then they showed a callous indifference to the sanctity of their environment. They junked their waste.

You heard me right. Nature's first daughters, her bacteria, were trash producers, litter scatterers, and garbage heap makers. They were industrialists without a conscience, mass-producers of detritus, ruins, and decay. They left old, played-out colony-sized homes behind. They trashed the place with their titanium oxide tubes and their layers of bedding, their sewage, and their rubbish. But their environmental indifference paid off. Their outrageous irresponsibility and their destructive, exploitive atrocities produced a gift.

With their garbage, the first generations of bacteria laid the base for life. They created the "soil" in which new generations would thrive. They provided the riches that would make radically new "microenvironments."[274] To quote Frances Westall, director of research at the Centre de Biophysique Moléculaire in Orleans, France, "As the glasses 'age', the autotrophs are replaced by chemoorganotrophs that use the organic carbon of the previous communities as a source."[275] Translation, please? The outrageous, ugly, materialist trash of the first generations was the treasure on which next generations would thrive.[276] But that's not all. Housing materials and layers of garbage—layers of discards, litter, and material waste, trash pits of manufactured goods and discarded technology—stacked on top of each other over time and generated end products like "microbial mats."

What are microbial mats? "Layered biofilms."[277] Generation after generation of materialist exploiters, of microbial growth-addicts, piled up their abandoned housing and remains and destroyed the nature of their environment utterly. Or, to put it differently, the microbial growth-addicts raped the virgin rock and desecrated the stoic purity at the bottom of the sea. But with their mass production of garbage, litter, and waste, they generated a revolutionary new bottom line. A big one—a gooey, mucky, dirty one. It's a substance that a fictional God would someday use to make a fictional Adam. The microbial rapists made mud.

Yes, mud would soon trash 70% of the seafloor of the world.[278] Mud would soon cover, bury, and dirty the seabed's pure and virginal face. Mud would soon be one of the most important forms of waste in the history of the cosmos. Mud would soon be one of the biggest makers of change.

The bacterial rapists would turn raw rock into mud flats that would eventually be six feet deep.[279] They would make so many layers of abandoned housing and sewage that eventually the pileup would despoil and recreate even the environment's geology. The multilayered garbage heaps would compact and make new kinds of rock—chert. Or the striped structures known as zebra rock.[280] Not to mention the stone that the geologist, sedimentologist, and editor of the *Journal of Paleontology* Brian Pratt calls "microbial biolithics" with "lufa-like structures of calcified microbial filaments."[281]

The undersea microbes raping the purity of the early oceans' bottoms would even make limestone, and that limestone would submerge beneath the seabed, heat, liquefy, and become the next asthenosphere, the next fluid base of liquid rock on which new tectonic plates would ride.[282] Yes, life would even remake the planet's geology. How dare it.

From dust to dust and from rock to rock—but not really. From rock to mud. And the difference would be huge.

***

What is going on here? Why all this despoliation, exploitation, desecration, and rape? Why all of this Franken-plunder? Why all this garbage? Why all this litter on an orbiting gravel pile that was still virginal and new?

Life was on a growth binge. Life was expanding from a teaspoonful to a handful, from a handful to a truckload, and from a truckload to a mountain or two. Of what? Of something that blows apart the previous rules. Of something improbable. Of something utterly impossible. Bacterial teams with persistence, hunger, and audacity. Bacterial teams with factory-like capabilities. Bacterial teams with an industrial base. Bacterial teams hell-bent on manic mass production. Bacterial teams indifferent to the existing "harmony." Bacterial teams insistent on remaking their environment utterly. Narcissistic bacterial teams. Bacterial teams obsessed with grabbing material stuff, "consuming" it, and making more copies of, guess what, themselves.

But these greedy, narcissistic bacterial teams were expanding a public works project. They were growing a global infrastructure, a collaborative initiative advanced by the efforts of every cell and colony. They were growing a new coat that would recostume an insecure and tempestuous gravity pile circling the core of a mid-sized[283] fireball, a mid-sized sun. They were growing a coat that with persistence would eventually swaddle and swathe the baldness of this gravity ball's stone. They were growing the garment that would soon cushion this planet's seizures of catastrophe. These molecule teams were growing something this solar system had never seen before—a biosphere. They were making nature into something that nature had never been. They were turning an orbiting slag-heap green.

***

To see how, let's go back to mud. Remember, mud was a toxic pollutant. It was the ultimate scar of consumerist and materialist desecration. It was waste. It was refuse. It was sewage and garbage. It was dirty. It soiled the virginal landscape. Hence another of mud's names: "soil." Yet mud would be the substance from which all that we call nature would arise.

Yes, pillaging microbes made the very soil in which others would thrive. They turned a mashing, crashing, angry, twitching ball of barren stone into the beginnings of a rock that today is named after its coat of trash, its muck, its garbage, its sewage, its litter, its dirt, its filth, its materialist, consumerist waste—its mud. Those looting microbes turned a planet of catastrophe into... an earth.

But we are getting ahead of ourselves.

CHAPTER 15

THE VIRTUES IN VICE: THE OUTRAGEOUS RULES OF LIFE

"A junkyard contains all the bits and pieces of a Boeing 747, dis-membered and in disarray. A whirlwind happens to blow through the yard. What is the chance that after its passage a fully assem-bled 747, ready to fly, will be found standing there? So small as to be negligible, even if a tornado were to blow through enough junkyards to fill the whole Universe."
— Fred Hoyle, *The Intelligent Universe*, 1984[284]

O ver 3.5 billion years ago, when our first foremothers, the first bacteria, appeared in the roiling, boiling seas of earth, the odds against the survival of the first teaspoon of life were trillions of trillions to one. Every sane calculation of probability shows that a final apocalypse, an end of life, was inevitable. Why? There were vastly more ways that life could fall apart than the tiny number of ways life could fall together. Yet life survived. Why?

Because there were commandments, algorithms, built into life at the deepest levels. Built into its biology. And those commandments were not what we think of as natural. In fact, they made a mockery of two laws that science believes are basic to the very nature of this cosmos: entropy and the law of least effort. And life's commandments went against the current notion that humankind is a cancer, a tumor that's about to destroy the planet.

To put it in blunt and ghastly human terms, life's imperatives were viciously immoral. And ethically unacceptable. Life's rules were colonialist, expansionist, ambitious, greedy, arrogant, acquisitive, and murderous. Those commandments demanded growth. They called for materialism, consumerism, and waste. And they would soon call for another sin, vain display. Why?

Rule number one of life was manic mass production. The Bible did a good job of summing it up: "Be fruitful and multiply, and fill the earth, and subdue it."[285] "Multiply your offspring as the stars of heaven."[286] Kidnap, seduce, and recruit every dead atom in sight. Weave it into the grand enterprise of life. Make as many copies of yourself and your communities as you can. Then make even more. Bulk yourself up. Increase your numbers. Expand the size of your societies. Expand the territory you control. Grow a macromolecular web of information exchange. Grow a global brain. But most of all, increase your total biomass. Increase life's share of the atomic weight of the planet.

How? Overpopulate the place. Every time you reach what seems to be the carrying capacity of your environment, open up new realms of nothingness. Reinvent every hell. Turn deserts and dire straits into new frontiers. Turn toxins into treasures beyond your ancestors' wildest dreams. Every time you reach the limit, invent a horizon beyond it. Invent an untapped landscape of abundance.[287] How? How do you invent resources?

In evolutionary biology, it's called "niche construction." You invent ways to mine the impossible. You invent new tools of exploitation. You invent new tricks. You invent new natural technologies.

- Nothing but rock in sight? Fine. Invent new ways to eat rock and turn it into food, fuel, and biostuff. Invent chemolitho-autotrophy.[288]
- The waters around you are impossibly acidic? Fine. Invent membrane pumps that turn an acid bath into a paradise. Invent acidophily.[289]
- Ghastly level of salt pollutes the water that you'd like to make your home? Terrific. Rejigger your internal chemistry to keep the salts out of your cells and to make over-salted seas your private paradise. Invent halophily.[290]
- Threatened by puddles of poisonous ammonia?[291] Terrific. Invent metabolic tricks that make ammonia your favorite energy drink.[292] Congratulations, you've created ammoniaphily.[293]

Which leads to the rule that makes all this possible: invent. Reperceive your surroundings. Remanufacture your environment. Bring things that have never previously existed into being. Turn catastrophes into opportunities. Turn poisons into gourmet buffets. Turn toxins into treats. Turn terrors into novelties. Turn every wilderness into a promised land and every danger zone into a delight. And invest in transportation. Be restless—spread, move, migrate. Take over every emptiness within reach. And stretch to lands and seas far beyond your parents' wildest dreams. Explore, grope, scope, test, and travel. Adventure. Take big risks and seek your fortune everywhere. But, again, don't just find new frontiers. Invent them.

And there's one more tiny trick. Harness the power of paradox. Harness the synergy of opposites joined at the hip. In this case, there is an extraordinary tool that creates new forms of teamwork. It's teamwork's opposite: competition. Insane, artificial, trivial competition. Just ask any NFL, NBA, or FIFA team owner. Create differences. Then flaunt them. Create face-offs. Again, generate competition and you will unleash cooperation beyond belief.

For example, gather your genes in a highly-structured nucleus, a central data library surrounded by fortress walls, and use the advantages of your armored nucleus to outcompete conservative cousins who insist on keeping the genes at the center of each of their cells free-floating in an unprotected necklace. The naked-necklace-of-genes adherents were bacteria. Prokaryotes. On the other hand, the creators of a fortress surrounding a library of genes were eukaryotes. And eukaryotes would be the ancestors of multicellular life. Eukaryotes would be the ancestors of you and me.

But there's more. To compete, grab hold of smaller bacteria, imprison them within your cell walls, and work out a deal. Give your prisoners protection and feed them what they most like to eat. In exchange, coax them to excrete their sewage in a form that is candy to you. Develop organelles. Develop the power sources called mitochondria. Develop the sunlight harvesters called chloroplasts.

Then use your newly-invented teamwork, your new form of group identity, to make your cousins look silly. Not just your old-fashioned cousins with the undefended rings of genes, but your newfangled cousins with nuclei. Go nuts setting up barriers and distinctions. Set up combats between us-versus-them.[294] Separate into the equivalent of Dodgers versus Red Sox versus Yankees. Better yet, separate into the equivalent of soccer versus football versus basketball. Separate into archaea, eukaryotes, and protists.[295] Separate to invent radically different group identities. Why? To increase your odds of finding whole new ways of making a living in the killing splatter, boil, and freeze. To increase your odds of researching and developing new genes and new "phenotypes," new cell bodies. Spiral twists, propulsive whips, rod-like bodies, thread-like bodies, spherical bodies, bodies that move by oozing, bodies that build crystalline boxes around themselves, bodies that glue themselves in place, bodies that scoot, scud, and race.

Competition is wasteful. It eats up time and energy. But do not be deceived.

Competition is a form of cooperation in disguise. Competition drives invention. Competition pits materialist communities against each other and forces them to build up surplus, then to expend it needlessly. Competition pits bacterial societies against each other to see who can throw the most away, who can afford to generate the most waste. One bacterial colony will use weapons of mass destruction, chemical weapons, to utterly exterminate another,[296] leaving the landscape littered with bacterial corpses. Leaving the landscape littered with lives that have been kindled in matter, then thrown away. Yes, bacteria make war. Like humans in war who leave behind vast numbers of destroyed tanks, planes, and dead young men. Competition is consumerism run amok, and consumerism is a curse. Right? Surely cooperation would accomplish more for less.

And with a whole lot less stress. But in the natural world, that is not the case.

If you are the first teaspoonful of life, how will you survive 142 mass extinctions? By gathering surplus. The more surplus, the more atoms you will kidnap, seduce, and recruit into the grand scheme of life.

And you will survive by using competition as the ultimate way to energize a global public works program—the ultimate way to explore the farthest nooks and crannies of possibility space and to bring your discoveries back to your companions and even to your competitors. To beat you, your competitors will ape you. They will learn every new trick that you've invented. Competition creates a massive information pipeline with your enemies. You spy on what they do and in some cases you imitate it. Or you improve on it. And they spy on you.

Meanwhile, competition increases the solidarity within your group. Competition boosts cooperation. This is the power of paradox. This is the power of opposites joined at the hip. Opposites are often the flip sides of the same process. So you—life—will survive by spending big time on competition.

You will survive by using yet another form of materialism, consumerism, and waste—gambling. You will survive by using one of the most grotesque forms of excess this cosmos has ever created. Over 3.5 billion years, you will survive by finding more than a trillion trillion trillion ways to place your bets.[297] You will create over a trillion trillion trillion organisms. Each one will be a feeler into the landscape of possibility space. Each one will be an antenna of a cosmos probing her potential. And each one will be subject to the ultimate form of disposability. The ultimate form of one-use throwaway. Death. The ultimate form of consumerist waste. A form that nature wallows in. And a form that we must oppose.

The result?

The next time a planetesimal the size of a city splats the sea with the force of 4.5 billion atomic bombs,[298] the next time a cloud of galactic fluff plunges the temperature into the deep freeze,[299] the next time the carbon dioxide goes up by a factor of two hundred and fries the place, and the next time a mega-volcano turns the world you're living in dark as night for three years at a time, you will be prepared.[300] You will be prepared not just to hunker down in energy-saving mode and tough it out. You will be prepared to rejoice. You will be prepared to be fruitful and multiply.

No light? Fine. You have found ways to make your energy from hydrogen sulfide or iron. Hot as Hades? Terrific. You have evolved microbial forms that thrive in the 700-degree temperatures of thermal chimneys in the deep sea. Galactic fluff plunges the planet into the deep freeze? Terrific. You have produced bacteria who get a kick out of living on the underside of ice.[301]

Put it all together, and you will have a force of nature that thrives via what some of us mistakenly call rape. Not to mention a force of nature that exults in gross disrespect for the status quo. A force of nature that exults in indiscriminate consumerism, industrialism, materialism, litter, and waste. A force of nature with which nature reinvents herself. You will have life.

CHAPTER 16

SHITTING EARTH

Remember, new resources are not found, they are made. They are invented. Three billion years after the first teaspoon of life spread out and turned the savage liquids of the sea into an organic soup and the dark and threatening bottom of the oceans into a fertile goop, some strange lifeforms set off on a voyage that once again would give a middle finger to the laws of nature. This time the rebellion would be against more than just toxicity. The uprising would be against one of nature's most fundamental commandments, gravity.

Above the seas there were two nothings: the air and the land. Both were impossible wastes, hostile, empty deserts. Places of danger and barrenness. Places not fit for life. What's more, going there defied a law so basic that even stars and moons have had the good sense to obey it: yes, gravity. Gravity makes a simple demand: "stay down." Down is natural. Up is not. Up is sin. Up is defiance. Up is heresy. Up is uppity. Up is rebellion against nature. And those who are uppity get slapped down. Right?

Going upward to the murderous emptiness beyond the sea would also defy another natural boundary, a barrier that protects life—the surface tension of the waters. Surface tension[302] is the sort of skin on the boundary between water and air that you can see at work when it surrounds a detached chaos of water molecules, pulls them together into a droplet, then keeps that droplet intact. As if that droplet was enclosed in a tiny plastic sandwich bag. To get beyond that skin of surface tension would be a struggle. Surely it was there for a reason. Surely nature was trying to tell life something. Something about the sacredness of the only existing biosphere 3.22 billion years ago,[303] the sea.

Which leads us back to the two torture-filled wildernesses beyond the womb of the waters. One was a vast emptiness where murderous radiation reigned supreme and temperature changes were appalling—the air. The atmosphere. The other was a barren waste with a face of naked stone. Empty rock. Rock where you could starve for lack of food and where you could dry down to a lifeless flake for lack of water. Rock where you could be poisoned by the downpour of ultraviolet radiation from the sun, where you could be pounded and washed away by the unpredictable rain of liquid from the clouds, and where you could be tortured by the massive climate changes of summer and winter, night and day. The rock face above the seas was no fit place for the living. That impossible height would someday be called land.

But life doesn't just find what's fitting. It makes things fit. It makes the impossible proper, the toxic tasty, and the barren rich. To repeat, life dared break nature's most basic law: gravity. Life thrived by aspiring high. Life thrived by aching for the sky. Life thrived by mounting space programs.

Roughly a billion years after life began,[304] a few brave bacterial search parties managed to struggle through the surface tension barrier, managed to break through the skin of the waves, and to rise up, making their way to the sterile and hostile rock face beyond the sea, rising from the nurturing embrace of the waters to the raw, wind-scoured, sunlight-battered stone of land. It may have been an accident. These pioneering teams of microorganisms may have been left behind when a flood ended and the waters pulled back. They may have been abandoned when puddles dried up. Left behind to die in a yawning empty space on high. But no matter how accidental, the move beyond the waters, the move up, the new space program, paid off.

On the sheer rock face, life showed once again how it creates new frontiers by inventing new ways to transform terrors into treats, new ways to rape old purities, new ways to desecrate the status quo. As geographer Denis Wood put it, the first bacteria, the first prokaryotes, to hit the land,

> slimed everything, their sheen was on every surface. They painted rock and river bank and mudflat and pond with crazy colors, with chemical colors, acidy and sharp, hallucinogenic oranges and aquamarines, and brilliant reds and greens.[305]

Why the flamboyance? Why the colors? I suspect it was for the sake of competition.[306] My community versus yours. Differentiation. Differentiation based on different strategies for eating your environment and for dodging its dangers. My community flashed one color. Your community flashed another.[307] And we competed.[308] But that's just a hunch.

Continues Wood,

> Some of the colors came from pigments bacteria had evolved
> to hook up the energy of the Sun. Others came from pigments
> they'd evolved to protect themselves from its ultraviolet ra-
> diation. The earliest photosynthesizers, anaerobic green and
> purple sulfur bacteria, came in all kinds of pinks and greens.[309]

Among these color-wearers were some of the greatest material
miracle-makers in the history of the cosmos, astonishing new
transformers of terrors into treats. These wonder-workers grabbed
photons of light machine-gunned from the sun, photons of the
sun's toxic waste, incoming bullets of radiation, and turned them
into energy sources. They used these bullets of light to start a flow
of electrons that powered the manufacture of new molecules,
molecules that defied nature—carbohydrates and sugars.[310] They
turned the toxin of radiation into the gift of light.

The result graffitied the natural landscape. Says Wood,

> Cyanobacteria, the earliest oxygenic photosynthesizers,
> bloomed scums bright and thick like oil paint. Some of the
> scums were green. Some were blue-green. Others were pur-
> ple.[311]

Desecrators of the natural landscape were innovators. What's
more, waste and garbage piles were life-makers. And lifesavers.
Explains Wood:

> Because the photosynthesizers ran off the energy of sunlight,
> they couldn't bury themselves in the mud like other bacteria
> to hide from the destructive energy of the ultraviolet. Instead

they shaded themselves with colorful mats made of the car-
casses of bacteria that had died from exposure to the radiation.
Or they learned how to tan.[312]

These first bacteria to invade the land were not nature-lovers; they
were nature-looters. They were industrialists massively mining the
environment. And changing it. Like stripminers, says Wood,

> They pushed trillions of tons of gases and soluble compounds
> around through the air and water.[313]

The bacteria who fed off the volcanic glass at the bottom of the sea
had shown no respect. And the microbes invading the land were
just as bad. They poked, probed, and opened cracks in the planet's
pristine rock face. They produced chemicals that penetrated the
tiniest orifices among the cracks and turned this third pebble from
the sun's outermost crust of rock, its pure and virginal skin, into
powder. And worse. Microbes defecated mineral particles from
which new rocks would be made.[314]

All of this savaging of the land was literally a big step up. Why? It
gave nature the finger. To repeat, it violated a basic law of nature:
gravity. And in the process, it changed the balance of life. It tilted
that balance from a majority of life in the sea, to a majority of life in
the forbidden realm on high.[315] A majority of life on the land.

CHAPTER 17

IS NATURE AN INVENTION ENGINE?

Let's rewind. 3.22 billion years ago,[316] the first microbes to test the land turned rock, light, and rain from dangers to delicacies, and turned chemical catastrophes into food, fuel, and shelter. They turned a nothing into an everything. They turned a terrain of tortures into a home.

Meanwhile, they laid the rock face waste and they laid their waste—their litter, trash, garbage, and sewage—all over the place. But that waste was not what it seemed. Yes, it was materialism, consumerism, and one-use throwaway culture run amuck. But it was not a desecration. It was a consecration.

It was a new layer of the sacredness we call green. A new layer of the biosphere. That microbial waste was the first hint of what we'd someday call topsoil. It was the first layer of what we'd someday see as fertile land. It was the first layer of what we'd someday call nature's resource base. To repeat, it was the first taste of what we'd someday call our planet, "earth." And it was something more. Bacteria's materialist, consumerist waste was the first layer of what we'd someday call... nature.

All of this came from a race to beat the life and death deadline of the next extinction. To beat the temperature deadline of the next nightfall, the freezing deadline of the next winter, the killing thirst of the next summer heat, the battering of the next rainstorm, the scouring forces of the wind, and the caustic deadlines of ice ages, global warmings, famine, and drought. All of this came from the race to beat climate change, the race to beat the lethal mood swings of Mother Nature herself.

***

The bludgeonings nature meted out were not easy to overcome. But there was a danger just as great as earthquakes and climate catastrophe. There was the threat from a garbage tossed away by life itself: pollution.

Especially pollution of the barren waste above, pollution of the atmosphere. As National Medal of Science-winning biologist Lynne Margulis and her son, ace science writer Dorion Sagan, write in their brilliant book *Microcosmos: Four Billion Years of Microbial Evolution,*

> No doubt in the first few million years of life's tenure, each "famine," change of climate, or accumulation of pollution from the microbes' own waste gases always extinguished some and probably sometimes almost all the patches of life on the face of the earth.[317]

What are Margulis and Sagan talking about? What do they mean when they write of the life-extinguishing power of "the microbes' own waste gases"? And what do they mean when they say gaseous wastes from microbes extinguished almost all the patches of life on the face of the planet? They are pointing to one of the most extreme examples of atmospheric desecration, of air pollution, this planet has ever seen.

It started with the invention of photosynthesis by cyanobacteria 3.5[318] billion years ago. The invention that allowed mere bacteria to snag bullets of radiation, photons, and to use them to turn the wheels of life. With every invention of a new biotechnology comes a problem. Organisms eat the bits of their environment that their new metabolic machinery turns into food. And they excrete what they can't digest. Often those excretions are a poison.

In the case of cyanobacteria, the first photosynthesizers, the new photosynthetic process spilled out a gas that was fatal to most forms of life. Each bacterium expelled the toxin from its system by farting[319] it out.

Now remember, bacteria are tiny. A single bacterial colony of seven trillion citizens contains more individuals than all the humans who have ever lived. Yet it is only the size of your palm.[320] And it is so thin that you cannot see it. So what's a little bacterial fart in an atmosphere of 31 sextillion cubic feet of gas? Nothing. Right? But over the course of the next billion years, those bacterial farts would build up. Finally, 2.4[321] billion years ago, the toxic pollution would reach a tipping point.

The toxic farts were so thick in the atmosphere that they precipitated one of the biggest species die-offs, one of the biggest mass extinctions, this earth has ever seen. Thanks to the waste gases farted by photosynthetic microbes, bacterial colonies died off left and right. Astronomer Phil Plait calls the result "an apocalypse that was literally global in scale, and one of the most deadly disasters in Earth's history."[322]

The name of the gas with which the cyanobacteria poisoned the atmosphere? Oxygen.

The oxygen-driven mass die-off is known as "the Great Oxygenation Event."[323] But it's more accurate to call it The Great Oxygen Catastrophe.

The sins of materialism, consumerism, and waste do not always work out to life's advantage. Or do they? Margulis won her National Medal of Science for proving what some microbes did next.[324] As you've seen before, they kidnapped smaller bacteria capable of turning the poisonous gas into a power source. They offered those smaller bacteria a deal—I will give you a safe home with all the food you can eat. In exchange, you will live inside of me, you will eat this toxic poison, and you will allow me to use what you poop out as my power source.[325] As you know, the toxic gas that fed the boarders was the new atmospheric poison: oxygen.

What did the bacterial kidnap victims use oxygen to poop out? ATP,[326] adenosine triphosphate, the energy-storing battery pack of life.[327] The new bacterial boarding houses were the ancestors of the cells that would someday make up multi-cellular plants and animals, animals like you and me. And their boarders, their guests, were the ancestors of the trillions of mitochondria energizing you at this very second with ATP.

The invention of the boarding house strategy was not the end of this apocalypse. There was a food vital to the photosynthetic process: carbon dioxide, CO_2, the greenhouse gas that protected the planet by keeping it warm. Over the course of more than a billion years, the eager photosynthesizers sucked up so much of that greenhouse gas that they triggered a global ice age.[328] Well, actually far more than a mere ice age. The CO_2 gobbling set off a period known as Snowball Earth.[329] A climate freeze you've already seen, the one in which the ice at the equator grew over half a mile thick. Yes, lovely green bacteria were behind that disaster.

There are two bottom lines. Number one: nature herself invented species extinction. In fact, she has used it so enthusiastically that today 99 percent of all the species that have ever existed are extinct.[330] Over four billion species of plants and animals have died out.[331] Over four billion splashes of cosmic creativity, of cosmic exuberance and flamboyance, are gone. And most of these extinctions happened long before the arrival of humankind.

Bottom line number two: species extinction has often led to something surprising: radical invention. Invention that changes the game. Invention that demonstrates something basic. Nature is an invention engine. Nature is a transcendence engine. She has the ability to utterly reinvent herself. To put it differently, nature's most potent power is not her ability to maintain a status quo. It is her ability to disrupt it. It is her ability to generate breakthroughs. It is her ability to utterly resculpt the nature of reality.

What does that mean for you and me? We will soon see.

CHAPTER 18

NATURE'S THIRD GREAT SPACE PROGRAM: SKYSCRAPERS OF GREEN

Nature advances by accomplishing the impossible. Nature reinvents herself.

In part, via space programs.

3.22 billion years ago, giant societies of single-celled organisms dared to tempt nature's wrath.[332] They embarked on the first in a series of life's great space programs. They set forth on a crusade to populate a toxic emptiness that lay above them. They defied the natural order and left the waters. They turned their backs on the very womb of life—the sea. They dared to pioneer that hostile, stony, doomscape we call "land."

These cheeky micro-beasts would have the audacity to live on seacoasts, in puddles, and in ponds. You've met these fearless creatures a minute ago. Formally these risk-takers are called cyanobacteria. We know them more colloquially as blue-green algae, as pond scum.[333] What was the risk these land-invading pond scum would face? Periodically their puddles and ponds would dry up. And these wee beasties could not live without water.

But some of the pond scum would not take the natural disaster of draught lying down. Over three billion years ago,[334] they harvested Armageddon. They learned to adapt to dryness.[335] Among other tricks, they invented housing. They surrounded themselves with the tough, Saran-wrap-like structure called a biofilm.[336]

And these widely-scattered, arrogant scabs of green dared use their new skills to do something suicidal. They left their ponds and puddles behind. And they built their housing on the bald and barren stone.

Two and a half billion years later,[337] these catastrophe-tamers, these pioneering land developers, evolved multicellular descendants, plants. The first land plants were bryophytes.[338] They were mosses and liverworts. But when it came to space programs, bringing a thin and tentative coat of green to an impossible wasteland of stone was not enough. The first mosses and liverworts took things a step farther. Like the first land-grabbing bacteria, they brazenly defied one of nature's most basic laws: gravity. They lifted tiny spore-shafts to the sky and formed a green shag a breathtaking inch or two high.

An inch or two doesn't sound like much, does it? But that's the human equivalent of erecting a building 28 miles high. And the new mosses and bryophytes erected skyscrapers like this all over the place. Remember, nature favors those who oppose her most.

The mosses and liverworts were tiny swatches on a vast and murderous rock face. Tiny huddles of green trying to stay intact and thrive on a landscape scraped and hammered by catastrophe. What's more, they were thwarted by death. To flourish, they had to multiply. But how? By using three of nature's favorite sins: materialism, consumerism, and waste.

First the mosses and liverworts upgraded an old technique that their ancestors in the sea,[339] bacteria, had invented—the spore. In sporulation—in spore making—you accomplish an impossibility. You pack your collection of genes, your genome,[340] into a tiny baseball-or-egg-shaped bundle. A bundle so small that it's invisible to the human eye. A bundle a mere 24,500th of an inch. Then you shoot, lift, or drop your micro-package of genes into a water current or a breeze and take your chances. You litter. You spread outrageous amounts of waste. All over the place. And you do it very deliberately. Why?

You are hoping for something that will prove crucial to life—transportation. You are hoping to hitchhike. And you are entering a lottery. The more tickets you buy, the more your odds of winning. So you gamble that two out of a billion spores will land in a corner rich in something to eat. You play the odds. And to do it, you make far more spores than can ever find a home.

You are materialist, consumerist, and wasteful. You, a single individual, crank out spores by the trillions.[341] You use those spores to spread out like finger tips, feeling for opportunities. But to pull this off, you throw millions of spores away.

So nature uses your materialism, consumerism, and waste to explore. To act as a search engine. To feel out her potential. To find unlikely possibilities.

But simply playing the odds isn't enough. You early land plants reach out for something more than mere survival. Evolution drives you to invent. You adopt a process introduced by a tiny single-celled sea creature known as "the last eukaryote common ancestor."[342]

You adopt...sex.

In fact, 99.9%[343] of you eukaryotic creatures will succumb to the addiction of sex. And sex is one of the biggest mysteries staring us in the face. A mystery that science may understand far less than it thinks. A mystery that may challenge our very notions of the way this cosmos operates. Why?

As you know, Aristotle is credited with saying that, "Nature operates in the shortest way possible." Nature always takes the shortest path. And Pierre Louis de Maupertuis backed that up with his principle that nature always takes the path of least action.[344] But sex proves that this is radically untrue.

Sex has become so ubiquitous that even de Maupertuis could not avoid it. In 1745 he wrote a book, *The Earthly Venus*. A book about sex. Austere as he tried to make his thinking, he could not evade the obvious, the urge that haunts our thoughts every day. And sex, he said, has a legitimate place in science. "There is no need for ridiculous prejudices to spread an air of indecency over a subject which has nothing indecent about it," he wrote.[345]

But sex has implications far beyond what de Maupertuis imagined. Implications that come from modern genomics. Sex proves that nature sometimes goes beyond the shortest possible path and invents not just new paths, but whole new highway systems, whole new knots, snarls, snags, and tangles. Such intricate snarls and tangles that even a hedge fund accountant could not keep them straight. If this were a thrifty cosmos, how could such a flamboyant dance of knots and strings possibly come to be?

What's more, sex may force us to modify Charles Darwin's idea that evolution is driven by a "struggle for survival."[346] As we will soon see, sex is not just a survival device. It is a macromolecular call for exuberance. It is an intricate ballet that exceeds your wildest dreams. And mine. The existence of sex implies that this cosmos is not in a struggle to merely hang in there. This cosmos is not in a struggle for mere survival. This cosmos is in a struggle for flamboyance. This cosmos is in a mad rush for the power to exult, to rejoice, and to do a victory dance. This cosmos is in a mad rush to turn impossibilities into new realities. This cosmos is in a mad rush to invent.

***

Yes, sex is the biggest entropy-defier and least-effort denier in this planet's history. To use sex, you, a land plant 450 million years ago,[347] do not take the simplest path. You do not simply live forever. Or you do not just split in two and make identical copies of yourself. And you don't make kids by packing spores with the simplest, thriftiest thing, a complete packet of your genes.[348] You don't just fill a spore with an everything-you-need-to-start-your-own-plant kit. A deed that in itself would be mind-exploding.

No, you don't take the shortest path. You don't bank on the tried and true. You don't simply keep the chains of genes you've inherited from your ancestors, then pass them down to your kids the way that you got them.

In fact, in sex you don't "reproduce." You don't make carbon copies of yourself. Instead, you perform some serious genetic engineering. Risky and expensive genetic engineering. You work to create something the cosmos has never seen before. Something utterly unique. Something totally untested and untried. Something that could help a curious cosmos scope out her next impossibilities. You generate one-of-a-kind offspring. You create extraordinarily different individuals. And to make those one-of-a-kind kids, you reshuffle your genes. In fact, you reshuffle your entire gene string.

Which means that the greatest example of flamboyance in this universe, the greatest example of nature's urge to splurge, is right here, teasing the back of your mind as you read this sentence—and teasing the back of my mind as I write it.[349] Yes, the cosmos' greatest display of materialism, consumerism, waste, and vain display is sex.

As you and I are about to see.

CHAPTER 19

WHY SEX CAN'T EXIST

It's impossible. It's improbable. And it breaks every rule in the universe.

You are the possessor of the most complex factory ever conceived. It's called your genome. And you have 100 trillion[350] of these jaw-dropping mass-production machines operating inside you as you read this sentence.

What's your genome? It's a tightly integrated collection of chains made from your complete collection of genes. All 30,000[351] of your genes. And you have a hundred trillion copies of these gene machines, these gene strings, these genomes. Why? Because you have one genome at the center of each of your hundred trillion cells.[352]

Yet your genome's existence is an impossibility. It utterly defies the odds. It flings a finger in the face of entropy.

Your genome is not a simple, straight necklace. It's not just the long boondoggle of a double helix that you've seen in pictures of DNA. It's not the shortest distance between two points. It's a confusing jewelry collection, a bag of beaded key chains. X-shaped key chains with a combined total of 3.2 billion beads, 3.2 billion base pairs,[353] per genome. Yes, 3.2 billion.

Think of trying to keep a gaggle of key chains with 3.2 billion links in order. Think of trying to keep them all straight. But your genome will someday perform a high-wire act in which there will be no room for mistakes. Get a bead or two out of order and your kids will be monsters. It's called teratogenesis—giving birth to monsters. And in theory—according to statistical probability and the concept of entropy—it should happen every time.

But it doesn't. Why?

The challenge is even tougher than it seems. The genome in each of your cells is a carefully organized team of 128 billion atoms. Yes, 128 billion.[354]

If the genome from just one of your cells were untangled and stretched out end to end,[355] the resulting chain of genes would be over six feet long.[356] Remember, that's the genome from just one of your cells. Which means that if all the genetic key chains from all the cells in your body were stretched out end to end, the resulting mega-chain would reach to the sun and back. Not just once, but 1,224 times.

However that's not the way your genomes are stored. Believe it or not, the chain of 3.2 billion base pairs in each of your cells[357] spends most of its time knotted and tangled in a tiny ball.[358] Like the wires of all the headphones you have ever owned thrown into a drawer and tangled together. Times a hundred trillion. In other words, the ball that is your genome looks like a total mess. But it's not. It is, in fact, very carefully and very precisely packed.[359]

Let's go back to you, the early land plant introducing sex to the naked plains of barren rock 450 million years ago. What do you do with your tangled key chains when it's time to have kids? The sensible thing would be to follow the path of the spore-makers in the sea.[360] Be narcissistic. Be conservative. Make identical copies of yourself. Or, more specifically, make identical copies of your DNA. Then be materialist, consumerist, and indifferent to waste. Mass produce those identical copies. And finally, send those trillions[361] of duplicates out in the world to do their thing. In other words, make copies of yourself. Reproduce. Be selfish. Be narcissistic. That, in itself, would be a stunning feat. But you, the land plant pioneering sex, are not content to leave well enough alone.

You are not in search of the simplest path. You are in search of the most challenging, the most audacious, the most ornate, the most complex, and the most unbelievable.

You use your tangled balls of genetic key chains as a library of blueprints most of the time. Under ordinary circumstances, you use 40%[362] of the material in your genetic key chains to tell your individual cells how to build themselves and once they are built, what role to play. But you set aside a few of your genomes and use them for nothing until it comes time to reproduce. Then you begin a process so hard to keep straight that very few of the brightest people on earth can remember all its details.

First you untangle your genome's mess. Not an easy job. Remember the last time you tried to untangle that ball of headphone cords in your drawer?

Then, when you have your beaded key chains isolated and straightened out, you end up with 23 of them per cell.[363] 23 key chains. But these are not just straight strings of genes. Each of the key chains is really two stubby strands of beads laid across each other and fused in an X.[364] Those beaded X-like key chains of genes are called chromosomes.

What's next?

A process so complex that it boggles the brain. You line up two similar X-shaped key chains next to each other. You line up two similar chromosomes. One has a lot of genes from your dad. The other has mostly genes from your mom.[365] With those two key chains you perform a trick very much like the shuffles a magician pulls off with cards. It's called crossing over.[366] You twist the key chains together and get a few genes to trade places.

Trade places? What does that mean? You get one or two beads[367]—one or two genes per key chain—to switch their allegiance from one key chain to the other. You get the beads to pull off this switch while staying in the precise location it will take to allow the entire jewelry collection of genes—the entire genome—to succeed in its task in life, making a brand-new organism.

Crossing over is like asking just a few pairs of passengers on a train to trade the window seat for the aisle seat. There's a girl with a flowery straw hat on the aisle in row twenty-five. Her date is a guy with a broken nose and tattoos all the way down to his wrists. He's in row twenty-five's window seat with his face mashed against the glass trying to see the view ahead of the train. Ask them to change seats. Now the girl with the straw hat won't be staring at the floor of the aisle between the seats anymore. Instead, she'll be next to the window taking in the view. But she'll still be in row twenty-five. And her tattooed date whose face was mashed up against the window trying to see the view ahead of the train will be on the aisle. Glowering at the black rubber flooring. But he, too, will still be in row twenty-five.

Meanwhile the well-tanned guy in shorts reading a paper on neurobiology on his iPad in row fourteen will have to stop leaning his shoulder against the window and switch to the aisle seat so that the elderly lady with shopping bags next to him can finally see the view from the window seat. And the elderly lady with the shopping bags and the neurobiology-loving guy in shorts, too, will switch seats without losing their places. They'll both still be in row fourteen.

Next comes one of the key tricks that make sex sex. You separate the folks in the aisle seats from the folks in the window seats. As if you were unzipping a zipper. You unzip the aisle-seat folks and set them apart on one single strand and you unzip the window-seat people and set them apart on a separate single strand. Now the muscular guy with tattoos from the aisle seat of row twenty-five is in row twenty-five of a single strand. He's on the same strand as the well-tanned guy with the iPad in row fourteen who you forced to take an aisle seat, too. But they are no longer seated next to their partners in the window seats. In fact, their window-seat partners have been unzipped on a separate bead chain far away.

You pull off this seat-switch with[368] roughly 75 of your 30,000 passengers. Seventy-five out of your 30,000 genes.[369] Which means that, thanks to the seat switching, each of your new single strands does not replicate your personal gene combination. You are not making carbon copies of yourself. You are not duplicating. Far from it. Each new single strand is a one of a kind. It's the only gene team in history to have the guy with the tattoos on the same string as the guy reading a neurobiology paper on his iPad. Which is why none of your kids will look like you or think like you. But more on that one-of-a-kind-ness in another chapter.

This is barely the beginning of sex's intricacies. Now you separate your opposite sides of the zipper. You put a distance between your single-stranded chains of beads. You pull your string of aisle-seat folks away from the string of window-seaters. You drag them away from each other and you shove them to opposite ends of the envelope surrounding them[370]—opposite ends of your cell membrane. Then you pinch your membrane in the center, and turn your cell from one envelope into two. You divide. And you make sure that each new envelope, each daughter envelope, each of your two new cells, has a complete single-stranded jewelry collection.

Believe it or not, sexual cells manage this mad shuffle with strings of up to 100,000 genes. Swapping seats in up to 100,000 rows.[371]

Confusing, right? Hard to understand. So think how appallingly difficult it must have been for nature to invent. Much less for nature to pull off millions of trillions of times.[372] Think of how tricky and energy-wasting it must be. Think of all the ways it could go wrong. Horribly, twistedly, in-the-wrong-row wrong. And think of how incredibly far it is from obeying the law of least action. In fact, it is the most-action process in the cosmos.[373] Yes. Literally.

What's more, if entropy and entropy's statistics ruled, crossing over would fall apart every step of the way. But your genome refuses to obey the law of entropy.

***

In this sexual minuet, what have you accomplished so far? You've gotten your cell to divide. Which means you've gone from having one cell to two. But inside each daughter cell is a collection of single-stranded key chains, single-stranded chromosomes, single-stranded X's of DNA.[374] Single strands slightly different from any single strands that have ever existed before. And those single strands are hungry. Hungry for company. Hungry for a neighbor in the next seat. Hungry to double up again. Hungry to zip together.[375] Which is where another of the tricks that make sex sex comes in.

To have kids, to have progeny, you, one of the first land plants, have to make your single strands double strands once again. You have to find your single strands a partner. Which means that you have to go out and find a mate. Yes, you have to find another creature of your species. One that's willing to entertain a collection of your lonely single-stranded key chains.[376] And possibly to even embrace it. Willing to zip together with it.

And if you think that's easy, you haven't tried it.

Yes, the REALLY hard part of sex is the mate hunt.

CHAPTER 20

GREEN HEARTBREAK: DISPOSABLE MALES

Yes, the REALLY hard part of sex is the primitive precursor of romance and heartbreak. The really hard part of sex is the mate hunt. Remember, you are a moss 450 million[377] years ago. You have ditched the path of least resistance and are taking your chances on the vast and murderous desert of rock that will some-day be called "land."

If there were Bernie-Sanders-style progressives around, they would probably say, "Let's take care of the sea first. Why waste precious resources on this land program? We are distracting our attention from what really counts. Let's take care of the species that are en-dangered and starving here in the oceans, here in the waters that gave us birth."

And these social justice warriors might be right. Your space pro-gram is an overwhelming gamble The persistence of life on land will be iffy. You, the moss astronaut, will be in a moss commu-nity[378] sitting precariously on a shoulder of rock that could shrug you off at any minute. You and your neighbors will be the merest speck of fuzzy green on fifty-seven million square miles[379] of blank and savage stone. Stone made only the tiniest bit hospitable by the cracks that bacteria have expanded, rock faces made only the barest bit livable by the powders, slime, housing, and mats that bacteria have left as garbage in their wake. Not to mention stone made only slightly welcoming by the bacterial colonies still thriving on the rock in biofilms.

What the bacteria left behind as garbage is food for you.[380]

But to compound your sins against nature, you, the plant invading the land and pioneering sex, are about to commit yet one more outrage against the principle of least action and the law of entropy. What sin could you possibly perpetrate next?

You could take another grotesque step toward making sex sex. A step filled with excess, overkill, and agony. A step that makes the law of least action look ridiculous. A step that flings a finger in the face of entropy. Yes, you could try to find a mate.

***

Remember, you are a moss, an early adopter of sex 450 million years ago.[381] In your sex cells, you have put your genome through the complication of "crossing over." You have made some genes switch seats. You have made new versions of your genome that are utterly unique. Then you've unzipped the double helix of your DNA and made single strands. One-of-a-kind single strands. Single strands hungry to double up again.

But you don't let these single strands find solace with another in their immediate vicinity. You don't allow them to take the path of least effort, the avenue of least action, the shortest path between two points. You don't let your masculine single strands pair up with their sisters, sisters only a tiny distance away. Instead, you inflict upon your males an agony.

Roughly 99.999%[382] of your cells with single-stranded key chains, your cells aching for a mate, have a precarious and potentially lonely destiny. They are male. They are sperm. And sperm are high-priced luxuries. They are outcasts. They are unbelievably intricate throwaways. Yes, maleness is the ultimate indulgence in materialism, consumerism, and waste.

Here's how the waste called maleness works. You, a single moss, will produce tens of millions of sperm in your lifetime. You will produce hundreds of thousands of sperm at a time.[383] Hundreds of thousands of extraordinarily intricate, whip-cracking, self-sustaining gene carriers driven forward by the corkscrew spinning of their bodies and their tails.[384] Gene carriers with a single strand of genes aching to zip together with another lonely single strand. Aching to pair up. Aching for a mate. And you, the moss, are a crazy gambler. You are about to do something insane. You are about to callously throw hundreds of thousands of these exquisitely engineered devices away.

As if you were to employ Foxconn workers to go through the toil and grief of manufacturing a few hundred thousand iPads, then as if you were to order the exhausted laborers to simply scatter the electronic tablets on sidewalks outside their Shenzhen factory in the hope that the iPads would find their way from a walkway in China[385] to San Francisco, Minneapolis, and New York.

Let's face it, sperm are as disposable as confetti.

But it turns out that sperm are a waste with a purpose. They are consumerist throwaways with a goal. They are the gambling chips of sexuality. They are the explorers of new possibility. Each sperm is a feeler and a scout carrying one lonely single strand of your genome to a distant mate with whom it can create offspring that are genetically unique. To find that mate, your sperm are built to go into territories far beyond your easy pickings. Far beyond your immediate vicinity. They are built to do things the hard way. They are not built to follow the path of least effort. They are built to follow the path of most effort.

How, pray tell, will you, an ancient moss 450 million years ago, send your male cells, your sperm out into the wild to see how they will do on the mate hunt? How will you send them out to make their fortune?

First, to repeat, you will defy the law of least effort. You will take a giant step to guarantee that your sperm do **not** take the path of least action. You will make sure that your sperm do not fertilize the eggs a tiny swim away from their birthplace. You will make sure that your sperm do not get together with their sisters less than an inch away, your eggs. In fact, you will hide each of your eggs in an expensive,[386] flask-shaped, protective vessel.[387] You will tuck each egg away in a sexual fortress. But that's not all. You will perfect a timing mechanism that shuts each egg-hoarding fortress while your sperm are venturing out. And you will sometimes back up the physical barrier of your fortress walls by making your eggs and sperm chemically incompatible.[388] To repeat, you will lock the law of least effort out of your reproductive cycle. You will freeze that law out. Utterly. You will ban the shortest path between two points. Viciously. You will follow the law of most effort.

What's more, you will make sure that your sperm are tossed as far from the easy pickings of home as possible. And you will be vicious and cruel in expelling those poor males. You will force them to risk their lives on territory utterly unknown to them. Territory they have literally never felt or seen.

But your sins against the shortest path will get worse. You will also defy nature's most basic law, the law of gravity. You will shoot some of your sperm[389] 5.5 to nearly 8 inches into the air.[390] Sounds like a mere nothing. But that's the equivalent of shooting you or me fifty-three miles into the stratosphere. That's a very unnatural form of travel for the cells of a two-inch-high plant whose ancestors formed green or black mats on the surface of the sea[391] and on the rocks of tidal pools[392] 1.2 billion years ago.[393] But you will go farther.

You will build some sperm to race over the landscape. You will make them double-engined microbial sports cars.[394] Microbial Lamborghinis. You will equip each of them with two long whips, two rotating[395] propellers.[396] But even bacteria with a single propeller will be able to swim a distance up to three to six feet.[397] That's the equivalent in human terms of swimming 2,272 miles—swimming from New York to Puerto Cabello in Venezuela. Just to up the challenge, your sperm will only be able to pull off this marathon swim if they are lucky enough to find water. Thank goodness for rain.

Your sperm—including the 99.999% you've built to throw away—are your bets on exploration. Your sperm are your wild, materialist, consumerist, waste-based gambles. They are your males. And males are your flagrant excess. Males are your probes on behalf of a cosmos feeling out her possibilities. And your consumerist, materialist, wasteful risk-taking is a strategy. But it's not your only strategy. Not at all.

You put your biggest bet on its very opposite: the conservative, stick-in-the-mud tactic of the tried and true, the strategy of the devil you already know, the strategy of the sure thing. The strategy of the land you are already familiar with. That's what the other 0.001% of the single-strand-carrying cells you've just manufactured are all about. Yes, that lucky 0.001% of your single-stranded cells, the cells you've mass produced using the sexual process, are your safety backup. Your fallback position. Your sure things. You will cherish them. You will make each one of them ten thousand times bigger than a sperm.[398] You will not carelessly shoot them into the air, scatter them to the wind, or let them trickle away in the rain. You will hug each one deep in your bosom. You will keep her protected. You will build that bottle-like fortress we talked about a minute ago in which you will keep her from harm.[399] Not to mention from incest. Why?

She is a female. And you do not waste tens of thousands or billions upon billions of females. You do not send them out to their death in the hope that for every million that die, one will strike pay dirt and survive. You do not waste them.

You ensure that when the males of another moss, that moss's wandering sperm, that moss's adventurers and knights, make their way to you, those males from a distant moss have already been tested for hardiness and for their ability to outdo a wicked fate. They have already been tested by the terrors of the landscape. Tested by dryness, puddles, obstacles, rivals, and those who would eat them. Tested by the mazes and dangers they have mastered to reach you. You will know that those who have managed to get to your bottle fortress are the clever few who were able to outwit death.

Ahhh, sex. How brutal. How picky. How it breaks hearts. How it holds us in its grip even when our deeds seem to have nothing to do with sex at all. How it drives us even when we are accomplishing the loftiest achievements. Like shooting 53 miles into the stratosphere. Just to find a girl.

***

But there's more. More to make sure that if a male does indeed find a female and enters her egg, the offspring will be unlike any creature who has ever lived before. There's yet another process that makes the genome of each child of sex utterly unique. It's called genetic recombination.

Let's say you are the sperm of a moss 450 million[400] years ago. You are a male. You have out-lucked your millions[401] of brethren and have made it all the way to the long, sausage-like cylinder in the fortress tower where I keep my eggs. You have swum down the throat of that cylinder and managed to make it to one of my eggs.[402] And that egg has chosen to accept you.[403] Congratulations. You have won the evolutionary prize. You have hit the jackpot in the sexual game.

But the sexual process is not finished making your offspring utterly unique. You are in for another shock: recombination. Remember your chromosomes, the X-shaped key chains of your DNA? An X is a cross formed by two single strands of DNA, strands aching to pair up again. But those lonely strands are in for a surprise. Thanks to the entry of your sperm into my egg, your X-shaped chromosomes and mine are now in one cell. A good place to get cozy. A good place to mix and mingle. Which is exactly what happens. Via the intimacy called genetic recombination.

What is genetic recombination? Imagine that you and I are two chromosomes, two X's of genes, tucked cozily into the bedroom of a cell. Our outstretched arms are the upper V of the X and our parted legs are the lower upside-down V. We go to sleep, let's imagine on a king-sized mattress. In our sleep, your leg crosses mine just below my knee. And your X does something very strange. When we separate, we have both been changed. You have my right leg from the knee down. And I have your left lower leg. You have a string of my genes and I have a string of yours. [404] In fact, now roughly 10% of your genes come from me and roughly 10% of mine come from you.[405]

But that's not all. Our swapping point could be anywhere on your leg. Not just from the knee down. From the thigh down. From the ankle down. Or from anyplace in between. One more step to guarantee that we will be one-of-a-kinds generating yet more one-of-a-kinds. And that when we generate new mosses[406], those tiny shrubs of green will grow to be something very different from you and me. Very different from their mothers and their fathers.

Guaranteeing that no matter how insignificant or glorious that infant moss's future, it will be an antenna of a collective intelligence, of a global brain. It will be a finger of an outstretched hand, a probe, a sensor helping the cosmos feel out her potential, helping her feel her way in possibility space. Helping her probe her future possibilities. Or invent them.

CHAPTER 21

WHY MALE AND FEMALE?

WHY DID NATURE INVENT GENDER?

It's all a matter of strategy. Search strategy.

The sperm of the earliest land plants, mosses, swam the equivalent of 2,272 miles just to find a female willing[407] to embrace them. For every moss sperm that managed to find a mate, millions[408] of sperm died.

That's materialism, consumerism, and waste on a grand scale. Whose materialism, consumerism, and waste? Nature's.

And it's an appalling cruelty. A cruelty meted out by nature herself.

What in the world could these mass murders of males be for? What does sex achieve with slaughter and pain? Three things.

Number one. The male-female combo is a way of tapping into the power of what game theorists call "mixed strategies."[409] It's a way of using a basic natural paradox, that opposites are joined at the hip.

When there are opposing strategies, we do best when we do *not* bet our lives and our passions on a single approach. We do best when we resist the temptation to pick just one. We do best when we pick both. When we pick two things that seem totally at odds with each other. When we pick fire and ice. When we pick day and night. When we pick up and down. When we pick recklessness and caution. When we pick exuberance and fear. When we pick reason and instinct.

Opposites are joined at the hip. And there is a power that comes from turning opposites into a team.

Let's get back to you, the moss 450 million[410] years ago. When it comes to multiplying, when it comes to being fruitful and flooding the empty earth with your offspring, one strategy would be to stay in place, to stick with what you know, and to be safe. The opposite gambit would be to take a chance, to roam, to explore, to pioneer, to adventure, to gamble on the unknown, and to ride the tides of the forces around you.

But there's a third option—don't just pick one, pick both. The combination of the two—the combination of exploration and homesteading, the combination of rambling and riveting yourself in place—has a name in evolutionary biology. It's called the fission-fusion strategy.[411] The reach-out-and-test-the-landscape, then-run-back-and-hug strategy. The explore-then-consolidate and compare notes strategy. The spread-out-then-clump-together-and-digest strategy. And this fission-fusion strategy is so widespread in nature and in human affairs that it's shown up in two of my previous books—*The Genius of the Beast*, a book about economics, and *The God Problem*, a book about cosmic creativity.

On the human level, the fission-fusion strategy is at the heart of the struggle between ideas and the clash of civilizations. It's at the heart of boom and bust. And guess what? The fission-fusion strategy is at the heart of gender. And gender is at the heart of something else: the search engine of the cosmos.

Gender is at the heart of a cosmos searching for her potential, probing for new peaks in possibility space. Poking around to find new pinnacles, not just new valleys. Feeling her way to new heights, not just to new shortest paths. And it's at the heart of a cosmos that sometimes manufactures peaks where formerly there were none. Peaks in possibility space. But more about possibility space in a bit. Let's get back to you.

You, the moss, are a sexual pioneer. So you do not put all your chips on staying in place and making a comfortable, predictable home. And you do not put all your chips on feeling out distant possibilities. You put your chips on both. You place your bets on a teamwork between opposing strategies. You bet on the team of male and female. You bet on the sexual team, the team of gender. You bet on what will turn out to be one of the most enduring and most breakthrough-generating forms of teamwork in the evolution of the cosmos. You vote for female and male.

But here's the trick. Each of your gender pairs—your male and your female—will pursue its extreme approach, its opposite approach, on the other's behalf. I, the male, will gamble my life and compete with hundreds of millions of others to let you, the female, take your pick. I will swim great distances and adventure to offer you new genetic possibilities from which to choose.[412] As if you were deciding on which fancy chocolate to take from a deluxe mixed sampler box. I will take risks and chances so that you don't have to. But you will get the benefit of my ramblings. Among other things, if I fail, you will be able to turn your back on me.

And you, by being female and sitting securely in a tried and proven place, will play out your strategy in a way that benefits me. You will give me a base I can count on. Even if you reject me and leave me miserable, suicidal, and alone, you will have favored one of my clones or cousins. And in doing so, you will have favored a rough copy of my genes. At the very least, you will have advanced the cause of the gene team that perpetuates my species. And yours.

What does this pairing of opposites achieve? I am a sperm. When genetic recombination made the one-of-a-kind gene string at my center, it put an audacious new feeler into the hidden potential of cells and DNA. And genetic recombination expanded the genetic bounds of possibility in you, the female, too. Put the experiments of crossing over and genetic recombination together and you get a new probe into the potential of cells and their genes. A new antenna scoping out the hidden possibilities of life. And the hidden possibilities of this planet. A new probe for the search engine of the cosmos.

You get a totally unique individual.

CHAPTER 22

WHAT IN THE WORLD IS POSSIBILITY SPACE?

There's a term you've now seen six times. It's possibility space. The cosmos is a search engine probing her next move in possibility space. But what the hell is possibility space?

Stuart Kaufman is co-founder of the Santa Fe Institute and one of the world's most influential theoretical biologists and complex systems researchers. In 1997, Kaufman introduced a useful concept.[413] He called it "the adjacent possible."[414] Those of us who have been influenced by Kaufman call it something else. We call it "possibility space."

Possibility space is an invisible landscape of hills and valleys. Invisible because it's the topography of all the things the future could possibly hold. It's all the possible implications of the past and the present. It's the terrain of all the things the cosmos could be next. And possibility space is the landscape of everything that you and I could do in the next minute, the next hour, or the next decade.

The valleys of possibility space are the easy paths. The hills are hard. And the peaks are the hardest of all.

The idea of possibility space implies that an entity like you or me will roll through the landscape of the possible like a pinball, sticking to the valleys and shunning the peaks. Following de Maupertuis' law of least action. But nature is not content with what's easy. She is constantly attempting what's hard. She likes a good climb.

Why? Because nature is hunting both the valleys and the peaks of possibility space. This cosmos is a search engine using its creatures the way you use your fingers when you spread them out to comb a shag rug in search of your lost contact lens. The universe spaces out us,[415] her momentary manifestations, like your probing fingers in the tufts of that rug, or like the members of a search party combing a field of chin-high corn for a lost child.

Each utterly unique gene string is a feeler into a previously unexplored corner of possibility space. Each is a potential discoverer of a paradise, a potential inventor of a breakthrough, a potential contributor to the cosmos' next supersized surprise. Each is a gamble on turning an impossibility into a reality.

And some, a very few, shun the valleys and climb the highest peaks. Or create them.

As we will eventually see with the story of Anne Boleyn and Henry VIII.

***

But let's get back to you, a moss 450 million years ago. You've un-snarled your ball of genes into separate key chains, you've pulled off a gene reshuffle, and, to repeat, you've made single-stranded strings that are unlike anything this cosmos has ever seen. You've managed to pull off some grittily demanding, enormously intricate chores. Chores riddled with room for error. Chores intolerant of the slightest misstep. And chores controlled with such precision that they very rarely go awry.

You've gone through processes with impossible names like meiosis, mitosis, and crossing over. You've used appallingly complicated tools like chromatids, chromatin, chromosomes, chiasmus, histones, haploidy, diploidy, metaphase, interface, and nucleotides.

Why did you go through all this toil? Why bother? How did your ancestors ever manage to invent a process so crazily concatenated, so bizarrely embroglioed, so fastidiously rigmaroled, and so finaglingly tangled?

If entropy were for real, none of this exquisitely precise molecular engineering would have been possible. To be "natural," your genome would have tumbled into the path of least effort. Your gene string would have fallen apart. Your key chains would have whoozled their beads into an arbitrary spew like a sugar cube dissolving in a glass of water. Which means that the processes at the heart of sex are radically unnatural. All of them.

Sex is unnatural? So why is it all over the place? Could it be that our ideas of "nature" are slightly off base?

***

Here's one of the biggest reasons of all to invent sex.

Nature is on the prowl for evolutionary accelerators, for creativity turbochargers, for search engine super-thrusters, for invention engines. Death and birth is one evolutionary accelerator. Sex is another.

Let's go back to Charles Darwin's "struggle for survival".[416] In reality the cosmos is in a struggle for both survival and for something hard to believe—for flamboyance. Sex is a victory dance over entropy, chaos, and stinginess. Yes, life is not just a struggle for continued existence. It's also a struggle for jubilation, a struggle for exhilaration, a struggle for exaltation, a struggle for exuberance. A struggle for surplus. A struggle for intricacy. A struggle for innovation. As we will later see in dinosaurs, fish, and hummingbirds.

Even the dust of space has done the very opposite of what Lord Kelvin's heat death predicted. Instead of falling apart in a random whizzle, instead of tumbling into a formless phmumph of entropy, instead of spreading out in a uniform mist like the molecules of your sugar cube, space dust has come together in galaxies. Then in stars, planets, moons, asteroids, and solar systems. Every swirl of a galaxy and ring-around-the-rosy of planets circling a sun is a victory dance over entropy. Every swirl shuns the lapse into uselessness. Every swirl defies the Second Law of Thermodynamics.

Extravagance is an evolutionary accelerator. A creativity super-thruster. Which means that it's time to give the Second Law of Thermodynamics a twin—the First Law of Flamboyance. But more on that law in a bit.

You, the land plant lifting your spore-carriers to the sky 450 million years ago, are a part of this quest of the cosmos to exult. You, the land plant, are pulling off very intense stuff. Not least-effort stuff. Most-effort stuff. Stuff so appallingly tricky and so basic to the fabric of this planet that one day the comedian Woody Allen could quip, "I don't know the question, but sex is definitely the answer."

***

OK, so sex slaps two basic scientific assumptions, entropy and the law of least effort, in the face. Is it time to ball out nature for her sins? Or is it time to change our scientific assumptions? We'll dig farther into this puzzle later. But first...

We've gotten a small taste of the way that nature uses sex to commit three deadly sins—materialism, consumerism, and waste. But in a few minutes, we'll see how nature created the most flamboyant sin of all: vain display.

And what she got out of it.

CHAPTER 23

NATURE LOVES THOSE WHO OPPOSE HER MOST

R emember, nature loves those who oppose her most.

Yes, merely moving onto the land 470 million years[417] ago defied one of nature's most basic laws: gravity. And by multiplying sexually, you—one of the first land plants 450 million years ago—are defying two other primal commandments: the second law of thermodynamics and the principle of least action.

What in the world are we going to do with you? You persist in breaking the rules. You persist in remaking the natural order. You persist in acting unnaturally. Yes you, the greenery of nature, you the mosses and liverworts,[418] you, the bryophytes, you who have brought sexuality to the land,[419] you who still sometimes do it the old fashioned way, without sex.

But you who so frequently splurge in this new thing called sex, you who dilly-dally with this incredibly risky business of sexual reproduction. You who risk shattering the fine-tuned order of the beads on your genetic key chains by making your genomes separate and do the twist, you who send out sperm on the trickles of water from dew and rain[420] in the absurd hope of finding an eager egg in a distant plant. You who indulge in this strange, iffy and agonizing proposition that will obsess life on land for the next 450 million years. You are radically unnatural! You are upsetting the applecart. You are a threat to the delicate balance of existing ecosystems. Yet someday humans will see you land plants as nature's very essence. As nature incarnate. How could this possibly be?

Could it be that you are following a natural law far more deeply seated in this cosmos than "respect for nature," "sustainability," and "harmony with the natural order"? Could it be that you are following an imperative more essential to this cosmos than entropy? Far more essential? Could it be that you are following a law that tells particles, stars, planets, and plants to rebel. To defy nature. To break her rules. To shatter her commandments. To go to the very edge, then to push, pounce, or bounce beyond the borderline. You, the first land plants to adopt sex, shrug off entropy and the struggle for mere survival.

You follow the First Law of Flamboyance. Why?

To invent. To flagrantly make new forms of order. New levels of complexity and amazement. New shape shocks. New twirls, twists, teams, group identities, and organizational personalities. Organizational personalities with the power to sustain. And with the power to produce progeny who, in turn, will rebel against what their parents have achieved. Youngsters with the power to produce new generations of rule-breakers, rebels who will bite the hands that fed them, new generations who will massively outdo their ancestors. New generations some of whose pioneers will cartwheel beyond the bounds of the status quo like a tumbler defying gravity with handstands, a tumbler exulting every cart-wheeled step of the way.

New generations who will outdo their ancestors in breaking the laws of nature.

Could it be that nature demands new generations who will manically mass produce their rule-shattering? Whether they multiply the way that stars do—by repeating a common pattern all over the place.[421] Or whether they multiply the way that lifeforms do—by piling up the innovations of millions of generations in their gene string, using that pileup to invent patterns that have never previously been seen, then recording the instructions for those rogue patterns in their gene strings, and finally manufacturing copies of those new rebel gene strings by the trillions. Manically mass producing. Doing it on an industrial scale. And pulling material stuff into the process like crazy. Kidnapping, seducing, and recruiting ever more raw matter into the hungry project of life.

Again, could it be that there's something deeply flawed in the "principle of least action"? Could it be that there's something wrong with "nature is thrifty in all its actions"? Could it be that the principle of least action's cousin, the Second Law of Thermodynamics, is deeply flawed too?

Could the claim that nature loves those who oppose her most be true?

Could one of the first laws of nature be: do something unnatural. Defy me. Do it in a way that is astonishing. Make me look silly. Shatter my existing order. Then put me back together in whole new ways. Take the path of most effort. Juggle me. Tumble me. Make me look terrific. Make me look like a magician and a sorceress. Build, assemble, knit together, and transform me. Reinvent me. Yes, invent.

Could one of the first laws of nature be, do not follow the Second Law of Thermodynamics. Follow the First Law of Flamboyance?

Could "stitch together a radically new reality" be Nature's real command? We are about to see.

CHAPTER 24

LIFTING *SEAS* TO THE SKIES: THE INVENTION OF AN OUTRAGE–THE TREE

W hat would soon inspire you, a plant, to invent vain display? What would inspire you to create advertising and to make an extravagant spectacle of yourself? You were driven by an obstacle. A threat. A dilemma.

That challenge would be one of nature's greatest space programs. One of nature's greatest rule-breakers and environment recreators. One of nature's greatest evolutionary-accelerators and creativity-turbochargers. One of nature's greatest followers of the First Law of Flamboyance. One of nature's greatest criminals against the natural harmony, one of nature's most astounding inventions: the tree.

Says the *New York Times*' master of science-reporting Carl Zimmer, 400 million years ago you land plants are still "little more than mosses and liverworts growing on damp ground."[422] You are a two-inch high coating of green spread in spotty patches across the landscape.

Then, 350 million years ago, 120 million years after you've left the waters, you come up with a radical invention—the leaf.[423] The leaf is a technological triumph, a sheet of gadgetry beyond belief.

It is a flat panel of tissue with specialized ventilation holes—stomata—to let carbon dioxide in but to prevent water from leaking out.

More important, it is a solar energy panel—a panel riddled with the tiny, round, green engines of photosynthesis—chloroplasts—green disks only three[424] times the size of a bacterium. Those tiny green polka dots pull off one of the most astonishing technological tricks in the history of the cosmos—like cyanobacteria, they grab the machine-gun wiggles of photons, turn them into a power source, then suck in molecules of carbon dioxide gas from the atmosphere, and use the photons and the CO_2 in an industrial process that builds carbohydrates and sugars.

The leaf breaks the rules of nature, gives a finger to the existing order, and makes the law of least effort look silly. Among other things, its plant has sophisticated plumbing. The leafy plant has veins to keep it supplied with water pumped from another invention, a root system beneath the ground. Which means that like all of life on the surface of the land, the leafy plant defies one of nature's most basic laws: gravity. But the leafy plant takes the defiance of gravity to radically new heights. It lifts water high above where water naturally "wants" to go. But pumping water on high is not the leafy plant's only anti-gravitational trick. The leaf's internal engineering allows it to be cantilevered above the old-fashioned, ground-hugging mosses and liverworts.

Then you leafy plants go a giant step farther. You go where no living thing should ever go. You mount nature's third great space program. The first was the rise of bacteria to the land. The second was the push to two inches high. And the third?

You invent new building materials: lignin and cellulose. Using these miracle stiffeners, you invent skyscrapers of a kind this cosmos has never seen before. You invent stalks and trunks. You invent wood. Why?

If you are a plant gifted with the new gadgets, leaves, and you take advantage of wood, you can get a competitive edge in the great grab for sunlight by reaching for the heavens, by lifting yourself above the competition. In other words, if you manage to rise toward the sky, if you manage to defy nature's most basic law, gravity, other plants can crowd and elbow all they want to monopolize the sunlight at ground level. But you can do them one better. You can spread a canopy of leaves high above them, capturing the sunlight before it ever reaches the ground.

Remember, new technologies open new frontiers. New niches are invented, not found. With your leaves and trunks, you create a whole new horizon and a whole new resource base for life...a second story that reaches to new heights. You invent leafy high-rises,[425] new towers that let you capture the light of the sun long before it can spill over to your competitors down below. You invent towers of your hard, stiff new building materials, lignin and cellulose. And you manically mass produce so many of these towers, these trunks, that cellulose and lignin—your new, woody super substances—will become "the two most abundant organic compounds on Earth."[426]

Even more astonishing, you go farther to defy nature. You go farther to defy gravity. You invent plumbing systems that can lift
11,000 gallons[427] of water to the sky every day. Yes, you lift 91,799
pounds of water. Per plant. Per day.

Your new plumbing systems will soon spread across the face of the
earth and will collectively hoist a total of over 61 septillion pounds
of water to the heavens every 24 hours. A thousand times more
than the entire Pacific Ocean.

Which means that you give one of the biggest fuck yous to gravity
in the planet's history. You defy nature. You invent trees. Cellulose
and lignin spires each of which can lift 200,000 leaves to the
sun.[428] Yes, 200,000 leaves per tree. And with trees, you spread a
new community high above the carpet of mosses and liverworts.
You invent a new niche. You create a new ecosystem. You invent
something violently unnatural—the forest.

To achieve these breakthroughs, you plants use nature's greatest
invention engine and evolutionary accelerator: sex. You trees use
the pollen and egg system.[429] And most of you use the honeypot of
flowers to purchase transport for your long-distance trysts. But the
flower honeypot is a story yet to come. So is the tale of pollen.

Your forest's roots will soon change the land. And the canopies of
your woodlands will blank out the heavens for those below. You
follow the path of most effort, not least. You do not roll down
the valleys of possibility space. You build new peaks. And nature
rewards you for it. Nature loves those who oppose her most. Nature
favors those who reinvent her.

Which presents ground plants with a problem. How do you compete for sunlight when the rays of the sun have been kidnapped high above you and grabbed long before their flood of light can reach the forest floor? You create. You innovate. You invent new niches. You invent new forms, new structures, new processes, and new communities. And you accelerate the speed of evolution.

You ground plants do your thing on the shores and borders of wetlands where forests haven't gotten a hold. First, you too adopt the use of leaves. But you up their efficiency. You invent new ways to turn scarcity into plenty. You up your leaves' productivity.

Like trees, you have already invented plumbing to lift water above ground level. You have already invented veins. Now you invent ways to make more of those veins. You produce leaves with "dense leaf venation"—lots of tightly packed veins.[430] More veins mean more water. So you are "able to dominate land by evolving more efficient hydraulics, or 'leaf plumbing.'"[431] And more water ups the rate at which your leaves can turn sunlight into sugars and carbohydrates, the rate at which your leaves can kidnap solar energy and turn mere electromagnetic waves, mere wiggles, mere photons of light, into food and fuel, into the stuff of life.

Your defiance of nature lets you open new horizons to the evolutionary race. And your inventions up the mass of the biosphere. They up the GAL, the Gross Amount of Life on this earth. Dramatically. From a tenth of an ounce to half a trillion tons.[432]

Congratulations. You have avoided the valleys of possibility space. And you are on your way to inventing new mountains, new peaks.

***

What other counter-tricks could you plants of short stature invent to survive the trees' monopoly on sunlight? What else could you do to speed the rate of your evolutionary creativity? First, more and more of your cousins, more and more short but massively veiny-leaved plants, show up all over the place.[433] Second, you up your productivity. Says Tim Brodribb of the University of Tasmania, one of the discoverers of this "cretaceous productivity stimulus package," "without this evolutionary step, land plants would not have the physical capacity to drive the high productivity that underpins modern terrestrial biology and human civilisation."[434] We'll save the astonishing manipulations that plants used to help birth human civilization for later. But first, let's get back to your cretaceous productivity stimulus package.

To solve your tree problem, you, the mid-sized plant, go beyond hyperpacking the density of your leaf venation. You take a risk. You take, in fact, what might seem like a giant step backward. Instead of taking the shortest path possible, you create a new longest path.

Your competitors, the trees, are locked in wood. You are not. So you place your bets on a childlike flexibility. You bet on a rela-tively woodless, naked, flexible green stem. You stay "non-woody at first," explains Linnaean Medal for Botany–winner Sherwin Carlquist.[435] And, adds Carlquist, you stay juvenile longer—young and reshapable.[436] That, says Carlquist, means you can come up with new water-conducting systems, new plumbing systems, to fit the circumstances.[437] And it means, says Carlquist, that you can invent "amazing new forms and wood formulas."[438] You can do ma-terials engineering. And you can experiment with new shapes and structures. Does this pay off? Says Carlquist, a hearty yes. It makes you the Ninja warriors of the botanical world—in Carlquist's terms, it makes you "the new weeds."[439]

Then, 250 million years[440] after you invent leaves with super networks of veins, you pull off an even more radical jump.[441] You pull off a great leap outside the box.[442] You pull off a change that will remake nature. A change that will remanufacture the status quo and will forever reinvent the way that the evolutionary game will be played. You invent your next law-of-most-effort move. You invent a new evolutionary accelerator. You bet on audacity. You indulge in the First Law of Flamboyance. You invent the sin of showing off. You invent the sin of vain display. You invent the flower.

CHAPTER 25

THE FIRST LAW OF FLAMBOYANCE: THE GREAT FLOWER TRICK

If you have allergies and want to blame your sneezes on someone or something, try sex.

Blame a plant invention. Blame the moonshot of an even longer path to the coupling of eggs and sperm. But a path that will produce a revolution. A path that will dramatically speed the pace of evolution.

This new plant breakthrough is pollen.

The first fossil sign of pollen shows up 270 million years ago in Tchekarda, Russia, "near Perm, the geological home of the Permian Era."[443] The journal *Science* explains that in Tchekarda in 1997 scientists found fossils of "bark lice, plant hoppers, and stone flies"[444] with half-digested food in their bellies. And what food was in the guts of these extremely ancient insects? The new stuff on the block— pollen.[445]

The fact that the insects had pollen in their intestines[446] was a clue, a tipoff to one of the greatest creative leaps in the history of this planet.

Let's go back to 450 million years ago, when you plants were early adopters of the male-female team. Recall that you kept your female cells home in a fortress for safekeeping. And you sent your male cells out to adventure and to take their chances in the great unknown. You sent your male cells out on the mate hunt. It took 150 million years of this, but eventually a few of you plants invented a massive upgrade in the sexual routine.

300 million years ago, instead of sending out your sperm naked,[447] you innovative plants built carriers, space capsules, small, round, tough-walled[448] vehicles with which to harness the transport powers of the winds.[449] Those space capsules looked like spores, but they were not. They were no longer tiny balls that housed the complete[450] genetic instruction manual for assembling a new plant. They were something far riskier. Something far more ingenious. Yes, they were tiny, spiked balls that contained only half a genome. An unzipped single-stranded half of a gene string hungry to zip up and become a double strand again. This was the male half of your genome. A genome half that was useless unless it hooked up with another genome half, unless it found a willing mate, unless it zipped together with the half of a genome in a female, in an egg. But the risk you plants took with these new capsules went farther.

Your new capsules were carriers of maleness, so you might think they'd have been packed with sperm. But no. They contained something far more astounding and far more risky: a sperm-making factory, an assemblage of nano-equipment set to manufacture sperm on the spot if it reached a welcoming destination. Yes, a miniature sperm-manufacturing machine. This was the law of most-effort taken to the nth degree. It showed utter disdain for the shortest path. In fact, it presaged the paving of entirely new cloverleafs, roundabouts, and overhead passes.

It was a manifestation of the First Law of Flamboyance. What is the First Law of Flamboyance? Shape shock in this universe is on a constant increase. New forms and new functions keep popping up all over the place. In other words, the cosmos is on a non-stop climb up an invisible staircase. A staircase to the next invention, to the next emergent property, to the next super-sized surprise. And sometimes to find her next step up, nature spurns the shortest path. In fact, she paves the longest highway.

***

Intricate as the new transport capsules were, you did not produce them in tiny numbers. Not at all. You used one of nature's favorite tools—manic mass production. You manufactured your tiny transport capsules in gargantuan quantities. You produced them in such overwhelming amounts that they formed a powder. You were materialist, consumerist, and wasteful. You were banking on the productive power of excess. You were banking on The First Law of Flamboyance.

With this new sexual cargo system, you, a plant, like so many before you, defied nature's most basic law, gravity. You lifted slender shafts on high with the powder of these micro-miniaturized transport capsules, the powder of the tiny new balls, on your tips, maximizing their odds of liftoff when the wind passed by. In the hope that a random breeze would slip over the tips of your upraised spires and would somehow carry your grains of powder to another sexually eager plant of precisely your species. An almost impossible proposition. What were the high-risk cargo carriers of sperm-making machinery in your powder? They were pollen.

Which means that 300 million years ago, you plants invented a dilemma. Pollen was great for riding the breeze. But its intricate machinery only sprang into action if a pollen grain splashed down on the wetness of another plant of your species. More precisely, on her female organ, a moist opening into the fortress—the ovular sack—where a plant walled off her eggs.[451] If your pollen grain managed to touch down on the sexual fluid[452] moistening the upturned lips of that sack, your pollen's sperm-making equipment went into action and manufactured a pair of sperm. But that's not all. Your pollen also contained a construction kit with which to instantly grow a long tube,[453] a bio-tunnel, a straw that would carry your pollen's newborn sperm to the egg down below.[454] A tube that would compete with the rapidly-growing tubes of any other pollen grains that managed to land on the target flower.[455] That managed to land on the upturned fortress lips of the girl of your dreams.

Does this sound like the shortest path to you?

What's more, there was a weak point—the reliance on something outrageously fickle, the breeze. The wind.

And there was another problem. According to the Smithsonian's Conrad Labandeira, "sawflies and wasps, moths, beetles" and "true flies"[456] noticed this new sexual powder—pollen. They tasted pollen and found it delicious. So they focused on you gymnosperms,[457] you plant pioneers of pollen,[458] and they plundered you.[459]

This was good for the insects, but not so good for you plants. It was hard to reproduce if all of your male reproductive carriers—your pollen grains—were eaten. Yes, some of your pollen may have survived the digestive process and may still have been ready to rock and roll when it was defecated out.[460] But it would, alas, have probably been in the wrong place. Far from a receptive female.

So what did you plants do? How did you plants deal with the ravages of the pollen gobblers? You invented.

***

Every crisis is an opportunity in disguise. To counter the plundering of the pollen-eaters, you shoved the horsemen of the apocalypse out of their saddles and rode their steeds. You turned a disaster into an opportunity. You used the usual tricks of maleness—materialism, consumerism, and waste. You mass produced so many grains of maleness that some of them would get past the eaters and feasters no matter what. You plants manically overproduced pollen.[461] And you did more. You showed why the path of most resistance, the path of most effort, can sometimes pay off. Mightily. You showed the First Law of Flamboyance in action.

You plants made an accidental discovery.[462] Overfeed the insects pollen, and you can get them to transport what they don't eat. You can get them to cart around the pollen that sticks to their bodies when they pig out.

But there's more. Why not take advantage of your plumbing? Why not take advantage of the liquid that carries nutrients from the photosynthesis factories in your leaves down to your stem and your roots?[463] Why not tap the sugar-rich sap called phloem?[464] And why not take advantage of the sugary drop[465] that many of you plants exuded to help smooth the meeting of incoming male pollen with your female egg?[466] The sexual fluids on the high tips of your egg-fortress towers, sexual fluids that gave an arriving pollen grain[467] the liquid it needed to fire up its germination machinery and manufacture sperm.[468] Why not take advantage of the robbery of these beverages by insects? Why not take advantage of your excess liquids, pump them up with extra sugars and amino acids,[469] and use them. To seduce.[470]

Conrad Labandeira calls the result of repurposing this waste an outright "bribe"[471]—a sugar-water so heavenly that the Greek gods were said to have drunk it on Mount Olympus.[472] Yes, you plants repackage your liquid as a libation that can even give immortality[473] to the gods.

In other words, you plants use an irresistible sugary drink to train the insects to favor your pollen. You invent nectar.

By offering a fast-food meal of pollen and nectar, you can get an insect group to specialize in your species and your species only. You can do this so powerfully that an entire insect species will evolve to focus on nothing but you and your kind.[474] Yes, you can get a bunch of insects to focus on you and your ilk so intensely that their focus will be written into their genes. Some insects will even develop specialized mouth parts to get at your fluids, including your "pollination drops"—the drops of liquid on the towering tips of your egg fortresses, your sexual drops whose purpose is to welcome pollen from other plants. But that's just the beginning. As you and I will see in a few minutes.

Meanwhile, once this training is done, these personalized insects will fly off to another plant of precisely your species for more pollen or nectar. Carrying your pollen on their body and their hairs. Carrying it directly to the tip of a target plant's sexual tower. Congratulations. You have turned your plunderer, the insect, into a postal worker. A cargo carrier far more precise than a mere gust of wind. In fact, you will reward your insect of choice so mightily that it will develop a fixation on just you and your kind. Which means that your pollen won't simply land on the ground in a random spot. Your insect will Federal Express it to a waiting female of exactly your sort. Yes, if you're in luck, the six-legged flier will carry your sexual package to the towering tip of an egg fortress, an egg flask, that your pollen will be able to penetrate.

Congratulations. You, the plant, have just invented a new longest path between two points. A new way to find a mate: commerce. You've just invented interspecies barter and trade. You've just created pay in exchange for services given. You've just invented Uber.

This is the germ of a great relationship. But what if it has further possibilities?

***

Let's take another look at the lay of the land. Roughly 300 million years ago,[475] you plants invent pollen.[476] Insects gorge on the stuff, plundering it and diverting it from its purpose,[477] forcing you to waste vast amounts of excess resources to make billions of back-up pollen grains. Then, wham, roughly 40 million years later, you green things do it. You turn your plunderers into postal workers.

But that's just the beginning. 125 million years ago, you plants come up with your great persuader.[478] You come up with an invention so new that it is startling—an interspecies communicator, a cluster of semaphore flags that insects can understand, a fistful of temptations insects cannot resist. You come up with a form of consumerist, materialist display so far beyond the law of least effort that it defies belief. You come up with your killer app.

Yes, you come up with the flower.

With the flower, you hit a motherlode. You hit an evolutionary jackpot. You invent one of the greatest evolution-accelerators in the history of life. One of the greatest invention engines. You invent the "sin" of vain display.

And that sin gives you flowering plants a staggering new power: instant evolution.[479] Roughly 130 million years ago[480] you flowering plants popped up in the fossil record and spread at astonishing speed.[481] As many as 369,000 kinds of angiosperms emerged from the womb of impossibility, as many as 369,000 new species of flowering plants.[482] Those species would someday include everything from peas to oak trees,[483] and from grass, rice,[484] and wheat to corn.[485]

But remember, nature is on the prowl for evolutionary accelerators, for creativity turbochargers, for search engine super-thrusters. For invention generators. And you flowering plants have come with a humdinger.

The fossil record shows that your flowers gave you the power to evolve so fast that it made a mockery of evolutionary "gradualism" and would someday shake an observer named Charles Darwin to his very core. But more about Darwin's big shakeup in a minute.

***

What if you, a pollen-producing plant, could make your new interspecies commerce even more efficient? What if you could hack your insects' passions? But how? The answer? "Meaningless" bling. The sin of vain display. Materialism, consumerism, and waste. The productive power of excess. The First Law of Flamboyance. Or, to put it differently, advertising.

All sins you were challenged to deploy to get around the catastrophe of the tree.

CHAPTER 26

HOW TO SUCKER
A TRUCKER: PLANT
PORNOGRAPHY

W ho invented showing off? Who invented spending vast sums to flaunt, dazzle, and to grab attention? And hopefully to get sex? Was it greedy capitalists? Was it the rich trying to suck blood and money from the masses? Was it big tech raking in the billions? Was it the advertising industry trying to shove consumerism down the throats of innocent proletarians? Was it rappers aiming to become billionaires? No, it was plants. In other words, it was nature.

Yes, one hundred and twenty million years ago,[486] nature herself was inventing the unholy transgression that the compilers of the Seven Deadly Sins,[487] from Saint John Cassian in 400 AD onward,[488] would characterize as pride and vainglory. The most deadly of all the seven sins.

Nature was inventing the vile offense that would someday be smacked down by sumptuary laws, laws against showy, expensive, and elaborate clothing, in ancient Greece, Rome, China, Japan, and the Islamic world. Not to mention in Medieval and Renaissance Europe.[489]

Nature was inventing vanity—competing to put on the most bodacious spectacle of what some social critics would have called pretentious, pointless superficiality. Indulging in a luxury that breaks the bank. A luxury so costly that it can soak up between ten percent and fifty percent[490] of a plant's total energy budget.[491]

Nature was inventing showing off. And she was doing it for a reason so devious that it defies everything we think of as natural.

Nature was inventing... marketing. Nature was inventing advertising. Nature was inventing promotion. Nature was inventing trade, barter, and commerce. You mere plants were inventing vain display.

You and I occasionally buy luxury goods—a gorgeous laptop, a Vuitton bag, a pair of Jimmy Choo knee boots, or Ferragamo oxfords. Or if we're rich enough, a Porsche, a Lotus, a Tesla roadster, a Lamborghini, or a yacht. Deep down we know why we do it. We want attention. We want admiration. If we're men, we are hoping to spark the envy of other men and to turn on the desire of attractive women. If we're women, we're hoping to make other women jealous and to get the attention of great looking guys.

Somewhere behind materialism, consumerism, waste, and vain display is the master-maker of invention—sex. But waste for the sake of attention and sex did not start with humans. It started with nature.

***

Let's go back to you, a land plant 125 million years ago. Trees are stealing your sunlight. And insects are stealing your pollen. How do you solve this problem? You do something radically unnatural. You defy thrift. You invent a new form of flamboyance.

In the process, you pioneer a new form of what the father of economics, Adam Smith, calls "the division of labor".[492] A new form of specialization. And a new form of cooperation. A new form of what Herbert Spencer calls "differentiation and integration."[493] A new form of what biologists and ecologists call "mutualism."[494] A new form of what these experts call something even more powerful, "co-evolution."[495] A new form of evolutionary acceleration. A new form of invention engine.

As you just saw, over 270 million years ago, you plants seduced, kidnapped, and recruited a species vastly different than your own: insects. You bribed them with nectar. But it's 125 million[496] years later. And you are no longer satisfied with that first audacious move. You push to go a step farther. To do it, you are about to go outside more than just one box. You are about to go outside of two. You are about to employ the Swiss Army knife effect. You are about to employ what 20th century evolutionary thinker Stephen Jay Gould called "exaptation"[497]—using something that evolves for one purpose to carry out a totally different task. Employing a device you've already got, but putting it to work for a function beyond imagining.

You have developed the genes for making leaves. But now you use these solar-panel genes to create another kind of panel entirely.[498]

In fact, you invent something that at first glance seems utterly useless.

Yes, it's a thin, extended sheet like a leaf. Yes, it has veins like a leaf. Which means it is expensive and uses a massive load of material resources. Like a leaf. But it has utterly lost its practical value. It only occasionally[499] has the high-cost ventilation holes that open to let in carbon dioxide and close to stop the evaporation of water. And it almost never has solar power generators—chloroplasts.[500] What's more, you use it for only a few weeks or months,[501] then you tell it[502] to die and you litter the landscape with its remains. It is big, flashy, and pricey, but you have the gall to throw it away. In other words, it's consumerist trash. It's pointless waste. Right? So what is it good for? Attention. Promotion. Persuasion. Recruitment. Vanity. Vainglory. Commerce. Social networking. Advertising. Vain display.

And what the hell is vain display good for? What's the point of vanity?

Remember your pollen transport problems? Remember how you were once forced to rely on the unreliable? How you were once forced to scatter your pollen to the wind? In the hope that a breeze would randomly take your male seed, your pollen, to a receptive female of precisely your species? Remember how you solved the problem by working out a deal with insects?

But that was not enough to satisfy you. You plants were ambitious. You were inching your way toward controlling an insect with precision, almost as if it were a drone. You were inching your way toward hacking its passions.

But how? Could you possibly use your new and seemingly useless leaves to kidnap, seduce, and recruit?

Could you find shapes, colors, and smells that an insect would not be able to resist? Could you modify your new, crippled leaves to fit the tastes, the pleasures, and the lusts of six-legged fliers with wings? Could you use the veiny panels that you originally invented for photosynthesis to do something so different it's ridiculous. Capturing an insect's attention. Feeling out an insect's motivational machinery. Feeling out an insect's passions. Could you use your panels to develop what author Michael Pollan calls a "botany of desire"?[503] Could you use your panels to mesmerize creatures of a very different kind? Could you use your panels as jumbotrons? Jumbotrons advertising... sex?

The answer in a universe where the most probable thing always happens would be no. Absolutely not. The answer in a universe where uselessness, chaos, and entropy are always on the increase is... forget about it. No way. But this is a universe where the least likely thing happens. This is a universe ruled by the First Law of Flamboyance: shape shock is always on the increase. Forms, functions, and amazements are constantly on the rise. The next material miracle is always just around the corner.

In fact, there are even more impossibilities teasing your drivetrain of natural ambitions. Could you refocus your panels from producing stomata and chloroplasts to producing a strange variation of the chloroplast, a strange variation on your photosynthesis-making polka dots of green? Could you repurpose chloroplasts as "chromoplasts"[504]—color makers. Could you use your disabled green photosynthesizing polka dots to produce reds and yellows to pull in butterflies? Ultraviolet patterns[505] and colors that contrast sharply with the background[506] to lure bees? Whites and pale colors to grab the attention of moths at night?[507] And could you go even farther?

Could you pay off insects with pornography? But how? How could you, a plant, pull off these wild impossibilities? How about upping the game of insect enticement with more than just your nectar. How about using your panels as impossible new inventions. How about repurposing them as billboards. How about reinventing them as, well, umm, petals.

How could such amazements come to be? Remember the Swiss Army knife effect? Exaptation? The discovery that something you developed for one purpose can be used to solve a very different problem? The time has come to use that trick. With your disabled leaves.

CHAPTER 27

THE ADVERTISING EXPLOSION

It's 275 million[508] years ago. Trees have defied gravity, scraped the skies, and with 200,000 leaves[509] per tree have hogged up all the sunlight. You, a land plant only a few feet high, are frantically trying to work around the edges of that light monopoly. To do it, you are attempting to invent new niches in which to survive. And to pull that off, you've conjured up a miracle drink with which to lure another great gravity defier, the arthropod who flies. The insect. Which is odd. Because insects are thieves.

What in the world is the point of attracting creatures who feast on your sexual essentials and force you to radically overproduce them? You plants "learn" something. You stumble into it tiny step by tiny step. Or do you "invent" it in a series of great leaps? In lightning flashes of biochemical creativity? We don't know.

But you "learn"[510] that by building flowers of specific shapes, you can fashion mazes. Mazes that will require an incoming insect to walk between walls of pollen balls and to allow those balls to attach themselves to the visiting insect's exterior. Mazes that will only give a payoff of nectar if I, the insect, rub myself all over your pollen in a specific way. Mazes you can use to get me to reshape myself. Mazes riddled with riches you can use to resculpt my physiology and my habits.

Meaning that if you lure me, an insect, to visit you, I will find you and your kind so richly rewarding that I'll let you coat me in your pollen, then I'll go off and visit another plant with a flower just like yours. And I will do it more efficiently than ever before. Thanks to the uniqueness of your species' flowers, spotting strangers of precisely your kind will be easy. Which means I'll carry your pollen on a non-stop private airline trip to your sexual ideal—a plant almost identical to you, but not quite. I'll become an even more efficient extender of your sexual organs. I'll become even more of what botanist Peter Bernhardt has called a "flying penis."[511] In the process, I'll increase the precision of your pollen's wanderings. What's more, over time I'll go even farther to reshape myself to fit only you and your kind. Meaning that I will even more effectively guarantee express delivery of your sexual material to a waiting mate of precisely the sort you ache for most.

So you, the plant driven to the margins by the sunlight monopolists, by trees, will spend between ten percent and fifty percent of your energy budget transforming a now useless leaf into what looks like a counterproductive luxury, a flamboyant signaling device, a materialist display as flashy as any building in Dubai. You will turn a deliberately disabled, repurposed leaf into a marketing banner that advertises radical upgrades on an old deal, an intensely colorful widescreen signboard that says, "Hey, don't waste your time and energy on my competitors. Don't diddle with other kinds of plants. Come over here and visit me. I've made extra pollen. Some pollen for my purposes. Some for yours. All I ask of you is a ride. I'm stuck here and can't move. But you can do more than just walk. You can fly. If you act as my cargo shipper and take my pollen to the sexual centers of other plants of exactly my kind, if you help me with my mate hunt, I will reward you mightily."[512]

By 140 million years ago, this is an old deal, but there's something you have never tried before: advertising it. And flagrantly bally-hooing your offer will produce a phase change so astonishing that it will trigger the evolution of entirely new insect species. It will generate moths, wasps, butterflies, and entirely new kinds of bees, [513] all species that have never before been seen. Not to mention new varieties of bugs, beetles, and flies. What's more, as you know, it will produce 369,000[514] new species of flowering plants.[515] More species than any other plant group on earth.[516] It will provide the species-richness of tropical rainforests. And, it will up the scale of animal biodiversity.[517] It will open the evolutionary paths that lead to orangutans, iguanas, flying foxes, giraffes, pigs, hippopotamus-es, camels, deer, antelopes, sheep, goats, cattle, honey badgers, and new kinds of turtles and lizards. It will produce pigeons, hum-mingbirds, and doves. Not to mention rats, mice, and primates.

It will trigger what University of Bristol paleontologist M.J. Benton calls a "reset" of the entire "Earth-life system on land."[518] An expansion of "the biosphere... to a new level of productivity." For the first time, this angiosperm explosion, this explosion of flowering plants, will populate the land with more species than there are in the sea. Despite the fact that land covers only a pitiful 29% of the globe's surface.[519] In fact, the debut of your brightly colored, disabled leaves will trigger what Benton dubs "a macroecological revolution."

Your big bet on the consumerist, materialist excess of vain display is, as you've guessed, the flower. And, yes, as you know, "flowers are more diverse than every other group of plants," according to Stanford University's Danielle Tucker.[520] Which means you flowering plants will invent 369,000 new ways to make a living. You will carve out 369,000 niches that never existed before. 369,000 new ways to suck up sunshine despite the ultimate sunlight hoggers, trees.

You flowering plants will defy the law of least effort with its shortest path possible. In fact, you will invent the longest paths imaginable. You will invent an evolutionary accelerator. A breakthrough generator. And, to repeat, you will invent another radical innovation: instant evolution. All thanks to coupling advertising with sex.

Oh, and one more tiny thing. Instant evolution, as you know, will someday flummox poor Charles Darwin. But Darwin's difficulty is a story yet to come.

***

Does sex's insistence on producing one-of-a-kinds help you generate your astonishing 369,000 new species? I suspect it does.

But evolutionary breakthroughs do not just fashion new kinds of legs, wings, and brains. They do not just generate new forms of individuals. They often create something we overlook when we obsess on the origin of species. They weave new group identities. And new kinds of teams, teams with new organizational personalities, are often the hidden goals of materialism, consumerism, waste, and vain display.

***

You are a land plant straining at the bit to outwit the light monopoly of trees. You have just invented the flower with its blaze of shapes and colors. And the flower has kicked off a great leap forward in your commerce with insects. Now you give your blossoms one more lure to tempt insects. You use the same chemical pathway that manufactures color pigments in your petals to produce, of all things, perfumes.[521]

Using odor-producing molecules is old hat to you. For a long time now, you've used smells to battle against plant-eaters. To fend off insect and microbial marauders. You've also used scents to coordinate defensive maneuvers against enemies, defensive maneuvers you pull off with a phalanx of other plants. For example, according to Purdue University's Natalia Dudareva,[522] when you were "infected with a virus," you used scent to mobilize your community.[523] You released "a volatile compound that signals other plants to set up defenses against the attacking microorganisms."[524]

You've also used offensive odors as repellants, pushing insects away. But opposites are joined at the hip. A molecule that cracks the code of another species' motivational machinery in order to chase its members away is the start of a communicative language, and a language can be used to repel or to attract. Switch strategies and reprogram the roughly fifteen genes[525] you use to produce the odors with which you send insects reeling, the fifteen genes you

use to produce the weapons of "essential oils" and "volatile compounds," and instead of repelling insects, you can reel them in.[526]

For example, you can mix and match your chemical building
blocks—building blocks like methyl benzoate, methyl salicylate,
salicylic acid, and linalool[527]—to smell like rotting meat to attract
flies. Or you can go for the biggest perfume breakthrough of all—
imitating the mating scent of an insect female. You can manufacture a perfume that promises an insect orgy to a male, a perfume
that will be your first step into insect pornography.[528]

Congratulations, you have hit the jackpot. Scent production will
prove so promising that eventually one of you new-style flowering ground plants, the orchid, will be capable of generating one
hundred different perfumes.[529] Sexual perfumes. But that may be
getting ahead of our story.

***

Yet another new kind of chemical advertising went into high gear
roughly 113 million years ago, a mere seven million years or so after
the evolution of you, the first plants with full-scale flowers. And
this new upgrade was kicked off by the evolution of a new desire
in female insects. A new passion. A new fad. A new fashion. A new
form of what economists call "demand." Demand in the females of
orchid bees.[530]

The female orchid bees' new passion demanded materialism, consumerism, excess, and vain display. From males. It demanded
new forms of waste. And it was driven by that old hand at one-use
throwaways, that old sinner against the law of least action, that old

creator of most action and then some, that old paver of the longest possible path, that old mistress of the First Law of Flamboyance, sex.

Faddish[531] orchid bee females became incredibly picky in their sexual appetites.[532] They would only pay serious attention to you, a male, if you gathered a back-breakingly hard-to-get perfume— one of 585 volatile compounds made primarily of terpenes and aromatics. A perfume mixed together from chemicals you, the orchid bee male, were forced to extract from

- the resin of tropical trees,
- from fungi,
- from rotting vegetation,
- from rotting trees, and
- from leaves.

Not just from one of these things. From all of them. Chemicals you males had to find, mix, match, and blend.[533] To perfection. Using a demandingly precise recipe.

This life-complicating perfume-obsession[534] spread among orchid bee females like a plague. Why? Why the demand that males go through this torment of chemical extraction?[535] Was it for a down-to-earth, practical purpose? To provide food and shelter? To guarantee the necessities of life? Was it to travel the shortest distance between two points? Was it to help the cosmos whizzle apart with entropy? Not a chance.

The female orchid bees demanded these rare scents for one purpose and one purpose only—to torture you, to test you. To give you the equivalent of a College Board Exam. To put you through the throwaway waste of sexual display.

The sort of materialist, consumerist, competitive display that allows British perfumer Clive Christian to charge $215,000 a bottle[536] for his Number 1 Imperial Majesty perfume and France's Baccarat to charge $6,800 a bottle for its Les Larmes Sacrées de Thebes (The Sacred Tears of Thebes).[537] At those prices only a tiny number of human males can afford to buy them. And to present them as gifts. In the mate hunt.

Like the women who demand that you present them with periodic bottles of Clive Christian's Number 1, orchid bee females demanded that you advertise your superiority. That you prove your powers. Or your lack of them.

Again, was this female orchid bee insanity the product of rabid capitalists creating false desires to enslave the proletarians? No, there were no capitalists anywhere in sight. This was nature on the wing. This was evolution doing its thing. This was Mother Nature's longest possible path between two points. Mother Nature's law of most-effort. Mother Nature's First Law of Flamboyance. Mother Nature's path of productive sin.

What did nature get out of inflicting an impossible perfume contest on you innocent males? Female orchid bees demanded this chemical feat to create an artificial and arbitrary form of competition. A competition that allowed the females to save their chastity for the males among them with the greatest ability to waste time and energy. Males who had found the richest food sources and had gathered enough surplus nutrition to be able to afford to play the perfume game.

The orchid bee females unknowingly followed the adage that many shall be tested, but few shall be chosen.[538] The perfume game was generated by females so they could be fastidious about males, could find those of exceptional ability, and could weed out the unworthy. And so that females could force males to do things nature had never seen before. So females could force males to push the envelope of possibility. The perfume game was a new twist on what Charles Darwin would call "sexual selection."[539] And making impossible things happen is often what sexual selection is all about.

Some of those new impossible things? As cliques of females developed their own unique tastes in perfume, they veered off on their own reproductive paths and developed brand new species. Brand new species of euglossine bees, orchid bees. Brand new paths of evolutionary possibility.

Yes, through female insects nature demanded surplus and luxury. Nature demanded inequality. She demanded competition. She demanded probes toward the peaks of possibility space. She demanded the invention of impossible things. She demanded the First Law of Flamboyance.

***

Twelve million years after the scent-obsessed female insects came forth from the maw of evolution, orchids hit on the same secret of advertising that we decry when marketers in Western society use it: yes, once again, sex. Pornography. But pornography with a purpose.

The female of the thynnid wasp let it be known that she was available by standing on the top of a plant and lifting her head to the sky. So the Australian hammer orchid built a flower that looked almost precisely like a female wasp in a mating position, complete with an upturned head that shined like the real thing and fuzz in the appropriate places.[540] Yes, the hammer orchid built a 3-D Playboy centerfold. And the orchid added one more lure with which to outcompete other plants in attracting Thynnid wasps. A chemical lure. The orchid's flower exuded the scent emitted by a female Thynnid wasp in heat.[541] The orchid's flower promised sex.

But most orchids went one giant step farther. They simplified the orchid bee's chemical chore. The orchids synthesized the ingredients of that rare chemical perfume that males had previously been forced to extract and blend from the combined resin of rotting fruits, tree bark, exposed roots, rotting wood,[542] and tropical trees.[543] Plus mushrooms.[544] If I, a male orchid bee, touched down on your elaborate surface, you, an orchid, rewarded me by letting me fill the hairs on my legs with the raw ingredients for the rare scent that I needed to seduce a real female.[545] Without forcing me to travel to four or five different kinds of sources to get it. The price you charged? That I also load up with your pollen.

You, the orchid spent a lot of energy on two ridiculously elaborate materialist lures that you would use for only six to ten weeks—a small fraction of the year—then would throw away:

1. the chemical perfume ingredients that performed no useful role in your internal economy;
2. and your flower.

In a world of least action and entropy, a world of thrift and parsimony, that should have been enough. In fact, it should have been way, way too much. But you, the flowering plant, the angiosperm, the orchid, were nowhere near finished lavishing energy, time, and material resources on capturing my attention and my pollen-carrying power. In fact, your elaborations, complexities, wild inventions, and new wastes of time and energy would defy belief.

If you put a real female wasp next to the orchid that imitates her, a male Thynnid wasp will spot the fake quickly, ignore the orchid, and go for the real female. The orchid to this day has still not perfected its pornographic lure. It has still not mastered the art of making a super-female[546]—a female more attractive than the real thing. But a long, long time ago, the orchid overcame this weakness. How? Timing.[547] It put out its flower in the weeks before female insects hatched from their pupae and emerged from underground.[548] It beat the competition—it foiled the real females.

Think of that for a second. Think of the plant "intelligence" it takes to discover a female insect's annual rhythm and to "see" how to use that rhythm to advantage. Think of all the trouble you, an orchid had to go through to beat the clock—spotting the fact that females emerged at a specific time of the year, building a biological calendar, a timing mechanism, that allowed you to accurately predict when the real females would sally forth, and to get ahead of that deadline. Think of the bio-machinery, the bio-technology it must have taken you to achieve timing with this precision.

Flowering plants responded to the challenge of the trees with a creative panic. And that panic showed up most in you orchids. You orchids were in a continuous competition, a continuous hustle to come up with new flowers, new perfumes, and new insect lures. You switched scents to lure different insects. Or you changed the part of the male insect's body to which you attached your balls of pollen. You came up with seven hundred different ways of attracting just one kind of insect, the aforementioned euglossine bee, a bee so tightly connected to orchids that it would become known as the "orchid bee."[549] Your small shifts in tactics resulted in seven hundred different orchid species. Seven hundred different kinds of flowers. Seven hundred different varieties of materialist consumerist, competitive display. All dedicated to orchid bees.

Not to mention what you orchids achieved in tweaking the genes of your former plunderers to produce 250[550] different species of orchid bees, euglossine bees. Two hundred and fifty species of orchid bees in the Americas alone.

The throwaway marketing displays of you orchids, your disposable billboards, seemed flamboyant beyond belief. But, in fact, they were profoundly practical. Like retail store windows, TV ads for blenders, or Amazon webpages, they were designed to both dazzle the insect and to help him or her to the goods that you flowers offered. Charles Darwin, who wrote an entire book on this subject, *The various contrivances by which orchids are fertilised by insects*,[551] says that the flower of every orchid includes an aircraft runway, a petal that "is larger than the others and stands on the lower side of the flower, where it offers a landing-place for insects." What's more, this specialized, easy-access petal, this landing pad, "secretes nectar for the sake of attracting insects, and is often produced into a spur-like nectary." Amazing!

Then would come more longest-path-possible complexities. For example, now that you, the flower have made a deal with me, an insect, to carry your pollen to a distant flower, you can go to extra lengths to block your own pollen from taking the shortest path possible. You can use an old moss trick. You can forbid your pollen from fertilizing your pollen's sisters, the eggs next to them. One of you flowering plants, an orchid found in southeastern Brazil, the *Notylia nemorosa*, makes extra sure that it can't fertilize itself. It lets its equivalent of sperm, its pollen, go out to seek their fortune for two or three days while keeping its female organ, its sexual tower, its stigma, blocked off. Tightly shut. Then this Brazilian orchid reverses itself, nails its remaining pollen in place, and opens its female sexual parts for business. Charles Darwin says that the Notylia nemorosa opens "a narrow slit...in the stigmatic cavity" allowing the pollen from distant suitors in, but shutting its own pollen out.[552] The orchid outlaws incest. It opens its female organ for an orgy only after putting its own pollen in lockdown.

The message? Don't take the shortest path. Don't take the path of least effort. Don't take the easy route to sex. Do not let the male material you produce mate with your own female offspring. What's more, the Notylia nemorosa orchid backs up its self-fertilization-prevention mechanism by using another old moss trick, producing pollen that is incompatible with its own egg. Why this extraordinarily expensive attempt to cancel out the path of least effort and to take the riskiest and most difficult route? The most wasteful path imaginable? And why produce a masterpiece of materialism, consumerism, and gaudy, useless display—your flower?

Why this absolute itch to take the path of most effort, to get out of your own sexual sandbox and to reach out to a flower way off in the distance? For the sake of one-of-a-kind offspring. For the sake of differentiation. For the sake of inventing as many forms of stems and leaves as you can, stems and leaves to suck up the sunlight that sifts between the greedy leaves of trees. And for the sake of a cosmos driven to explore her potential. A cosmos feeling out her possibilities. A cosmos on a quest. A cosmos that's a search engine of possibility space. A cosmos using you as an antenna to feel out the flamboyant, the ornate, and the impossible. A cosmos that invents.

A cosmos that loves those who oppose her most. A cosmos faithful to the First Law of Flamboyance.

CHAPTER 28

DID THE FLOWER CREATE THE BEE?

One hundred and twenty million years ago, you flowering plants went through wild gyrations to reshape yourselves as insect attractors, insect manipulators, and insect payday providers. But you did more.

Your ancestors remade the mouthparts[553] of the 270-million-year-old insects found at Tchekarda in Russia, refashioning those mouth parts to specialize in eating just your species' pollen.[554] Roughly one hundred and fifty million years later, you flowering plants went even farther. You reshaped more of the insects' physiology, its sexual specializations, and its society. In fact, you helped fashion an insect with its own promotional communication display, a figure-eight dance. A social insect whose entire lifestyle would be built around your pollen and your nectar. What was that insect made by you, the flowering plant? The bee. Or, more accurately, the blindingly inventive group identity we call the hive.

The bee hive is a group intelligence[555] in which hordes of conformist bees follow the crowd and plunder a popular flower patch with extreme efficiency while searcher bees, non-conformist bees, go out and wander, self-indulgently following their whims. These un-conventional bees seemingly waste their time, and, what's worse, waste the precious food it takes to keep them alive. They ramble an extraordinary ten miles[556] a day. But this meandering is not as purposeless as it seems. The scout bees' "aimless" roaming allows them to find the next candidates for hot flower patch of the day, the next contestant for the patch of blossoms that will be needed when the flower patch of the moment runs out of pollen and nectar. Five or six of the thousand or so searcher bees[557] beeline back to the hive and announce their finds of new flower patches by stepping inside, and entering a breakdance competition on the hive's inner wall.

On that wall, a crowd of "discouraged" worker bees have gathered, bees whose flower patch has run out of nectar and pollen, bees who no longer seem to have a purpose in life, bees who crawl instead of walking upright, bees whose body temperature I suspect is low,[558] meaning they are what we would call depressed. Thomas Seeley, one of the world's leading bee researchers, calls these aimless fliers "unemployed."[559] And these workers in a tailspin come to the hive wall just inside the entrance literally looking for excitement. These dejected conformists check out the instructions the break dancing non-conformist scouts convey with their figure-eight waggle dances. If an unemployed bee gets hopped up by one dancer, she follows the instructions coded in that dancer's figure eight and flies out to check on that searcher's discovery. Then, if the fact-checker is thrilled by the find, she joins the scout bee as a backup dancer, figure-eighting her enthusiasm.[560] Finally, the crowd picks the explorer bee with the greatest number of backup dancers, and follows her instructions to the next ripe flower patch.[561]

In other words, the hive is an exploration engine that beats a life-and-death deadline: collect 60 pounds of honey[562] in the six weeks of blossom-time to carry the hive through winter. If the group makes the wrong decisions about which flower patches to plunder, every member, the entire hive, will die in the season of snow and ice. Why? Because in December, January, or March its fuel, its honey, will run out.[563]

Which forces the bees to maximize the number of flowers whose pollen they can carry to distant blossoms. The number of flowers whose sexual ambitions the bees can serve.

Who helped shape the social intelligence of the hive? Who helped make it a servicer of as many blossoms as possible? You flowers. You angiosperms. Your blossoms have helped fashion a whole new kind of group identity.

The hive.

***

Bees live solely on your pollen and your nectar. They are creatures that you flowering plants bred for a simple purpose. To service your sexual needs. But bees are more than individuals. To repeat, they are a dramatically new kind of team. They have a breathtakingly new form of group identity. A whole new kind of organizational personality. Once again, they have the hive.

How did you plants pull off this radical rewiring of another creature's genome? This elaborate creation of a new form of living thing? And this astonishing invention of a new form of insect society? How did you mere plants rise up against the status quo, rebel against the principle of least action, mutiny against the shortest path possible, disturb the existing harmony, reengineer another species' genome, and give the finger to entropy?

Through the process of evolution. But we know a lot less about evolution than we think. As Charles Darwin acknowledged when he wrote to his friend Asa Grey, a botanist at Harvard, "the thought of the eye made me cold all over...[and] The sight of a feather in a peacock's tail, whenever I gaze at it, makes me sick!"[564] Why was Darwin sick? Because he could not account for the cosmos' ability to generate breakthroughs, for the cosmos' contrivance of mind-blowing complexity. Yes, Charles Darwin couldn't account for how the cosmos creates. He couldn't account for how the cosmos invents.

Meanwhile you, the flowering plant, invented a new niche for insects. Then you rewarded those who went along with the deal you offered. And you did not reward those who ignored your signals. You were as picky as humorist P.G. Wodehouse's Lord Emsworth, a pig enthusiast whose goal in life was to breed the pig that would win the Fattest Pig Prize at his local country fair.[565] Or as picky as a pigeon fancier trying to breed an all-white bird. More on that pigeon fancier when we get back to Darwin.

Which means that you flowering plants reshaped the very gene string of insects simply by increasing the odds that those who were lured by your elaborate billboards would be well fed, well equipped with courtship perfume, and would produce more young insects. And you reshaped insects by increasing the odds that the fliers who loved your flashy displays would eventually be so numerous that they could dominate. You became an instrument of evolutionary pickiness. You became a natural selector.

Did your wild plunge into the materialism, consumerism, and excess of nectar, pollen, and flowers pay off? Did it prove worthwhile to spend huge amounts of energy on flamboyant, gaudy, materialist displays? Displays that you would use for a few weeks or a few months, then throw away? Displays that would litter the floor of forest margins with discarded petals, with white and pink one-use throwaway waste? Did your investment in advertising prove to be worthwhile? You be the judge.

An Indiana University press release on the research of paleobotanist David L. Dilcher and three of his colleagues in Europe says that you flowering plants "supplanted ferns and gymnosperms in many regions of the globe."[566] You flowering plants—you angiosperms—pulled this off, says the Indiana University press release, by mounting worldwide "invasions."[567] Invasions of patches of land where trees were scarce.

Yes, evolution often exults in using you "invasive species." First you flowering plants invaded the wetlands around lakes 130 million to 125 million years ago. Then you invaded "understory floodplains," the flatlands between forests and rivers, "between 125 million and 100 million years ago."[568] Finally, you flowering plants invaded "natural levees, back swamps, and coastal swamps between 100 million and 84 million years ago."[569]

And when you flowering plants invaded, you took over. First you occupied the tropics and nearly wiped out the native cone-bearers and ferns. You were one manifestation of nature nearly exterminating another. Then you flowering plants moved north and south, into the temperate zones. In the end, you forced the cone-bearers and ferns to eke out an existence in the cold of the far north and the far south.[570] And in the chill high up on mountain sides.

You angiosperms—you flowering plants—you who wasted vast amounts of resources on throwaway displays, elbowed the old, conservative, far more economical cone-carrying plants and ferns out of niches all over the globe. And you did it with unaccountable speed.[571] Your flamboyance won out over the pine[572] trees' thrift. Your law of most effort won out over Pierre Louis de Maupertuis' law of least effort. Your law of the longest distance between two points won out over the law of the shortest path.

What's more, you flowering plants were evolutionary turbochargers. The explosion of wildly different forms of you flowering plants took place so fast that a global bouquet of your flowers seems to appear from nowhere in the fossil record. The profusion of forms and species was so wild and so instant in geological terms that Darwin's critics claimed it proved that his theory of evolution was wrong. There was no step-by-step, slow-but-steady process of change, the critics said. No Darwinian gradualism. Declared Darwin's attackers, the fossil record showed clearly that flowers had been created all at once, in one giant whoomph. Which was unequivocal evidence, they said, of a giant Whoompher—a God. A Creator.

Darwin took this argument seriously. Very seriously. To counter it, he "wrote three volumes on plant reproductive biology."[573] One was an entire tome to counter the flower argument against evolution, a book we bumped into a minute ago: *The various contrivances by which orchids are fertilised by insects.*

Yes, Charles Darwin believed that a new species could evolve only slowly. But you flowering plants, with your wasteful ways and your gaudy displays, were able to churn out more than 300,000 new breeds, new species, in what seemed like no time.[574] As you know, an agonized Darwin called this explosive defiance of his theory of slow-but-steady natural selection an "abominable mystery."[575]

How did you flowering plants pull this "mystery" off? For one thing, you were able to make the most out of minor changes. The difference between wildly varied flowers could be as little as one or more extra copies of a gene.[576]

But your real secret was, brace yourself, waste. Yes, you heard me right. Your real secret was materialism, consumerism, vanity, and the throwaway. Your real secret was the productive power of excess. And the First Law of Flamboyance. At least if you believe what Wageningen University ecologists Frank Berendse and Marten Scheffer wrote in 2009 in the journal *Ecology Letters*: "Fossil evidence suggests that by the end of the Cretaceous the angiosperms had spectacularly taken over the dominant position."[577] One secret to that takeover was what Berendse and Scheffer called "litter." Yes, litter.

The cone-bearers that you flowers shoved aside were exactly what we are told nature demands. They were sustainable. They were as-cetic. They were thrifty. They "respected" their environment. They lived in harmony with their surroundings. They honored the law of least effort. They were parsimonious. They were puritanical. They threw very little away. To repeat, they were sustainable. But thrift is not always a virtue. Especially in evolution. Especially in the world of nature.

To quote Berendse and Scheffer, the cone-bearers, the gymnosperms, "kept the soil poor - with their poorly degradable litter."[578] Think of pine cones and pine needles. They are hard and unyielding. They are Scrooge-like hoarders. They resist becoming food for the bacteria that make things decay.

You flowering plants, on the other hand, were literally decadent. You courted decay. You made a rich litter. Very rich indeed. What's more, say Berendse and Scheffer, the petals and leaves that you flowering plants tossed away were litter of a very special kind. Unlike tough, woody pine cones, they were "easily degradable."[579] Easily digestible. They were a banquet. In fact, your garbage was a superfood. For bacteria.

Your waste was generosity. You used it to kidnap, seduce, and recruit bacteria into your ecosphere. You used it to expand your social network, your interspecies weave. And you used it, as a *Science Daily* report on Berendse and Schaffer's work put it, to remake the world to suit yourselves.[580]

Remember all those flower petals you saw on the ground last spring just after the magnolia, apple, and cherry trees had shed their blossoms? And how rapidly the thick drift of white, purple, and pink petals disappeared? Why did your path's coating of petals melt away so fast? Because it was a bacterial feast. And what bacteria eat, they turn into the raw stuff of soil. Rich soil.

Here's the result according to Berendse and Scheffer: there was a "shift in Earth vegetation" on a "massive scale."[581] Underlying that shift was the radically revolutionary initiative on the part of you flowering plants, your campaign to enlist the services of other species, from the bacteria that cleaned up your mess to the animals who would someday eat your fruits, carry your seeds in their guts, and deposit those seeds in new potential homes. Not to mention your campaign to recruit insects.

According to Berendse and Scheffer, you flowering plants did not live in harmony with your environment. You reinvented your environment to suit your ambitions. And that environment included animals. Animals that, as you know, you plants actually reengineered. As the *Science Daily* article puts it, "the improved edibility of the leaves and fruits of the flowering plants led to a tremendous increase in the number of plant eaters on the Earth, which opened the way to the rapid evolution of mammals, and finally to the appearance of humans."

Yes, you flowering plants sinned against the status quo. You shattered the harmony. You rebelled against nature. Instead of respecting her and living at peace with the existing order, you flowering plants remade that order. You rejiggered the zoo of microorganisms and animals on this pebble of a planet.

Instead of saving, hoarding, and conserving, you produced a wild profusion of materialist excess. Then you indulged in waste. You threw it all away.

In the process, you increased the number of atoms and molecules kidnapped, seduced, and recruited into the grand enterprise of life.[582] More on how you did that in a second.

Meanwhile, you flowering plants—you angiosperms—helped fashion the humans of today. At least if Berendse and Scheffer have got it right.

Maybe there is a point to vanity after all.

***

Oh, and one more tiny thing. Vanity gave you wasteful, flaunting, flouncing, flowering plants one more superpower. Sixty five million years ago, when a meteor "the size of a small country,"[583] the meteor called Chicxulub, would wipe out 75% of the species on earth including the dinosaurs, it wouldn't even faze you flowering plants. You would dance unscathed through the Cretaceous-Paleogene Extinction Event. Why? Your exuberant variety gave you survival power.[584] It gave you over 3,000 different strategies with which to cope with new realities. And it gave you the ability to evolve and adapt almost instantly. Thanks to your blossoms. Thanks to your scents. Thanks to your partnership with insects. Thanks to your acceleration of evolution. And thanks to the First Law of Flamboyance.

***

Vanity works because seduction pays. In fact, making an offer too good to refuse can produce a bigger bonanza than blood-sucking exploitation or bone-chomping predation.

Commerce outperforms violence. Even among plants.

**

In the original version of the angiosperm contract 125 million years ago, you plants worked your energy budgets off on insects' behalf. In exchange, both you plants and the insects you seduced became fruitful and multiplied, taking over the earth. All for the sake of sex.

That angiosperm sex contract would lead to the deal humans would someday make, allowing domesticated plants and animals to reshape human habits, societies, and genes. In the agricultural arrangement, you plants would get humans to find the choicest soil for you, to clear the land, to irrigate and fertilize it, to plant your seed, and to defend your offspring from nibblers and competitors. You would get humans to ensure that you could be fruitful and multiply in ways that nature had never previously conceived. And you would get humans to toy with your genes and to tease forth entirely new forms of descendants, new variations on your species. In exchange, you plants would feed your humans. In this agricultural contract, you plants would use humans as your new insects. But once again, we are getting ahead of ourselves.

**

The angiosperm contract, the insect-plant alliance, was a food-for-transport arrangement. A food for **sexual** transport. And that bargain made the world of birds, bees, and flowers that we humans love today. The world of birds, bees, and flowers that we naïve humans worship as nature. But in its day 140 million years ago,[585] you flowers on this ball of stone were as unnatural as a Madonna concert at a convention of nuns. Complete with an opening act: an abortion performed onstage.

So why did nature allow you, the flowering plant, to survive? Why did she do more? Why did she allow you to thrive?

CHAPTER 29

IS SEX SEVEN~DIMENSIONAL CHESS?

Why did nature allow you, the gaudiest and most wasteful organisms in history, you flowering plants, you angiosperms, to flourish? Why did she favor you despite your sins of materialism, consumerism, waste, and vain display?

Because your sins were blessings in disguise.

With your sins, you harnessed two opposites joined at the hip— cooperation and competition. Remember, competition with ferns and cone-bearers played a critical role in your success. You flowering plants were vicious, greedy, and unjust. You drove the ferns and cone-bearers to the hard-scrabble margins, to the cold lands high on mountainsides or in the extreme north and south.[586] Then you flower-bearers competed with each other. Frenetically.

Remember, competition is not a capitalist invention. It is a tool of nature. Nature is green in tooth and claw.

But despite your victory over the ferns and cone-bearers, you flowering plants did not just triumph at the expense of others. You also succeeded because you, a plant, genuinely looked out for my needs—the needs of an insect—in exchange for my looking out for yours.

When cooperation and competition worked together to evolve you, the flowering plant, cooperation and competition also worked together to increase the amount of living matter on the planet. Cooperation and competition worked together to spur the expansion of the biosphere. Cooperation and competition worked together to increase the GAL—the Gross Amount Of Life—on this planet. And the GAL in the very cosmos.

Opposites are joined at the hip.

Yes, cooperation and competition worked together to increase the percentage of the atoms on this planet woven into the bio-mesh, stitched into the eco-net, the percentage of atoms knitted into the tissue of the living. Thanks to cooperation and competition, you plants and we insects flashed together like knitting needles, stitching a growing weave of green.

The flowering-plant-and-insect explosion happened not because we struggled against each other to divide a tiny pie. It happened because we worked together to grow that tiny pie into a wedding cake. And it happened because we worked together to turn that cake into a bakery. Even our competition contributed to that larger enterprise. Which means that the four deadly sins of materialism, consumerism, waste, and vain display are teamwork makers. Group identity crafters. And biosphere expanders.

But behind it all, behind the flower and insect partnership, was sex. Why?

What does the wild success of the flower say about our current assumptions in science and popular culture? What does it say about basics like the law of least effort and the Second Law Of Thermodynamics, the concept of entropy? What does it say about the nature of the cosmos within which evolution operates? What does it say about a nature that supposedly pursues harmony, sustainability, and thrift?

And what does it say about the contribution of bling?

***

When you are pondering nature's supposedly pinch-penny ways, her insistence on bare bones asceticism, her push toward doing things in the most economical and energy efficient manner, her stress on harmony, sustainability, and balance, put one simple fact at the forefront of your brain.

Nature invented death, the ultimate materialist throwaway. The ultimate one-use indulgence. The ultimate waste. Yes, death is nature's invention, not the invention of capitalists, industrialists, paternalists, or even of humankind. Since life began over 3.5 billion years ago, more than twenty quintillion individual animals have died. Twenty billion billion.[587] And if you add the deaths of plants and microbes, the number of individual organisms who have died skyrockets.

Just to make things worse, nature invented pain. What does that say about nature's benevolence? What does that say about her tendency to do things in the most harmonious way?

Death is a tipoff to a brutal and disturbing fact: that nature is deeply addicted to manic mass production, then to manic mass destruction. Nature is hooked on industrialism. Worse, she is hooked on the mass production of something that is ephemeral beyond belief—life. Life that is priceless to you and me but to nature is cheap, cheap, cheap. Death makes one-use throwaways out of you and me.

Think of all the energy, matter, and time that's gone into making you. And you and I will be discarded someday. Thrown away. By who or by what? By nature. That is excess. That is materialism and consumerism. That is waste. Worse. It means we live in a cosmos of cruelty.

Why death? Because it is apparently an evolutionary accelerator, a creativity turbocharger, a search engine super-thruster. An expander of life's ability to kidnap, seduce, and recruit more dead atoms into the grand enterprise of life. Hate our mortality though we may, death is apparently an expander of the cosmos' ingenuity.

What a truly horrible thought.

***

Is life in a world of death worth living? Is part of what makes life worthwhile beauty, a beauty that has sex at its core? In other words, is part of what makes life worthwhile the wild waste of sexual display?

More important, does nature get breakthroughs from sex's vain and flashy throwaways? Does nature harvest innovation from mindless bling?

Let's reframe our question. Why do females of roughly 30 million[588] species, including ours, put males through artificial games, competitions that seem very far removed from the basics of life— very far removed from food, clothing, and shelter? Competitions that waste vast quantities of material goods, energy, labor, and ingenuity? Like the competitions of orchid bee males[589] to show off their talent by blending the most compelling perfumes? And the competition between aristocratic human females to show off the sophistication of what they demand from males. The competition to show off their taste. And why do these arbitrary and artificial competitions, these high-priced mating games, work? What do they contribute to the evolutionary process?

Sexual competitions produce elaborations, ornamentations, and dance steps that are really inventions in disguise. In the case of dinosaurs, for example, sexual contests generated horns, plates of armor, heads like snowplows built for ramming, heads ballooned up like bowling balls built for bashing,[590] long lines of upright backbone scales standing like rows of tombstones, skin that flashed bright colors, skin that photo-luminesced,[591] and a game-changing innovation we'll get to in a few minutes, the feather.[592] Not to mention martial-arts moves and songs. Elaborations and ornamentations that often did the seemingly impossible. Elaborations and ornamentations that broke nature's rules. And in the process of breaking her rules, ornamentations and elaborations that helped nature give herself new upgrades, new tools. New powers, powers that punched open her envelope of possibility.

Elaborations and ornamentations that helped nature work out the unseen potential in her current state of play. Elaborations and ornamentations that helped nature find new moves. Moves that sometimes flew off the game board and created entirely new games. Moves that helped nature advance her position in the biggest game of all. The innovation game. The game of self-reinvention. The game of non-stop self-re-creation. The game of producing the next shape shock, the next supersized surprise.

As we will soon see. But first...

***

For an example of the way in which even insects are seized by materialism, consumerism, waste, and vain display, go a bit deeper into the evolution of those expensive luxuries we looked at a minute ago, perfumes. Perfumes allowed female insects to dictate new competitions, new sports, new games. They allowed females the luxury of following their tastes in whole new ways.[593] They allowed females the luxury of aspiring high. The luxury of going for perfection. The luxury of reaching for genetic optimization. The luxury of reaching for the sky.

And differences in female taste allowed insects to form groups that walled themselves off from each other and went down different genetic paths. The paths that led to breakthroughs. The paths that led to new genomes—to new gene teams. The paths that led to new modifications in bodies. The paths that led to new habits.

New tactics and new strategies. New ways to turn a poisonous environment into a paradise. New species.

Orchid bees, euglossine bees, for example, invented two hundred and fifty different tactical approaches—two hundred and fifty different species. With different tongue lengths, different mouth parts, different colors, and different recipes for the perfumes that put males through hoops.[594]

And honeybees invented wax, the honeycomb, the egg laying queen, and the factory-like teamwork of the hive. All to satisfy the plant and insect bargain. All to reshape themselves to satisfy the ravenousness of plants. The hunger of plants for sex.

CHAPTER 30

DOES NATURE REALLY HATE WASTE?

Who are the test pilots for invention? Males.

Which brings us back to waste. In the form of male disposability.

In many species, the very existence of the male is an example of materialism, consumerism, waste, and vain display. Maleness, like the flower, is an example of wild excess. Maleness is an incarnation of the First Law of Flamboyance.

A male in a bee hive—a drone—does nothing of value. And he's expensive. He needs to be fed, housed, and tended to. That's why his very name, a drone, has come to be associated with high-cost, lazy human males who slump on the sofa, drink beer, watch football games on TV, expand their waistlines, and seemingly do nothing but demand continuous care and feeding. So why do beehives have males? For genetic mix and match. For sex.

At one brief time of the year, the males of up to a hundred neighboring bee colonies are sent out of their hives to the air above a meadow to compete with each other and with the males of other hives. Hundreds, thousands, or tens of thousands of males.[595] What prize are these males competing for? Intercourse. To get their semen into the sperm-storing sacks of queen-bee wannabees. Would-be queens who will have sex with roughly seven males each,[596] and will store these seven kinds of sperm for future use.

When a male succeeds, when he manages to mount and penetrate up to ten females, what's his reward? His semen-carrying equipment explodes. This internal shrapnel is fatal. Winning is deadly. The very sex act kills the triumphant male.[597] And nature literally tosses the male's body and whatever its living spark may have been away. At the ritual mating meadow—"the drone congregation area"—to which as many as a hundred neighboring hives have sent their sexually ripe males and females, their drones and would-be queens—by the tens of thousands,[598] to compete, the bodies of males literally pile up on the ground.

They are thrown away.

Yet the hive gets enough out of this competition to make drones worth a steady fifteen percent of the hive's yearly budget. Why? To innovate. To invent 2,000 new ways of making a living, 2,000 new species of hive-making bees.[599]

Nature wallows in waste and luxury to reinvent herself. She produces waste to give the finger to entropy. She produces waste to achieve her most astonishing feat—moon shots, shape shock, supersized surprise. Emergent properties. Intricacy. Invention.

She uses sex to increase the percentage of dead atoms and molecules kidnapped, seduced, and recruited into the grand enterprise of life.

And more. Nature uses sex to find her way up the next mountain peak in possibility space. And if there is no mountain, nature uses sex to build one. Nature uses sex to innovate. She uses sex to invent. She uses the coupling of sex and waste to follow the First Law of Flamboyance.

***

And what do you flowering plants get out of your expensive transport economy, the insect-bee partnership, to which you've committed between ten and fifty percent of your budget?[600] What do you flowers get out of increasing the distance your sperm can travel and the precision of its targeting? What do you flowers get out of long-distance mating?

You are a flowering plant. Why not just insert your pollen into your blossom's own remarkably convenient stigma, your female organ, and have done with it? Or, better yet, why not do away with sex altogether and simply send out shoots and let them become new plants, new plants that are your precise genetic clones, something plants from dumb canes[601] to strawberries[602] do? Or why not be immortal and do away with reproduction altogether? Why not take the shortest path, the path of least effort?[603]

To put it differently, why is nature horny? Why does nature have a libido? Why does she put so many resources into sex? Because she has a special place in her heart for gene shuffling. She has a special place in her heart for one-of-a-kinds. She has a special place in her heart for oddballs.

And oddballs are what gene shuffling produces. You remember gene shuffling—the guy in row 25 with tattoos all the way down to his wrist and the well-tanned guy in shorts reading a paper on neurobiology on his iPad in row fourteen. Each of them was forced to switch from the window seat to the aisle seat. When all of the aisle seats were unzipped from all of the window seats, both of these gentlemen became part of the same new string of single genes. Thus making that new string colorfully unique. Making it the first of its sort ever seen. Making it a one-of-a-kind.

Raw duplication of your self is not why nature drives you to find a mate. You are not driven by what superstar Oxford zoologist Richard Dawkins calls selfish genes.[604] In fact, you are driven by remarkably unselfish genes. Genes that step aside and let others take the spotlight. Genes that relinquish their positions and switch seats.

As you know, none of your kids will look, walk and talk like you. Thanks to gene shuffling, every child you have will be a wild mix. A shot in the dark. A weave of something old making something new. Sex produces what author and spiritual philosopher Marc Gafni calls "unique selves."[605] Sex produces one-of-a-kinds. Sex produces oddballs.

But if your freak, geek, or rebel children find others of their sort, they will hang out together and these outcasts will have kids. Kids whose shared oddities will be a new normal. Kids whose shared peculiarities will produce new subcultures. Kids whose shared oddnesses may just possibly produce a new species. Kids whose shared strangeness may produce new swirls in the hurricanes of history. Kids whose shared eccentricities will be new antennae for the search engine of the cosmos.

***

The shortest path indeed. Nature shuns simplicity and plays seven-dimensional chess. In fact, she invents seven-dimensional chess. Then she finds seven dimensions beyond even that, and she invents entirely new games.

CHAPTER 31

WHY DO FEMALES FALL FOR DEATH SLAYERS?

There's a hidden reason that you and I are addicted to sex. A hidden reason that we obsess over this insanely intricate sin against entropy and against Pierre Louis de Maupertuis' law of least effort. Nature is driven by the First Law of Flamboyance. Nature is driven to invent.

Please bear with me while I repeat, even where it counts the most, way down on your gene string. Sex is about making gene combinations that are utterly unique. And torturing you with romance is one way that your genomes, your gene strings, pull their one-of-a-kinds off.

But what, pray tell, is the value of this mind-walloping uniqueness? Each new living being that sex produces is a new feeler for the search engine of the cosmos. A whole new finger with which she can poke beyond the current limits of the possible.

In fact, sex tempts the impossible into the realm of reality. Sex makes nature an ever-expanding realm of former impossibilities.

Take the notothenioid fish of the Antarctic. Between 15 million[606] and 25 million years ago,[607] these finny wrigglers invaded a territory whose year-round use was previously out of bounds to fish, the underside of the ice floes around the South Pole.

The seas beneath the Antarctic ice warmed up during the summer, when the winter ice melted and drifted away.[608] These warmed waters provided fish with food.[609]

But when winter came, the waters dipped below the temperature at which blood freezes. They became much too cold for animal life. They became an impossibility. At least for existing animals.

Remember the power of oddballs? New gene combinations can turn hellscapes into riches. And niches. Between 15 and 25 million years ago, some pioneering fish evolved an impossible technique to overcome the cold. They were notothenioids. And they generated sugary molecules, glycoproteins, with a special property. A property captured in the name by which scientists know them: anti-freeze proteins. The anti-freeze proteins lowered the temperature at which the fish's blood froze.[610] In other words, at winter temperatures that would freeze the blood of a normal fish, the blood of the "freeze tolerant" notothenioid fish just kept flowing.[611] The Howard Hughes Medical Institute's Biointeractive.org explains that, "as the antifreeze proteins circulate through the blood, they bind to ice crystals and prevent them from growing. The fish's blood thus does not freeze and continues to flow normally."[612]

The anti-freeze invention created a niche. It turned an impossibility into an opportunity. It turned a wasteland under the winter ice of the Antarctic into a wonderland. New niches are invented, they are not found.

It appears that once male notothenioids had worked out their anti-freeze trick, they were able to stake out territories, to build nests,[613] to attract females, and to produce baby fish, baby fish with the anti-freeze-making process built into their genes. He who turns a terror into a treasure gets the girls. And the kids.

This turning of a death trap into a bounty opened a vast new horizon to not just fish, but to life. Today, anti-freeze-making fish constitute 90% of the fish biomass in the Southern Ocean.[614] Thus increasing the resources available to the living. And thus, increasing the GAL, the gross amount of life, on the planet.[615]

Each new living being that sex produces is a feeler for the search engine of the cosmos. A feeler into the impossible. Once notothenioids had settled down, they apparently set up cliques. Cliques whose ladies led their clans down separate paths, paths to eating the formerly uncatchable[616] or the formerly indigestible. For example, phytoplankton ate sunlight and turned it into biomass. Despite the Antarctic darkness, they photosynthesized. So some notothenioids scaled down to the size of minnows and ate that phytoplankton. Other notothenioids grew bigger and ate, guess what? The minnow-sized notothenioids. In other words, the notothenioids evolved new species to maximize the ways in which they could turn the meager stuff of the Antarctic into a feast. Today there are roughly 283 species[617] of notothenioid fish supping in the darkness below the ice of the Antarctic.[618] Which means the notothenioids have invented 283 new ways to make a living. 283 new ways to turn nothings into niches. And niches into riches.

Then there's the search power of sex between similar oddballs, between similar one-of-a-kinds. One-of-a-kinds cluster in cliques. And their females drive males down separate paths. Paths paved by different ways of showing off. In Eastern Africa less than 12,400 years[619] ago, a small number of look-alike fish found its way to Lake Malawi.[620] It didn't take long for the finny explorers to overpopulate the place. As food became harder to find, squabbles and serious fights[621] apparently pushed these cichlids to square off in spatting gangs. So did the pickiness of females, who went for their clique's[622] taste[623] in the colors, sizes, smells, social status, croons, and wiggles of courting males.[624]

Kelly Kissane, of Lassen College, shows how things like colors, sizes, smells, social status, croons, and wiggles can help drive animals down the paths of invention, the paths to new species. She points out that the females of two groups of water spiders on opposite sides of a pond will develop different fashions in their courtship dances. Take an outrageously-popular male superdancer from one side of the pond to the other, and the local females will regard his choreography as alien and shun him. They will stick with the close-by spider males who give them the dance they have come to prefer. Thus the two groups of spiders will segregate genetically. Walled off by dance steps. Walled off by spider pop culture. Walled off by the fads of females.[625]

This sort of differentiation has a name, schismogenesis.[626] And schismogenesis even shows up in the sperm whales of the Pacific

Ocean, whales that wall themselves off in seven massive clans of up to 20,000 whales each. Clans led by females. Clans also separated by pop culture, separated by different "vocal dialects," separated by their styles of song.[627]

So it's not surprising that the same principle shows up in cichlid fish. Female cichlid fish go for five things: a male's swaggering standing at the top of the pecking order. And a male's smell, courtship song, moves, and color. To quote Stanford University's *Stanford Report*, "A dominant male attracts choice females to his territory by dancing seductively... waggling its tail and quivering."[628] Whether a female swoons or not over this courtship display determines the next nano-step in the evolution of her species, the next probe of her genetic tribe, and the future success of the entire cichlid clan.[629]

This female snootiness, this fickleness based on animal fashion—this choosiness that Darwin called sexual selection—apparently walled the cliques of cichlid fish in Lake Malawi off from each other. The farther the groups grew apart, the more different they became.[630]

The result?[631] The cichlids of East Africa's Rift Lakes rapidly went[632] from a single species of fish to 1,000,[633] each equipped with a crowbar to pry open opportunities others had missed. Some evolved mouths wide enough to swallow armored snails. Others generated thick lips to yank worms from rocks. One diabolical coven acquired teeth like spears, then skewered its rivals' eyeballs and swallowed them like cocktail onions. In the geologic blink of 12,400 years, what had begun as a small group of carbon copies in

Lake Malawi became a myriad of separate species—a carnival of diversity, a carnival of invention, a carnival of oddballs.[634]

Strangely, a gene called bmp4, Bone Morphogenetic Protein 4,[635] was what shifted the shape of cichlids' jaws and mouths.[636] And bmp4 was the same gene that had crafted the many shapes of beaks in Darwin's Galapagos finches.[637] It was the same gene[638] that had patterned the brains of insects, and that would someday serve half a dozen functions in shaping humans like you and me,[639] including fashioning your teeth, your face, your bones, and your brain.[640]

In fact, the diversification of genes shaped over 1,700 cichlid fish species in just three of Africa's Great Lakes.[641] In a mere 50,000 years.[642] A blink of the eye in evolutionary time. And how did these 1,700 species act as probes of a search engine? Each poked its way into new nooks and crannies. Each invented a new niche. And together they maximized the use of Africa's Great Lakes. They pushed the envelope of the possible. They increased the ways those Great Lakes could add to the planet's biomass. And to the amount of dead stuff kidnapped, seduced, and recruited into the grand scheme of life.

What gave the cichlids the ability to pull this off? Oddballs, the finicky tastes of females. And sex.

**

Yes, sex favors probes for the search engine of the cosmos. Sex favors discoverers. For example, when sex drives males to set out and seek their fortune, the fervor to impress females sometimes spurs those males to open the treasures of whole new territories.

Sex drives males to open the realm of the impossible. To see how, let's zoom in on the recent exploits of creatures who first appeared 300 million years ago, damselflies.[643]

Damselflies, explains the Encyclopedia Britannica, "are graceful fliers with slender bodies and long, filmy, net-veined wings." They look like dragonflies, but they are "smaller, more delicate" and can't match dragonflies in the strength of their flight.

But don't be fooled by their iridescent blue and green colors and by their delicate elegance. Damselflies are predators. They eat "flies, mosquitoes, and moths and some eat beetles and caterpillars."[644] Damselflies are wickedly successful. There are 2,600 species of them worldwide.[645]

How did damselflies manage to evolve 2,600 new upgrades, 2,600 inventions, 2,600 new species? The damselfly females optimized the genomes of their kids by being picky. By mating most with males who did remarkable things. Males who amassed a surplus of resources. An overflow of riches. Riches they'd acquired by deeds like opening new territories. Opening whole new niches, whole new resource jackpots. Whole new realms of the impossible.

In southern Europe around the year 2000 AD, just a few decades ago, the dainty blue *C. scitulum* damselflies were confined to the area near the Mediterranean Sea—Spain, Sardinia, Sicily, and the southern tip of Italy.[646] But the climate was warming. And some males were restless, rebellious, and bold. Those with the strongest wings flew outside the boundaries of their species' territory and headed north to Germany,[647] Holland, and France.[648] They explored the possibilities for their kind in the standing waters of distant bogs and fens.

He who finds a jackpot gets the sex. He who found food attracted females. And the damselflies formed permanent colonies in the new homes. This was the discover-then-consolidate strategy, the fission-fusion search strategy, come alive. But the discovery was the job of the males.[649] The consolidation, the settling down, was the job of the females. So was the job of setting challenges and being picky about those who would succeed in mastering them. That, too, was a female task. If this sounds "overly gendered,"[650] please complain to nature.

Oh, and the children of these damselfly parents were born with unusually powerful wing muscles. Why? Because their fathers had been restless explorers whose main tool of exploration was flight.

***

Males are waste with a purpose. They are probes for a search engine of the cosmos.

And who were the explorers of new territory among the cichlids, the damselflies and the anti-freeze fish? Who were the seekers of the impossible? Oddballs. Adventurers flying or swimming off the beaten track. Taking big risks. Death defiers. Death slayers. Disposable males.

Not to mention one-of-a-kinds. The sort of one-of-a-kinds produced by guess what? Sex.

***

The restless explorations of anti-freeze fish, the sexual song and dance of cichlid fish, and the itch for new territory of damselflies reveal the evolutionary search engine of the cosmos at work. They reveal a hungry nature turning ever more of the inedible into an hors d'oeuvre. And turning ever more abiotic dead stuff into threads in the weave of life.

The anti-freeze fish,[651] the cichlid fish, and the damselflies'[652] extraordinary "adaptive radiation" reveals a cosmos restlessly probing to find her next moves in possibility space, hunting for her next wiggles beyond awe, reaching for her next radical impossibilities. Her next supersized surprises.

What's more, life's gamble-placing process, her exuberant exploration, her insistence on breaking the rules of the status quo, and her obsession with invention have just begun. You are part of it. You are a wager, a probe, an antenna. You are life's way of feeling out the next impossible landscape. You are nature's way of feeling out the cracks in rock and testing the highest limits of the sky. Whether you are a moss shooting your sperm five inches higher than life has ever gone before or you are a human hungry to read the latest reports from telescopes in space that hunt for livable planets softballing around distant suns. You are an extension of the first teaspoon of life's itch to girdle one poison pill of stone, then to garden an entire solar system and to green an outstretched galaxy. You are an extension of the cosmos' itch to move up. Her itch to penetrate the impossible. Her itch to reinvent herself.

***

Let's get back to you, the flowering plant. Why did you, a plant obsessed with sex, take the path of most effort? Why did you use the ultimate team of rivals, male and female? Why did you take huge risks and shuffle genes in a process so complex that even a Mensa member has trouble following it? Why did you gamble your chips on sexuality's materialism, consumerism, and waste? And why did you go hog wild for vain display?

Remember your earliest ancestor? Remember that first teaspoon of life on a hostile planet? Remember how your forebears were born in shock, disaster, and catastrophe? Remember how even the change from day to night and back again was a cataclysm that repeated every three hours? Remember the massive climate changes of summer, fall, winter, and spring? Remember the Milankovitch cycles that shook the climate every 22,000, 41,000, 100,000, and 413,000 years? Remember the sun's 240-million-year journey around the heart of the galaxy, with that trek's unpredictable clots of space dust and arbitrary storms of climate-shocking cosmic rays? Remember the miscellaneous end-of-the-world scenarios that produced mass die-offs every 26.5 million years?

How did your great, great foremothers survive? By turning poisons into pistons in the engine of life. By turning disasters into delights and cataclysms into fields of dreams. By putting feelers into as many puddles, sinkholes, cracks, chemicals, currents, storms, and stews as possible. By stretching out, searching, tunneling, soaring, spreading, pirating, and inventing. By seizing a toehold in every inconceivable kind of place. By kidnapping, seducing, and recruiting as many dead atoms as possible into the grand enterprise of life. And by inventing new ways to surf the waves of change.

Your animal ancestors survived by following the First Law of Flamboyance.

***

There have been roughly 142 mass extinctions since life first began. And today, life's race against disaster is far from over. Global warmings and ice ages are not just things of the past. They are things of the future. Sure things. Things that may be going on at this very moment. Things that may not arrive tomorrow or twenty years from now. But things that are guaranteed to arrive someday. Things that we may have caused with our carelessness. Things that might have happened without us. Even a return of snowball earth—the total freeze that has smacked this planet at least two times[653]—is possible someday. Remember, this is a planet of climate catastrophe. Climate change is as certain as the change from day to night.

Sex is expensive. Sex defies the laws of nature. But sex is the ultimate survival device.

Because sex is the ultimate inventor of the next big thing. As we are about to see in the case of the loony dinosaurs who flew.

CHAPTER 32

THE *STORY* OF THE LOONY DINO*S*AUR*S* WHO FLEW

We are all a little weird and
Life's a little weird,
And when we find someone whose
Weirdness is compatible with ours,
We join up with them and fall in
Mutual weirdness and call it love.
—Robert Fulghum[654]

Can gaudy displays of bling really lead to breakthroughs? Can flamboyant ornamentations and shows of useless stuff really upgrade, empower, and uplift the grand project of life? Can strutting your stuff really produce invention? You bet. But to understand how, you have to understand the birds of a feather effect.

Nature uses the genetic shuffle of sex to generate one-of-a-kind creatures, oddballs. Then she gathers oddballs of the same sort together in flocks, herds, species, subcultures, nations, and civilizations. The birds of a feather effect is nature's way of gathering similar eccentrics and turning them from solitary outcasts into movements, into social forces. Into forces of evolution. And into forces of history.

Here's how the birds of a feather effect works. Take a yellow natural sponge. Run it through a sieve into a bucket of water. A sponge may look and feel like just one solid handful, one squeezable thing. But, like you, it's actually a community of microscopic beings. It's actually a society of cells. Cells that can live on their own. But cells that choose not to. As you are about to demonstrate. With your sieve, you tear those cells away from each other and turn them into a yellow cloud in your bucket. A yellow cloud of solitary cells swimming around and aching for company.[655]

Now take a red sponge and run it through your sieve into the same bucket of water. It forms a pink cloud. Swirl the water with your hand to make sure that the two clouds—the yellow and the pink—are thoroughly mixed. Your water is now orange. You would expect that if the sponge cells in your bucket glom together again, they would create an orange sponge. But that's not what happens. The red cells find each other in the fog of cells and grab hold of each other. The yellow cells find their way through the sponge cell mist and find each other, too. In the end, the cloud disappears, the water clears, and what do you have at the bottom of your bucket? Two sponges. Two societies. Two group identities. One is red. The other is yellow.[656] If these sponge cells were human, we'd call this apartheid.

The separation of the cells in your bucket comes from the birds of a feather effect. The red cells want to hang out with other red cells. And the yellow cells want to cozy up to other yellow cells.

The birds of feather effect can reshape genes. How? Through the process of sexual pickiness.

An Israeli researcher, Gil Sharon, of Tel Aviv University, divided a swarm of identical fruit flies in two and gave each a different diet.[657] One half got molasses. The other half was fed starch. Animals like you and me have a population of bacteria in our gut, a microbiome, with hundreds[658] of species of microorganisms that help digest our food and nourish us with necessities like thiamine, folate, biotin, riboflavin, pantothenic acid, and Vitamin K.[659] So do fruit flies.

The fruit flies feasting on molasses developed one sort of intestinal menagerie. The fruit flies gorging on starch developed a different microbial zoo in their guts. When it came time to mate, molasses-eating females preferred to hook up with molasses-eating males. And starch-eating females went for males who had also eaten starch. The birds of a feather effect.

What did the biomes in the fruit flies' guts have to do with it? Everything. When Sharon gave the fruit flies antibiotics and killed the microbial zoos in the fruit flies' intestines,[660] starch-eating females were suddenly interested in molasses-eating males again and molasses-eating females enjoyed the attentions of starch-gobbling males. In other words, the microbes had erected a barrier of sexual prejudice. A barrier that, over hundreds of generations, would have made the genes of the starch eaters different than the genes of the molasses guzzlers.

But the tendency of oddballs to find each other is not limited to sponges and fruit flies. Butterflies and moths[661] mate with others who have similar patterns and colors on their wings. Fish, from

Amazonian tetras to salmon, mate with others who are like them in size.[662] And a 2023 study of flamingos found that these elegant, pink birds "tend to spend time with other birds with personalities similar to their own."[663] You read that right. The researchers said that the flamingos "assort by personality." It's the birds of a feather effect again.

For another peek at the birds of a feather effect, check out this squib stolen from my book *The Lucifer Principle: A Scientific Expedition Into the Forces of History*:

> We mammals are uncannily good at gravitating toward those who share our hidden joys and woes. This talent for emotional homing crops up among beavers, wolves, and even deer.[664] In the macaca mulatta and rhesus monkeys Harry Harlow studied[665] [from 1931 on in his Psychology Primate Lab at the University of Wisconsin] it's particularly astonishing. When it came to mating, those who'd been raised in isolation fell for others also brought up in quarantine. Those who'd spent their youth in cages wooed other victims of captivity. Now here's the topper. Some of the monkeys had been lobectomized. Though none were handed pictures of each other's brains, those with similar neurosurgery managed to sniff each other out. And to pair up and mate. So subtle were the differences detected by the simians that even researchers couldn't spot them without a careful study of medical and rearing charts.[666]

Humans are much the same.[667] Children whose gifts or disabilities make them seem bizarre, for example, manage to find each other and to congregate.[668] Among our kind it's called validation. Without others on our wavelength, the strangeness of our emotions can make us feel we're losing our minds.

But when we gather with others of our kind, our weirdness becomes a new normal.

In 1903, a term for the results of the birds of a feather effect appeared in the journal *Biometrika*, "assortative mating."[669] Eighty three years later, in 1986, premiere primatologist Frans de Waal called the tendency to gather with others like you the "similarity principle." De Waal showed it at work in rhesus macaque monkeys.[670] But in evolutionary biology, it's the term "assortative mating" that has stuck.

The result of the birds of a feather effect? The result of assortative mating? Over time, oddballs mate with other oddballs who share their peculiarities. And those peculiarities hammer themselves into strings of genes.

Yes, oddballs find each other and go from being alone and feeling lost to forming a clique, a movement, a subculture, a nation, a civilization...or a new species. And when it comes time for sex, oddballs have sex with others who share their oddball qualities.[671] Thus planting their oddness in an army of offspring. And making that oddness their offspring's new social reality. Making that oddness their offspring's new normal.

Is there an evolutionary power to these gatherings of oddballs? To these gatherings of one-of-a-kinds? You bet.

Once upon a time, roughly 144 million[672] years ago, 19 million years before plants began to flower,[673] sex's genetic shuffling led to an oddness in dinosaurs. Some were born with quills, quills decorated with weird flattened planes of fluff. Today's scientists speculate that the planes of fluff were useful. They kept dinosaurs warm.[674]

It is probable that the strange males with fluffy quills were shunned. Conformists are not kind to those who are weird and deformed. And it is likely that these male outcasts found others who shared their deformity. Including females made monstrous by fluffy quills.[675] Like red sponge cells finding other red sponge cells in a cloudy bucket of water. Yes, the males deformed by quills seem to have found females deformed by quills and to have had sex.

How do we know? Researchers speculate that planes of fluff on quills were useful for something more than just warmth. They were useful for getting attention. For display. For sexual display.[676] For showing off for females of your kind, primping and preening for females who have also been born with quills and planes of fluff. In fact, researchers believe that the males tried to outdo each other in quill and fluff displays.[677]

Piloerection is the phenomenon that makes your goose bumps. Your skin breaks out in tiny peaks. In animals with quills and panels of fluff, piloerection can spread those quills and fluff out in mantles of magnificence, making the quills stand erect and making a dinosaur look twice its size, a good way to rouse fear in the heart of a rival male.

Another good way to awe a rival into submission is to have quills and planes of fluff on your forelegs,[678] then to stand on your hind legs,[679] and to spread your forelegs out to maximum width. Displaying an intimidating wall of feathers. Brightly colored feathers.

And sometimes impressively striped colored feathers.[680] What's more, standing on your hind legs lets you do something else that makes you a winner in male showdowns—getting above your rival. Being able to stare him down. By breaking nature's most basic law, the law of gravity, and rising on high. Remember, nature favors those who oppose her most.

In species from lobsters and crayfish to lizards and dogs, females favor the male able to stand the tallest. The male tall enough to look down on his rivals. The male most able to break one of nature's most basic laws, the male most able to defy the law of gravity.

Even hormones like serotonin[681] favor the male who rises the highest. In crayfish, the male able to lift his head the highest and to defeat a rival is rewarded with the hormones of victory. An intoxicating blend of chemicals that makes him strut. And that makes him eager to keep other males in their place with more height showdowns, more altitude contests.

On the other hand, the males who can't get it up are numbed by the hormones of defeat. Instead of standing tall, after their defeat in a height contest they crawl. Researchers studying crayfish showdowns say that it's as if each of the males in the competition receives an entire brain transplant. Based on whether or not he was able to rise to the heights of glory.[682] And the same thing happens to lizards,[683] the relatives[684] of dinosaurs. Winners have one blend of hormones. Losers have another. Winners have the hormones of victory. Losers have the hormones of defeat. In fact, you can see those hormones in green anolis lizards from their color. The winners turn a triumphant green. The losers turn a defeated brown.[685]

Researchers on feathered dinosaurs take it for granted that these beasts went up against each other in display contests.[686] Why? Because in species after species the male best able to rise up high gets the girls. Which means the dinosaurs who could rise the highest almost certainly got sex's grand prize—the privilege of inserting the genes for height and fluff displays into the maximum number of dinosaur chicks. In other words, each new generation of oddball dinosaur chicks were given the genes for standing on their hind legs and making a grand display of height. Sex between birds of a feather locked the genes for quills, fluff, and the aspiration to the heights into place. Yes, sex was the tool for genetic change and genetic upgrade.

Sexual display is also the father of invention. It is the maker of impossibilities. What if you could get even higher than mere legs and neck muscles could lift you? What if you could truly stare down your rival from the space above his head? What if you could rise up into the air and look down your snout, beak, or nose at him?

If one of these quill-and-fluff covered dinosaurs stood as tall as he could and still could not look down on his rivals, he could try another trick. He could hop,[687] flap his fluff-covered forelegs, and hover in the air for a few seconds. Those who could pull off this trick could rise above even their haughtiest non-hopping competitors. And these hopping oddballs could win the girls and insert the genes for hopping into the genetic line.

Then, speculate those who research these dinosaurs, some went even farther in the competition to see who could rise the highest. Some climbed trees, jumped, and used their planes-of-fluff covered forearms to glide.[688] This favored a modification in the fluff. If your genes disciplined your fluff into aerodynamic blades,[689] you could out glide everyone else around. And you could, once again, get the prize of inserting your peculiar genes for blades of disciplined fluff into as many females as possible. You could win the prize of sex.

As many females as possible? Is that true? Is it true that he who rises the highest—he who most successfully breaks nature's law of gravity—gets the privilege of a harem? And the privilege of spreading his genes? The privilege of sex? Yes.

E.O. Wilson, in his paradigm-shifting book *Sociobiology*, describes the following experiment. Put six lab rats in a cage,[690] three females and three males. What happens? The males go into a series of face-offs to determine who's on top and who's not. One male rat wins the most showdowns.[691] He beats the other two male rats into submission. Then he courts and consummates. He gets the ultimate prize, sex. How can we be sure?

Once the pups from this rodent fever of lust appear, researchers test the rat pups' DNA. And what do they find? The pups of all three females carry the genes of just one male—the dominant male, the alpha male, the male who came out on top. That is true among species from crickets and flies[692] to lobsters, lizards, and human beings. And it was apparently true in the oddball dinosaurs.

So the genes that help you rise the highest in a pecking order are the ones that take over your tribe.

Note the use of terms like top male, standing tall, and staring your rivals down. All of them are clichés in our common vocabulary. What do these catchphrases have in common? Defying nature's most basic law, gravity. Rising to new heights. Yes, nature favors those who oppose her most.

And rising to new heights is what the oddball dinosaurs did. They learned to flap their forearms and fly.

Let me give this to you from a different point of view. Here's a script I wrote on the topic for a National Space Society animation:

> once upon a time,
> 125 million years ago,
> a loony bunch of you dinosaurs
> came up with a weird idea:
> flying.
> if dinosaurs could speak,
> their eco-conscious, nature-loving parents would have
> pooh-poohed the entire notion.
> don't you get it, their parents would have said,
> look up above your head.
> there is nothing up there but empty space.
> the earth is your mother.
> every good thing in your life is down here on her breast—
> food, shelter, and company.
> not to mention greenery.
>
> up there there is absolutely nothing, now listen very carefully
> and look up above your head.
> what do you see?

absolutely NOTHING.

clouds and at night stars.

you can't eat stars and clouds.

you can't make nests in them.

there is nothing but danger above your head.

if you fall from a great height, you're dead.

but a strange thing happened.

when you walked out of your house this morning, how many of the earth-and-nature loving dinosaurs did you see? None. But how many of the loony dinosaurs who flew did you see? dozens.

the dinosaur conservatives with a love of nature and a deep commitment to the earth died out 65 million years ago.[693]

and the nutty loons who wanted to loop and play in the empty space above their heads are called birds.

what's more

there are twice as many species of birds

as there are of us nice, conservative, ground-walking mammals.

meaning that the fliers have found twice as many ways of making a living in the emptiness of the sky.

and fliers,

be they birds or flying mammals like bats,

live roughly 60% longer than us groundlings.[694]

is nature trying to tell us something?

is there another empty space above our heads waiting for us to ply?

So flowers are not the only gaudy display mechanisms that evolution has produced for the sake of the sexual game. With flight, nature has shown yet again what she first showed with the evolution of the tree—the impulse to rise on high. The impulse to take to the sky. And evolution's use of oddballs, invention, and sexual display to achieve impossibilities.

***

Why does nature favor those who oppose her most?

Why is nature obsessed with sex? Is it because she is obsessed with reinventing herself? Is self-upgrade what evolution is really all about?

Does nature use you and me to rise above herself? Does she use us mortal creatures to soar in whole new ways? Does nature crank out oddballs to elevate her very nature?

Is that why she uses sex?

Here are a couple of hints.

CHAPTER 33

FROM GREENHOUSE TO ICE HOUSE: THE REAL THREAT OF CLIMATE CHANGE

Nature is a killer. And one of her murder weapons is climate.

Our fears of human-caused global warming seem to be on target in the short term, but may be as nearsighted about the long term as Yossarian's well-intentioned rush to stop the bleeding in his gunner's stump of a leg. Remember, the gunner who Yossarian was working frantically to save died just after Yossarian was sure he'd succeeded in saving the injured man's life. Why? Yossarian was rushing to stop the bleeding from the gunner's severed leg. But it turned out that the real problem was the blood loss from the gunner's eviscerated gut. Like Yossarian, we are looking for the cause with admirable zeal and commitment. But we might be looking in the wrong place.

Surely we are the cause of all that is wrong with the world. Surely if we change our behavior, if we offer up sacrifice, if we worship nature, we can usher in nature's petting zoo. Right? But a petting zoo is a highly unnatural state.

Every Garden of Eden has its winter. Or its dry and wet season.[695] Every Garden of Eden has its floods, droughts, and forest fires. Every Garden of Eden has its death and pain. Without the transgressions of human beings.

For example, Mother Nature's five biggest mass extinctions were all driven by climate changes in the range of 9.36 degrees Fahrenheit.[696] And all of those mass extinctions happened long before there were human beings.

We are not the only ones to trigger global warming. 92 million years ago, roughly 52 million years after the evolution of the loony dinosaurs who flew, nature lashed the earth with a massive global warming called the "Cretaceous Hot Greenhouse."[697] It looks to those who research this period of shake and bake that volcanoes became hyperactive and spewed carbon dioxide into the atmosphere.[698] As you know, carbon dioxide is one of the ultimate greenhouse gases.[699] A gas some of us seem to think that only we can emit. We are wrong.

All animals exhale carbon dioxide.[700] What's worse, so does the planet. In volcanic upheavals, the earth coughs out carbon dioxide in massive quantities.[701] The result: in the Cretaceous Hot Greenhouse there was three to five times[702] more carbon dioxide in the atmosphere than there is today.[703] And there were fewer plants to grab that carbon dioxide and to pack it away. In other words, volcanoes were generating climate change. And volcanoes are not man-made. They are nature.

The result 92 million years ago was a massive global warming. The average temperature on earth shot up to 80 degrees year-round in the areas close to the north and south poles. At the tropics, even the ocean was between 82 and 111 degrees Fahrenheit.[704] All the ice at the south[705] and north poles melted. All of it. The earth was so warm that forests of "warm-temperature" trees "thrived near the South Pole."[706] Earth was so warm that crocodile-like Champsosaurs[707] lived in what we think of as the frozen Arctic. In fact, there were so many seemingly tropical animals making the Arctic and Antarctic[708] their home that the curator of earth sciences for the University of Alaska Fairbanks Museum of the North, Patrick Druckenmiller, calls an entire klatch of them "polar dinosaurs."[709]

These dinosaurs in the polar regions included the bizarrely armor-headed, horned, rhino-bodied Pachyrhinosaurus, the ten-foot tall, T-Rex-like Hadrosaurus, the 30-foot long duck-billed, ancient grazer Ugrunaaluk, the 20-foot long, 2,000-pound "polar bear lizard," Nanuqsaurus, and the six-foot-long, bird-like, big-brained Troodon.[710] A testament to the genetic inventiveness of oddballs.

To repeat, a massive global warming held the earth in its grip and there were no tailpipes, smokestacks, or capitalists in sight. The generator of this global bakeoff was nature.

But global warmings are not the worst of nature's climate tortures. The real natural mass killers are ice ages. The earth has been locked in ice ages for at least 415 million years of its 4.5 billion year existence.[711] According to the Utah Geological Survey, "At least five major ice ages have occurred throughout Earth's history: the earliest was over 2 billion years ago, and the most recent one began approximately 3 million years ago and continues today."[712] What's worse, according to Australian National University Professor of Ocean and Climate Change Eelco J. Rohling, we may be overdue for our next ice age. Writes Rohling, according to one theory of the planet's long-term climate changes, that ice age should have arrived over two thousand years ago.[713] But we have fought it off by inadvertently emitting carbon dioxide. We have fought it off by emitting the gases that produce the greenhouse effect.

Even a minor freeze can be deadly. Let's imagine that you are an animal 34 million years ago.[714] You are a Hyaenodon. It's 31 million years after the asteroid Chicxulub smashed into what is today the Gulf of Mexico and wiped out the dinosaurs. Thanks to that three-mile-wide[715] asteroid, this is now a planet of mammals. But not all mammals will make it through the next major extinction. An extinction called The Eocene-Oligocene Transition. An extinction caused by, of all things, too little carbon dioxide. Let me repeat that: too little carbon dioxide. Yes, too little.

The climate change that will threaten you Hyaenodon will apparently be pushed along by yet another of the many things that makes earth the planet of climate catastrophe: the peculiar tilts and wobbles of the planet as it faces the sun. The result of these twitches are the Milankovitch Cycles.[716] Cycles of climate apocalypse this planet is permanently sentenced to. Then there's a see-saw between what geoscientists call greenhouse and icehouse.[717]

Again, you are a Hyaenodon[718] 34 million years ago. You look like a hyena, but you're not. Like a hyena, you are a carnivore. A dangerous carnivore. Meat is your only option in life. Which means that killing is your only way to stay alive. Why? Because all you have are sharp, ripping teeth similar to canine teeth. From the front of your jaw to the rear. Self-sharpening teeth.[719] Flat molars are good for crushing and grinding plants and seeds. But you don't have flat molars.[720] All you have are wickedly sharp teeth good for one thing and one thing only—ripping into flesh.[721] And to make sure you can break necks and crush skulls, your bite reportedly has a force of 1,300 pounds.[722] You Hyaenodon come in a wide variety of sizes, from killers the size of polar bears to slayers the size of house cats. But your teeth mean that you will starve if you don't kill.

Your strategy of meals via murder is successful. Very successful. The asteroid Chicxulub has wiped out a fearsome range of dinosaur meat eaters. That has left a lifestyle of hunting and breaking necks open, a niche you Hyaenodon have seized and made your own. But you will have a problem in your future. This is a poison pill of a planet. This is a planet of climate catastrophe. This is a planet addicted to swinging back and forth from greenhouse to icehouse.

Flash forward to 25 million years ago. You Hyaenodon have been around for nine million years, ripping open the hides of the fellow animals you have managed to bring down and turning your victims' muscle into meals. But the next awkward wobble of the planet you live on is coming, the wobble that changes this gravity ball's ability to suck power from the sun every 23,000, 41,000, 100,000, and 413,000 years. The planet is about to take its next Milankovitch punishment.[723]

What's more, the success of flowers, plants, and trees has set you up for catastrophe. Up until now, the planet you've called home has been kept tropical and green by carbon dioxide levels four times higher than they are today: 1700 ppm[724] versus today's 420.[725] But the wild profusion of flowering plants and trees has slowly sucked in that carbon dioxide and imprisoned it in stalks, leaves, petals and trunks. And now the plants' eager carbon capture has reduced the carbon dioxide level in the atmosphere[726] down to a fraction of what it was when you Hyaenodon first evolved.

Reduce carbon dioxide and you reduce the power of the planet to capture sunlight and to keep it trapped—trapped by the greenhouse gas effect. An effect you, the Hyaenodon species, need to survive.

Something else is happening that will lead to cataclysmic change. It's been going on 9,700 miles to your south. The South Pole was long an empty sea. But the tectonic plates of land mass on this planet have undergone a long, slow evolution of their own. A long, slow migration. Shortly before you Hyaenodon first arose, a portion of what had been South America detached and slid into a new location, a location that covered the South Pole with its first-ever land.

Put two and two together and what do you get? The success of plants in capturing carbon dioxide is cooling the planet. So is the latest Milankovitch twitch in the planet's curtsy to the sun. Meanwhile, ice has slowly grown on the new continent[727] positioned at the South Pole, Antarctica.[728] As the ice of Antarctica sucks up water, the lands on the continents[729] that you Hyaenodon live on grow drier. Swamps you depend on are disappearing. Drought is becoming common. And the sea level is going down by as much as a staggering 270 feet.[730] Leaving new lands open to trees and flowering plants. Leaving new lands open to carbon capturers.

Meanwhile, the ice at the South Pole does not do what the dark soil of land does. It does not suck up sunlight and hold on to its warmth. Ice is white. White reflects sunlight and sends it back in the direction it came from. Back to space.[731] Black heats. White cools. It's called albedo. Remember, the lack of carbon dioxide is already cooling the planet. So, it appears, is the Milankovitch effect.[732] Now add to that a massive sheet of white ice at the South Pole, ice ambitiously reaching out to cover all of the new continent of Antarctica.[733] Ice reflecting light and heat back beyond the stratosphere. Which means that you Hyaenodon are hanging onto life in a whole new world, a world vastly different than the warm, tropical planet you evolved on.

The average temperature is falling by between nine and thirty-six degrees Fahrenheit. Yes, the temperature is dropping by as much as thirty-six degrees.[734]

The trees that have covered the planet you've known, palms,[735] have long, broad leaves that do a great job of capturing sunlight, but those leaves turn out to be a burden when winters grow harsh,[736] and when even the summers are less comforting than they used to be. Those tropical trees are forced to pull back from the latitudes that used to be so sunny and warm.[737] They are replaced by trees able to survive the winter[738] using a clever trick. The new trees put out leaves in the sun of spring, use those leaves all summer, then kill them off and toss them away when winter comes. And when having leaves is no longer an asset, it's a liability.[739] The new trees make use of materialism, consumerism, and waste. They drop their leaves in winter. To adapt to climate change. These new trees with disposable leaves are cousins to flowering plants. They, too, are angiosperms.[740]

Meanwhile, the ice of the South Pole, the ice of the Antarctic, is not alone. The new cooling of the atmosphere has given birth to glacial ice sheets in the North Atlantic.[741] The high and low latitude lands that have been abandoned by the palms and ferns have not all become forests that throw their leaves away when the weather chills. Or meadows of flowers. Grasses can also survive the new frigid winters. And they have taken over vast swatches of territory.

Those grasses have created a new source of riches for species that can adapt to grass-eating. And, indeed, grass eaters rapidly evolve to take advantage of these vast green, rippling savannahs, plains, veldts, and steppes—horses, deer, camels, and elephants.[742] Which sounds perfect for you meat-obsessed Hyaenodons. But it is a bit too perfect.

It is such a rich opportunity for steak and chop eaters that other mammals evolve to compete with you for the meat. Other animals evolve from the production of oddballs and the spread of their oddities via the birds of a feather effect. Your new carnivorous competitors include dogs, saber-toothed, cat-like Nimravids, bears, weasels, and raccoons.[743] And these new meat eaters eventually defeat you. Why? Among other things, some of them have a new genetic twist. Their legs are better built for high-speed running.[744] Far better built.

By the end of this new refrigerated age, you Hyaenodon have gone extinct. Killed by an age of glaciation. Killed by a switch from greenhouse to icehouse. Killed by a freeze generated by overeager carbon sequestration. Overeager slashes in earth's primary greenhouse gas. Overeager cuts in the atmosphere's carbon dioxide.

But there's more than just the plants' enthusiastic imprisonment of carbon dioxide to blame. The earth itself kidnaps carbon dioxide and jails it. Tectonic plates have collided and caused whole new mountain chains to rise: the Rocky Mountains in what would become the West of North America; and the Himalayas[745] in what would become Asia. Raw rock can grab carbon dioxide,[746] pack it away in water and send that water down mountain rivers to the sea, where it is snagged by coral polyps, clams, oysters, and snails,[747] and turned into reefs and shells that hug the sea bottom and that cage the carbon dioxide in carbon-loving rocks like limestone.[748]

Then there is the occasional meteorite.[749]

But the bottom line for you Hyaenodon is this: you have been killed by the earth's see-saw from greenhouse to icehouse and back again. You have been killed by the lack of carbon dioxide in the atmosphere. You have been killed by your own planet.

CHAPTER 34

ELEPHANT *SEALS*: TO HE WHO HATH, IT *SHALL* BE GIVEN

Are you, I, Hyaenodon and the creatures we are about to meet, elephant seals, really probeheads, antennae of a search engine? A search engine with which the cosmos feels out her potential, with which she climbs her next mountaintop in possibility space, with which she hunts for her next supersized surprise?

Is the cosmos really a learning machine? An invention engine?

Let me put you in my place once again. In your 1995 book, *The Lucifer Principle*, and your 2000 book, *Global Brain*, you treated all of life—from microbes to manta rays and from Gila monsters to human beings—as a world-spanning interspecies learning machine. Like a neural net. Like the massively parallel system that makes a supercomputer super. And that makes a large language model like ChatGPT intelligent.

But in 2013, when you began to write this book, *The Case of the Sexual Cosmos*, you tackled something far, far bigger. You wanted to know if the entire universe was a learning machine. You wanted to know if this is a cosmos searching for her potential. A cosmos using time, space, quarks, atoms, and clots of space dust as feelers, fingertips, and probes. Probes into possibility space. And, surprise, when you dug back all the way to the Big Bang, the cosmos *did* show signs of being a learning machine, a discovery and invention engine. But, with this idea of the cosmos as an exploration device, a learning machine, an invention engine, you were out on a limb. Alone.

In 2022, that changed. Bobby Azarian, a cognitive neuroscientist educated at George Mason University, sent you an advance copy of his astonishingly cross-disciplinary book *The Romance of Reality: How the Universe Organizes Itself to Create Life, Consciousness, and Cosmic Complexity*. In it, Azarian, too, proposed that, thanks to information, the cosmos extracts energy, replicates, adapts, and learns. In other words, says Azarian, information turns the universe into a learning machine constantly climbing toward "higher complexity."[750]

Then came two articles in academically respectable outlets. One, in the advanced math and physics pre-print source ArXiv.org, said point blank, "We present an approach to cosmology in which the Universe learns its own physical laws."[751] The authors proposed a mathematical model of the universe as "a learning machine." And not just any learning machine. The sort of learning machine you had described in your 1995 book *The Lucifer Principle* and again in your 2000 book *Global Brain*—a complex adaptive system, a neural net.[752]

Next an article appeared in the *Biological Journal of the Linnean Society*, an article that said, "we propose teleonomic... behavior" [753] in this cosmos. Teleonomic behavior is not just behavior pushed by the causes in its past. It is behavior pulled forward by its future. Pulled forward by the lure of possibility space. Pulled forward by what the Greeks called a "telos," an "end, purpose, ultimate object, or aim." [754] A goal. To talk about a goal-driven universe—a teleological universe—out loud in the scientific community had been forbidden for a century and a half. Mentioning the role of the future in what happens today could destroy your career.

Why? Because teleology—the idea of the future as a cause that effects the present—was ruled out of science in 1855 by German physiologist and physician, Ludwig Büchner, in his massively influential book *Force and Matter, or Principles of the Natural Order of the Universe.* [755] Büchner was certain that regarding the future as a causal agent, regarding the future as a goal pulling us forward, was religious. It threatened the secular edifice of science. Being yanked into the future by what comes next has been taboo ever since.

Scrub teleology and you rule out the idea of the cosmos as a search engine probing her next step in possibility space. Why? Because possibility space **is** the future. The future seducing the present. Luring it. Beckoning it. Whispering to it.

And if teleology is forbidden, then the idea of possibility space— the range of all the future things you could be—is also out of bounds. But Wesley P. Clawson, a neuroscientist at the Levin Lab at Tufts University and Michael Levin, director of the Tufts

Center for Regenerative and Developmental Biology, the authors of the *Biological Journal of the Linnean Society* article "Endless Forms Most Beautiful 2.0: Teleonomy and the Bioengineering of Chimaeric and Synthetic Organisms," went even farther out on the very limb you'd been perched on since 2012, alone and without company. They said,

> We suggest that a multi-scale competency architecture facilitates evolution of robust problem-solving, living machines. [756]

A "competency architecture"? What, pray tell, is that? A competency architecture is a tool for handling challenges. It is a learning algorithm. One that builds on experience to fashion new ways to survive. And to thrive. An algorithm that may someday give birth to new ways to be materialist, consumerist, wasteful, and flamboyant. An algorithm that may give birth to even more of the longest paths possible. An algorithm that may someday produce amazements.

Alas, Clawson and Levin's learning algorithm is limited to "living machines," to the realm of life. But you believed that may be far too narrow a point of view.

OK, let's get back to the basic question you were trying to answer. Is the very cosmos itself a learning machine? Is there a learning algorithm for the entire universe, from the Big Bang to what's going on in your brain as you read this sentence? Is a cosmic learning rule of that sort anywhere in sight?

Yes, there are two of them. And they are right under your nose and mine.

- Learning rule number one is summed up by Jesus in the Book of Matthew. "To he who hath it shall be given. From he who hath not, even what he hath shall be taken away."[757]
- Learning rule number two is explore-then-consolidate. Spread out to discover new possibilities, then pull back together and digest what you've learned. Use the spread-out-and-probe, then huddle-together-again-and-process strategy. Use what evolutionary biologists call the fission-fusion strategy.

Why spread out? To probe the possibilities beyond the boundaries of the known. Why clump together? To digest what you've discovered.

Spread-out-and-discover, then clump-together-again is a learning rule.

But wait, isn't Charles Darwin's natural selection a basic rule of this evolving cosmos? Yes. But natural selection is just another way of summing up rule number one, "To he who hath it shall be given, from he who hath not, even what he hath shall be taken away." "To the victor go the spoils." He who shows the most panache at gathering surplus stuff and inventing new resources wins sex. He or she who reproduces and who ends up with the most kids, takes over the future.

And he or she who promotes ideas that stick may remain an influencer long after he or she is gone. Like Jesus, Muhammad, and Marx. But more on that when we get to Anne Boleyn.

In the case of stars, galaxies, planets, and black holes, he who grabs the most stuff wins. This is what Charles Darwin initially called the "struggle for existence." Until he found a phrase he said was "more accurate," a phrase suggested to him by Herbert Spencer: "the survival of the fittest."[758] But more about Herbert Spencer's contributions in a bit.

You can see the rule of "to he who hath it shall be given, from he who hath not, even what he hath shall be taken away" at work in the first eleven minutes of the Big Bang when neutrons hit a deadline. If they had managed to hook up in a social pairing with a proton, they survived. If they were still alone, they died. Yes, if they were still isolated, still solitary, still one of those who hath not, they went through neutron decay. But neutrons that had buddied up with a proton survived.[759] To he who hath it shall be given. From he who hath not, even what he hath shall be taken away. Indeed.

380 thousand years later, you could see the rule of to he who hath it shall be given at work once again in the Great Gravity Crusades. Remember, when two gravity balls faced off against each other, the bigger always won. And the bigger swallowed the smaller whole. The winners bulked up farther with each victory, with each smaller gravity ball they digested. Eventually most of these roly-poly gravity balls were swallowed by the biggest gravity balls of them all, suns, planets, moons, galaxies, or black holes. To he who hath it shall be given, from he who hath not, even what he hath shall be taken away.

And you could see materialism and consumerism, the greed for material stuff, in these wars between gravity balls, in these races to see who could bulk up the fastest.

You could also see the cosmic obsession with waste. A single black hole burped out flares with more light than a thousand trillion suns.[760] Stars were almost as bad. They reveled in tossing out a non-stop sewage, a toxic spill of radiation—light.

But in the quartet of sins—materialism, consumerism, waste, and vain display—vain display would not arise until the ascendance of the horny reptiles called Aetosaurs,[761] then dinosaurs, dinosaurs who showed off the bowling ball shapes on their heads[762] or on their tails, not to mention their upright scales and their feathers. Then 10 million years later would come the ultimate champions of vain display, plants that showed off. Flowers.

13.765 billion years after the first to-he-who-hath battles were over and after the triumphant gravity balls of the sun and its planets had settled into a stable hierarchy, elephant seals would show how the rule of "to he who hath" was still alive and kicking in mammals. Yes, elephant seals.

***

It's 20 million years ago. Roughly five million years after the Hyaenodon have gone extinct. Another new species has nosed its way into existence, evolving to take advantage of the riches of the North Pacific Ocean. Riches under the water. Like the Hyaenodon, this new player is a carnivore. Yes, it's the elephant seal. [763] And elephant seals show the two basic learning rules of the cosmos at work in remarkable ways. Rule number one, to he who hath it shall be given. And rule number two, fission-fusion—fan out to explore; then clump together and digest what your explorations have taught.

To peer into the dramas and daily lives of these newcomers to the sea, you can look at the patterns of elephant seals today. And what do you see? The spread-out-and-explore phase of the fission-fusion search strategy. The National Park Service says that among elephant seals who winter[764] off the coast of California, "Males return" each summer "to the same feeding areas off the Aleutian Islands," an 1,100-mile chain of islands that arcs from the coast of Alaska west across the Bering Sea to Russia. And females feed farther south,[765] "in the open ocean of the northeast Pacific."[766] To complete their round trip, males journey 13,000 miles. Females get off easy. They travel a mere 11,000. That is fanning out. That is exploring.

In these distant dining areas, males dive over a mile deep for food.[767] As behavioral ecologist[768] Lee Dugatkin puts it, male elephant seals fatten up during their "months at sea, diving thousands of feet below the surface to feed on ratfish, dogfish, eels, rockfish, and squid, beefing up for their stints on the sand dunes, where they will eat nothing at all."[769]

In other words, you elephant seals expend a lot of energy on the search-and-explore phase of your lives. The fan-out-and-poke-around-to-find-food phase. The phase in which you bulk up like gravity balls, accumulating as much body fat as you can. Why all this feeding and fattening? Very soon you will need that surplus fat to make it through the cosmos' greatest sorting mechanism of them all, the one that makes the most use of "to he who hath," the one that comes with the clump-together phase of fission and fusion: the tournament for sex.

When winter arrives,[770] you outspread elephant seals will come together on one small island in the Pacific off the coast of California. You will enact the fusion part of the fission-fusion search strategy. You will cluster and clump. And on your home island, your herd will sum up what it's found. How? With competitions.

The competition between you males is fierce. It is the ultimate user of consumerism, waste, and vain display. If you are a male and you survive your first year of life, you will put as much material surplus as you can on your bones for six years. You will bulk up to 5,000 pounds,[771] considerably more than an all-wheel-drive Buick SUV. You will shoot for a length beyond 13 feet. Your bulk will sum up your success at sea. How? In the winter of your sixth year, all hell will break loose. Says elephant seal researcher Caroline Casey, you will become a cast member in "a living soap opera."[772]

You will spend four months of winter on the dunes of a home island like the one in Ano Nuevo State Park, off the coast of Northern California. It's three weeks before the females arrive. Your goal will be to provoke one-on-one tournaments with each and every one of the other 149 males on the island in the three weeks before the return of the ladies. You will find another male, stand as high as you can on your front flippers, inflate your balloon-like nose to its maximum size, and bellow your name. You will make a fierce call that in tempo and timbre is unique to you. Meanwhile, you will slam your body to the sand hard enough to shudder the ground. And to spit sand in the face of your opponent. Then you and your rival, still bellowing, will charge each other until it becomes obvious that either you or your competitor is just not big enough, strong enough, or persistent enough to cut it. The loser—you or your opponent—will skulk away, defeated, and go through this all over again with some other hulking male. Most of the time these showdowns will be definitive—one seal will win and the other will hunch away a loser. But if neither of you male elephant seals are willing to concede defeat, the battle will get bloody. The two of you will bite. Until finally one of you yields.

After you've done your showdowns with most of the other 149 fellow male members of your sand dune community, it will become obvious that seven or eight males have won the most battles and subdued the most of your neighbors. Hopefully one of these champions will be you. And you and the five or six other winners will get all the females. Yes, all of them. Very much like the winning rat who mated with every female in the cage.

Of all the elephant seal males born, less than one percent ever manage to mate.[773] Yes, only one percent. Ninety nine of every hundred never get a crack at the ultimate male elephant seal prize, sex. They never get a crack at the reward of passing their genes to the next generation.

Where is materialism, consumerism, waste, and vain display in this picture? Males are waste with a purpose. 32% of baby elephant seals will never make it to their first birthday.[774] That's materialism, consumerism, waste, and cruelty. But there's more. The material and consumerist waste of 142 5,000-pound males who will never get to mate. The waste of nearly 710,000 pounds of living beings. The waste of six years of bulking up. The waste of over eight million hours of seal effort.

And the waste of all the million or more ratfish, dogfish, eels, rockfish, and squid you've eaten. The waste of the million lives you've taken. All for the sake of vain display. Vain display of your size, your strength, your ferocity, and your ability to persist. Vain display that acts as a sorting mechanism for the females. Vain display that powers a competition in which less than five percent[775] of you males come out on top. To he who hath it shall be given indeed.

Then there's your use of the explore-then-consolidate strategy, the fission-fusion search strategy, in which you males and females spread out in the sea for eight months of the year,[776] traveling a combined total of 3.6 million miles hunting for food. Then clustering together again on your sand dunes[777] during the winter and summing up what you've accomplished with a contest to see which of you males managed to eat the greatest number of other living beings and convert them into the greatest stores of fat and muscle. Fat, muscle, and something immaterial. Passion and persistence. The fighting spirit.

Natural selection bulks up 600,000 pounds of living stuff just for the sake of a competitive display that will allow nature to throw 142 males away. A competitive display with which nature will digest what she's learned about which males are the so-called fittest. Not to mention which elephant seal one-of-a-kind gene combinations worked out best. Nature uses materialism, consumerism, waste, and vain display as probes of possibility space, antennae in her search engine, feelers poking into the invisible landscape of the future.[778] Nature uses materialism, consumerism, waste, and vain display to power her learning machine. And to fuel her power to invent.

Not to mention her power to waste something beyond measure— lives.

Why? To feed two rules of the cosmic learning machine:

- To he who hath it shall be given. From he who hath not, even what he hath shall be taken away. And
- Spread out and explore. Then rush back together again. Hug and merge what you've learned. Rebel then embrace. Or compete. Follow the fission-fusion strategy.

CHAPTER 35

HOW BUTTERFLIES SOLVED CLIMATE CHANGE

Remember, this planet has had five major mass extinctions and roughly 137 smaller ones. This loving earth has killed off over five billion species of plants and animals.[779] That doesn't include microorganisms, of whom it is said there were 100 million species "before animals invaded the land."[780] All of these savageries have been produced not by humans but by nature.

And remember something else. The earth has been locked in ice ages for at least 415 million years of its 4.5 billion year existence.[781] Tilting the planet back to ice age would be easy. Which means that we need to decrease the current level of carbon dioxide in our atmosphere. But we also need to make sure that we don't go overboard and toss the planet back into one of its favorite states, a deep freeze.

Our goal of a climate stabilized at its pre-industrial temperature range is unnatural. Our climate goals fly in nature's face. They are anthropogenic. They are man-made. But they may be wise. And they may be achievable.

Our next challenge may be to do something as startlingly audacious as what monarch butterflies have done to deal with climate change. Remember, these fragile orange flutterers use the planet's annual changes of climate to their advantage.

You are a monarch butterfly. Butterflies emerged 150 million years ago.[782] And your species—you monarchs—evolved over a million years ago in Mexico. Twenty-four million years after the death of Hyaenodons. And roughly nineteen million years after the emergence of elephant seals. You monarchs coevolved with the plant you loved the most, milkweed.[783] A plant from which you harvested a poison to ward off butterfly eaters.[784] Twenty thousand years ago, when the last glaciers of the Pleistocene ice age retreated from North America, the milkweed were gifted with a bonanza, a whole new world to conquer—the rich lands of the newly-ice-free North American continent. The milkweeds increased in number dramatically as they shouldered their way from Mexico[785] into the North American territory that the glaciers had left behind. You monarchs followed. And you invented something new—an extraordinary annual migration. You traveled between three thousand and nearly six thousand miles, from the warm south to the cold north, then back again.

Your migrations were the fruit of wild invention. The fruit of bio-technologies. You monarchs used "a bidirectional time-compensated sun compass for orientation."[786] To help that sun compass make its calculations, you told time with a "time-compensating circadian clock that resided in your antennae." As a backup, when clouds covered the sun, you used "a light-dependent inclination magnetic compass."[787] Between 35 million and a billion[788] of you monarchs migrated in an annual North American swarm using these built-in navigation devices. Devices just possibly invented by the productive power of oddballs

And what did you monarch butterflies employ all this bio-equipment for? To adapt to one of Mother Nature's favorite tortures, climate change. To turn a crisis into an opportunity. To invent a new niche. To take advantage of the fecund milkweed of the northern United States, Canada, and the northern Midwest[789] during the summer. Then to fly back home to Mexico for the winter.[790] And what did you do once you were south of the border? You went into suspended animation—diapause—to wait out the cold. And you didn't just settle for any old location in Mexico. You zeroed in on the tops of "a few mountain ranges in the center of the Transverse Neovolcanic Belt of Michoacán Mexico."[791] You headed by the millions[792] to a pinpoint of territory less than 15 acres in size.[793] Your ancestral home.[794]

It doesn't matter whether you millions of butterflies had spent your summers in the Eastern United States or in the Midwest. All of you ended up on the same peaks in Mexico. What's more, by today, you've been doing this for 20,000 years. Which means you have a collective memory built into your genes, complete with the timing mechanisms that tell you when to go north or south and the equivalent of mental maps of your routes.[795]

Imagine that you are a newly birthed monarch butterfly. You've never seen anything but the inside of your cocoon and possibly the milkweeds on which you fed in your caterpillar days.[796] Yet you are expected to travel thousands of miles over land you've never seen to a destination you've never been to before. Surrounded by companions who, like you, have never experienced long-distance flight or seen their destination. Ever.

Adding one more layer of invention, you butterflies take roughly two generations to get to Mexico and roughly another two generations to fly to the American north. In the spring, one generation of you travels north from Mexico, finds milkweed[797] in the American south, and settles down. That generation's children continue the journey to the northern United States and Canada. What sort of collective memory does it take to pull this off? Is that map-sense stored in navigation genes?[798] Alas, there is no definitive answer.[799] Yet.

We won't even go into the diabolical cleverness it takes you monarchs to work out becoming a caterpillar at one point in your life, then building yourself a transformation capsule, a cocoon, and changing into another body entirely, your flying, traveling morph. In other words, you monarchs pull off the intense invention it takes to have two radically different bodies in one lifetime. And to build your own housing, your cocoon.

To repeat, these innovations are adaptations to climate change. Nature's climate change, not man's. These innovations are climate hacks. And we'll dive farther into climate hacks in a few minutes.

Meanwhile, you North American monarchs will be so successful with these hacks that you will manage to plant colonies in "Southwest Europe and North Africa...Guam, Palau, and Taiwan."[800] Populations that will forget how to migrate. And the territorial invasions of your North American traveling hopscotch, like the migrations of damselflies and anti-freeze fish, will make it glaringly obvious that colonialism is an invention of nature, not of human beings.

You monarch butterflies will use learning machine rule number two: spread out and explore the North American continent. Then pull back together again and consolidate in Mexico. Spin apart then rush together again. Do the Big Bang Tango. Fizz then fuse. Use the fission-fusion strategy.

But in the human case, our answer to climate challenge will hopefully allow us to stay at home. The answer for humans may lie in what we do best: giving birth to new technologies. Specifically, climate stabilization technologies.

What the hell are climate stabilization technologies?

***

We will soon see. But first, a story of how sex's gene-shuffling, oddballs, the birds of a feather effect, and the generation of the longest path possible works its ways on human history.

CHAPTER 36

HURRICANES OF HISTORY~ THE PEACOCK`S TAIL AND ANNE BOLEYN

Moonlight and love songs
Never out of date
Hearts full of passion
Jealousy and hate
Woman needs man, and man must have his mate
That no one can deny
It's still the same old story
A fight for love and glory
A case of do or die
The world will always welcome lovers
As time goes by
–Herman Hupfeld, 1931

Evolutionary breakthroughs do not just fashion new kinds of legs, wings, and brains. They do not just generate new forms of individuals. They often create something we overlook when we obsess on the origin of new species. They create new networks of interaction, new communities, new teams. New kinds of organizational personalities. New group identities. Like the hive, the tribe, the nation, or the civilization.

And here's a dirty little secret. Even a seemingly isolated individual like a plant, an animal, a you, or a me is a form of teamwork, a teamwork between twenty trillion cells in a plant and a hundred trillion cells in you or me. Not to mention the 39 trillion bacteria[801] inside you that digest your food for you and protect you from microbial invaders in your nose and throat. Yes, you are a network, a mesh, an ecosystem, a team, a community. So am I. But that's not where the teamwork stops.

You know that you are a you, a person with a name, an address, and a life story. But that "you" is a network bigger than you think, a social mesh that binds you, a seemingly individualistic organism, to your colony mates and relatives, your friends and enemies, the plants and animals you eat, the people who built the floor beneath your feet and the roof over your head, and your environment. Not to mention a culture left to you by 10,000 generations of ancestors. And your obligations to generations yet to come.

You, an individual, are a knot in a network of relationship that pulls in even the energy of a sun 93 million miles away and the photon flows of stars over four billion light years away, stars that chill you with awe. Stars that the creatures around you use as navigational signals in the night. Stars that even indigo buntings and robins use when they migrate.[802]

You and I are nodes in what Princeton University philosopher Manual DeLanda calls "meshworks."[803] And the evolution of new meshworks is as crucial as the evolution of new organisms. The evolution of new kinds of societies, from the society of atoms that we call a molecule to the societies of atoms and molecules we call galaxies and solar systems, to the gathering of cells in organisms, organisms like you and me. Not to mention the societies called bands, clans, tribes, nations, and civilizations. All of these when they first emerged were brand new forms of social organization. Brand-new meshworks. Brand-new kinds of group identities. Brand-new forms of organizational personalities. Brand-new emergent properties. Brand-new inventions. Brand-new supersized surprises.

Which leads us back to another evolutionary secret. Materialism, consumerism, and waste are teamwork makers. They are social integrators. They are meshwork fabricators. They are evolutionary breakthrough creators. So is sex.

***

Sex is not the shortest path between two points. In fact, it's the longest path this cosmos has ever conceived. And the most twisted. Yet it pulls together the most astonishing things. Things of enormous power and scope. Like bacteria's construction of biofilms, titanium tubes, and bitumen. Like the rise of the loony dinosaurs who flew. Like nature's invention of wings, insects, birds, butterflies, and elephant seals. Like the creation of new group identities—hives or tribes—and new partnerships like the commerce between insects and flowers. But that's just the beginning. Sex can also shift and change the forces of history.

Take the genome from just one of your 100 trillion cells. It's knotted, spooled,[804] tangled, and balled up with unbelievable precision. It's far too small for you to see. But undo the knots and tangles and stretch your genome, your string of genes, out in front of you and it spans, as you know, over six feet, more than two full meters. What's more, it's a community, a team, of roughly 128 billion atoms,[805] sixteen times as many as the number of humans on earth. All working together to pull off miracles every minute of every day. The miracles that keep you and me alive. That genome, like the New England Patriots, is a group identity.

Inflicting sex on this genome is not the most efficient way of doing things. It is not spare and streamlined. It does not minimize the amount of energy involved. In fact, it's the very opposite. It's the longest way to achieve something, the most intricate way that nature has ever devised. Which hints that in the future even more flamboyantly complex tricks of nature are to come.

Or, to put it differently, sex is the very opposite of thrift. It is the most expensive process in the universe. The most ornate process nature has so far been able to invent. And the genetic component of sex is just the beginning. Why? Because context counts. Sex is a catalyst. Sex is a choreographer of new social combinations. Sex is a harnesser of hurricanes of form. Sex is a rider and a changer of more than just species. It is a shifter of the evolution of group identities. It is a shifter of organizational personalities. It is a shifter of the forces of history.

Let's look beyond the first moss, the first trees, the first dinosaurs who flew, the first flowers, the first orchid bees,[806] the first fish with anti-freeze in their veins, and shoot forward a million years after the birth of monarch butterflies to look at a human love story, the story of the love between Henry VIII and Anne Boleyn. A story of sex. A story of wooing and winning that took seven years of Henry's blood, sweat, and tears to consummate.[807] And seven years of Anne's seductions and negotiations. Seven years on the surface. In reality, it took over 700. And it influenced the 500 to come. It was a multi-generational project. It was a civilization changer.

What's more, it cost Anne her head.

***

In theory, the tale of Anne and Henry could have been short, simple, and sweet. The sort of thing that would have brought a smile to the lips of those who believe in the law of least effort. Why?

Henry VIII could have simply lived forever. Or Henry could have done what our bacterial ancestors did. To reproduce, he could have split in two. Then the two new Henrys could have split and produced four more. And Henry could have skipped wooing, aching, and mating altogether. He could have skipped sex.

But that is not the way nature works. She demands the sexual process. With all its obstacles, twists, and agonies.

There's another way Henry could have dodged the torments of his seven year quest for Anne Boleyn. Henry VIII had almost unlimited access to sex. He could have snapped his fingers and bedded almost any woman in the kingdom.

What gave Henry this unlimited access? Like the one rat in the cage who impregnated all three females, he was an alpha male. One of the most powerful alpha males of his time. And how, pray tell, did Henry get his alpha status? It was not through a simple display of magnificence like a loony dinosaur's display of outstretched arms of feathers. It was the product of over 700 years of work. Work by a small army of ancestors.

How do we know that Henry's status had been achieved by a multi-generational team? By 850 AD, 641 years before Henry VIII was born, his ancestors had already fought their way to the top of the social heap in Wales. Ancestors like Elfyw ap Môr were listed as "Lords of Brynffenigl," as fabled warriors, and as protectors of Rhodri the Great, King of Gwynedd and founder of one of the Fifteen Tribes of Wales.[808] Which means that like Hyaenodons, Henry's ancestors were fabled killers. That was way back in 855 AD.[809]

But coming out on top can be costly. It involves materialism, consumerism, waste, and vain display. And it is driven by the birds of a feather effect. The obsession with finding a mate who is on your frequency. Who is on your wavelength. Who is precisely your kind of oddball.

Let's go back for a second to the modern descendants of the loony dinosaurs who flew. You know the story of the peacock's tail. Darwin first used it in 1871 in his book *The Descent of Man, and Selection in Relation to Sex* to illustrate one of his most brilliant insights—that evolution is not just shaped by natural selection. It is also shaped by sexual selection.

Evolution is shaped by the pickiness, tastes, and fashions of females.

And females tend to fall for the male who can make the greatest show of materialism, consumerism, waste, and vain display. For example, male peacocks flaunt astonishing tails and compete to see whose tail is the most magnificent. The one with the most outrageously gorgeous tail gets the girls. He gets to have sex. He gets to reproduce. Yes, the peacock with the grandest display of materialism, consumerism, waste, and vain display gets sex. He gets to pass his genes to generations yet to come.

To get the girls, a male peacock has to pull off the impossible. Or, as Darwin put it, "The peacock with his long train appears more like a dandy than a warrior."[810] The price the male peacock pays just for a slim chance of sex is enormous, says Darwin. Darwin writes that "we ought not to accuse birds of conscious vanity; yet when we see a peacock strutting about, with expanded and quivering tail-feathers, he seems the very emblem of pride and vanity. The various ornaments possessed by the males are certainly of the highest importance to them, for in some cases they have been acquired at the expense of greatly impeded powers of flight or of running."[811]

Yes, peacocks' tails make them poor runners and fliers. Poor evaders of predators. Birds with flamboyant tails, continues Darwin, "must be much more liable to be struck down by birds of prey. Nor can we doubt that the long train of the peacock...must render them an easier prey to any prowling tiger-cat than would otherwise be the case. Even the bright colors of many male birds cannot fail to make them conspicuous to their enemies."[812]

So an awesome tail increases your odds of a violent death from the talons of a hawk or the teeth of a tiger. To get sex, a male peacock has to lay his life on the line. To get sex, a male peacock has to do the impossible.

But you've heard the story of the peacock's tail before. I wanted a story about the cost of sex that would give you something new.

When I was hunting for a good example of just how hideously expensive sex really is, I wanted to know what peacock's tail human males used in the great romances of history, and just how costly their sexual displays were. In other words, I wanted to show you just how expensive sex can be. I wanted to hunt down a saga of romance in which the sheer price and complication was beyond belief. In which the law of least effort was defied outrageously. But the cost turned out to be far more gargantuan than I had expected. Why? Because in the most expensive acts of lust and love, the future of entire civilizations can be in play.

**

The stories I looked into and tossed aside were instructive. For example, the story of the Taj Mahal is one of incredible cost spent on love. But, ironically, Shah Jahan, the Mughal Emperor of India, did not build the Taj Mahal to be an architectural peacock's tail. He did not build it to woo and win the love of his life, Mumtaz Mahal. He built the Taj Mahal to preserve her memory. He lost her in child birth when she was 38. And he apparently couldn't take the pain. The Taj Mahal was an extremely expensive soother of his grieving soul. An expense he was privileged to be able to afford.

Privileged by what? Privileged by a multi-generational project. Privileged by a bloody Muslim conquest of India that put Shah Jahan, a Muslim, on the top of India's largely Hindu heap. A bloody conquest that gave him access to the wealth of every road, bazaar,[813] and bit of land under his control.[814]

How did this lofty status come to be? Starting in 636 AD,[815] 917 years before Shah Jahan, Muslim invaders killed tens of millions of Indians—Hindus, Sikhs, Jains, and Buddhists.[816] Muslim warriors raised mountains of Indian skulls in Delhi, Khanua, and Chanderi.[817] The Sikh website, SikhNet, points out that the ruler and conqueror Akbar the Great "ordered a general massacre of 30,000 Rajputs after he captured Chithogarh." And the Bahamani Sultans set a quota of killing 100,000 Hindus a year.[818] SikhNet calls this "The Biggest Holocaust in World History."

The Taj Mahal was an advertisement of Shah Jahan's power.[819] His status. His ability to extract taxes and contributions from Hindus, Jains, and Buddhists. And it was an advertisement of Shah Jahan's ability to show off, to spend, and to waste more than any other ruler on earth. It was a demonstration of Shah Jahan's ability to pull off the impossible. So the Taj Mahal was Shah Jahan's peacock's tail after all.

***

I asked Albert Einstein College of Medicine neuroscientist Lucy Brown,[820] one of the world's premiere researchers on love and sexuality in the human brain,[821] if she could refer me to research on just how expensive sex can be. Lucy asked what I meant.

Here was Lucy's note:

> What do you mean by "expensive"? Literally, like how much money it can cost? How much time it takes? The mental cost of rejection to the individual? The social/mental cost, like murder because of rejection, which happened just yesterday on

[New York City's] City Island in broad daylight at lunchtime and in public at the base of the City Island Bridge?

A woman who holds the stop/slow sign for a long-term construction project on the island was shot on the job by a jealous man who had brought her lunch a few times and then found out she had a boyfriend. The boyfriend was there and chased down the murderer, who was on a bicycle.

I had exchanged smiles and waves with the woman in the morning. A troubling reminder of the "costs" of love and courtship and how we never know if we will make it until the evening. Let me know more what you are thinking about.

It turns out that I was still figuring out what I was thinking about. And the further I got into it, the bigger and more expensive romance and its driver, sex, seemed to be.

I finally settled on a well-known tale, but one with hitherto unseen surprises, the story you glimpsed a few minutes ago, the saga of Anne Boleyn and Henry VIII. What was at stake in their case?

Sex, it turns out, harnesses the hurricanes of history.

✳✳✳

As you know, history records more than 700 years in which Henry's family had been attempting to achieve an alpha position. And succeeding. Then there's the multi-generational effort that had given Anne's family a position close to Henry's. A position high in the English aristocracy. So the positions of Henry and Anne were the result of multi-generational projects.

But in the romance between Henry and Anne, a whole new form of human organization was at stake, a form attempting to be born. A new form of group identity. A new form of organizational personality that the scramble of the Tudor and Boleyn families for alpha position helped birth. A nation state.

And at stake in the romance between Henry and Anne was more than the shape of the ears, nostrils, or brains of any baby that might emerge from a successful romance. At stake was the fate of Western Civilization. Literally.

***

In 1519, Christendom was threatened by the Muslim Turks,[822] who seemed unstoppable in their conquests and who cast terror into European hearts.[823] But Europe was also about to be gripped by a subcultural struggle whose results would be massive. That battle would be between the Catholics and an explosively growing new subculture of heretics. A subculture that would be eager to replace the Catholic Church: the Lutherans. This would be a sumo wrestling match between a Church that said it was the only interpreter of God's word and a movement that said you could read God's word yourself—in the Bible.[824] It would be a rebellion of oddballs. And a clash of group identities. A clash of organizational personalities.

The love affair between Anne and Henry would also be a key flashpoint in a literacy revolution—yes, a literacy revolution. A democratization of reading that was about to shrink the supreme authority of the papacy. In fact, the struggle in Europe would be over the ownership of that literacy revolution. And over the ownership of God himself.

In other words, the romance between Anne and Henry would harness the turbulent birth pangs of new group identities. Potentially massive group identities. Yes, the romance between Anne and Henry would harness the hurricanes of history.

***

What's a hurricane of history? A hurricane is nature in her most naked form. A hurricane happens when opposites meet: a warm spot in the tropical sea and a cold thrust of wind from the north.

But there's more. Nature abhors an equilibrium. She also abhors entropy. She abhors random disorganization. And she is obsessed with the climb to higher degrees of order. She is obsessed with creating new forms of social aggregation. She herds together clans to form tribes. She herds together tribes to form nations. And she herds together nations to build civilizations. She births new group identities. She births new organizational personalities. Nature follows the First Law of Flamboyance. But guess what?

Nature even builds group identities among gases.

Yes, gases in the atmosphere are social. They shun the randomness of entropy. They gather in the armies of molecules that we call breezes, clouds, warm fronts, cold fronts, and storms. But the grandest of all the group identities of gases on this earth are hurricanes.

Opposites are joined at the hip. A typical hurricane begins when opposites meet over East Africa—when a cold wind from the north meets a mass of warm air[825] in the south and generates a whirlwind of form, a spiral of air. That air is corkscrewed by the Coriolis effect.

It is twisted into a spiral by the pull of the earth's thousand-mile-an-hour[826] rotation at the equator and by the drag of the earth's slower rotation farther north.

The swirling society of atoms from East Africa travels 5,000 miles across the African continent to the Atlantic Ocean, keeping its spiral pattern as it goes. Keeping its group identity. Crossing the Atlantic westward toward the Americas, the whirlwind may get lucky. Near the equator, scudding across the Atlantic, it may run into a hotspot of tropical ocean 80 degrees or warmer,[827] a hotspot whose air is saturated with water. Hot air rises. And the society of moist air can be as much as 21 degrees warmer than the air around it. So the swarm of hot, moist air rushes upward, rocketing roughly ten miles high.[828] But when that jet of warm air hits the cold air of the upper atmosphere, when it slams into the frigid air at roughly the 7.5-mile-high mark, the warm air's water precipitates in new group identities: clouds or thunderstorms.

A lucky African whirlwind recruits this ten-mile high speed-rush, this jet-turbine of hot ocean air, as a center. An eye. If it's even luckier, this African-born wind-twist harnesses the circle of thunderstorms generated by the ocean's hot spot.[829] The African-born whirlwind herds the newborn thunderstorms like sheep into its growing spiral swirl, a corkscrew that can be over 1,300 miles[830] across. In other words, a hurricane is a group of smaller gas societies kidnapped, seduced, and recruited into a massive new social force. A force that can churn out 200 times the total amount of energy generated by humanity.[831]

The hurricane is the First Law of Flamboyance on steroids: things don't fall apart, they fall together. The hurricane is a group identity. An organizational personality.

Wherever she can, nature turns turbulence into group identities. Organizational personalities. She even forms the swirls of a permanent hurricane 10,159 miles across—1.3 times the size of the earth—on the planet Jupiter. It's Jupiter's Great Red Spot, the biggest hurricane in the solar system.[832]

The romance between Anne Boleyn and Henry VIII would harness hurricanes of new social structure aching to leave the world of possibility and to enter the realm of reality. Their love would harness a struggle between opposites, a struggle between forces of social organization based on opposing strategies.

The Catholic movement would call for obedience. Obedience to a holy hierarchy. One single hierarchy intent on ruling the world. And the Lutheran movement would call for independence. It would call on you to read and interpret God's word on your own. With the help of a new clergy that produced at least eight different Lutheran views to choose from.[833]

What would be the role of sex in this battle? And would this contest do what Pierre Louis de Maupertuis decreed—finding the shortest distance between two points? Or would sex do something far, far more ambitious?

CHAPTER 37

ANNE BOLEYN AND THE GATHERING STORM

Again, did the romance of Anne Boleyn and Henry VIII do what Pierre Louis de Maupertuis decrees—finding the shortest path between two points? Or did it do something more outrageous? Did it create the longest path between two points? Did it create an entirely new highway system? Is the story of Anne and Henry simple and thrifty or expensive, flamboyant, intricate, and vain? You be the judge.

When Anne Boleyn was sixteen,[834] Europe was in turmoil. Religious turmoil. A Church run in Rome had held Europe in a headlock since 323 AD,[835] when the Roman Emperor Constantine had converted to Christianity.[836] Romans who believed in pagan gods like Jupiter, Juno, and Mars had called the head of their pagan College of Priests the "Pontifex Maximus," the maximum high priest. The newly Christian emperor took that pagan title, Pontifex Maximus, for himself, thus declaring himself the high priest of the Christian religion.[837] Over 1,200 years later, when the Renaissance sparked a fascination with ancient Roman history, popes would resurrect that pagan title—Pontifex Maximus—and give it to themselves.

And they would give themselves the power the title implied. The power over an earthly empire.

Then a whole "new world" with two continents previously unknown to Europeans and Asians was discovered in the 1490s. Suddenly, the Catholic Church not only had a grip on all of Europe, but was eager to go global.

Nineteen years after the first voyage of Christopher Columbus, a 28-year-old monk from the tiny copper and silver mining town of Mansfeld in Germany headed to Rome to get his first glimpse of the control center of his religion.[838] At that moment, the church whose bureaucratic system he was entering[839] seemed anxious to demonstrate its position as the number one connection with God on the planet. One reason the church may have wanted to make a statement: the rapidly expanding Islamic empire of the Turks was threatening to invade and to run Christendom off the map.[840]

The Church had chosen a tool to tattoo its grandeur into human minds: a spectacle of materialism, consumerism, waste, and vain display beyond what humans had ever seen. A project so expensive and so elaborate that it would take a hundred years and five generations of architects to complete. Among those architects would be Michelangelo.[841] The very Michelangelo who had carved the daringly nude statue of David we saw early in this book.

The Church's megaproject would be St. Peter's Basilica. A project designed to achieve what nature herself demands—a great leap beyond the boundaries of the possible.

One goal may have been to outshine any other religious complex of its kind, whether that competitor be Istanbul's Hagia Sophia mosque[842] in the capital of the Islamic empire,[843] the Grand Mosque of Mecca in the Arabian Peninsula, the Huēyi Teōcalli temple[844] in the heart of Aztec Mexico, or the Temple Of Heaven[845] in Beijing. This was an astonishing new expansion of the peacock's tail. It was a stunning display of a group identity determined to rise. Determined to secure top position in the pecking order of social organizations, the pecking order of group identities, the pecking order of religious groups, the pecking order of civilizations.

And it tapped into a basic trick of evolution. Whether you are an animal or a religion, if you want the sort of attention that will put you on top of the heap, do something impossible. Nature has built awed attention to the impossible into our biology. It's a little tweak she has planted in us to expedite her inventiveness.

In 1505, Pope Julius[846] had the plans, but he needed the money. Fortunately for him, the church had invented a money-raiser par excellence a thousand years earlier.[847]

The Church held the keys to your afterlife, to your endless residence after death in the sunlit clouds of heaven or in the flame-flicked torture chambers of hell. Why not cash in on the fear of hell and its waiting room, purgatory?[848] Why not sell certificates[849] that would reduce your customer's time of torture in the afterlife's nether rooms? Certificates with which the church and the pope guaranteed to get you time off from the tortures of eternity?[850] In exchange for a contribution. A contribution that would help pay for Pope Julius' wild wallow in architectural extravagance.[851] His wild splurge in materialism, consumerism, waste, and vain display. His pole jump beyond the boundaries of the possible.

The funding certificates were called indulgences. And Pope Julius invented a new indulgence to rake in the loot.[852] Priests like Johann Tetzel, Grand Commissioner for indulgences in Germany,[853] would become "hawkers" [854] for these afterlife-alleviation certificates. They would become salesmen. And these certificates disturbed the monk from Mansfeld. One line in particular bothered him. As Hayley Nolan puts it in her book *Anne Boleyn: 500 Years of Lies*, the monk from Mainz was horrified when the super-salesman of the new certificates, Tetzel, preached that the new indulgences were so powerful they could even get "a sinner who had violated the Virgin Mary"[855] out of hell's barbecue pit. Yes, you could seduce or rape the Virgin Mary herself and this mega-indulgence would pay off your time in hell. It would set you free from having your earlobes set aflame and your derriere sautéed in a gigantic cast-iron frying pan.[856]

But the monk visiting Rome for the first time would eventually declare that "indulgences are nothing but knavery and fraud."[857]

So in standard histories, it is said that the monk, Martin Luther, wrote 95 theses[858] objecting to what he saw as corrupt church practices and pinned those theses to the door of the cathedral in the German town of Wittenberg. Hayley Nolan writes that Luther was actually less dramatic. "He sent" his theses "to the Bishop of Mainz for approval," she asserts. "And he accompanied his theses with a very polite letter."[859] One way or the other, Martin Luther's 95 theses went viral.

Printing was a new technology. Christopher Columbus, an arch pamphleteer, had come back to Europe from Cuba and Santa Domingo just twenty-four years earlier, in 1493, and had promptly shown how you could change the way Europeans saw their world by publishing printed fliers about your discoveries.[860] Twenty-four years later, the printers of pamphlets spotted the glimmer of a profit in Martin Luther's complaints and spread them with wild abandon.[861] Like the millions of spores spread by a moss, Luther's criticisms took off. Explosively. They became the talk of Europe nearly overnight. And Martin Luther's protests against the Church were about to change Europe's destiny.

***

Why? Europe's rulers had battled with the church for more than four hundred years over the control of their kingdoms' bureaucrats.[862] A long series of popes had insisted on appointing all of the bishops and abbots under a king, thus leaving you, a ruler, stuck with a team of church officials who owed their loyalty to the pope in Rome, not to you. And those officials were the most literate men in your kingdom. They were the backbone of the team with which you ran the place. So you needed their services.

Rulers like the Holy Roman Emperor Henry IV objected. But the Pope always won out in the end.[863]

Yet what if the grievances of Martin Luther, the furious monk, were on point? What if they communicated the will of God more accurately than a college of cardinals[864] and a Pope? You, a monarch, could adopt Lutheranism, shake off the shackles of the Vatican, and appoint your own bureaucrats.

So prince after prince rebelled against the church and adopted the religion of protest, Protestantism.[865] They adopted Lutheranism. The result would be 132 years of war[866] between the rebelling rulers and the potentates who held steadfast to the papacy. Yes, the result would be 132 years of war between the Protestants and the Catholics.

Henry and Anne's romance would enter the scene just as those wars between the Protestants and Catholics were about to start.[867] And just before the battle for hearts and minds of kings and princes was about to reach full boil.[868] In fact, Henry and Anne's sexual interest in each other would arrive just as the winds of group identities were gathering to create a hurricane.

**

Remember, in a hurricane, two different societies of molecules, two opposites, meet: a warm center and a cold flow—a center of moist hot air near the equator and a flow of cold air coming from the direction of the earth's poles. The tropical center, the eye of the storm, sucks[869] in the cold air, heats it, saturates it with water evaporated from the sea, and sends it rocketing into the sky. Meanwhile, new cold air rushes into the center to fill the void. And it too is heated, saturated with water, and sent to the heavens. The clash of opposites does not create chaos. It does not create a random spew. It does not create entropy. It creates a spiral form with immense power. A group identity with massive force. An organizational personality.

Conflict in the air produces a higher form, a form that did not exist before, a vortex up to 1,300 miles across, a hurricane.[870] Conflict in the air follows the First Law of Flamboyance: things do not fall apart, they fall together; shape shock is always on the increase. And shape shock sometimes happens with clashing societies of human beings.

CHAPTER 38

ANNE REAPS THE WHIRLWIND AND THE WHIRLWIND REAPS ANNE

"There is a tide in the affairs of men."
–William Shakespeare

Anne Boleyn was plunged into the battle between Lutherans and Catholics at the age of sixteen, when Martin Luther's 95 theses first went viral.

When she was just twelve years old,[871] Anne's father, an English diplomat with spectacular international connections, had sent Anne to the French province of Burgundy[872] to be an attendant to the most powerful woman in Europe, Archduchess Margaret of Austria.[873] Margaret's court overflowed with intellectuals—poets and scholars. Says Hayley Nolan, "Anne spent only a year at the court of Margaret...before" moving to France to serve Mary Tudor.[874] Mary Tudor was the sister to Britain's king, a sister who was married briefly to the king of France, Louis XII. Yes, confusing.

But to put it more simply, Anne moved from the court of the most powerful woman in Europe to the most powerful court of any kind in Europe,[875] the French court at Versailles.[876] As one of only four maids of honor to Queen Mary. Mary Tudor. Now brace yourself for a bit more confusion. France's Queen Mary was the younger sister of a British ruler new to the throne, the 23-year-old king Henry VIII. Mary had been only eighteen when her father had married her off to France's King Louis XII.[877] King Louis, on the other hand, had been 52. So Louis, king of France, died three months after the wedding.[878] And young Mary Tudor went back to England. But Anne Boleyn stayed at the French court,[879] at the pinnacle of European power, for another six years.[880]

Which means that Anne's courtly training included more than just etiquette. Anne was steeped in the machinations of kings and queens for seven years. And those machinations were fierce. They were the eye of the storm.

When Martin Luther's heresies, his 95 Theses, smacked Europe up the side of the head in 1517, Anne, as you know, was a sweet sixteen. And the court of France, in the words of Tudor historian Hayley Nolan, "was rife with reformers."[881] In other words, the French court was roiling with rebels against the Catholic Church. Frothing with Lutherans. Seething with oddballs. And Anne caught the Lutheran fever.

Or did the Lutheran fever catch Anne? Did Lutheranism harness Anne Boleyn the way flowering plants harness insects? Or the way hurricanes recruit thunder storms? Once again, you be the judge.

***

Little did Anne know it, but she would soon have an opportunity to infect an entire nation with Martin Luther's heresy. And she would soon get the chance to turn the order of that nation on its head. Thanks to sex.

After her nine years in Europe,[882] Anne returned to England at the age of 21 to marry her cousin, James Butler. But that marriage would never happen. Why? Like a cichlid fish, Anne would prove to be picky about her men. First, she fell in love with a dashing military officer, Lord James Percy.[883] Whose family felt Anne was beneath them.[884] But it turned out that Anne could do better.

In early March[885] of 1522, Anne was in York Place for a Shrovetide Pageant[886] that lasted "several days."[887] In fact, she was in York Place at a prized location, the sort of location she'd been used to for nine years. She was at the court of the British king.

In the royal Shrovetide pageant,[888] Anne was one of sixteen young women playing the parts of characteristics like Beauty, Honor, and Kindness. Anne played the part of Perseverance.[889] An omen of the perseverance about to come.

Acting the role of an attacking knight in the pageant was a figure no one could miss. He was over six feet two inches tall at a time when the average height was between five foot five and five foot eight.[890] Wrote Thomas More, "Among a thousand noble companions," he "stands out the tallest, and his strength fits his majestic body."[891] A Venetian diplomat visiting the English court in 1515 gushed that this mystery man was, "the handsomest potentate I ever set eyes on; above the usual height, with an extremely fine calf to his leg, his complexion very fair and bright, with auburn hair combed straight and short, in the French fashion, and a round face so very beauti-ful, that it would become a pretty woman."[892] Thomas Cromwell, the future chief minister to England's king, summed the man act-ing a knight up as an "Adonis."[893] That prince was in his sexual prime. He was 31 years old, ten years older than Anne.

But he was an oddball seized by the birds of a feather effect, deeply yearning for a mate on his frequency.

It is almost certain that this "Adonis" saw Anne. After all, Anne's part in the pageant put her on display. And it is absolutely certain that she saw him. Why? He was the king of England. He was Henry VIII. And wherever he went, he made a splash.

What's more, King Henry was apparently ripe for love. Stitched on the trappings of his horse when he rode out for the Shrovetide jousts was his personal motto for the event, "Elle mon coeur a na-vera," in English, "She has wounded my heart."[894] But this man with the wounded heart wasn't available. He was married. Was the woman who had inflicted a wound his wife? Or was it, as Claire Ridgway, writes, Anne's sister, with whom the king was starting an affair?[895]

We do not know what happened in the next two years.[896] But by December 21, 1524, everything had changed. Henry had been mar-ried for fifteen years[897] to a wife who had brought him the most powerful ally that England could possess—The Holy Roman Em-peror,[898] head of the first transoceanic empire in history.[899] Henry's wife's name was Catherine of Aragon. But there was a problem. A sexual problem. Catherine had not given birth to an heir. Yes, she had given birth to a son, but that son had died within weeks of birth.[900] The only child of Catherine's to reach adulthood would be her eight-year-old daughter. And in English history, trying to put a woman on the throne had led to disaster.[901] So Henry had stopped sleeping with Catherine and seemed intent on getting out of the marriage.

There may have been another reason that Henry had left his wife's bed. Some dedicated Anne Boleyn fans—and there is an international pack of them—think that Henry had fallen in love with "the Boleyn." With Anne.

***

Henry was athletic and liked to show it. He was particularly big on jousting. So he staged a tournament in the Christmas season of 1524 at which he appeared in the tiltyard of Greenwich Palace on horseback furiously unseating and thrashing opponents[902] with an image of a heart in flames above the motto "Declare Je Nos"—declare my love I dare not—stitched on his costume. Some interpret this to mean that he was consumed by a love he dared not reveal. And that his overwhelmingly powerful attacks on his opponents were meant to impress a lady. Anne.

Meanwhile, Anne was in high demand. She had that new love in her life who we glimpsed a minute ago, Henry Percy. The two were secretly engaged.[903] Yes, Anne was engaged with Henry Percy while she was apparently still engaged to James Butler.[904] Very messy. But it meant that Anne had lots of choices. Like an insect requiring her males to carry precisely the right scent, a cichlid fish requiring the right social status, a spider insisting on the right dance, or a female peacock demanding an extraordinary tail, she could afford to be picky.

In early 1525, King Henry went to Cardinal Thomas Wolsey[905]—one of those high-level bureaucrats appointed by the pope—and asked him to "break up the relationship between Anne and Percy."[906] The King's excuse? The love was unseemly. After all, Anne was already betrothed to her cousin.

But George Cavendish, who wrote about these events way back in 1641, a mere seventeen years after they happened, believes there was more than a sense of propriety at work in Henry's heart. Wrote Cavendish about the king, "he could hide no longer his secret affection, but revealed his secret intendment unto my Lord Cardinal in that behalf."[907] In other words, the king was forced to confess to the cardinal that he was in love with Anne Boleyn.

To make sure that Anne's potential engagement to Percy was thoroughly stomped out, Anne was sent to her family home, Hever Castle, 30 miles outside of London, where she "was left to simmer and sorrow... for a year or more."[908] To simmer and sorrow in anger. As Cavendish describes it, she was so furious that she "smoked."[909] When she returned to the king's court, it was as an attendant to, ironically, Catherine of Aragon,[910] the wife Henry was apparently trying to ditch.

In the autumn of 1526,[911] when Anne was still at Hever Castle, Henry began to bombard her with love letters.[912] Letters filled with something that you know from your own experience. The keenest price of love is not material. It is emotional. Sex harnesses hurricanes of the heart. Sex stirs tornadoes of emotion. And Henry was thoroughly tornadoed. The result was emotional torture. Henry wrote to Anne:

> In turning over in my mind the contents of your last letters, I have put myself into great agony, not knowing how to interpret them, whether to my disadvantage, as you show in some places, or to my advantage, as I understand them in some others, beseeching you earnestly to let me know expressly your whole mind as to the love between us two. It is absolutely necessary for me to obtain this answer, having been for above a whole year stricken with the dart of love, and not yet sure whether I shall fail of finding a place in your heart and affection.

Note the emotions: agony, feeling stricken, feeling brutalized by insecurity. Remember the last time you fell in love? Just how agonized you were, how uncertain about whether the object of your affections loved you? And how every minute of uncertainty felt like an hour of pain? And every hour felt like a year? If you had texts or emails from your beloved, you went through them over and over again trying to find clues to whether that beloved loved you or was simply being polite. That's what Henry VIII told Anne Boleyn he was going through. He was paying one of the bitterest prices of love. An emotional price. A price created by, guess who? Nature.

By 1528, Henry's pain had grown even harsher. He wrote to Anne that the torture of her absence at Hever Castle[913] was unbearable. "My heart and I surrender ourselves into your hands, beseeching you to hold us commended to your favour, and that by absence your affection to us may not be lessened: for it were a great pity to increase our pain, of which absence produces enough and more than I could ever have thought could be felt."

Yes, Henry VIII, a king, was feeling more pain "than I could ever have thought could be felt." Was this a manifestation of a universe that always takes the shortest path between two points? That always finds the path of least effort? Was this a manifestation of entropy? Of a universe constantly falling apart? Or was it a manifestation of a cosmos that is constantly falling together?

Despite being separated from Anne, Henry swears that, "by absence we are kept a distance from one another, and yet it retains its fervour, at least on my side; I hope the like on yours, the pain of absence is already too great for me... it would be almost intolerable, but for the firm hope I have of your unchangeable affection for me."[914] Henry said point-blank that he was being dragged through a pain beyond endurance. By the agonies of the mate hunt. By torments like those of the moss sperm who swam the human equivalent of over 2,272 miles to find a willing egg. Henry was being assaulted by the tortures of sex. By the tortures of the longest path imaginable.

But Henry had certain privileges that you and I will never experience. Kings can have both wives and mistresses. In Easter of 1527,[915] two years into the relationship, Henry made Anne an offer—to take her as his one and only mistress.

> If you please to do the office of a true loyal mistress and friend, and to give up yourself body and heart to me, who will be, and have been, your most loyal servant, (if your rigour does not forbid me) I promise you that not only the name shall be given

you, but also that I will take you for my only mistress, casting off all others besides you out of my thoughts and affections, and serve you only.[916]

A generous offer if you take into account the fact that Henry was a king and could put Anne at the very top of England's social pecking order, England's dominance hierarchy, England's elite. And if you realize that Henry could not only have his pick of women, but could have as many women as he wanted. But Anne turned the mistress offer down.

This is where the peacock's tail comes in, the display of materialist, consumerist luxuries. The fanning of your feathers in vain display. For the sake of sex. "Henry ordered four gold brooches from his goldsmith: one representing Venus and Cupid, the second of a lady holding a heart in her hand, the third depicting a man lying in a lady's lap and the fourth of a lady holding a crown."[917] The one-of-a-kind jewelry was a materialist expense of love. And Henry had more to expend than almost[918] anyone else in England.

Then Henry gave another flick of his peacock's tail. He sent Anne "my picture set in a bracelet."[919] A painting no normal mortal would be able to afford in a bracelet that no one like you or I could ever have purchased.

Anne reciprocated with a piece of jewelry that Henry himself described as a "fine diamond and the ship in which the solitary damsel is tossed about." But Anne gave another present of far greater value to the poor, love-tortured king. She accepted his love and returned it.

In a letter to Anne, Henry said that, "The demonstrations of your affection are such, the beautiful mottoes of the letter so cordially expressed, that they oblige me forever to honour, love, and serve you sincerely, beseeching you to continue in the same firm and constant purpose, assuring you that, on my part, I will surpass it rather than make it reciprocal, if loyalty of heart and a desire to please you can accomplish this... assuring you, that henceforward my heart shall be dedicated to you alone."[920]

But keep your eye on that piece of gold jewelry Henry had sent with "a lady holding a crown." Henry had asked Anne to be his mistress. But Anne had apparently wanted a different sort of hold over the crown. So she had turned him down. It appears that she was fishing for something better.

***

In fact, Anne was so successful at holding out yet tantalizing Henry that his courtship of her would last, as you know, a full seven years. Anne was doing her own version of inventing the longest path between two points. And she was demonstrating the prerogative nature gives to females—the right to be picky.

But there was another reason for the delay: there was a fly in the ointment. Or, more precisely, a wife. Henry had petitioned the pope to annul his marriage to Catherine of Aragon. The pope had turned him down. Over and over again. So Henry wrote to Anne, "henceforward my heart shall be dedicated to you alone. I wish my person was so too."[921]

Only "God can do it," said Henry, "if He pleases, to whom I pray every day for that end, hoping that at length my prayers will be heard." And God, indeed, was about to give Henry an escape route from his existing marriage. And a way to override the Pope.

A path that was definitely not the shortest distance between two points. A path that would make entropy look silly.

Through harnessing the hurricanes of history. And through Anne and Henry's choice of which hurricane to ride.

CHAPTER 39

ROMANCE AND REAL ESTATE

Put a one-of-a-kind person into a one-of-a-kind circumstance and you can move the evolution of humanity up a step. You can edge into the realm of the impossible. You can invent new niches. You can launch great leaps forward in social organization. You can elevate group identities to whole new levels of astonishment.

Let's go back to the role of sex in harnessing the hurricanes of history. Let's go back to Henry, Anne, oddballs, and the birds of a feather effect.

In 1527, Henry VIII showered Anne Boleyn with jewels. And she, in return gave Henry a gift that would change the course of British history.

But this gift did not consist of diamonds or gold. Instead, Anne introduced Henry to a mere wisp. A wisp with the power to upend group identities and to remake organizational personalities. A wisp with the power to change nations. An idea. She introduced Henry to the work of a heretic.

The Catholic Church had kept the Holy Scriptures hidden in an obscure language—Latin—for 1,100 years.[922] And only Catholic priests and a few aristocrats could read Latin. But Anne told Henry about a sinful new version of the Bible,[923] a Bible translated into, of all things, English.[924] Yes, plain, simple English. Which meant that Henry no longer needed his pope and his bishops to tell him the word of God. He could read that word of God for himself.

The forbidden Bible that Anne was talking about had been translated by William Tyndale, a wanted man who would later be strangled, then burned at the stake for his crime of translation. In 1527 you forfeited your property for merely owning an English translation of the Bible.[925] For being the translator, you forfeited your life.

And, in 1529, two years after Anne made Henry aware of the forbidden English Bible, she introduced the king to something even more subversive, another of Tyndale's works, "The Obedience of the Christian Man and How Christian Rulers Ought to Govern."[926] In fact, Anne did more than just hand over the new book. She pointed out a few stunning passages. Passages explaining that according to the Bible, "a ruler is accountable to God alone and that the Church should not control a monarch."[927]

As Hayley Nolan explains it, "Tyndale's book" prompted "Henry VIII to have an epiphany: as king of England, he was answerable to no man on earth. No, not even the pope. Only God."[928] But that would be true if, and only if, Henry pried England out of the Catholic Church. With the English translation of the Bible, Tyndale, and Martin Luther as his crowbars.

Oh, one more little detail. If Henry broke out of the shackles of the Roman church there would be three additional bonuses:

- Henry would not need the permission of the Pope to annul his marriage to Catherine of Aragon.
- He could marry and have sex with the woman he loved and who loved him, a woman who, during six years of courtship, had refused to yield her body to him.
- And, finally, the kicker: up to a third of the land of England belonged to the Church.[929] It was being used for buildings, farms, and gardens that housed nuns and monks. If Henry yanked England out of the Catholic Church, he could take possession of that third of the land of England. He could take possession of a fortune in real estate.[930]

Yes, Henry's way out of his marriage might be to ditch the pope and to do his own version of what the Lutherans were doing in Northern Europe—establish a new church. With a twist that only a tiny number of other European rulers had tried.[931] He would be the new church's head. He would be his own pope.

Thanks to the gift of Tyndale's words, Anne moved a giant step toward really holding Henry's crown in her hand, a step toward allowing her to change the belief system of the entire English nation. And a step toward changing England's group identity. Its organizational personality.

What's more, Anne was proving over and over again that Henry's emotions were on target. He and Anne were on the same wavelength. Raging rapids of seemingly inchoate emotion were pulling them ever closer. Because of the birds of a feather effect. And because of the power of paired oddballs to invent.

***

Says Hayley Nolan, during their seven years of courtship, "it's clear that" Anne and Henry "bonded over their shared goal of fighting for the country's independence from Rome, and would often have felt that it was ' us against the world.'"[932] A battle of us against the world that would last until Anne was 32 years old, Henry was 42, and Henry could undo the bonds of the Church in Rome and establish a church of his own.

A battle of this kind can hold two people together with an intensity that normal matings simply do not possess. It's what sociologist Norbert Elias calls "functional bonding"[933]—bonding over a common project. And bonding by facing common obstacles.[934] What would that bond produce?

Thanks to sex, something remarkable.

***

In 1529, Anne Boleyn was 28 years old and was growing impatient. According to Eustace Chapuys, the envoy of the Holy Roman Empire to Henry's court, Anne complained to Henry, "I see that some fine morning...you will cast me off. I have been waiting long, and might in the meanwhile have contracted some advantageous marriage, out of which I might have had issue, which is the greatest consolation in this world; but alas! farewell to my time and youth spent to no purpose at all."[935] Four years later, a very long four years, in 1533, Henry finally got the first seeds of his new church up and running,[936] rid himself of his wife, Catherine of Aragon, and began the process of elevating Anne to the level of queen.

First, Henry made Anne Duchess of Pembroke in a Windsor Castle ceremony where Anne wore so many jewels that you could barely see her ermine collar and her crimson velvet bodice.[937] Says contemporary chronicler Edward Hall, Anne entered:

> under a rich canopy of cloth of gold, dressed in a kirtle of crimson velvet decorated with ermine, and a robe of purple velvet decorated with ermine over that, and a rich coronet with a cap of pearls and stones on her head.[938]

But that was not the only outrageous display of materialism, consumerism, waste, and vain display in the ceremony. Anne was the center of attention in a procession that included:

> all the monks of Westminster going in rich copes of gold, with thirteen mitred abbots; and after them all the king's chapel in rich copes with four bishops and two mitred archbishops, and all the lords going in their parliament robes, and the crown borne before her by the duke of Suffolk, and her two sceptres by two earls.

This was not a manifestation of a thrifty cosmos. It hinted strongly that the cosmos has an itch toward excess, an impulse toward flamboyance.

But Anne and Henry were still going through the elaborate social obstacle course that finally allowed you in those days to have sex. And to obtain sex's product: children. Legitimate children. Children with the right to inherit a crown.

Meanwhile, Henry was also performing yet another delicate dance of a very different kind. But this one was geopolitical. For a quarter of his reign, Henry had been at war with France.[939] But now he needed France's support to switch wives. Remember, Catherine of Aragon, the woman he worked so strenuously to throw away, was the aunt of the Holy Roman Emperor.[940] And the Holy Roman Emperor was the most powerful man in Europe.[941] Only France's king was his rival for power.[942] If Henry planned a move that would offend the Holy Roman Emperor, he would need an alliance with another superpower. And the only other superpower in Europe was France.

Anne had spent seven years at the French court.[943] Says author Olivia Longueville, she "was probably more French than English."[944] She knew France's power players well.[945] So Henry planned to take Anne to France to get the approval of his new queen from France's king. Meanwhile, why did Henry give Anne the privilege of holding court as queen? Henry apparently wanted to give Anne additional standing and to guarantee that any children of their mating would have respect, money, and royal heft. Despite the local scandal and the international uproar[946] over tossing his first wife away.

In 1532, Henry and Anne—who were still not married—sailed to Calais for a week, including four days with the French king, who greeted the pair with a display of regal splendor[947] that included bull-baiting, bear-baiting, wrestling,[948] elaborate dinner parties, and dances. In other words, the king of France approved of the couple's engagement. He showed it with materialism, consumerism, waste, and vain display. And finally, Anne and Henry had sex.

Out of wedlock. But just a few weeks before their wedding. Yes, finally.

Does this sound like the shortest distance between two points to you? Does it sound like things falling apart in entropy? Or does it just possibly sound like things falling together?

***

Then came two weddings, one in secret, and one in public. And Anne became pregnant[949] with what Henry hoped would be a son. In September of 1533, Anne's baby was born. To Henry's dismay, Anne Boleyn had given him exactly what he'd gotten from Catherine of Aragon, a daughter. In fact, Henry's dismay was so keen that he had Anne beheaded in 1536 so he could try again with a younger wife. Little did he realize that he had planted two unique seeds, two one-of-a-kinds, two oddballs, two daughters, each of whom would harvest and be harvested by the hurricanes of history.

***

Henry VIII believed that a woman could not rule England. The experiment had been tried almost 400 years earlier, in 1139, with King Henry I's daughter, Matilda. And it had failed. In fact, it had led to a period of civil war so nightmarish that it was called "The Anarchy."[950]

When Henry died in 1547, his daughter by Catherine of Aragon—Mary—would prove Henry wrong. She would expand the envelope of possibility. She would be the first woman to rule all of England. Yes, the first woman. A stunning achievement.

But Mary's mother, Catherine of Aragon, had raised her as a Catholic, and Catholicism was the historical cold front that Queen Mary would choose to ride. Or that would choose to ride her. To eradicate Henry's Protestant legacy,[951] Mary would put over 300 prominent Protestants to death. Painfully.[952] 280 were burned alive. And Mary's efforts to return England to the Pope would earn her the nickname "Bloody Mary."

Abraham Lincoln would say three centuries later, "public sentiment is everything. With it, nothing can fail; against it, nothing can succeed. Whoever molds public sentiment goes deeper than he who enacts statutes, or pronounces judicial decisions."[953] The current of British public sentiment was against Mary.[954] More important, that tide of public opinion was against Catholicism. If the public had loved the pope, every death of a famous Protestant would have been cheered. Instead, every execution produced a martyr. A person whose blood acted as social glue. A person whose death fueled a cause. The cause of Protestantism. A nationalist Protestantism. A Protestantism that was assembling two new group identities: Henry VIII's Church of England. And a new sense of the British nation.[955]

Mary would rule England for five years. But she would die at the age of 42 during an influenza epidemic. And she would be succeeded by her half-sister, Elizabeth.[956] Elizabeth, the daughter of Anne Boleyn. Elizabeth, who had been raised as a Protestant.[957] And whose father had believed that no woman could rule.

Remember, sex does not use you to merely reproduce. Your kids do not come out exactly like you. Far from it. Sex does not carbon copy you. And sex doesn't use you to make kids who are identical to each other. Sex uses your comingling with your mate to make odd-balls, to make one-of-a-kinds. To make individuals who are utterly unique. The upheavals of romance twist together strings of genes that have never been twisted together before. Genes following the First Law of Flamboyance. Genes in search of unique opportunities. In search of niches they can invent.

What sort of unique mix and match had the chromosome-dance of sex created when Henry VIII and Anne Boleyn finally mated? Their daughter, Elizabeth, was intellectually gifted. She was given the same education as a young prince, and that education took. It is said that she had a command of seven languages.[958] She was also good at music. Which may be why she was good at sensing the rhythms and melodies of public sentiment. A necessity in a political leader who would stay on the throne for 45 years. And who would ride the tides of conflict so well that she would produce something precious and rare: peace.

Under Queen Elizabeth, England would flourish in whole new ways, giving us, among other things, the plays of William Shakespeare. Under Elizabeth, England would defeat the Spanish Armada, expand Britain's naval activity to the New World, and would produce the scientific thinking of Francis Bacon. Yes, under Elizabeth, England would experience what historians called its "golden age."[959] Who were Elizabeth's parents? Henry VIII and Anne Boleyn.

And what were the cold and warm fronts of the time? Catholicism and Protestantism.

***

Put a one-of-a-kind person into a one-of-a-kind circumstance and you can move the evolution of humanity up a step. You can invent new niches. You can launch great leaps forward in social organization. You can elevate group identities to whole new heights. You can turn impossibilities into new realities.

To his dismay, Henry had no sons. All he had were two daughters. But those daughters vastly exceeded his expectations. Each was a unique individual. And each was a powerful rider on a different current of history. Thanks to the power of romance, sex, genes. and something else.

Circumstance. Context. An evolving environment. Flickers and twists in the spirit of the age—flickers and twists in the zeitgeist. Flickers and twists in the tide in the affairs of men and women that William Shakespeare wrote about. Flickers and twists in the hurricanes of history.

***

Flickers and twists are totally immaterial things. Yet flickers and twists are realities. Powerful realities. Monumental realities. Realities whose mysteries we will dive farther into in a bit.

CHAPTER 40

IS SEX A TRIGGER OR A GUN?

S ex is the enemy of entropy. And at the heart of sex are genes.

It's been over 70 years since we first detected[960] genes, and we still know less than we think about how they work.

We don't know, for example, how a string of genes manufactures a box. We don't know how a knotted thread, a genome, builds the walls and interior scaffolding—the cytoskeleton—of a cell. We don't know how a hundred trillion gene strings make the tower of 100 trillion boxes, 100 trillion cells, stacked together as you or me.

Yes, we are getting a handle on how genes make proteins. And we are beginning to see how genes turn those proteins into teams.[961] But protein-making doesn't give you the walls, ceilings, and floors called cell membranes. Or the architecture that turns these membranes and the cytoskeleton beams within them into muscle and bone. Into high-rise towers of cells like us *Homo sapiens*. Into organizational personalities. Into group identities.

Proteins on their own just give you a soup. The next step in genomics may be to see how protein teams harness the hot and cold fronts of their environment. And how the resulting hurricanes power the "self" and the "soul" of the group.

But there's more. We also have only the faintest clues about how and why the cosmos invented sex. Why go through the agonies of materialism, consumerism, waste, and vain display—not to mention what Shakespeare calls "the expense of spirit", the emotional tortures— that Henry VIII expressed in his letters to Anne Boleyn? What are emotions, anyway?

What did nature get out of the romance, agony, sex, and gene-mixing that produced two one-of-a-kinds: Bloody Mary and her half-sister, Elizabeth I?

A lot of what genes achieve has to do with context. And context changes. Constantly. Environments evolve. And as they change, they offer up new possibilities. Put a one-of-a-kind gene string into the right one-of-a-kind position among the hurricanes of history, and you can change the perceptual lens of a nation.

New ways of seeing lead to new ways of being. Elizabeth I helped literally change the English mind. Above all, she promoted her father's rebellious form of Protestantism, a whole new way of interpreting the relationship between humans, God, and the state.[962] She also changed the reality of everyday English life.

Under Elizabeth, a new class without a name emerged. A class that was neither aristocracy nor peasantry. Today, we call it the middle class. What's more, being able to read was no longer just a privilege of the upper crust. Schools and literacy spread.[963] Theater grew in London. The town fathers of the city were convinced that this new form of entertainment promoted sex and violence and that it "corrupted" the young.[964] So they tried to shut the city's playhouses down. Elizabeth stopped these theater bashers in their tracks.[965]

Among the playwrights Elizabeth protected was a real-estate investor[966] from Stratford, William Shakespeare. The son of a prosperous glove-maker,[967] a son of the new and still nameless middle class. Literary critic Harold Bloom says that Shakespeare invented a whole new notion of what it means to be human. Instead of being born to your lot and pinned there from birth until death, you could change. You could have an inner life. You could have contradictory impulses. You could have entire "new modes of consciousness."[968] You could make choices. Like to be or not to be.

On the material level, clothing for the new middle class showed off an ebullient blaze of colors—"gooseturd green (yellow green), dead spaniard (pale grey tan), orange tawney (orange brown), incarnate (red), lusty gallant (vivid red) and maide's blush (rose)."[969] Thus opening up even more the sort of personal choice that Shakespeare was portraying in his plays.

But it was the perceptual lens of

- England's unique, national Protestantism,[970]
- England's greatest playwright,
- and of England's greatest theater supporter, Anne Boleyn's daughter, Elizabeth,

that changed everything.

Shakespeare raised a crucial question. Before Henry and Elizabeth, the Church in Rome had told us English men and women how and what to think. But now that Henry and Elizabeth had freed us from the headlock of the Church, says Stony Brook University English Professor Amy Cook, Shakespeare forced us to ask, "how do we know what we think we know?"[971]

That question, implies Cook, came from plays like *Hamlet* and from characters like Iago. Iago who poisoned the mind of the general Othello by altering his perceptual lens. Shakespeare, says Cook, produced a "reconceptualization of cognition and intellection." And that reconceptualization, Cook says, influenced the seminal scientific thinkers Francis Bacon, Robert Boyle, and even the French philosopher René Descartes to change "how data were gathered and examined." In other words, the actions of Henry and his daughter, Elizabeth, led to the invention of our secular way of thinking. And to our modern science.

If you feel you are in control of your life, you may owe that to Shakespeare. And to a protector of his work, Elizabeth I. Daughter of the romance between Henry VIII and Anne Boleyn.

***

Put a one-of-a-kind person into a one-of-a-kind circumstance and you can move the evolution of humanity up a step. You can invent new niches.

Put a unique genome into a unique hurricane of history, and you can catalyze whole new forms of human social organization.

The perceptual changes in the Golden Age of Elizabeth I contributed to the evolving group mind of Western Civilization. To the evolving group mind of all humankind. And to the evolving group intelligence of all life. The perceptual changes in the Golden Age of Elizabeth I contributed to the global brain. Not to mention to the tool kit with which the cosmos creates.

***

Under Henry VIII and Anne Boleyn, things defied the dogma of entropy. They followed the First Law of Flamboyance. To repeat, things did not fall apart. They fell together.

125 million years before Mary and Elizabeth, there had been a pay-off for nature's lavish expenditure on flower petals. That payoff? A turbo-charge of evolution. A fast track to new one-of-a-kinds and their innovations. A boost to the speed at which nature runs up, not down. And a boost to the First Law of Flamboyance. An acceleration of the invention generated by the most tangled course of action nature has ever knotted together: sex.

Which means that sex demands something of you and me. A radical rewrite of the way we see this cosmos. A radical rewrite of the rules that science takes for granted. But if we're going to toss out the old rules, what in the world would the new rules be?

Before we answer that question, we need to follow another story fueled by the First Law of Flamboyance, another story of the longest path between two points: the story of how the idea of entropy came to be. And the story of another clash between opposites, another hurricane of history. The tale of how the subculture of entropy would pit itself against the subculture that would give us Charles Darwin.

CHAPTER 41

THE BIRTH OF A MIND DESTROYER: ENTROPY

The villain of this book is entropy. But how did this mere wisp, this idea of entropy, come to be? And how did it get a headlock over you and me?

Flowers and Anne Boleyn have something in common. Their vain display sped the pace of evolution.[972] As you know, flower species exploded so fast that their step-by-step development of variations is not in the geological record. On one fossil layer, flowers don't exist. On the next, flowers appear in massive profusion. Darwin was perplexed. Remember, he called this sudden explosion of flowering species an "abominable mystery."[973] Yes, there is no step-by-step record of how the thousands of flowering plant species came to be. The reason? The flower-insect partnership accelerated the evolutionary process dramatically.[974] It led to instant evolution.[975] It gave a big screw you to entropy.

Ideas produce that kind of evolutionary acceleration among humans. Ideas are among humans' new billboards of display. Like flowers, ideas massively goose the speed of change. Like flowers, ideas are evolutionary turbochargers. Like flowers, ideas give the finger to entropy.

And the story of entropy, like the tale of Henry VIII and Anne Boleyn, reveals how the hurricanes of history and the birds of a feather effect sweep one-of-a-kinds together. One-of-a-kinds on the same wavelength. One-of-a-kinds who reproduce not just their oddball genes, but their oddball theories. And whose theories sometimes ride history's whirlwinds. Or are ridden by them.

The story of entropy reveals an intellectual struggle that's been hidden for over a hundred years, the saga of one of the biggest battles of ideas in the history of science. A battle that very few even realize exists. A battle with major implications for your understanding and mine of the cosmos and of our place in it. And a battle that has been written out of scientific histories.

A battle that shows how nature pits scientific subcultures against each other. Just as nature did with Protestants and Catholics in the days of Anne Boleyn. Why? To climb the mountains of possibility space.

There were two contending scientific subcultures in the mid-1800s. What were they? Opposites joined at the hip. Like the swirl of hot and cold fronts in a hurricane. The reductionists versus the holists. The specialists versus the multidisciplinarians. The professors versus the polymaths. A contest in which the holists, the multidisciplinarians, the polymaths, would lose. And would be consigned to anonymity. Despite the fact that they would get one of the biggest predictions of the nineteenth century right.

***

Yes, entropy would be a weapon in the battle between the holists and the reductionists. But what, pray tell, is a reductionist?

You recall that in 335 BC, Aristotle laid out a plan for a future field, science. In it, he said that to understand things, you must break them down to their smallest bits, their elements. Then you must understand the "laws" of those elements. And with that understanding, you would know everything that you need to know.[976] We are still following Aristotle's dictum. At a cost of close to seventeen billion dollars, in 2024 we were planning CERN's follow-up to its Large Hadron Collider—CERN's proposed Future Circular Collider[977] on the border between France and Switzerland. To pull off an Aristotle, to break down the particles named for Aristotle's idea of "elements"—elementary particles—into smaller bits. That Collider was a project put together by reductionists.

The modern battle between holists and reductionists began 2,200 years after Aristotle and 214 years after the death of Anne Boleyn. And, like Anne Boleyn's life, it ended in a form of beheading.

***

From 1812 to 1896[978] one-of-a-kinds in the scientific community were drawn together in what may have been one of the most consequential multi-national scientific quests ever, a profoundly reductionist quest—the quest to understand the nature of heat. Why heat? Because of evolving niches in an evolving environment. Because heat was regarded as the not-so-secret ingredient in something new, the steam engine. And the steam engine was transforming Europe.

How did steam engines come to be? There was something unique in the nation that Henry VIII, Anne Boleyn, and Elizabeth I had birthed. England's individualism.[979] And that individualism had led to a wild proliferation of new skills, new concepts, and new technologies:

- "The invention of an English sense of national identity."[980]
- Copper mining.[981]
- Harbor building.[982]
- The rise of practical mathematics as a discipline and a hobby.[983]
- The development of precision mathematical instruments.[984]
- The application of math to navigation.[985]
- The creation of navigation manuals.
- And "an ideal of justice, what we would now call liberalism, in politics, law, economics, citizenship, class, and gender."[986]

All of this led to the individualism, secularism, and use of numbers in Andrew Marvell's attempt to seduce a girl, his poem "To His Coy Mistress."

But the father of the concept of entropy would be the steam engine.

And the steam engine struggled to be born earlier than you might think. Englishmen like Edward Somerset, 2nd Marquess of Worcester, had been dreaming about powering engines with steam since 1655, a mere 52 years after the death of Elizabeth I and six years after "To His Coy Mistress." Somerset proposed a steam-powered engine, and in London, Thomas Savery actually built and patented what he called a "fire engine" in 1698. But because the level of British metal-working was not yet up to making valves and seals, Savery's steam pump was inefficient and had a tendency to blow up.[987] The first steam engine ready for prime time was invented by an engineering apprentice and ironmonger[988] named Thomas Newcomen, in Devon in 1712.[989] Sixty-two years before the American Declaration of Independence, Newcomen called what he had created an "atmospheric engine."

Engines powered by steam were invented to solve a problem. England had a resource crisis. It used trees for fuel and for shipbuilding. But ever since Henry VIII, the British Navy had been growing. So had the size of its warships. And it needed timber to build these massive battleships. So did the craftsmen who used wood to make the charcoal that turned raw ore into iron. So did blacksmiths, who used charcoal fires to work this iron. So did glassmakers. And so did ordinary people trying to keep their fireplaces going with wood.

Remember, Britain is a cold and rainy country. Body heat won't warm your house to a tolerable temperature in winter. And it won't cook your food. Then there was the demand for wood to build castles and cathedrals. But there was a problem. Trees were running out. In the 1570s, England entered a series of timber crises[990] that would continue for 210 years. In the days of Queen Elizabeth I, the crisis was so severe that more than 30 bills were introduced in Parliament to deal with the wood shortage.[991] None succeeded in solving the problem.

The time was crying out for the invention of a new niche, a new resource base. Were there any candidates? Perhaps. There was this sooty black refuse littering the soil of English lands like Northumberland, Durham, and Yorkshire. Up until now, it had been what preachers called "the devil's excrement."[992] But in the face of the multiple tree crises, England invented new uses for it, new uses for coal.[993]

In 1618, just fifteen years after the death of Elizabeth, the illegitimate son of Edward Sutton, 5th Baron Dudley of Dudley Castle, Dud Dudley, left Balliol College to join his father's iron-making enterprise 135 miles northwest of London.[994] Dudley noticed that the alleged 20,000 blacksmiths within ten miles of Dudley Castle were running out of wood, and came up with an answer, the stuff that would later give the surroundings the name the Black Country,[995] coal. Dudley adapted the technique used for turning wood into charcoal and turned it on coal. The secret was heating coal without oxygen. The process turned out a spongy substance that was almost all carbon—coke. Coke produced more heat than coal with less smoke. Dudley patented his coke-making technique in 1622. Then he wrote a book promoting coal for iron-making in 1665—*Mettallum Martis, Or, Iron Made with Pit-coale*.[996] New niches are invented. They are not found.

But coal could do more than smelt metal.[997] It could also fuel your fireplace and your oven. It could be used by glassmakers[998] and blacksmiths.[999] What's more, it would eventually be used to make a breakthrough metal—steel. But there were obstacles. Coal was off in the distant countryside, far away from the cities where it was needed. And, worse, when you tried to mine coal the way Europeans for over eight thousand years had mined copper,[1000] you ran into a problem. The water table of a land frequently shrouded by fog and rain was high. Dig a hole to get at coal, and your hole filled with water.

The pump had been invented in 1746.[1001] But powering enough pumps to keep your coal mines dry would have taken more horses on treadmills than England could possibly raise.[1002]

Newcomen invented the first workable machine to solve the problem, an engine that could pump the troublesome water out of the growing coal mines, an engine driven by steam. The first Newcomen engine was installed near Dudley Castle, Dud Dudley's former abode. But pumping water out of mines was all that Newcomen's engine could do. Why? First of all, the new engine was the height of a five-story apartment building.[1003] And it was as immovable as an apartment building. Second, the Newcomen engine burned a hideous amount of coal. It was staggeringly inefficient. Which wasn't a problem. The engine was built above a coal mine. Though it burned through coal like a dragon's flame burns through lighter fluid, it allowed the digging of far more coal than it consumed. So if you wanted a Newcomen engine, you had to put it on top of its fuel source, cheek-by-jowl with its coal supply.

But Dudley and Newcomen had invented a new niche.

In 1782, when the revolutionaries in America were negotiating a peace treaty with England,[1004] the potential of the steam engine changed dramatically. The steam engine was transformed by two friends of Charles Darwin's illustrious and scientifically influential grandfather, the holist to beat all holists, Erasmus Darwin. The friends got together with the elder Darwin and a remarkable group of buddies in Litchfield near the new industrial city of Birmingham[1005] once a month[1006] when the full moon allowed them to find their way home late at night. These five "Lunar Men" included Joseph

Priestly, discoverer of ten gases in the atmosphere including oxygen and nitrogen, Josiah Wedgwood, founder of the Wedgwood pottery company, and two steam-engine startup founders, James Watt and Matthew Boulton. Watt produced a host of steam-engine innovations that reduced the fuel a steam engine burned to one-fourth of what the old Newcomen engines had consumed.[1007] When Watt perfected his radically upgraded steam engine, he patented it. And when Boulton joined Watt's business, he got Parliament to extend Watt's steam-engine patent by seventeen years.[1008] An amazing feat. Meaning that if you wanted a steam engine that burned a far more reasonable amount of coal, you had to buy it from Boulton and Watt. You had to buy it from the friends of Erasmus Darwin.

Matthew Boulton,[1009] Watt's business partner, then used Watt's steam engine to power a new creation of his own, a modern industrial factory. A factory complete with "executive development programmes, sickness benefit schemes and welfare programmes."[1010] A factory that mass produced, guess what? The components of steam engines. Interchangeable parts. Using a precursor of an assembly line.[1011]

Despite Boulton and Watts' mass-produced parts, your steam engine still had to be custom-built on your chosen location and was still huge. Yet the spread of steam engines in England was explosive. They were used to grind grain, to pump water, to power looms, and to drive massive hammers that shaped iron and steel.[1012] By 1800, there were 2,500 steam engines in England. By 1907, that number was up to nearly ten million.[1013]

To see how the steam-engine revolution turned crafts into industries and utterly changed daily life, focus on fabric. Cloth was one of the most important items in your life. It was one of your few climate stabilization technologies, one of your few protectors against the cold. And against the lacerations your skin would endure while you were working in the field. The lacerations meted out by the knife-like edges of grass blades. But clothing was expensive. The cheapest shirt you could buy, a shirt for the lowliest laborer, took 579 hours to make. That included spinning nearly half a mile of thread and weaving that thread into 45 square feet of cloth. Then came the sewing together of the final garment. If you paid today's minimum wage for that 579 hours of time, the shirt would have cost you over $5,000. But wages were low in 1775, so the least expensive shirt cost you a staggering $3,500.[1014] No wonder when your clothes were so thoroughly worn that they embarrassed you, you didn't throw them away. You sold them to a cloth vendor[1015] who, in turn, sold them to someone else.

Why were clothes so expensive? And so prized? Because the thread used to weave cloth was the product of a painful process, spinning. Spinning thread and yarn with distaffs and spinning wheels. Says economic historian Gregory Clark, it "took well over a week to spin a pound of yarn." Then the threads and yarns were painstakingly woven into fabric on hand-operated looms. Hand or foot-powered looms. These were processes normally done by folks working long hours in their cottages in the city or in the countryside[1016] with their own primitive cloth-making equipment. Insists Clark, the machines were "all human powered and required enormous inputs of labor."[1017]

What's more, the most desirable thread came from the cotton plant. And cotton was grown in places like the West Indies[1018] and South America.[1019] A long and expensive trip by sailing ship. A 4,185 mile trip.

But starting in 1750, people like James Hargreaves, inventor of the spinning Jenny,[1020] Richard Arkwright, inventor of a spinning machine powered by water, [1021] and Edmund Cartwright, inventor of the powered loom,[1022] developed machines that could do the spinning and weaving with far less labor.

Arkwright and Cartwright's spinning and weaving machines ran into opposition. They put people out of work. And those displaced workers were not happy. Which led to protests like the Luddite Riots that started in Nottinghamshire in 1811.[1023] The Luddites broke into the new factories and destroyed the machines. The British National Archives says, "They also attacked employers, magistrates and food merchants." And they fought with soldiers.[1024] Not exactly the shortest distance between two points. But a step on the tangled path to the dogma of entropy.

Up until 1783, when Richard Arkwright first used a steam engine to power a textile mill in England's Derbyshire,[1025] automatic cloth-making machines were powered by water. Mills were equipped with paddlewheels to harvest the force of gravity, the force that pulled the water of streams and rivers downhill toward the sea. But the Boulton and Watt invention freed the new operations to install the breakthrough cloth-making machines in any location they wanted. These new workplaces were called "factories." And Boulton and Watts' steam engines liberated these high-efficiency cloth-making facilities from a location next to flowing water. Yes, the new engines of Boulton and Watt could be erected just about anywhere.

Inventing new niches opens new riches. As damselflies and notothenioid fish will tell you. Thanks to this wedding of cloth-making machines and steam engines, England's economy went from producing 1.9% of the world's gross domestic product in 1700 to 25% in 1870.[1026] Britain became known as the "workshop of the world."[1027] And once England began mass-producing steam engines—yes, using steam engines to make more steam engines—England's workers became the best paid on the planet. Their wages doubled in just the 32 years from 1819 to 1851.[1028] England's gross domestic product per person soared to 50 times what it had been three thousand years before.[1029] Just one city, Manchester, produced 40% of the world's textiles.[1030] The result? A shirt went down in cost from $3,500 to $135.[1031] And England became mighty. It climbed to the top of the pecking order of nations. It commanded a global empire on which it was said the sun never set. But more on that empire in a few minutes.

Mainland Europe's military superpowers, France and Germany, were envious. They wanted in on the steam revolution. But the details of cloth-making machinery and of steam engines were British state secrets. It was illegal for steam engine erectors in England to move to another country. And it was against the law to take drawings and plans of cloth-making machines outside the boundaries of Britain.[1032]

In other words, the early 1800s were the years of the steam engine. And the English had a monopoly on steam engines. Which, as you know, put England on top in the pecking order of nations. But France regarded Paris as the global capital of science. So did most of the rest of Europe. Which led to a dilemma. How could France counter this British steam-engine leadership? There was an answer. Ever since Descartes and "I think therefore I am" in 1637, France had been known for the quality of its abstract thought. So France's scientific thinkers could try to come up with the ultimate steam engines. And the ultimate theoretical way of optimizing the energy these engines derived from their steam. France could produce Sadi Carnot.[1033]

Why did a Frenchman focus on optimizing engines that his nation did not possess? France did not have Britain's rich store of coal. It couldn't afford to burn the black stuff in massive amounts the way the English did. If it was ever going to catch up with England and exceed it, it would need a whole new generation of steam engines. Steam engines maximized for efficiency. Which is where Sadi Carnot came in.

Sadi Carnot was born into a new kind of group identity. A new kind of organizational personality. Based on a new pecking order. He was born into the very top of a new French elite. In 1789, the French had had a revolution to rid the country of its king and its aristocracy and to replace these blue bloods with ordinary working men and women. In 1793, France had beheaded its king. In 1794, it had beheaded 40,000 of its aristocrats in public executions using an efficient and precise new technological invention, the guillotine. Then the French had beheaded some of the very revolutionaries who had led the overthrow of the old system. France had become the eye of a new hurricane of history.

Meanwhile, the surviving revolutionaries tried to replace the worship of God, Christ, and the Pope with the Cult of the Supreme Being and the Cult of Reason.[1034] And someone came along to make order out of the anarchy into which the Revolution had descended, a very short,[1035] very well read man.[1036] A man who had started his military career as an artillery officer and had been a master of the math needed to aim cannons. This math-and-books obsessive was Napoleon Bonaparte.

One of the few revolutionaries to survive the guillotine madness was named Lazare Carnot.[1037] Carnot was a military engineer and mathematician who had made contributions to the field of fortifications. He had not only survived the times of troubles, he had surfed them. He had become a member of the bloody revolution's Committee of Public Safety, the Committee that ordered executions by the thousands.[1038] In 1795, Carnot had also come up with the military strategies that had defeated a seemingly unbeatable alliance of England, Prussia, Austria, and Russia. An alliance pulled together to crush France's Revolution. What's more, Lazare's triumph had allowed France to occupy Holland. For all of that he was known as "The Organizer of Victory."[1039] Lazare had also directed the creation of France's renowned science and math academy, the École Polytechnique. Then Carnot had been appointed minister of war in Napoleon's cabinet.[1040] Five years later he had topped that by publishing a book on *The Geometry of Position*, a book that had made major contributions to projective geometry.[1041]

In a sense, Lazare Carnot's son, Sadi Carnot, inherited his steam-engine fixation and France's steam-engine envy from his father. His father had written two works on the theoretical principles of machines—"An Essay on Machines in General,"[1042] and the "Fundamental Principles of Equilibrium and Movement." Lazare had been concerned about machines like capstans, pulleys, and mill wheels. But he had turned the stream that powers a watermill into an abstraction. He had reduced machines to three abstract factors: time, force, and velocity. Forces that Isaac Newton would have approved of. Forces that allowed you to "achieve mechanical advantage."[1043]

Time, force, and velocity were factors you could translate into mathematics. Into Lazare's geometry and trigonometry. Lazare's principles of machines could in theory be used to get the most work possible out of a water-powered mill. But only in theory. Alas, Lazare Carnot's work was too abstract for actual watermill builders to understand.

Sadi Carnot, Lazare's son, was another abstract thinker. But Sadi took on the task of imagining how to maximize the efficiency of a new generation of machines, steam engines. Why all this abstract thinking about real world devices? First, there was the opportunity to win fame as the new Archimedes. Archimedes in 250 BC[1044] in the Greek colony of Syracuse had turned the machines of his day—the lever, the pulley, and the screw—into mathematical abstractions. Using the math of his time, geometry.[1045] And he'd gained everlasting fame for the feat. What's more, as you know, ever since Descartes, the French had been known for the quality of their abstract thinking. And England's Isaac Newton, who we'll get back to in a few minutes, had shown how you could use abstract thinking to discover "the laws of the universe." How you could use abstract thinking, astronomical observations,[1046] and mathematics to explain the movements of the heavens. Which may explain why Lazare and Sadi Carnot forgot about actually building water mills and steam engines and transferred these machines from the real world to the land of imagination. The land of mathematics.

For over a hundred years, under leaders from Louis XIV to Napoleon, France had fought to be the central culture of the globe. Lazare's son, Sadi Carnot was one of those aching to outdo the British in steam-engine development. So he wrote a paper on "The Motive Power of Fire." Full title: "Reflections On The Motive Power Of Fire And On Machines Fitted To Develop That Power."[1047] Yes, machines. Steam engines.

In his paper, Sadi reduced a steam engine to a series of thought experiments,[1048] to what he called a "mechanical theory,"[1049] very much like what his father had done with capstans, pulleys, and water mills. Sadi came up with a theory of "heat engines,"[1050] a theory that focused on what he saw as the steam engine's driving force, the difference between two opposites joined at the hip, hot and cold, the heat transfer from a boiler and hot steam to a cold sink, to the steam engine's condenser, to the cooling mechanism that reduced the steam to water. A choreography of opposites very similar to the massive dance of hot and cold air that creates hurricanes.

Yes, of all the factors making a steam engine work, Carnot zeroed in on just one: heat. In the words of a later expert on Carnot's work, Rudolf Clausius, who we shall meet in a few minutes, "Carnot...

regarded production of work as the equivalent of a mere transmission of heat from a warm body to a cold one, the quantity of heat being thereby undiminished."[1051] Carnot was a reductionist.

In the opinion of Cornell University professor of mechanical engineering Robert Henry Thurston, writing in 1897, Carnot was also a "genius."[1052]

Another thing: Carnot noticed that no steam engine would ever be perfect. If you hooked a steam engine producing heat to a steam engine running in reverse, an engine that produced cold, a refrigerator, and used each one to power the other, you would not have a perpetual motion machine, said Carnot. Why?

Because there are untested assumptions in science. Traditional ideas. Ideas that are seldom challenged. Scientific dogmas. It had become a dogma in the science of the time to say that a perpetual motion machine can't exist. And Carnot adopted that dogma without questioning it. Meaning that no matter how efficient you made a steam engine, there would always be some wasted heat.

Keep that wasted heat in mind. It will prove crucial.

✳✳✳

Carnot's paper promptly went into oblivion. And Carnot himself was killed off by cholera at the tender age of 36.[1053] Or was he? One author says that Carnot's brother made up the cholera story to cover up for the fact that Sadi died in an insane asylum.[1054] One way or the other, when Carnot died in obscurity, it seemed certain that his work would be forgotten. But that was not to be the case.

In the United Kingdom's northern realms, there was a science prodigy.[1055] His father had been professor of engineering and mathematics in Ireland at the Royal Belfast Academical Institution, so this remarkable child had been born in Ireland.[1056] And born into academe. His mother had died when he was six. Then when the prodigy was eight, his dad had gotten a job as a professor of mathematics at the University of Glasgow in Scotland, where the gifted boy, William Thomson, would spend the rest of his life.[1057] Thomson's father had taken advantage of the social mobility of the post-Elizabethan age and the age after the French Revolution. He was a farm laborer who had educated himself. He was determined that a laborer's status would never befall his son William. So William's father "took him unto himself" so thoroughly that he slept with him at night. And before William was six, his father had begun to teach him history, geography, Latin, and mathematics.[1058] Not just any mathematics. Mathematics so new that few citizens of the British Isles had mastered it.[1059]

Some biographers say that James Thomson raised his son to be a great, a scientific genius. And Thomson's father succeeded.

Young William Thomson was regarded as a scientific prodigy from the age of ten, when he first started taking courses at Glasgow University. When he was fifteen, Thomson expanded his studies to include astronomy and chemistry, then he studied magnetism, electricity, and heat. Yes, heat. In Thomson's time this study of magnetism, electricity, and heat was considered just one part of something bigger, "natural philosophy." Four or five decades later it would be walled off, turned into a narrow specialization, and re-named physics.[1060]

Thomson had more than just the advantage of his father's mathematics. While Thomson pursued one of his father's two disciplines, math, his one-year-older brother went after their dad's other specialization, engineering. However James was as science-obsessed as his younger brother William. And he would soon be as obsessed with abstract theory, with the flow of heat, and with the imaginary engines of Sadi Carnot. But instead of restricting himself to high-flying theory, James preferred improving steam engines in the real world. He had presented his first practical plan for improving the efficiency of steam-boat paddle wheels when he was fourteen. And his father had made sure that the fourteen-year-old's idea was taken seriously.[1061] Only to discover that a similar idea had been patented just a few weeks earlier.

Not only were the two brothers—William and James—close, they were a pair, solving problems together.

***

Meanwhile, Glasgow was a ship-making town. It cranked out one huge steamship a week. Sailing ships had taken as long as four months[1062] to carry 200 passengers[1063] across the Atlantic to North America. But the new steamships being constructed in Glasgow could carry 400 passengers across the Atlantic in a breath-taking three weeks.[1064] With far greater odds that the passengers would live through the experience.[1065]

But back to William Thomson. When he was fifteen, William continued to astound. He wrote an Essay on the Figure of the Earth. It won him a gold medal from the University of Glasgow. When he was sixteen, one of Thomson's professors turned him on to a work from the hated French. That work was *The Analytical Theory of Heat*, by the French multi-disciplinarian Joseph Fourier.[1066] Yes, a mathematical theory of heat. Says Thomson, "I took Fourier out of the University Library; and in a fortnight I had mastered it—gone right through it."[1067]

Fourier's paper made an impression that would last the rest of Thomson's life. Writes Paul Sen, who studied engineering at Cambridge, then wrote a terrific history of thermodynamics:

> Fourier's aim was to mathematically describe how heat behaves and, in particular, how it flows. An example is a metal bar that is hot at one end and cold at the other. Experience tells us that heat will diffuse from hot to cold until the bar's temperature equalizes. Fourier showed how this kind of diffusion can be described mathematically. His approach was unusual for the time, and Fourier had critics.[1068]

Critics or not, Fourier's mathematical approach fascinated Thomson. But that was just the beginning of Thomson's fascination with the math of heat and of its flow. Thomson threw himself into the new French use of complex math like differential equations,[1069] despite these equations political unpopularity.[1070]

The reason for the equations' unpopularity? According to Paul Sen, they were "French math."[1071] The British and the French had been at odds for centuries. Since long before the age of Henry VIII and Anne Boleyn. And only 25 years earlier, the French—led by Napoleon—had threatened to take all of Europe plus all of the possessions of the British Empire, including Egypt and the jewel in the crown, India.[1072] Britain had faced down the French in a life-and-death struggle at the Battle of Waterloo in 1815. Yes, like individual lobsters and lizards, armies of humans face each other down. And in William Thomson's day, the nasty memory of Napoleon's threat lingered. So did the terror of the French Revolution. Which meant that the mathematics of the French was frowned on. Thomson disregarded that political prejudice. Over and over again.

Then, when he was still sixteen, Thomson came across an 1837 book by a professor from his own university, the University of Edinburgh, a professor with the additional prestige of also being a Fellow at Cambridge. The book said that Fourier and his math were wrong. Thomson, despite his youth, disagreed.[1073] So, writes Paul Sen, "Aged sixteen, Thomson published a detailed defense of the Frenchman's methods in the scholarly Cambridge Mathematics Journal."

How did a teenager break into a major journal? His father pulled strings. Nonetheless, quite a feat.

In 1844, when he was 20, Thomson made a trip to France to be in the land that had given the world math geniuses like Fourier,[1074] Poisson, Coulomb, LaGrange, and Laplace. Then William's father played politics and landed him a position as "Glasgow University's next professor of natural philosophy."[1075] In other words, a professor of what would soon be called physics. But the position wouldn't start for a year. William's dad had enough money to assure that William would not have to work for a living. So what was William going to do with the spare time?

To succeed as a professor of natural philosophy, especially in Glasgow, you needed to put on public demonstrations of natural phenomena. You had to show these physical phenomena so vividly that you could fill an auditorium with 500^{1076} spellbound spectators. William's father was concerned that his son did not yet have the chops to pull this off. Which is why he sent William back to the country whose physical demonstrations, whose public science shows, he believed were the best on earth—England's traditional enemy, France.

In the words of Paul Sen, William Thomson's father "urged his son to obtain letters of introduction to eminent French savants, travel to Paris, and get his hands dirty once he'd received his Cambridge degree." In Paris, Thomson worked as an assistant to an experimental physicist, Victor Regnault, who had been funded by the French government to study the thermal properties of, guess what? "Steam."[1077] The state had funded Regnault to crack the secret of the steam engine.

Despite Sadi Carnot's death in anonymity and despite the obscurity into which his 1824 booklet had fallen, the 20-year-old Thomson read about Carnot's 45-pager[1078] in the late summer of 1843 in an article by a French fan of Carnot's who had helped the Russian czar build a new system of roads and who would later help study the feasibility of the Suez Canal. That Carnot fan was dead serious about Carnot's work. He translated it into a language that was delicious to William Thomson, mathematics. This translator was the French engineer Émile Clapeyron.[1079] Clapeyron's article laid out Carnot's ideas in newly-invented charts. And Thomson began fervent conversations with his steam-engineer older brother on the ideas of the mysterious Carnot. Discussions his brother could use to improve the efficiency of the steam engines that he was helping build for the ships that his city, Glasgow, was cranking out at the rate of one a week. Remember, the more efficient the engine, the farther a ship can go and the less coal it needs to carry. The less coal, the more paying cargo. So any improvement James Thomson could achieve would make a difference.

But to William's frustration, he couldn't find a copy of Carnot's paper. So when he got to France, he searched madly for a version in the bookstores of Paris. Recalls Thomson,

> I went to every book-shop I could think of, asking for the Motive Power of Fire by Carnot. 'Cainot? I do not know this author. Ah! Ca—rrr-not! Yes, I have his work', producing a volume on some social question by Hippolyte Carnot [Sadi's brother], but the Motive Power of Fire was quite unknown.[1080]

Not having a copy of Carnot's Motive "Power of Fire" did not stop William Thomson. In 1848, when he was 24, he presented a paper to a live audience at the Cambridge Philosophical Society "On an Absolute Thermometric Scale Founded on Carnot's Theory of the Motive Power of Heat." Thomson had invented our modern scientific temperature system, with the coldest the universe could go, minus 273 degrees Celsius, as zero degrees Kelvin;[1081] the freezing point of water as plus 273 degrees Kelvin; and the boiling point of water as 373 degrees Kelvin. Yes, temperature points on the thermometric scale would be named after him.

But why would these temperature points be called "Kelvin" instead of "Thomson"? Thomson would be knighted eighteen years later in 1866 by Queen Victoria for helping turn a technological fantasy, the transatlantic cable, into a reality.[1082] Then, in 1892, he would be elevated yet another step up. He would be raised to the peerage and given the title Lord Kelvin. That's how absolute zero would become zero degrees Kelvin.[1083]

But in 1848 William Thomson credited the inspiration for his invention—the thermometric scale—to Sadi Carnot. What's more, Thomson had found Carnot's ideas so compelling that he had published an article about them without ever having seen Carnot's original words.

Or, as he put it,

> Having never met with the original work, it is only through a paper by M. Clapeyron, on the same subject, published in the Journal de l'École Polytechnique, Vol. xiv. 1834, and translated in the first volume of Taylor's Scientific Memoirs, that the Author has become acquainted with Carnot's Theory.

Finally, in 1849, Thomson managed to obtain a copy of the phantom Carnot booklet. Through the generosity of one of his colleagues. He recalls,

> A few months later through the kindness of my late colleague Professor Lewis Gordon, I received a copy of Carnot's original work and was thus able to give to the Royal Society of Edinburgh my "Account of Carnot's Theory."[1084]

Carnot's theory became one of the pivot points of William Thomson's life. The sort of pivot point that is vital to the inventiveness of the cosmos.

✳✳✳

In fact, Thomson was so fascinated by the Carnot paper that in 1849 he wrote an English translation and summary, then reperceived Carnot through the lens of the math he had learned from his study of Fourier and the research conclusions from his teacher in France, Henri Victor Regnault. Thomson titled his 1849 paper on the topic "An Account of Carnot's Theory of the Motive Power of Heat—with Numerical Results Deduced from Regnault's Experiments on Steam."

Topping it all off, in this 1849 paper, Thomson referred to the steam engines that obsessed Carnot and himself as "thermo-dynamic engines." Thermo means heat. Dynamic means motion. Thermodynamic means the motion of heat. So William Thomson came up with the term "thermodynamics."[1085] The article in which he did it was his "An Account of Carnot's Theory of the Motive Power of Heat." This would soon prove to be a giant step forward for, guess what? Reductionism.

But there's more. Remember, Carnot had noticed that heat engines, steam engines, have a problem: wasted heat. That waste, too, would have a major consequence. A major consequence for something that had just begun to acquire a name in 1833, when William Whewell coined the term "scientist."[1086] A major consequence for "natural philosophy" as it evolved into "science."

CHAPTER 42

LORD KELVIN`S COSMIC DOOM

The waste produced by steam engines would lead to one of Thomson's most disturbing contributions to science: a whole new form of scientific pessimism. Scientific doomism.

In 1852,[1087] Thomson asked in his paper "On a Universal Tendency in Nature to the Dissipation of Mechanical Energy," about "the loss of power experienced by steam in rushing through narrow steam-pipes." And, he said "for the best steam engines...at least three-fourths of their work...is utterly wasted." Wasted in friction. Then he took a huge leap and generalized this to the entire "material world." And he proclaimed that, "There is at present in the material world a universal tendency to the dissipation of mechanical energy." That's a big jump, from the pipes of a steam engine to the entire material world.

But that's not the end of William Thomson's big jumps. He concluded his paper with the claim that, "Within a finite period of time past, the earth must have been, and within a finite period of time to come the earth must again be, unfit for the habitation of man."

In other words, the earth will become uninhabitable. And plants, animals, and humankind will be out of luck. They will die out.

Why? Because of "the dissipation of mechanical energy."

So the prodigy rapidly becoming the most august scientist of his time, William Thomson, soon to be known as Lord Kelvin, predicted apocalypse. Because of waste heat. The friction and exhaust of the cosmic steam engine. Not a cheerful conclusion.

Thomson did not have a name for the creeping cataclysm of this "universal dissipation." Naming it would take thirteen years. That's when Rudolf Clausius would christen the universal tendency to dissipation "entropy." As you and I will see in a minute or two.

Meanwhile, one of the causes of Thomson's scientific pessimism was this: Thomson was a reductionist. And reductionism was contagious.

Two other figures in the scientific community—two other oddballs on the same wavelength—got excited about Thomson's paper. And about its conclusion. The end of the world.

The two scientific thinkers were Scotland's William John Macquorn Rankine and Germany's Hermann von Helmholtz. It was Von Helmholtz who came up with another perfect name for Thomson's scientifically proven end times: "heat death."[1088]

What's heat death? In 1852, William Thomson laid the base for it.[1089] But much, much later, the standard example of heat death would become the good old sugar cube[1090] in a glass of water. Hold a sugar cube between your index finger and thumb. It is a well-ordered mass. Flat sides. A surface roughened by crystals. A nice, uniform white color. Sharp, raspy edges, unlike anything else you've ever touched. Your sugar cube has a hard and fast group identity. A group identity you can literally grasp.

Now drop the cube into a glass of water, leave it, come back in an hour or two, and what do you have? Just water.

What happened to the sugar cube? The molecules of sugar have ceased to be a community, they've abandoned their unity, they've left behind their team, their cube, their group identity, their organizational personality, and they've dissolved.

You recall that Thomson scaled up this idea in the last paragraph of an 1852 paper[1091] when he declared that, "Within a finite period of time past, the earth must have been, and within a finite period of time to come the earth must again be, unfit for the habitation of man." The good things of the earth itself, Thomson suggested, would be whizzled apart like the sugar cube by "the dissipation of mechanical energy."

Then thirteen years later, in 1865, Germany's Rudolf Clausius gave that random whizzle a name. A name soon to be famous. Clausius proclaimed his Second Law of Thermodynamics,[1092] the idea of entropy. The notion that the entire cosmos is constantly falling apart.[1093] Dissolving like the sugar cube. But more on Clausius' epiphany in a minute.

Now that entropy had two names, it needed a final touch. And that touch came in 1884, eight years before William Thomson would be named Lord Kelvin. That finishing touch came from

Vienna's Ludwig Boltzmann, who pulled together a mathematical justification for the catastrophic claims of entropy.[1094] In fact, Boltzmann came up with the radical improbability of everyday reality.

Here's the essence of Boltzmann's reasoning.[1095] Statistically, the number of ways your sugar molecules could hang together in a cube is small. On the other hand, the number of ways those sugar molecules could randomly move around in the water in your glass is vast. According to Boltzmann's mathematical upgrade of entropy, in the end the more probable outcome always wins. And the number of ways your sugar cube could lapse into chaos outstrip the number of ways your sugar cube could hang together. The probability of chaos overwhelms the probability of a cube by zillions to one. So probability dissolved your sugar cube. Or, more accurately, improbability.

And probability will do the same thing to the cosmos. Order is improbable. Intricacy is statistically unlikely. And the odds of flamboyance are nearly nil. But there are gazillions of ways to achieve disorder, to achieve randomness. Which means that the universe, like the sugar cube, will eventually break down in a random whizzle. It will relax into its most probable form. It will turn into a random gas. A useless gas. A spent gas. A gas whose energy can no longer be tapped. That's heat death. That's entropy.

But there's a problem. Heat death started with an idea from William Thomson. And William Thomson would become such a lofty figure in the world of science that in 1892, as you know, he would be lifted to the peerage and would take the name Lord Kelvin. For solving many of the thorniest engineering problems in one of the world's most improbable achievements—the transatlantic cable.[1096] For solving those problems brilliantly. For expanding the boundaries of possibility. For handing the cosmos a few new inventions.

But lord or not, William Thomson was not always right. Thomson thought that the world was between 20 million and 400 million years old.[1097] In fact, he was sure he'd proved it. Unequivocally. Scientifically. Using the newly discovered melting temperatures of slate, sandstone, garnet, and granite.[1098] Using Fourier's fancy equations[1099] for the flow of heat. Using reductionism. In the end, Thomson would prove to be wrong. Very wrong. And his mistake would have consequences.

***

Today we believe that the planet is considerably older than Thomson "proved"—240 times older. Today we believe that the earth is 4.5 billion years old. And we believe that the cosmos is an age Thomson would have ridiculed—13.8 billion years old. Yet we still have faith in Thomson's heat death. And in its cousin—the Second Law of Thermodynamics, the theory of entropy. Why?

Remember, underlying the concept of entropy were two metaphors: the water mill and what had replaced it as a power source, the steam engine. In the flowing brook that powers the water mill, a current of liquid moves downhill, pulled by gravity. When the water reaches the level of the sea, you can no longer use it to power a paddlewheel. It has become useless. That's entropy.

But, to repeat, for the last 13.8 billion years, the cosmos has not been running downhill like a stream. It has been running uphill. What's worse for entropy, the universe has been building new hills and creating new mountains from which water can run.

Then there's the steam engine. When steam comes from the spout of your teapot, that's entropy. It's useless energy. But invent the steam engine, and steam can pull a railroad train. New niches are not found, they are invented. The creators of the idea of entropy looked at the spent gases coming out of the steam engine's exhaust chimney and concluded it was useless. Then they proclaimed that the entire cosmos behaves like that spent gas.

They failed to realize that the steam powering their engine had also been useless just a few decades earlier, before the invention of the steam engine. They failed to see that waste is a resource waiting to be tapped. Entropy is opportunity in disguise.

What accounts for this cosmos' defiance[1100] of entropy, a law so basic that one of the greats of 20th century science, Sir Arthur Eddington, the legendary explainer of Einstein's relativity, says,

> If someone points out to you that your pet theory of the universe is in disagreement with Maxwell's equations—then so much the worse for Maxwell's equations. If it is found to be contradicted by observation—well, these experimentalists do bungle things sometimes. But if your theory is found to be against the Second Law of Thermodynamics I can give you no hope; there is nothing for it but to collapse in deepest humiliation.[1101]

Yes, you will be humiliated in science if you admit that the Second Law of Thermodynamics, the law of entropy, is wrong. You will be mocked and scorned if you proclaim that the Second Law Of Thermodynamics simply does not fit the universe that's revealed by your timeline of the cosmos, a cosmos that evolves from the Big Bang to scientists in lab coats. Scientists shunning those who dare to say that the Second Law of Thermodynamics is wrong.

But why will you be shunned? Could it be because the scientific community is a club? And clubs have initiation rituals, senseless rites of belonging? For the last 160 years, a key rite in science has been something the White Queen in the sequel to *Alice in Wonderland* would have approved of. Said Alice to the White Queen, "one can't believe impossible things." To which the White Queen replied, "Why, sometimes I've believed as many as six impossible things before breakfast."[1102] The White Queen's words are perfect descriptions of blind faith, the belief in something impossible that demonstrates that you have handed control of your mind over to the club. Swear you believe in the Second Law of Thermodynamics and you're in the scientific community. Criticize the Second Law of Thermodynamics and you're out. You're out in "deepest humiliation."

Or you're erased. The way Alexander von Humboldt and Herbert Spencer would soon be erased from the history of science, a history to which they would make enormous contributions.

***

Entropy is an idea Thomson, Clausius' and Boltzmann—three birds of a feather—got from imagining the universe as a steam engine.

Could something be wrong with using the metaphor of a 250-year-old technology, a steam engine, to understand this particular cosmos?

CHAPTER 43

HOW THE UNIVERSITY CREATED THE SCIENTIST

Meanwhile, the most militarily powerful kingdom in Germany, Prussia,[1103] was even farther behind England in steam engines than France and knew it. So the Prussian government did three things.

Starting in 1810, the government established universities that combined equal amounts of the arts and of something brand new to institutions of higher learning, natural philosophy, later to be known as science. These "Humboldtian" Universities were named for a man whose astonishing brother you will meet in a few pages. They were named for Wilhelm von Humboldt. Wilhelm had invented the unitary school system that included something we'll bump into in a second, Prussia's gymnasiums. And, even more important, von Humboldt had created a new form of university. The first form of university that hired full-time professors of natural philosophy. Full-time teachers and practitioners of what we know today as science. The first universities that gave these natural philosophy professors labs in which to experiment. And the first universities to demand that these professors produce a constant flow of original research.

Humboldt's universities would change the very nature of natural philosophy. Among other things, Humboldt's new universities would radically resculpt natural philosophy's group identity. Its organizational personality. Thanks to a battle of opposites—the competing subcultures of the holists and the reductionists.

Meanwhile, the Prussian government increased its spending on these new universities fivefold.[1104] And its demand for original scientific research made German universities the leading scientific centers in the world.

Writes Dartmouth historian of science R. Steven Turner, Germany's emphasis on research produced a whole new kind of scientist. Erasmus Darwin—a holist beyond all holists—had been a doctor practicing science in his spare time. And Sadi Carnot—a reductionist—had made his living as a military engineer. But thanks to the new German university system, "By midcentury the career scientist was almost invariably a professor."[1105] In other words, a scientist was a full-time teacher in a cloistered world, the world of the new, Humboldtian University. The world we now know as "academe."

One of the new crop of full-time scientist/professors would twist the path of scientific thought. You bumped into him a minute ago. His name was Rudolf Clausius. And thanks to his full-time job as a professor, Clausius may have been one of the most colorless personalities in scientific history. His father had been a pastor and a bureaucrat—a Councilor of Prussia's Royal Government School Board. But Clausius' dad stepped beyond mere bureaucracy, founded a small private school, and briefly enrolled his sixth son in it, Rudolf.[1106]

In his book Einstein's Fridge, Paul Sen explains that the French Revolution 29 years earlier had left a lasting impact on the culture of England and of Europe. The revolutionaries had taken the focus off of educating aristocrats and had put the spotlight on cultivating the hidden genius of the common people. The revolutionaries had felt that children of the common folk deserved a rare privilege that had formerly been limited to the elite—schooling. So schools for kids whose parents did not have titles sprang up all over Europe. That cleared the ground for the schools that Clausius' father supervised as a councilor of the Royal Government School Board. And for Clausius' father's own private school.

When Rudolf was roughly ten, he moved from his father's school to another new Prussian phenomenon,[1107] another invention of Wilhelm von Humboldt, a gymnasium, the Gymnasium in Stettin. To us, a gymnasium is the basketball court of a school. But thanks to von Humboldt, in Prussia it was a rigorous educational institution that took in kids at the age of ten. Its purpose was to prepare them for entry to a university.

At sixteen, Clausius left the gymnasium and entered the University of Berlin, another institution established in 1810, a mere six years after Napoleon declared himself emperor, a Napoleonic act some historians feel ended the French Revolution.[1108] The University of Berlin was the first "Humboldtian" university. A university devoted both to traditional subjects and to full-time "natural philosophy." Rudolf wanted to become a historian. But then his mind was swayed by something he found more exciting: math and what would later be called physics.

Clausius was not a child wonder like William Thomson. Clausius wrote his doctoral thesis when he was 24. A thesis that used mathematics to explain why the sky is blue. And his use of math went far beyond what others had ever achieved in considering the topic. The result? He received his doctorate. With distinction.

There was only one small problem. The math was extraordinary. But the explanation of why the sky is blue was wrong. A problem that would dog Clausius again at the peak of his powers.

MacTutor is the leading online source of biographies of mathematicians. It's curated by the School of Mathematics and Statistics at Scotland's University of St Andrews. And MacTutor says that Clausius' explanation of why the sky is blue was "a good illustration of how physical problems drive the development of mathematics even when their physical basis is unsound." But is math that proves to be "unsound" really an achievement?

The big puzzle of the time was radically reductionist, it was the flow of heat. A problem that another oddball, William Thomson in Glasgow, was working on. In other words, the big problem was the efficiency of the steam engine. Not the efficiency of the real steam engine. But the efficiency of Carnot's thought-experiment steam engine. An imaginary steam engine. A "Carnot heat engine."[1109] A steam engine translated into mathematics.

When he graduated from university, that problem was what Rudolf Clausius tackled. But to get at it, he had to confront the theory of heat itself. What in the world is heat? Antoine Lavoisier, the "founder of modern chemistry"[1110] was the man who had named oxygen and hydrogen 67 years earlier. He was also the man who had brought numbers[1111] to chemistry. And Lavoisier had come up with an answer to the heat problem. A reductionist answer. In 1787, Lavoisier had argued that heat was a substance called caloric. Caloric, he had preached, was "made up of tiny, weightless particles" released from heat sources like fires.[1112] Yes, heat, said Lavoisier, was a particle. Aristotle would have been pleased.

Lavoisier had also kick-started what would eventually be called the First Law Of Thermodynamics. In 1785 he had proposed the "Law of Conservation of Mass."[1113] He had declared that matter can neither be created nor destroyed.[1114]

But, alas, lives can be created and destroyed. In 1791, Lavoisier would become a French Revolutionary. Despite the fact that he was an aristocrat. And in 1794, his own Revolution would guillotine him. For making his living as a tax collector. Lavoisier lost his head to a hurricane of history.

But fifty years later, Lavoisier's caloric theory of heat—that heat was a particle—was proving incapable of explaining heat's flow and diffusion. It was especially poor in explaining how electric currents in something new, dynamos and copper wires, produce heat. Does electricity somehow create the heat particles called caloric? Impossible. Remember, Lavoisier had decreed that matter can neither be created nor destroyed.

1848, the year Clausius was getting into gear for his 1850 paper "On the Moving Force of Heat", was a year of revolutions.[1115] Revolutions hit over 50 nations worldwide.[1116] Demanding the end of monarchy. Demanding the privilege of what two revolutionary nations—the United States and France—had achieved. The French had had a republic, then had lost it when Napoleon lost the battle of Waterloo and the French royal family, the Bourbons, had been returned to the throne. But the Americans still had their republic. Complete with a kingless democracy. In other words, the Western world was in turmoil over competing forms of group identity, over competing forms of organizational personality.

Clausius, despite his colorlessness, was a revolutionary in another realm: science. He toppled the theory of heat that had dominated natural philosophy for 65 years, Lavoisier's theory of caloric particles. Clausius pulled this off with his 1850 paper "On the Moving Force of Heat." A paper that came just one year after William Thomson had introduced the phrase "thermo-dynamic."

Clausius wrote, "many facts have lately transpired which tend to overthrow the hypothesis that heat is itself a body." In other words, new facts have emerged to topple the idea that heat is a particle called the caloric. Instead, said Clausius, new "facts" have proven that heat "consists in a motion of the ultimate particles of bodies."[1117] In essence he said that heat is not caloric. It is not a substance. It is motion. It is the motion of what Clausius would later call "atoms" and "molecules."[1118] Yes, heat is not a particle, it is the motion of particles. Brilliant. In fact, astounding.

But where did Clausius get the bizarre idea of "the ultimate particles of bodies," atoms and molecules? Where did he get reductionism?

Remember, in roughly 335 BC, in his Posterior Analytics, Aristotle had told us that if you break things down to their smallest parts, if you break things down to what he called their elements, and if you understand the "laws" of those elements, you understand everything. To repeat, it was Aristotle who had laid out the scientific program of what is today called reductionism.

The program Clausius and William Thomson were working on would prove to be one of the most successful reductionist scientific initiatives in history. Reducing things to their smallest parts. Says the MacTutor biography of Clausius, "In his 1850 paper, Clausius states clearly that the assumptions of the caloric theory are false and he gives two laws of thermodynamics to replace the incorrect assumptions."[1119] These first two laws of thermodynamics were muddy and wordy. In fact, Clausius would not be able to express his two laws in just a single sentence each for another fourteen years. Then would come his 1864 paper "The Mechanical Theory Of Heat, With Its Applications To The Steam-Engine And To The Physical Properties Of Bodies."[1120] In it, Clausius would come up with this:

> ...the two fundamental theorems of the mechanical theory of heat.
>
> 1. The energy of the universe is constant.
> 2. The entropy of the universe tends to a maximum.

Decades later, Clausius' first law would be rephrased using wording borrowed from chemistry's founder, Antoine Lavoisier, nearly 80 years earlier, way back in 1785:

Matter and energy can neither be created nor destroyed.[1121]

But keep your eye on that second law: "The entropy of the universe tends to a maximum." That law would become a blindfold for modern science.

CHAPTER 44

BREAK IT DOWN TO BITS

Let's get back to the tale of how Rudolf Clausius, William Thomson, and their oddball friends were in the process of advancing the ultimate reductionist theory, the notion that matter is made up of particles called atoms and molecules.

The idea of atoms had been around since 400 BC in Greek philosophy. Towering thinkers like Greece's Democritus in roughly 400 BC had tried to account for a paradox, a pair of opposites joined at the hip—stability and change. To explain how stability and change, those two opposites, could co-exist, Democritus proposed that the material world was made up of tiny unchangeable "particles." The tiniest parts into which you could divide a tiny part. Democritus called these ittiest bits atomos—atoms. The unchangeability of atoms was a base of stability for all that was. But those immutable particles were constantly coming together and spreading apart in new ways. That's why we see change.

The concept was so striking that 340 years later, in roughly 60 BC,[1122] the Roman poet Lucretius put it into a six-volume poem, "On the Nature of Things." Six years later, in 54 BC, the iconic Roman orator Cicero was bowled over. He wrote to his brother that Lucretius' work stunned with its "flashes of genius."[1123]

Nearly 1,700 years later, in 1620, a Dutch scientific thinker, Isaac Beekman, revived the Greek term "atoms." Then he added another word, "molecule."[1124] And Beekman came up with "the first molecular theory of... matter."[1125] *The Journal of Molecular Structure* calls this "one of the major achievements of the... Scientific Revolution of the 17th century."

Then 30 years after Beekman, in 1650, Lucretius' poem resurfaced in Europe, was translated into French,[1126] was printed, and was popularized by a French priest, scientific thinker, and experimentalist, Pierre Gassendi.[1127] Gassendi was out to prove that atoms were created by God.

However, the idea of atoms and molecules would stay out of use for the next 200 years. The really big stone to ripple the pond of science was yet to come.

Thirty eight years later, Isaac Newton changed science utterly when he published his *Philosophiæ Naturalis Principia Mathematica* in 1687 and argued that you could understand the movements of the heavens with a radical change in the way you viewed those heavens. And a radical application of math.[1128] From that point on, Europe's leading intellectuals were on a hunt for natural laws of Newton's kind, laws to make sense of everything including what the founding fathers of the United States called the "natural laws" of human behavior.[1129] Mathematical laws.

In Germany there was a towering philosopher and scientific thinker who doubled as an international diplomat. He was a master at massaging group identities and organizational personalities. In fact, he had helped put a German ruler from the electorate of Hanover on the throne of England, King George I. But that was the least of his achievements. He was also a master of math. He had co-invented calculus. And he had invented binary numbers.[1130] He was Gottfried Wilhelm von Leibniz, and he came up with a name for what Newton had achieved. Leibniz called Newton's clockwork solar system a system of "dynamics,"[1131] a word that would later stimulate William Thomson to call the study of heat and its movements "thermodynamics." But that would be another 140 years down the line.

Meanwhile, Newton-envy seized the sciences. As early as 1812, the French mathematician and philosopher Pierre-Simon de Laplace, one of the new science's most influential thinkers and the man who helped put the idea of determinism on the map,[1132] wrote,

> The regularity which astronomy shows us in the movements of the comets doubtless exists also in all phenomena. The curve described by a simple molecule of air or vapor is regulated in a manner just as certain as the planetary orbits; the only difference between them is that which comes from our ignorance.[1133]

In other words, Laplace felt that we could discover Newton-like mathematical laws underlying "a simple molecule of air or vapor." You could comprehend the entire material world by boiling things down to billiard balls and cannonballs, the metaphors associated with Newton.[1134] You could grasp the entire material world by boiling things down to particles. To atoms and molecules. Now that's reductionism.

But there was a problem with this billiard-ball-like predictability. It was subject to probability. Subject to the same math that the landmark French philosopher Blaise Pascal had conceived in 1654,[1135] 158 years earlier, to help solve a problem in gambling— how do you divide up the pot of winnings in a dice game that hasn't been finished.

Extrapolating from dice to the cosmos seemed interesting and potentially useful. And in 1830, the astronomer John Herschel showed how the math of probability could be used to put together observations of stars taken over many days and months with a shaky telescope to find the actual position of a star.[1136]

Like Newton's mathematical explanation of the solar system, probability, the math of dice, was reductionist. And it was a way to grab hold of the rest of the cosmos with numbers. But it lacked a certain something. It misread the tendencies of the cosmos. It was blind to the cosmos' seeming headlong lust for the intricate and the ornate. It was blind to the cosmos' love of invention. It was blind to the First Law of Flamboyance.

Probability was blind to the fact that in this universe, it's not the most probable thing that happens. It's the most improbable.

* * *

But the really big test of the new reductionism, the really big battle over wisps, would be over William Thomson's ultimate reductionist triumph, William Thomson's mathematical calculation of the age of the earth.

How did that battle begin? One of Thomson's contemporaries, like Thomson, had graduated from Cambridge. Near the top of his class.[1137] But unlike Thomson, he had left Cambridge as a lost man. An oddball. His passion was collecting specimens of pebbles, plants, and animals. Like his grandfather. But his father had dictated that he must be one of two things, a doctor like his dad and granddad, or a priest.[1138] This lost graduate had gone to school at the University of Edinburgh to study medicine, had been traumatized by witnessing surgery without anesthetic on a child, and had switched schools to Cambridge to study theology.[1139] Neither medicine nor theology seemed to kindle his enthusiasm. But there was another role that had seized Europe's imagination ever since Christopher Columbus' first letter about his discoveries in 1493,[1140] the role of explorer.

In 1797, 34 years before this student's confusion, global exploration had been wrangled into science by a Prussian intent on using all of the sciences at once and going out to sample, measure, and test all over the world. Yes, the real world. Not the imaginary world of Carnot and his steam-engine abstractions. And not the world of the university lab in which science would be reduced to highly unnatural simplifications, reductionist simplifications. The real world. Its complications. And its unknowns. In 1799,[1141] this Prussian fired by curiosity had boarded a sailing ship and had spent five years traveling, mostly in South America. He had covered 6,000 miles. The hard way. By sail and on foot. Then he'd made a dogleg to North America, where he'd met Thomas Jefferson and Jefferson had gathered some of the new nation's top minds to meet him, folks the Smithsonian identifies as "leading American artists, writers and scientists." Says the Smithsonian, this high-energy, highly persuasive European traveler shaped "their views of an emerging national identity linked to nature."[1142] Yes, this scientific adventurer contributed to a new nation's emerging group identity. To its emerging organizational personality. To the emerging identity of the U.S.A.

With him this scientific explorer had carried 75 scientific instruments.[1143] And he had carted those instruments around in wooden cases lined with velvet,[1144] even when he had climbed what was thought to be the highest mountain in the world, Chimborazo in the Andes. Even when his hired hands, the men who carried his equipment, had been convinced that this adventurer would meet his death and had refused to go any farther up the mountain with him. He still had shouldered his scientific instruments himself, had taken samples of plants and insects, and had taken measurements "assiduously"[1145] as he climbed.

Even when the mud in the valleys had sucked off his shoes, when the insects had bitten and stung him, and when his traveling companion, a student of medicine, botany and zoology, Aimé Bonpland, had come down with typhoid and had been forced to spend close to a month in bed,[1146] the indomitable young Prussian had still struck out for new territory as soon as Bonpland had gotten better, taking samples and measurements every bit of the way. Translating what he had seen into numbers. "Numbers banish disorder," he had said.[1147] This determined scientific explorer, this virtuoso of the scientific expedition,[1148] was Alexander von Humboldt. Wilhelm von Humboldt's younger brother. The younger brother of the man who would invent the modern university.

Alexander von Humboldt's scientific adventures made him a global superstar. They made him so famous that 400 species would be named after him, and 1,600 towns and streets[1149] all over the world would be called Humboldt in his honor. Not to mention two celestial objects, the carbonaceous asteroid 54 Alexandra and the sun-orbiting asteroid 4877 Humboldt. We humans celebrate doing the impossible. And Alexander von Humboldt's adventures had yanked new swatches of the impossible into the fabric of the every day.

Von Humboldt was not a specialist. He was the ultimate multi-disciplinarian. What's more, he was the very opposite of a reductionist. He was the ultimate holist. The ultimate omnologist. He was out to use every science available to piece together the biggest picture possible. In fact, he invented a term for that big picture. He called it a "cosmos."[1150] And *Cosmos* would be the title of the last book series he would write, the five-volume "Cosmos: A Sketch of a Physical Description of the Universe."

Von Humboldt was such a seminal figure that his modern biographer, Andrea Wulf, calls him the man who invented nature.[1151] What's more, von Humboldt would play a role in an agricultural revolution that would feed the world. But more on that later. Back to our lost new Cambridge grad.

It wasn't the role model of doctoring from his father that this uncertain 22-year-old graduate from Cambridge wanted to follow. Instead, he wanted to follow the path of Alexander von Humboldt. The path of a scientific explorer. So he signed on for a two-year stint[1152] as naturalist[1153] and "gentleman companion" to the captain of the exploratory sailing ship the Beagle. A ship out to map the South American coast.[1154] His two years on the Beagle would become five. And he would carry with him the three-volume version[1155] of the memoir written by, you guessed it, Alexander von Humboldt.

The Beagle's captain, Robert Fitzroy, was a geology enthusiast. When the young man boarded his ship, Fitzroy gave him a "welcoming gift."[1156] That gift was a book that would change this perplexed young man's life, the first newly published volume of Charles Lyell's *Principles of Geology*.[1157]

Many would have regarded this lost puppy as unqualified to do science. He was not proficient in math.[1158] He did not do experiments in a lab. He was not a university professor. And he was not a reductionist. He was not intent on breaking things down to their smallest bits. But he had two other tools in his kit. He had the example of a role model he denied influenced him: his grandfather.[1159] And he would soon have three words from another 19th-century thinker who would use all the sciences at once and would make major contributions to sociology, philosophy, psychology, biology, anthropology, and economics: Herbert Spencer. The words from Spencer would be "theory of evolution." The lost young man's name was Charles Darwin.

Charles' grandfather, Erasmus Darwin, had been a contemporary of Alexander von Humboldt. In fact, Erasmus Darwin had been admired deeply by von Humboldt's intellectual circle in the German town of Jena.[1160] A circle that included German intellectual superstar Johann Wolfgang von Goethe. Why had this group looked up to Erasmus Darwin? Erasmus Darwin had done what von Humboldt's friends ached to achieve. Erasmus had combined all the sciences with an unusual plus, poetry. Erasmus Darwin had written six scientific books, two of them in rhyme.[1161] And Erasmus had put forth a timeline of the universe that went from a beginning in a big explosion to the emergence of suns, planets, plants, animals, human beings, and human societies. [1162]

One other figure would influence Charles Darwin's way of thinking, Charles Lyell, the geologist. Remember, the captain of the Beagle, Robert Fitzroy, had given Darwin a copy of Lyell's *Principles of Geology* when Darwin had boarded his ship at the age of 22. Darwin had read it while he was traveling on the Beagle. Then, five years later, when the Beagle had returned Darwin to England, Darwin had met Lyell in person. In London in 1836. The two Charleses—Darwin and Lyell—had become friends. And Lyell's ideas would change Darwin's mission in life.

The two bonded over shared ideas. Like Anne Boleyn and Henry VIII bonding over the emotions that would eventually impel them to found their own church. And like William Thomson, Rudolf Clausius, and Hermann von Helmholtz bonding over an equation-powered reductionism. Darwin and Lyell bonded thanks to sex's production of oddballs and to the birds-of-a-feather effect.

Twelve years earlier, in 1824, a futuristic machine called a "steam carriage" had debuted. It was a steam-engine-powered transport device built to carry coal from the countryside to the port of Stockton, where it could be hauled to London by ship. To lay a path flat enough for this newfangled "carriage," you had to cut away the sides of hills and mountains.[1163] And when you did, you saw in the raw rock face a stack of stone that looked like one pancake laid upon another. You saw layers of rock. Layers rich in fossils. In 1824 Lyell had visited the Black Country of Dud Dudley. Coal country. It was six years into the construction of the railroad line for the first train that would ever make a long-distance, 25-mile trip. And it was a year before that train's inaugural run. This is where Lyell had seen firsthand the cut flapjacks of stone revealed by railway construction. And Lyell, like a tiny number of others before him, had believed that each layer represented a different era. He had been certain that what you saw when you sliced into a hill was a history of the planet, a timeline. And since fossils of strange creatures were embedded in the layers, he'd been convinced that these odd beasts were not just the pranks of a playful God, but were creatures who had lived before us.[1164] Each dated by the flapjack of stone—the layer—in which it was found. And Lyell had believed something farther. He had been certain that the earth was not 6,000 years old,[1165] as the Christian religion had said. It was far, far older. Hundreds of millions of years older.[1166]

This was a holist view. An omnological view. A view the reductionists would soon try to demolish.

And it came from a scientist who, when he did his most important thinking, did not have a university job.[1167]

Neither did Darwin. Darwin adventured. He learned how to throw the bolas,[1168] the boleadoras, the three balls connected by rawhide cords that the South American cowboys slung from horseback at escaping cattle to tangle their hooves. He rode with those cowboys, the gauchos. He negotiated safe passage with a general who was taking an hour off from leading a bloody revolution.[1169] And as he traveled, Darwin began to build the kind of theory that Herbert Spencer would talk about over and over again—a theory of evolution.

When Darwin got home from his travels on the Beagle he had a collection of 1,500 species.[1170] And he had a far bigger collection of experiences in his fourteen notebooks.[1171] He wrote up his emulation of von Humboldt's journeys in an 1839 book, *The Voyage of the Beagle*. He then spent twenty years corresponding with like-minded people, with birds of a feather, with oddballs all over the world, collecting evidence, piecing it together, and fitting it into his emerging theory. The concept Darwin would call "the theory of natural selection."[1172]

Why "natural selection"? Let's go back to pigeons. Breeding these birds was an English passion. According to the *New York Times*' Carl Zimmer, Darwin took up the hobby in 1855, building a dovecote in the garden of his house in the countryside of Kent, fifteen miles south of London, and raising "pouters, carriers, barbs, fantails, short-faced tumblers, and many more."[1173] Darwin saw first-hand how pigeon breeders can start with birds speckled with gray and white and, over the course of generations, could produce all white or all black pigeons by what he called "modification by selection."[1174] What we call "selective breeding."

If you wanted a pure white pigeon, you took advantage of oddballs, of one-of-a-kinds. And you were picky. In each generation you selected the male and female pigeons with the most white, let them breed, then selected the whitest of their chicks and bred them, too.

In the end you produced the pure white birds you were after. By careful selection of which pigeon oddballs to breed. By careful selection of which pigeon oddballs, you gave the privilege of sex.

But in the world beyond the pigeon cage, said Darwin, nature was the selector. Yes, nature. Nature culled out the birds who couldn't meet the challenges thrown their way and granted outsized reproductive rights to those birds who could make the most of their circumstances. Nature gave the winners outsized access to sex. Nature gave the winners the privilege of siring the next generation. To he who hath it shall be given. From he who hath not, even what he hath shall be taken away.

After twenty years of patient fact-gathering, theorizing, and dodging the explosion of outrage he felt his ideas would produce among religious folks like his wife, Darwin in 1859 would finally introduce his new theory in a book called *On The Origin of Species*. The book Darwin was afraid would trigger a storm of opposition sold out the first day it went on sale.[1175] Then Darwin would write fourteen[1176] more books expanding his theory. But even in *Origin of Species*, Charles Darwin would only use Herbert Spencer's word, "evolution," once.[1177]

Darwin's form of science was radically different from the science of Sadi Carnot, William Thomson, and Rudolf Clausius. It was not abstract. It was not mathematical. It was observational. It relied on mountains of evidence from all over the world. The real world. Not the highly artificial world of the labs in the new Humboldtian universities or the even more abstract world of "ideal" steam engines, steam engines that existed only in the realm of the imagination. Only in the realm of mathematics. Darwin's science, to repeat, was not mathematical. It did not use equations. Darwin's science was observational. And it was what Herbert Spencer called "synthetic."[1178] It used as many sciences as fit the problem. Welding them together into one.

And one of its main tools was not the lab in a university. It was the scientific expedition. The scientific adventure.

Darwin's science was a form of what in 2001 I would name Omnology, a science that uses as many disciplines as the scientist can master. Here's the Omnologist Manifesto:

> We are blessed with a richness of specializations, but cursed with a paucity of panoptic disciplines—categories of knowledge that concentrate on seeing the pattern that emerges when one views all the sciences at once. Hence we need a field dedicated to the panoramic, an academic base for the promiscuously curious, a discipline whose mandate is best summed up in a paraphrase of the poet Andrew Marvell: "Let us roll all our strength and all Our knowledge up into one ball, And tear our visions with rough strife Thorough the iron gates of life."
>
> Omnology is a science, but one dedicated to the biggest picture conceivable by the minds of its practitioners. Omnology will use every conceptual tool available—and some not yet invented but inventible—to leapfrog over disciplinary barriers, stitching together the patchwork quilt of science and all the rest that humans can yet know. If one omnologist is able to perceive the relationship between pop songs, ancient Egyptian graffiti, mysticism, neurobiology, and the origins of the

cosmos, so be it. If another uses mathematics to probe traffic patterns, the behavior of insect colonies, and the manner in which galaxies cluster in swarms, wonderful. And if another uses introspection to uncover hidden passions and relate them to research in chemistry, anthropology, psychology, neuroscience, history, and the arts, she, too, has a treasured place on the wild frontiers of scientific truth—the terra incognita in the heartland of Omnology.

Let me close with the words of yet another poet, William Blake, on the ultimate goal of Omnology:

> To see a World in a Grain of Sand
> And a Heaven in a Wild Flower,
> Hold Infinity in the palm of your hand
> And Eternity in an hour.

The field of Carnot, Thomson, Clausius, and Boltzmann was peephole science. Reductionist science. In studying the steam engine, it did not look at coal, fire, or steel. Or at the ingenuity of the steam engine's makers. It looked only at heat. Yes, it produced crucial research on that heat. But it worked primarily in the world of imagination. The world of mathematical abstraction.

The territory of Carnot, Thomson, and Clausius was reductionist science. Carnot, Thomson, and Clausius followed Aristotle's dictum: reduce things to their smallest parts, their particles, their elements, find the laws of those elements, and you will understand everything.[1179] Reduce things to atoms and molecules.

That's not what Darwin tried to accomplish. Instead, Darwin followed von Humboldt's implied imperative, an omnological imperative—see as much as you can of the planet. See as much as you can of nature. See as much as you can of human affairs. Use every science available to you. Go out into the real world, immerse yourself in it, experience it up, down, backwards, and sideways. Then take your ideas from what you see. Look for the big picture.

Which form of science is right? Reductionist science or holist science? Mathematical science or observational science? University science or real-world science? Narrowly specialized science or omnological science? Both. Each one needs the other. Opposites are joined at the hip.

***

But there's something more. Remember, in science, if your prediction fails, you have to either abandon the theory it came from or alter that theory until it can make accurate predictions. And when it comes to the prediction business, holistic science, omnological science, real-world science, big-picture science was about to beat reductionist science at one of the grandest predictions of them all.

CHAPTER 45

OMNOLOGY VERSUS REDUCTIONISM: THE BIG TEST

Williﬞam Thomson was 35 when Charles Darwin hit the book stores with *On The Origin Of Species*. Thomson waited until the third edition in 1863 before he went through it.[1180] He apparently didn't like what he read.

So he tried to disprove it. He used Fourier's math[1181] on the travel of heat in an iron rod from the red-hot tip to the cool handle. He used the newly discovered melting temperatures of slate, sandstone, garnet, granite,[1182] and alkali feldspar.[1183] He factored in brand-new observations of underground rocks in Southeast London's Greenwich.[1184] He ran the equations. And he came to the conclusion that the earth was between 20 million and 400 million years old.[1185] Now that's a lot compared to the Christian belief in an earth only 6,000 years old. But, he said, it's too little time to allow for what geologists James Hutton[1186] and Charles Lyell had proposed, and what Charles Darwin had picked up and run with.[1187] It's too little time for an evolutionary process so slow that it would be called "gradualism."[1188]

Remember, William Thomson did mathematical and experimental science. University lab science. Reductionist science. Charles Darwin did holist science, science that looks at the whole ball of wax instead of just its tiniest parts. Holism would flourish in the 19th century with greats like Charles Darwin's grandfather Erasmus Darwin, Alexander Von Humboldt, Charles Darwin himself, and Herbert Spencer. All of them were omnologists. And these holists were some of the most famous and admired thinkers of their time. They were giants.

Herbert Spencer was considered the greatest philosopher of the nineteenth century.[1189] He made major contributions to "evolutionary theory, philosophy of science, sociology, and politics," not to mention fields "like sociology, anthropology, political theory, philosophy, and psychology."[1190] In fact, Mark Francis, Professor of Political Science at the University of Canterbury in New Zealand, credits Spencer with "the Invention of Modern Life." What's more, Spencer was idolized as far away as India,[1191] China,[1192] and Japan.[1193]

And, as you've begun to see, Alexander von Humboldt was one of the century's most famous men. You'll glimpse one more surprise of that fame in a minute or two.

But Thomson wanted to wipe one key to holist science out—"uniformitarian geology."[1194] The geology that implied that evolution had taken place over a very long period of time. And Thomson had an ulterior motive, his religion.[1195] He would eventually be appointed an elder in the Church of Scotland and would become Chairman of the Christian Evidence Society,[1196] an organization founded in

1870[1197] to counter "atheism." So Thomson wanted to eradicate the evolutionary timeline. On behalf of his faith. On behalf of his God. And he felt he'd done it by proving that the earth was too young to support the long, slow process that Darwin had proposed. William Thomson—the future Lord Kelvin—felt that he had demolished Darwinism.

Darwin himself feared that Thomson was right. Not about the age of the earth, but about reducing Darwin's theory to an object of scorn. Darwin felt he was at war with what he called "the physicists."[1198] In his letters to friends, Darwin admitted that:

> "I am greatly troubled at the short duration of the world according to Sir W Thomson."

> "Thomson's views on the recent age of the world have been for some time one of my sorest troubles."

> "Then comes Sir W Thomson like an odious spectre."[1199]

Darwin called Thomson's objections to his theory of natural selection, "one of the gravest as yet advanced."[1200]

And Darwin tried to defend his theory from Thomson's ideas by writing,

> I can only say, firstly, that we do not know at what rate species change, as measured by years, and secondly, that many philosophers are not as yet willing to admit that we know enough of the constitution of the universe and of the interior of our globe to speculate with safety on its past duration.[1201]

In other words, said Darwin, despite all of Thomson's achievements, science was still too primitive to measure the age of the earth.

But in the end, it was Charles Darwin who would be right about the age of this planet, not William Thomson. Thomson, the man with the infallible tools of equations and lab science, Thomson, the professor of natural philosophy at a major Humboldtian university, was off by 4.1 billion years. An overwhelming error. And Darwin, the man who went out into the real world, who observed, took samples, and who used intuition, reason, and idea-testing, Darwin, the scientist without an equation, the scientist without a lab, the scientist without a university chair, Darwin, the seeker of a massively big picture, would be right. The holist would triumph over the reductionist. The polymath would triumph over the professor. The omnologist would triumph over the specialist. Or would he?

Darwin's holism might have proven massively more accurate. But Thomson's successors would triumph where Thomson himself had left off. They would deep-six holism. They would make the sciences that dared to meld all of the disciplines invisible. They would erase the names of Alexander von Humboldt and Herbert Spencer from the history of science. And they would kidnap Charles Darwin's theory of evolution and would imprison it in a new reductionism, caging it in what they would call the Modern Synthesis.[1202] They would mangle Darwin's evolution to fit a new mathematical framework. A framework that in 2012 would be proven wrong.[1203] By one of the greatest omnologists of the 20th century, Harvard's E.O. Wilson. The man who founded the field of sociobiology.

In 1982, the pope of the Modern Synthesis, Oxford zoologist Richard Dawkins, would call the new holy writ "the central dogma of individual organisms working to maximize their own success."[1204] Yes, the specialists would turn the holistic evolutionary theory of Charles Darwin into a new form of reductionism. A mathematizable form of reductionism. They would boil evolution down to a new breed of particles: what Dawkins, as you know, would call "selfish genes."[1205] And they would reduce evolution to something called the "rational choice"[1206] of the individual. Each of us individuals, says rational choice theory, makes our choices based on what will selfishly benefit us. An idea Darwin's books radically disagreed with.[1207]

Remember, in science if your predictions prove wrong, you have to modify or abandon the theory from which those predictions came. Yet science never rejected William Thomson's reductionism. Or its product, Clausius' concept of the Second Law of Thermodynamics. Science would never abandon Thomson's cosmic pessimism, his apocalyptic vision, his heat death. And it would never abandon the idea that Thomson had laid the base for, entropy.

Sometimes science is less scientific than it claims.

***

All of this happened thanks to the hurricanes of history. All of it happened thanks to opposites joined at the hip. All of it happened through the birds of a feather effect. All of it happened through pulling oddballs together into packs. All of it happened through pulling together one-of-a-kinds into group identities, organizational personalities, movements, subcultures, and mind-tribes. Darwin was a one-of-a-kind, an oddball. Lyell was an oddball of the same sort. When they came together in 1836, they formed the nucleus of one new kind of mind tribe. The gradualists. Thomson, Clausius, and von Helmholtz were oddballs of another sort. They, too, found each other in roughly 1850 and started a movement. A mind tribe. A group identity. An organizational personality built around thermodynamics.

The battle of these mind tribes—the battle of the holists versus the reductionists[1208]—resembled the conflicts between the Lutherans and the Catholics in the days of Anne Boleyn. Like the Lutherans and the Catholics, the holists and the reductionists took up opposite sides in a culture war. A culture war in which the winners would erase the losers from the history books. Wiping out the tale of one of the most important hurricanes in the history of science.

But a hurricane about whose opposing forces you can say at least one positive thing: this equivalent to Protestants and Catholics never stooped to violence.

***

In a world ruled by entropy, in a world where everything is falling apart, is the evolution of opposing worldviews championed by two different scientific subcultures, two different mind tribes, likely to have consequences? The scientific tribe that reduced heat to the motion of particles opposed the scientific tribe of the evolutionary timeline. Is this kind of competition the shortest distance between two points? Or is it nature weaving the longest paths possible? Using opposites joined at the hip. Knitting together hurricanes of thought. Vortexes that abjure a straight line. To create something new? To invent?

Is it time to get beyond the metaphor of the steam engine? Is it time to get beyond the mechanical metaphor that's produced the ideas of heat death and entropy? Is it time to get beyond the tyranny of the concept that got the age of the earth radically wrong? Is it time to revive observational science, holistic science, the science of real-world experience, the science of what seminal anthropologist Robin Fox calls the participant-observer?[1209] The discipline of the scientific expedition? The science that can look at big picture astonishments like group identities, organizational personalities, hurricanes of history, and mind tribes?

Is it time to revive the science of the overview? The science that got the age of the planet right? Is it time to pursue Omnology, the science that urges you to follow all of your curiosities at once? Is it time to pursue the science that encourages you to know as many fields as your mind can master and that invites you to use those sciences rigorously to pursue the grandest sweep you can conceive?

In reality, it's not an either-or proposition. Opposites are joined at the hip. Science needs both. It needs holists and reductionists.

But which does nature encourage you to do, to be generous to yourself and follow your curiosities? Or to be stingy to yourself and try to squeeze yourself into just one narrow discipline? Which does nature favor, your passion or your thrift? Your self-indulgence or your self-control. Good question. The answer is both.

CHAPTER 46

BUILDING MOUNTAINS FROM MOLE HILLS

L et's do a quick review.

Mother Nature isn't nice. She is not kind and caring. In fact, Mother Nature loves disasters. In the early days of life, natural disasters were all over the place. Yes, natural disasters. From earthquakes and volcanoes to this planet's favorite—climate change. All of these took place without humans, without smokestacks, and without capitalists.

But here's the trick. More things ran uphill than ran down. More things were created than destroyed. More shockingly new group identities popped into existence than exited the scene. More forms of organizational personalities emerged. A fact that makes two sacred tenets of today's science untenable.

Holy doctrine number one comes from the work of William Thomson, Rudolf Clausius, and Ludwig Botlzmann. It's the Second Law of Thermodynamics. Remember, the Second Law of Thermodynamics says, "The entropy of the universe tends to a maximum."[1210] *Die Entropie der Welt strebt einem Maximum zu—* those were the words of Rudolf Clausius in 1865.

Emeritus Professor of Physics at Boston University Samuel J. Ling expresses another traditional facet of entropy. Entropy, says Ling, is "the most disordered state."[1211] But the 13.8 billion years of cosmic history in the timeline that you began to assemble at the age of twelve seem to prove entropy wrong.

Look again at cosmic dust. In the 13.8 billion years since its birth, has cosmic dust sifted randomly through the universe? Has it dissolved like the molecules of sugar in your glass of water? Has cosmic dust collapsed in chaos? Has it become formless? And useless? A random mist? No. As you know, the dust of the cosmos has done the very opposite. It has assembled in new teams. It has come together in new group identities. The dust of the cosmos has come together in gravity balls, galaxies, stars, planets, black holes, and ultimately in you and me. Mere dust has assembled astonishments.

In this cosmos, energy does not whuffle into uselessness. The invention of new ways to harness energy has been on a constant increase. In fact, seemingly useless energy has been harnessed with each upward step this cosmos has taken. Space, time, quarks, atoms, galaxies, and stars have all harnessed spills of energy in brand new ways.

Life harnesses more than just energy. Life harvests disasters. Roughly 3.5 billion years ago—when the cosmos invented life—wind, rain, light, and darkness were all death traps. But a short time later,[1212] the bacteria[1213] known as blue-green algae would use the toxic flood of radiation from the sun to make sugars and carbohydrates.[1214] And those sugars and carbohydrates would be woven together to produce, not chaos and disorder, but new generations of bacteria.

At the same time, an entire menagerie of bacteria would turn the murderous darkness at the bottom of the sea into a pleasure dome.[1215] And three billion years later, the first mammals—rodent-like eutherians[1216]—would use darkness to hide their movements as they hunted for food at night. The eutherians would turn the curse of night into a resource.

Life thrived by turning poisons into powers. Life thrived by eating its environment.

No, the cosmos is not running down. It never has. It is running up. The law of entropy is nonsense in a self-upgrading universe. The law of entropy is gibberish in a cosmos that revels in breaking her own laws, pulling herself up by her own bootstraps, and inventing new stairsteps up.

For what purpose does nature shatter her own commandments? To reinvent herself. To resculpt the very landscape of reality. To push the envelope of the possible. To rise above herself. To soar above herself quite literally. Nature is a search engine probing her potential. Nature is a transcendence engine. Nature is an invention engine.

Nature abhors the Second Law of Thermodynamics. She follows the First Law of Flamboyance. And she does it most flagrantly with sex.

***

Then there's a second scientific mistake it's time to call out. It's Pierre Louis de Maupertuis' law of least action, which, as you already know, says that "Nature is thrifty in all its actions." De Maupertuis had the backing of Aristotle himself, one of the greatest philosophers of all time,[1217] who, you recall, is alleged to have said 2,080 years before de Maupertuis, "Nature operates in the shortest way possible."[1218] Nature always finds the shortest path. Nature does everything by taking the most penny-pinching route. Nature, implies de Maupertuis, is obsessed with energy conservation. Nature is obsessed with what de Maupertuis calls thrift. And what we call sustainability. But is she? Apparently not.

How has the cosmos herself given the law of least action the finger? Through the First Law of Flamboyance. Through this universe's restless, itchy, obsessive, and unstoppable creativity. Through this cosmos' unabashed lust for novelty. Through her thirst for inventions and breakthroughs. Through her love for materialism, consumerism, waste, and vain display. Through her adoration of those who oppose her most. Through nature's insistence on reinventing her very nature.

CHAPTER 47

WHICH DOES THE COSMOS LOVE THE MOST: LOCKDOWN OR LUXURY?

There are yet more lessons from the tale of Anne Boleyn and the story of the rise of entropy. Group identities compete. Organizational personalities compete. Groups of humans once formed around genes, but ever since the Axial Age[1219]—the age that gave us Confucius, Buddha, Zoroaster, the Hebrew prophets, and the Greek philosophers—they've formed around ideas. Ideas are like the flowers of plants. They are markers of identity and accelerators of speciation, accelerators of schismogenesis,[1220] accelerators of differentiation, accelerators of invention. And invention is evolution's obsession. In fact, invention is apparently evolution's purpose in life. To achieve it, she pits groups against each other and offers them a path to winning. That path is coming up with something new. Even if the something new is something old being radically reperceived. Like the idea of god in the era of a new invention, the printing press. An era in which books were no longer just preserving information for information specialists—

monks and priests. An era in which books were suddenly used to disseminate knowledge to the broader public. And to spread something even more crucial to the acceleration of invention—the right to interpret knowledge. The right to reperceive it. A right the Bible-translator William Tyndale was denied. Well, not quite denied. He got his radical reinterpretations out. But he was punished for it. He lost his life.

Remember, one totally unique individual can find another individual whose total uniqueness fits hers and catalyze a movement of those on their wavelength. As Anne Boleyn did with Henry VIII to create the Church of England. And to create a new English group identity, a new English organizational personality, a new sense of what it meant to be English. And to generate what literary critic Harold Bloom (not a relative) calls "The Invention of the Human."[1221]

The oddballs of a feather finding each other and creating something bigger also showed up in the 1800s when William Thomson synched with men he had not yet met in person, men he communicated with by snail mail and journal articles—Sadi Carnot, a man who, alas, had found no others to resonate with. Plus Rudolf Clausius, and Hermann von Helmholtz. The result was a new group identity—the thermodynamicists, the reductionists.

The oddballs of a feather finding each other and creating something bigger also showed up when Charles Darwin found geologist Charles Lyell and later, his "bulldog," his promoter[1222] and defender, Thomas Henry Huxley. Not to mention Darwin's resonance with Herbert Spencer. That resonance with others, that ability to speak what others have only felt, to give words, concepts, and social structures to others' emotions, was like the dance of cichlid fish and water spiders. Among fish and spiders, resonance—being on the same wavelength—gave rise to new species. New species whose differences were spelled out in genes. And in humans, the oddballs finding each other and creating something bigger led to new species of ideas. And to new groups that formed around those ideas. Groups that competed. Yes, like the Protestants versus the Catholics. And the reductionists versus the Omnologists. Sometimes, these resonances of oddballs and the competitions between oddball groups do not dissipate in uselessness. They power the forces of history.

But let's get back to Pierre Louis de Maupertuis and his law of least effort, his law that nature always takes the shortest path. I've been saying that it's wrong. And I've been pointing out that it's utterly shattered by the longest path nature has been able to conceive to date, sex. But, and this but is a big one, opposites are joined at the hip. De Maupertuis law is at one end of a continuum. The lavish, the expensive, the outlandish, the impossible, and the ornate are at that continuum's opposite end. And guess what? Nature uses them both. What in the world do I mean?

The law of least effort calls for parsimony. It calls for thrift. So does another standard idea in science—Occam's Razor—which says that the simplest explanation, the thriftiest, is always the best. But there was nothing thrifty in the seven-year romance of Anne Boleyn and Henry VIII. And if you've seen her costumes after she ascended the throne, you know there was no thrift even in the everyday outfits of Anne's daughter, Elizabeth I. To awe the British nation, Queen

Elizabeth had to indulge in thrift's opposite, splendor. On a daily basis. And what is splendor? A materialist, consumerist, wasteful, vain display, a display of excess that only a queen can afford. A display designed to create awe and obedience. By showing off the achievement of the impossible. By showing off a surplus bigger than anyone has ever seen. By flashing your group's peacock's tail.

By putting up a display that shows that the group identity you control and that controls you can pile up an impossible excess of resources, a massive surplus. A surplus that puts a ruler on top of the heap. On top of the pecking order of her nation. And hopefully on top of the pecking order of something new—her nation-state.

In fact, thrift and flamboyance are different facets of a single strategy. A search strategy. A spread-out-and-explore, then consolidate strategy. A reach-out-boldly-and-adventure, then-pull-together-again-in-fear strategy. A strategy that increases the percentage of atoms on this planet entrained in the uber-enterprise of life. The fission-fusion strategy.

Keep that reach-out-then-clump-together strategy in mind. You'll find it all over the cosmos of dust, stars, and galaxies. And you'll also see it wherever there are living beings.

Yes, opposites are joined at the hip. Thrift and extravagance are Siamese twins. Don't pick one, pick both. Take, for example, communities that have been around for over three billion years,[1223]communities you've run into before, microbial mats. Microbial mats are green multi-species colonies of bacteria and algae layered on each other until they look like squares of rug. Until they look like big green squares the size of couch cushions. During the Antarctic winter, microbial mats freeze-dry and go into cryostasis, a suspended animation that stops all their life functions. Then when the weather warms in spring, the mats come back to life again.[1224] Microbial mats see-saw between thrift and extravagance. To outwit a nasty twitch of the planet: climate change. The climate change of the seasons.

To repeat, nature couples thrift to extravagance. This coupling is one of nature's favorite tricks.

Another example: the tiny "water bears" called tardigrades, creatures who evolved 550 million years ago,[1225] 2.5 billion years after microbial mats and 80 million years before the first land plants would arise. Tardigrades have to know which to use when—penny-pinching or flamboyance.[1226] When conditions are bad, these "water piglets" become one of the greatest virtuosos of thrift this cosmos has ever seen. They pull their legs into their bodies.[1227] They look freeze-dried. They go into what's called "cryptobiosis."[1228] They become tiny dried up ovals named after the beer barrels they resemble, "tuns."[1229] For all practical purposes, it looks as if these micro beasts are dead. They use no energy. They need no water. They insert a sugar called trehalose into their cells where water used to be. Very much like the strategy that would later be adopted by anti-freeze fish. And they cut their metabolism down to one-thousandth of the norm.[1230] They become super savers. Yes, they become the ultimate puritans. The ultimate misers. They take thrift to the nth degree. De Maupertuis would have been proud.

The result? These penny-pinching tardigrades can stand temperatures from minus 457 degrees Fahrenheit (almost absolute zero) up to plus 300 degrees (way past the boiling point of water). Tardigrade tuns can survive in the zero pressure of a vacuum and in the 1,200-atmosphere mega-pressure of the sea at the bottom of the Mariana Trench seven miles below the surface of the Pacific Ocean.[1231] Tardigrade tuns can even keep it together under radiation levels of 5,000 to 6,200 grays.[1232] Five grays is enough to kill you or me. Hiroshima-atom-bomb victims were killed by less than ten grays.[1233] 6,200 grays, the radiation a tardigrade in Puritan-mode can handle, would kill you and me more than a thousand times over.[1234]

According to Harvard researchers, tardigrade tuns can even survive the rigors of outer space. And more. Harvard-Smithsonian scientists say that tardigrade tuns can weather "asteroids, supernovas, gamma ray bursts," and hang in there for ten billion years.[1235] One Oxford University and Harvard-Smithsonian Center for Astrophysics paper claims that tardigrades will be around until the sun dies. In fact, the Harvard Gazette speculates in a headline that tardigrade tuns could someday be "the last survivors on earth."[1236] That is thrift.[1237]

Or is it? The flipside of the tardigrade's thrift is flamboyance. And nature's most flamboyant process is sex. Even for tardigrades.

When water arrives and the temperature is right, the seemingly dead tardigrades burst into life. In fact, if things look really good, a female tardigrade becomes a spendthrift. She goes for materialism, consumerism, and waste. She generates 30 eggs and opens herself to romance. As many as nine male tardigrades trundle slowly toward her, crowding around, hoping to climb on her back and inseminate her eggs. Or the hopeful males march around the external envelope she exudes with her eggs in it. Males are waste with a purpose. These excess males, these ambulating mini-barrels of materialism, consumerism, and waste, hope for the chance to penetrate one of the female's openings, preferably "the posterior end of a female exuvia containing eggs."[1238] They want sex. In fact, they compete[1239] for sex. The lucky male that gets to hook up with the female can spend an hour ejaculating. That is a huge expenditure of time and energy in a beast whose entire lifespan is between two months and two years.[1240] It's the equivalent of 3.3 days and nights in human time. Having sex. Copulating. With just one female. Without sleep or food. That's indulgence. That's flamboyance. That's the expenditure of a surplus.

Then there's another display of materialism and waste—that mass of male suitors who come calling. Nine is eight more than a female tardigrade needs if all she's after is reproduction. But she's apparently after something more. She's after perfection. She's after upgrade. She's after self-improvement. She's after genetic optimization. She's after the impossible. That's what she gains from the competition between males, a competition that will leave many on the sidelines, without sex, without reproduction, the incels of the tardigrade world. The living throwaways of materialism, consumerism, waste, and vain display. But the female tardigrade is not alone in seeking a step up.

So does nature. Nature drives her species to produce an excess of sexual choices. In order to improve herself. In order to invent. In order to find her next step up. And excess is the opposite of thrift.

***

Like bacterial mats and tardigrades, hummingbirds switch from thrift to exuberance and back again. The hummingbirds of the Andes mountains are among the descendants of the loony dinosaurs who flew. To deal with the climate slam from day to night and back again, they go into what researchers call a near-death state. They do it every night to wait out the darkness. They become living incarnations of thrift. But when daylight returns, they burst into a wild expenditure of energy. They flap their wings an average of 53 times per second.[1241] Yes, per second. That's exuberance.

Then there's the flamboyance of the hummingbirds' sexual role. Or, to be more specific, its two sexual roles. Hummingbirds hover like helicopters over the sexual parts of a flower, drink in the flower's nectar, and accidentally get the plant's powdery pollen on their beaks and heads. When the hummingbirds move on to another flower of precisely the same species, they do more than drink more nectar. They help consummate the plant's sexual act. They deposit the first plant's pollen on the new flower's sexual parts. They deposit male pollen on a distant plant's female organ—the receiving plant's stigma.[1242] Yes, like insects, the hummingbirds act as a plant's flying penis.[1243] Like insects, they are pollinators.

Then, when night comes, the hummingbirds swap extravagance for thrift once again and go back into an energy-saving, near-death state. Like the microbial mats and the tardigrades, the hummingbirds switch from thrift to flamboyance and back again. Why? To deal with the massive climate change from day to night. Not to cope with man-made climate change. But to deal with a climate change inflicted by the planet itself.

Servicing the sexual hungers of plants is a hummingbird's first sexual role. The second role? Servicing the sexual hungers of his or her self. To impress females, a male pushes the envelope of possibility. Male hummingbirds can't woo mates by strutting around. They have no ankle joint.[1244] They can't walk, swagger, and prance.[1245] They can't dance. Instead, they demonstrate their mastery of the impossible by showing off in the sky. They display their ability to accumulate surplus energy, surplus food, and surplus skill. They shoot 150 feet into the air, then dive bomb the ground, pulling up just before they crash. An extraordinary achievement from the descendants of loony dinosaurs who flew. A literally death-defying achievement.

Females fall for death slayers. For the slayers of death. Females fall for the conquerors of the impossible.

But dive bombing is not enough to woo a mate. Males gather in a spot where they can compete—a lek.[1246] They show off the iridescence of their feathers in the sunlight. They sing. They flutter their wings at more than the normal 53 beats per second, at 4,000 beats per minute,[1247] in order to raise the tone of their hum. And most of all, they show off the abundance of flowers in their territory.[1248] After all, a hummingbird female will need to drink her own weight in nectar every day just to survive. Not to mention to lay three clutches of two eggs each during the summer.[1249] She needs access to as many flowers as a territorial male can offer.

That is the First Law of Flamboyance at work. Nature does not use sex to make things easy. She uses sex to make things hard. She uses sex to make things increasingly lavish, increasingly extravagant, increasingly ornate, increasingly expensive, and increasingly demanding. She uses sex to show how gaudy she can be, how much surplus she can crank out, how flagrantly she can display that excess, how materialistic and consumerist she can be, and how much she can afford to throw away. How much she can drive males into the realm of the impossible. How far she can drive males to perform feats no female has ever seen. And in the process, how much innovation nature can create.

How much nature can invent.

Nature uses sex to generate the longest path between two points. She uses sex to push the envelope of the possible. She uses sex to upgrade herself. Yes, nature uses sex to reinvent herself.

***

Thrift and flamboyance are not an either-or. They are Siamese twins. They are opposites joined at the hip.

Opposites that have evolved to help species survive the opposites of nature, opposites of climate change—night and day, summer and winter.

The pairing of opposites is among the new rules we will need to replace the old mistakes of entropy and the law of least effort. And the pairing of opposites is among the laws that you and I will need to focus our own lives. But there's more.

CHAPTER 48

THE FISTS OF GHOSTS

Remember, sex is the longest distance between two points. Sex makes rules like entropy and the law of least effort look silly. But despite its expense and its byzantine tangles, sex sure must do something major for nature, because sex evolved between one and two billion years ago.[1250] And nature has gone hog wild with it ever since.

Why? Sex, as you know, helps nature feel out her next moves in possibility space. Sex works because billions of one-of-a-kinds scattered in the right place at the right time can produce evolutionary leaps. Leaps like the loony dinosaurs who flew. And leaps like England's Elizabethan age, an epoch in which the range of human possibilities soared. Sex helps nature expand her repertoire. Sex helps nature add to her toolkit. Sex helps nature reinvent herself. Sex helps nature satisfy the First Law of Flamboyance.

Sex increases the percentage of atoms on this planet kidnapped, seduced, and recruited into the lofty enterprise of life.

Sex also uses oddballs and the birds-of-a-feather effect to resculpt genes. But among humans, sex is not just a matter of whose tail feathers are the most spectacular. It's not just a matter of genes. It's a matter of another evolutionary accelerator. It's a matter of what super-evolutionary biologist Richard Dawkins calls "memes"— ideas reproducing themselves in a sea of human minds the way an annoying pop song takes over minds by the millions. Or the way a religion exerts a hold over humans all over the planet. Like Anne Boleyn's Protestantism versus Bloody Mary's Catholicism. Like William Thomson's reductionism versus Charles Darwin's holism, Charles Darwin's natural selection, Charles Darwin's evolution. Or like a few modern madnesses we'll get to in a minute.

In humans, sex is guided by shared perceptual lenses, shared worldviews. Worldviews like Islam, Marxism, Christianity, evolution, and entropy. As you know, worldviews are the human equivalents to flowers, to the water spider's dance, and to the hummingbird's death-defying dives to the ground. Worldviews are ways we humans show off. What's more, worldviews are the ways we justify another of nature's learning-machine tools, intergroup tournaments.

Intergroup tournaments—competitions between clans, kingdoms, mind-tribes, and civilizations—can go on peacefully. Like the competitions between holists and the reductionists. Or intergroup tournaments can produce the mass murder of war. And the origins of war go all the way back over 3.5 billion years when armies of billions of bacteria used chemical weapons to produce carnage and slaughter among their enemies—among rival bacterial colonies.[1251] Was this conflict unnatural? No. War was one of nature's creations. Remember, it was nature who invented death and pain. It was nature who gave generously to winners and plunged losers into poverty, loneliness, and misery. Which may make nature happy, but is morally unacceptable to us human beings. Nonetheless, remember, we are not stuck with war. Nature uses us to reinvent herself.

Meanwhile, the blunt fact is that intergroup tournaments enrich the cosmic invention engine. They sort out winning ideas. And winning forms of social organization. They also strengthen organizational personalities, group identities. In fact, intergroup tournaments are at the heart of each group's soul.

How do worldviews like Catholicism, Protestantism, Islam, Marxism, holism, and reductionism glue together individuals and sculpt group identities? Argue for the right idea, and you're one of us. Argue for the wrong idea, and you're one of them. Argue for the wrong idea and you will be tossed out "in deepest humiliation." Your ideas are marks of the group you belong to. Your ideas are the equivalents of the color red or the color yellow in reassembling sponges. Or the equivalents of red, blue, and gray in 19th century military uniforms.

You could see the way ideas set groups against each other in 624 AD when Allah granted Mohammed the right to make war to spread an idea, Islam.[1252] That form of idea-driven war was jihad. And its goal was global conquest. Global supremacy.[1253]

You could see how ideas set groups against each other again in 1555[1254] in England under Bloody Mary when hatred flared in the struggle between Catholics and Protestants and when 280 prominent Protestants were burned at the stake.[1255] The blood of the victims bonded the Protestants more tightly together.[1256] Protestants regarded the executed men and women as martyrs.[1257]

And you could see the way ideas set groups against each other again in 2022. In that crucial year, if you professed enthusiasm for Donald Trump and his view of the world you were a member of one group and the enemy of another. Argue for a woman's right to abortion and you were a member of a totally opposite crew. Why?

In 2022, America had two different tribes of patriots living in two very different realities. Two different tribes of people who loved the United States but who saw America in radically different ways. Two different tribes living in the bubbles of two very different worldviews. Two different tribes who saw reality through two radically different perceptual lenses. Two different tribes who viewed their world through the lenses of radically different ideas. Two different mind tribes.

Republican candidates before the 2022 congressional election sent thousands of emails telling their readers that the United States was on the brink of annihilation. On the brink of falling into the hands of a dedicated band of Communists, Marxists, and socialists. A band of people who hated America. That band of Communists, said the Republican candidates' emails, was the Democratic Party. And mailers from former president Donald Trump himself said that "deranged" Democrats were "thugs,"[1258] "communists," "Marxists," and "tyrants" out "to destroy American democracy."[1259]

The believers in QAnon took this dark vision even farther. We are under threat from a globalist elite, said QAnon. The globalists worship Satan, glorify sex with children, and sacrifice these children to the devil. But soon there will come the Storm. And in the Storm, all of those guilty globalists will have their wrists zip-tied behind them, will be dragged out to the streets, and hung. Or executed by firing squad.[1260] On live television.[1261]

Among them, will be House Speaker Nancy Pelosi and the Democratic members of Congress who opposed, attacked, or criticized Donald Trump.[1262] Many of the participants in the insurrection of January 6, 2020, were hoping that the day of the Storm had arrived. That's why some had white, plastic zip-tie handcuffs hanging from the back pockets of their jeans as they climbed the capitol steps. And that's why others built a gallows. Vice President Mike Pence, who refused to disqualify the election of Joe Biden, was only the first of many the gallows makers hoped to hang. They also wanted to execute all the Democrats in Congress and the Senate.

According to QAnon,[1263] there was only one savior who could sweep away the Deep State and free us of the Satanic Democrats. His name was Donald Trump.

I'm a Democrat, and I can tell you that at least 95% of this dark view of Democrats was hogwash. Less than 5% had elements of truth.

Meanwhile Democrats like me saw Donald Trump as the biggest threat to America's democracy since the Civil War over 150 years earlier. A close reading of Trump's actions and speeches showed that he was a worshipper of Vladimir Putin and an admirer of China's dictator Xi Jinping. Trump's words and those of his minions hinted that The Donald was trying to create a one-party system in the United States and to install a Trump dynasty like the dynasty of a family Donald admired, the descendants of North

Korean dictator Kim Il Sung, whose family line had kept North Korea in an iron grip for three generations, for 74 years.

I know because I covered every day of the presidential campaign of 2016 between Donald Trump and Hillary Clinton for 545 radio stations via America's highest-rated syndicated overnight talk radio show, Coast to Coast AM.[1264] And I've been following Trump daily and in depth ever since.

Yes, Donald Trump's own words and actions indicated that he wanted to be dictator for life. Then that he wanted to be followed in the White House by his sons, Eric and Donald Jr., and possibly his daughter, Ivanka.

That's the struggle of the gene pool of one family to take power in the USA. A very selfish-gene sort of thing. But the primary instrument that the Trump family's selfish genomes used was not a 20-ton weapon. It was not a tank or a bomb. It was a wisp, a phantom, a perceptual lens, a tight mesh of memes, a worldview. A mesh of ideas that engages your passions. And mine.

A worldview does something vital to bond us humans together into a group identity. A worldview works its magic by telling you who you have permission to hate. Yes, politics is permission to hate. Tenant groups give permission to hate landlords. Progressives give permission to hate billionaires. Republicans give permission to hate progressives. Organized groups of rebellious teenagers give permission to hate the supporters of the established way of doing things. "Patriots" hate globalists. And globalists sometimes give permission to hate "patriots." Worshippers of a fearless leader give permission to hate the enemies of that fearless leader. And anti-fascist groups give permission to hate that "fearless leader's" followers.

How did Donald Trump provide a new perceptual lens?

When he descended an escalator in the lobby of Trump Tower on Fifth Avenue in New York City on June 16, 2015, to announce that he was entering the presidential race, real-estate developer and reality TV star Donald Trump introduced the MAGA worldview, the Make America Great Again worldview. Complete with folks you had permission to loathe and despise. Immigrants from Mexico, Trump said, are "bringing drugs, they're bringing crime, they're rapists."[1265] You are free to look down on them. You are free to abhor them. Democrats aim to take your freedoms and your guns away. Please feel free to hate them too. There is one and only one savior from the immigrant hordes and from the Swamp of the Deep State. His name is Donald Trump. That was the MAGA vision.

Hate is a tool of social construction. Hate bonds us in tight-knit groups.[1266] In organizational personalities, in group identities. Groups define themselves by who they hate.

Hate gives us scapegoats. And scapegoating gives us pleasure. In 1962, experimenters Roger Ulrich and Nathan Azrin put eight rats in a box with an electrified floor.[1267] When they turned the current on, it sizzled the rats' feet. Painfully. How did the rats react? Did they huddle together to give each other comfort from the pain? Not at all. It appears that the seven strongest rats felt out who they were and mauled the weakest rat among them. They apparently turned the pain of their foot shocks into a bonding experience.[1268] For a gang of seven. And they chose the eighth as a target. They marked out a rat to hate. Then they mauled him mercilessly. Beating up a bottom rat, a scapegoat, was apparently social glue. Which is one reason a worldview gives permission to hate.

There's more to a worldview's bonding power. Donald Trump unveiled the MAGA view of the world when he descended the golden escalator of Trump Tower in 2015.[1269] The MAGA idea unleashed the power of the birds of a feather effect. Research shows that human females seek out those who share their views of the world and who share their group loyalties.[1270] Which means you have to share another person's group affiliation and his or her worldview if you're going to mate. Yes, the MAGA worldview provided a new way to help girls get guys and men get mates. And like the different dance steps of spiders on opposite sides of a pond, worldviews wall us off from each other. In theory, that wall, over a long period of time, can make mind tribes genetically different. As genetically different as klatches of cichlid fish.

No wonder Henry VIII and Anne Boleyn were attracted to each other. They shared a newly emerging worldview. They shared what the Germans call a weltanschauung. They also shared several group identities. They were British. They were of the ruling class. And eventually, as if they were flowers spreading their pollen, they would spread their new worldview and change the group identity of England. They would change England's organizational personality. And they would change England's perceptual lens. Its ideas of who to hate. They would give permission to hate the Papists, the Catholics.[1271]

Why is a worldview that tells you who you have permission to hate so gut-grabbing? Because of what researchers call "altruistic punishment."[1272] We get a kick out of hating those who are our group's designated scapegoats. Or our group's chosen enemies. We get a jolt of dopamine.[1273] The neurotransmitter that gives us the pleasures of cocaine and heroin[1274] highs. In other words, when Donald Trump descended the escalator in Fifth Avenue's Trump Tower in 2015 and told the folks in the audience that they had permission to hate immigrants, it was as if he was handing out trays of cocaine.

But there's more. Love and hate are emotions produced by a single neurotransmitter in your body—oxytocin, sometimes called the cuddle hormone. Oxytocin warms you to love your groupmates. But it also produces a cool flipside: your distrust[1275] of outsiders.[1276] The grounds for hate. Opposites are joined at the hip.

Why would evolution plant such powerful rewards in your brain? Why would establishing an us-versus-them give us humans a survival advantage? Because the competition between groups is vital to evolution. Vital to nature. Vital to invention. And vital to you and me. Just as in the days of the Great Gravity Crusades, when human groups compete, the big eat the small. The leader who can mobilize the largest and most committed group wins. The idea that can mobilize the biggest pool of true believers reigns.

To he who hath it shall be given. From he who hath not, even what he hath shall be taken away.

How has group size and group efficiency helped organisms like you and me survive? If your group is wiped out, you will be wiped out too. If your group thrives, you are likely to do okay.

A clan based on genes will be limited in size. But a group based on ideas, a group based on a worldview, a group based on a view of who to hate, can take over huge swathes of humanity.

What's more, leaders eventually die. But worldviews can survive. Worldviews can organize societies for thousands of years. And the longer societies hang in there, the bigger and more powerful their worldviews are likely to become.

For example, from 622 AD to 750 AD, Islam pulled together the biggest empire the world had ever seen. An empire eleven times the size of the conquests of Alexander the Great, five times the size of the Roman Empire, and seven times the size of the United States.[1277] The influence of those conquests has survived. In 2015, there were more than 1.9 billion Muslims.[1278] That's more than the entire human population of the planet in 1910.[1279] It's nearly six times the size of the entire population of the United States. This vast sea of believers is held together by a worldview that has stayed strong for roughly 1,400 years.

And at the core of that worldview is a vision of who to hate. Unbelievers. Infidels and Jews. Apostates and hypocrites. Scapegoats and enemies.

Eight hundred years after the explosive military expansion of Muhammad and his followers, 800 years after the continent-spanning land grabs of the Empire of Islam, England yanked together an empire on which the sun never set. For the first time in 1,200 years, there was an empire bigger than Islam. But barely.[1280] The British Empire ruled over 458 million human beings at its peak in 1922, 23% of the world's population.[1281] And today 1.35 billion people speak English. Seventeen percent of the people on the planet.[1282] But that's over half a billion smaller than Islam's 1.9 billion. Yes, smaller.

In those nations that were once part of the British Empire, the ideals of democracy and the language of the English live on. In other words, the contributions of Anne Boleyn and Henry VIII still survive. And the worldview of Islam—complete with Muhammad's holy law of sharia—is vividly alive today in the 57 nations of the Organization of Islamic Cooperation. In fact, the Muslim worldview is trying to take more territory as we speak, fighting to swallow nations like Nigeria,[1283] Kenya,[1284] Ethiopia,[1285] and even Trinidad and Tobago in the Caribbean,[1286] 6,643 miles away from Islam's birthplace, Mecca.

But there's a competition between empires that's seldom spoken of, the competition to last, the competition to thrive and survive over eons of time. The competition to see which worldview can keep its hold on minds the longest. The competition to succeed as multi-generational projects. The British worldview stays alive in democracies in Europe, the United States, Canada, Australia, and India. It has been alive in the hearts and minds of the folks in these nations for roughly 400 years.[1287] But that's a mere piffle in the competition for longevity. The Muslim worldview is alive in the hearts and minds of a quarter of the world's population.[1288] And it has been throbbing in hearts and minds for over 1,400 years. Meanwhile, modern Chinese claim that their concept of Tianxia, in which the emperor rules all under heaven[1289] and the barbarians bow down to him, has been pulling together their group identity for 3,000 years. [1290] And Tianxia often gives permission to hate barbarians.[1291]

Which means that strange as it sounds, sex, one of the most private things in your life and mine, is not just a matter of individual variation. Sex provides the atoms and molecules of organizational personalities. Of group identities. And it provides the clusters of oddballs that make each of these organizational personalities different. Like the organizational personality of the Anglicans versus the group identity of the Catholics in England in 1558. Like the group identity of the Americans versus the group identity of the British in 1776. Like the group identities of the Europeans and the Americans versus the group identities of the Russians and the Chinese in the 2020s. Like the group identity of Islamists versus all those who adhere to any other forms of belief from 629 AD to today. Like the group identity of North Koreans hating the "warmonger imperialist," [1292] "gangster-like," "blood-stained"[1293] Americans. And like the group identities of the Free World versus the alliance of authoritarian states in the early 2020s. Not to mention the group identities of MAGA Republicans versus Democrats in 2022.

For at least the last 4,125[1294] years, the hurricanes of history have been powered by these competing meshes of ideas. By these worldviews. Worldviews that enclose a group in a bubble the way a membrane encloses a cell. Or the way surface tension encloses a drop of water. Worldviews that act as the perceptual lenses of a people. And the perceptual shells. Worldviews that give permission to hate.

But worldviews are challenged and altered every century. And newcomers like Donald Trump are among the challengers.

Which means that ideas are evolution's new accelerators. Ideas are evolution's new flowers.

***

Which brings us back to another of the biggest modern hurricanes of all. Another of the biggest competitions between worldviews and group identities. The intergroup tournament between climate change activists and climate deniers.

CHAPTER 49

TAME THE TORNADO: WHAT THE HELL ARE CLIMATE *STABILIZATION* TECHNOLOGIES?

"Do you seriously mean to tell me that you want to solve a
problem caused by technology with... technology?"
Daniel Pinchbeck, author of *Breaking Open the Head* and
2012: The Return of Quetzalcoatl
"Yes." Howard Bloom

D r. Luke Kemp at the University of Cambridge's Centre for the Study of Existential Risk says it's time to stare a simple fact in the face. We run the risk of "global societal collapse or human extinction."[1295] The University of British Columbia's William Rees agrees. He says that this collapse is inevitable. It will lead, he insists, to the deaths of at least five billion people, and it will leave the few humans who survive using waterwheels, draft horses, mules, and oxen.[1296] All because of man-made climate change.[1297]

Meanwhile I've hinted that we have ways out of climate change—we have climate hacks. Climate stabilization technologies. Again, what the hell are climate stabilization technologies?

To nature, climate engineering is old hat. Bacteria force the formation of clouds in the Arctic. They circulate from the ocean depths to the surface, are blown into the air, and act as seeds for the ice crystals that bring clouds to life. Clouds are vital to the Arctic climate in which the bacteria thrive. Says Jessie Creamean, an atmospheric scientist at Colorado State University, the climate-engineering bacteria "regulate the surface and atmospheric temperatures, affecting sea ice, ecology, shipping, Arctic climate and weather."[1298]

Rainforests engineer the climate to suit their needs. They produce the clouds that hover above them permanently.[1299] And they produce the rain that keeps them moist.

There's even a chance that bees may be in this climate-hacking club. Says Ellard Hunting, a biologist at the University of Bristol in England, his research revealed that a swarm of "bees had a [electric] charge density that was about eight times greater than a thunderstorm cloud and six times greater than an electrified dust storm."[1300] The *Scientific American* speculates that this charge could impact "things such as weather events, cloud formation and dust dispersal."[1301]

What's more, as you know, we humans have been inventing climate stabilization technologies since we first used a stone tool to slice the hide off of a bear and turn it into a fur cape roughly 300,000 years ago.

Why are animals born with fur coats, but we are forced to steal their pelts to survive the cold? Why can your dog or cat play in the snow or chase squirrels in the heat of summer with nothing but the fluffy stuff on its back? Why are animals born with thermoregulatory abilities we simply do not have?

Because we are children of our tools. Because we've been selectively bred by our technologies.

We were born to eat meat. How do we know? When you wolf down a steak or a chop, a vital chemical goes off in your gut[1302] and travels to your brain.[1303] It's cholecystokinin. And it tells your brain and mine to bond to the lovely folks who are sharing their barbecue with us.[1304] Those people may be a winning team. They could up our intake of two vital nutrients: protein and fat.

But if we're born to eat meat, why were we born without the tools we needed to bring down prey? Why were we born without fangs and claws? Because we were born to invent new fangs and claws. Or, to put it differently, our inventions reinvented us. They reshaped our biology.

We invented stone tools 3.3 million years ago.[1305] Then our tools reinvented us.[1306] First the new tools favored those of us who had already discarded natural fangs and claws[1307] and who had made artificial meat-ripping machinery. Mega fangs and monster claws. Radically unnatural things. Fangs and claws of stone.

Despite our nakedness,[1308] a million years after we invented stone tools, our cousins,[1309] *Homo erectus*, became the first hominids to migrate into the cold. They migrated north from tropical Africa[1310] to China 2.1 million years ago and to Europe 1.4 million years ago, taking their chances on the climate tantrums of the north. First they settled in the familiar warmth of subtropical China. Then, 1.8

million years ago,[1311] they invented our first climate stabilization technology. They tamed fire.[1312] And 1.15 million years ago, our *Homo erectus* relatives were finally able to take their chances on land plagued by winter, the temperate Chinese Loess Plateau.[1313]

Why were we early hominids able to finally set up shop in lands of winter frost? Because of our climate stabilization technologies. Think of it. We migrated into the deep freeze just about the time we lost our fur.[1314] We were literally naked to the cold. But the use of fire, which apparently started with just a few of us, became more and more widespread.[1315] Fire was our first external thermoregulator. Our first technological attempt to deal with this planet's perpetual climate change. Our first attempt to create a climate bubble. Our first climate stabilization technology. And the campfire gave those who gathered around it an evolutionary edge. Especially in latitudes of frost and in ages of ice. Like the Quaternary[1316] glaciation that began 2.58 million years ago and continues, believe it or not, to this day.

We tamed fire, and fire tamed us.[1317] It gave us a new invention, cooking. And back home in Africa,[1318] cooking shrank the size of our jaws and made room in our skulls for bigger brains. In Africa, fire and cooking changed our genes,[1319] gave us smaller guts,[1320] helped us stand up straight,[1321] and helped turn us into runners.[1322] Oh, and fire kept us warm. Something it would do even on the edges of sheets of ice. Sometimes when we radically remake our tools, our tools radically remake us.

Then came another climate hack: the hearth. The hearth dramatically upped a fire's heat, made it safer, more controllable, and kept your cave or camp ground cleaner. 800,000 years ago the first hearths appeared at campsites in the Jordan Valley. [1323] It appears that the hearth users gathered around their flames to make stone tools and to feast on "fruits and seeds, as well as...turtles, elephants, and small rodents." Three hundred thousand years ago, a hearth at Qesem cave in modern Israel would be massive—twelve feet wide and six feet deep.[1324] Probably in the hopes of warming the entire cave.

The age of the hearth brought a second great migration that was more daring. It came during an age of glaciation, between 600,000 and 400,000 years ago.[1325] But we weren't quite in modern form yet. We were *Homo erectus, Homo heidelbergensis,* and the muscular folks who evolved roughly 200,000 years later, Neanderthals. *Homo heidelbergensis* and Neanderthals were burlier than modern humans.[1326] And all of these early hominids were equipped with stone tools, fire, and hearths.[1327]

Then a hundred thousand years before we became fully human, we had two more breakthroughs. 400,000 years[1328] ago we sharpened straight tree branches to invent something nature alone had not conceived—the spear. And 100,000 years later we fashioned bear hides[1329] into the first fur capes and fur bedding. Now we were on our way to having our own, personal, private climate bubbles. Climate hacks that would keep us warm even when we stepped away from the fire.

And we would need these coats of animal fur. Remember, our first great migrations into the lands of frost came over 1.15 million[1330] years ago, when our cousins, *Homo erectus,* headed for the Loess

Plateau[1331] of China and 780,000 years ago when they traveled to Europe.[1332]

Then we actual *Homo sapiens* departed Africa on the next great ripples of hominid migration to Asia and Europe between 100,000 and 40,000 years ago.[1333] But by now, we were modern humans, *Homo sapiens*. And this time, we were bolder than the *Homo Heidelbergensis*. Like Neanderthals, we were "periglacial." We lived on the edges of ice sheets. We made ourselves at home on the edges of glaciers.[1334] Yes, despite our wretched hairlessness. We could afford to pull this off in part because roughly 120,000 years ago we had upgraded from fur pelts thrown across our shoulders to actual fur coats.[1335] Then 45,000 years ago,[1336] we invented bone needles and actually sewed those furs together, tailoring the new fur garments to our bodies. Thus providing a massive upgrade in our personal, portable climate stabilization technologies.

Roughly 30,000 years ago,[1337] when we were working in well-coor-dinated[1338] groups to bring down animals of massive size—something Neanderthals[1339] had also been very good at—we upped the scale of our climate stabilization bubbles dramatically. We erected frameworks of the tusks and ribs of up to sixty mammoths at a time, frameworks forty feet in diameter, then covered them with mam-moth hides.[1340] We made tents. Big ones. Big enough for between three[1341] and fifteen[1342] families. With three separate hearths, three separate fire places, per tent. These mammoths-hide super-tents were climate stabilization bubbles far bigger than the space within a fur cloak or around a fire.

Eleven thousand years ago, we invented two even more substantial climate stabilization advances. First, we pried obstacles out of the ground—boulders. We rolled these massive rocks to a common destination, chiseled their surfaces with stone hammers and bone scrapers so that they would fit together, and invented the stone wall. With the wall, we built the first city—Jericho. And Jericho was a human hive of man-made permanent homes, round straw and clay dwellings sixteen feet across.[1343] A human hive of climate stabilization technologies.

But that's not all. Two thousand years later we upped the game. We took a toxic substance, the mud of the rainy season, mud that could suck you into a swamp and never let you out again, patted it into a shape that no human or animal had ever seen before—a perfect, flat-sided rectangle—and took advantage of the dry season to leave the mud-patties in the sun to harden and dry. We made ten million of these impossible new devices.[1344] And we built an entire city of two-story-high apartment buildings with three-room flats and a nice, warm fireplace—a hearth—in each. A new hive of climate bubbles. Another triumph for climate stabilization technologies. And another niche we'd invented, a niche in which we could compete like male elephant seals. A niche in which we could evolve[1345] and thrive. The first city of brick. Çatalhöyük in Turkey.[1346]

This was the first time humans had ever seen perfect flat surfaces, right angles, and accommodations for something else new—not a hodgepodge of fellow tribe members, but a nuclear family. A nuclear family living within a three-room flat. Enjoying a radical new invention—privacy.

In the process, our technology reshaped our genes and turned us into a brand-new species.[1347] Not just *Homo sapiens*. But *Homo technologicus*.[1348] Humans the technology makers. Humans remade by our own technology.

From fire to the fur coat, the mammoth hide tent, and the brick wall, all of these were climate stabilization technologies. Technologies to create envelopes of space enclosing the internal climate we chose. Technologies to ensure that we would not have to follow the path of the monarch butterfly and migrate. Technologies to ensure that we would be able to stay in place.

But now it's time to scale things up. Time to ratchet up our climate stabilization game.

***

First, we need to beat man-made greenhouse gases in the atmosphere. And there is a way to do that.

Harvesting solar power in space and transmitting it to earth is an idea first proposed in a 1941 science fiction story by Isaac Asimov.[1349] In 1968, space solar power was turned into a serious scientific project by a private engineering consultant who had worked on America's Apollo moon program, Dr. Peter Glaser.[1350] Then in the 1970s, space solar power was advanced[1351] by Princeton's architect of imaginary 500-square-mile space colonies,[1352] Dr. Gerard O'Neill, who inspired a raft of students including future space entrepreneur and Amazon founder Jeff Bezos.

I cut my teeth on solar energy in 1981, when I put together the first public service radio ads for solar power in conjunction with

James Young, guitarist and vocalist from the band Styx. Then I was educated on space solar power in the late 1990s by Paul Werbos, Program Director for Control, Networks & Computational Intelligence at the National Science Foundation, the Vatican of American science. Paul converted me to solar energy harvested in space. He converted me so famously that when I had my first phone conversation with Elon Musk in 2005, he already knew where I stood on the issue. The first thing he blurted out was, "I don't believe in space solar power."

Then, in 2010, I put together a meeting on Skype between astronaut Buzz Aldrin—the second man to set foot on the moon—and India's eleventh president, Dr. A.P.J. Kalam, the most popular politician in South Asia. The topic? Space solar power. Dr. Kalam was known as "the people's president." He was voted one of the two most trusted men in India in an Indian Reader's Digest Poll.[1353] He won this poll in a nation that deeply distrusts politicians. Not only was Dr. Kalam one of India's very few beloved political leaders, he was also the rocket scientist who had put India's space program on the map.

The goal of our Skype call was a global initiative on space solar power. A few months later, Buzz ran into conflicting interests at the Obama White House and dropped out. But Dr. Kalam and I continued to work together for space solar power until his death five years later. Kalam was the man who called space solar power "harvesting solar power in space." Like harvesting two descendants of the first flowering plants, wheat and rice. And he was sure that space solar power could lift two billion people out of poverty.

What is space solar power? It's solar power harvested in space and transmitted to receiving stations on earth using the sort of harmless microwaves your cellphone relies on. To harvest solar power in space, you construct solar farms, vast sheets of solar panels five miles long and five miles wide. Preferably parked in geosynchronous orbit.[1354] Which means that your space solar power farm continuously hovers 3,586 miles[1355] above the receiving stations of your city. Or you use small, modular solar-power-harvesting tiles orbiting together in constellations.[1356] Orbiting like SpaceX's unfolding 42,000-Starlink-satellite constellation. No matter which technique you use, you convert the solar power to microwaves and aim those microwaves at rectennas. Rectennas are 4.3-mile by 8-mile[1357] geometric grids of wires. On the earth. Grids of antennas beneath which you can raise crops, graze cattle, or rewild the land. And from the rectennas, the power can go to your city, to communities in Africa or Asia that have never had electricity before, or into the grid.

Space solar power is far more grid-friendly than ground-based solar farms. The power production of solar panels on the earth's surface is unreliable. Ground-hugging solar panels are robbed of sun by clouds and by night. On the other hand, space solar power is 24/7. Space solar power produces what the power industry calls "base load" power. Steady, reliable, hour-by-hour, constant power. The precise kind of power the grid can lap up and distribute. And unlike ground-based solar farms, the rectennas don't wipe out the local environment on which they are strung. Remember, you can rewild, farm wheat, or graze cattle beneath their wire receiving grids.[1358]

Phys.org reports that "173,000 terawatts (trillions of watts) of so-lar energy strikes the Earth continuously. That's more than 10,000 times the world's total energy use."[1359] And the amount of solar energy in the space beyond the earth's atmosphere is even greater. Roughly 44 times greater. So, for all practical purposes, space solar power is infinite.

Most important, space solar power is a climate stabilization technology. How? With it we can end the use of fossil fuels. Completely. We can reduce our emissions of greenhouse gases for energy production and transportation to zero. In other words, we can bring ourselves close to net-zero and eliminate the bulk of the man-made contribution to global warming. We can help achieve the Green New Deal.

Is space solar power a practical idea? Yes. We've been harvesting solar power in space and transmitting it to earth since 1962, when the first commercial satellite, Telstar, went up. Telstar looked like a beach ball covered with medallions. Each of those medallions was a primitive photovoltaic panel, converting sunlight into electricity. Electricity that the satellite beamed to earth. We've been harvesting solar power in space, beaming it to earth, and using it as the backbone of the telecom and global media business ever since. In fact, space solar power has driven an industry that was already worth $447 billion a year way back in 2020.[1360] That's nearly half a trillion dollars. So space solar power is a proven technology.

But that's just the beginning of space solar power's promise for climate stabilization. I was speaking in Kobe, Japan, in 2014 at a conference on space solar power when Isabelle Dicaire, of the Advanced Concepts Team at the European Space Agency,[1361] stopped me in my tracks with her speech. To harvest solar power in space, we will need farms of photovoltaic panels up to 25 square miles in size. Dicaire's idea was to equip these huge solar farms with massive lasers. Why? If the weather service sees a hurricane headed for New Orleans, focus a solar-powered laser beam on one of the hurricane's edges. Heat that edge, and steer the hurricane away from populated areas, sending it out to die in the Gulf of Mexico. A brilliant climate stabilization technology

But the real climate stabilization goal comes from realizing that every catastrophe is an opportunity in disguise. Every disaster is an energy source waiting to be tamed.

What does that mean? In Egypt, 5,000 years ago, the land was a catastrophe. A climate catastrophe. There had been 5,000 years of monsoon rains.[1362] Then the downpours had retreated, turning lush, green savanna to desert. That was climate disaster number one. And the Nile River was climate disaster number two. Every year it overflowed its banks and flooded the desert lands around it. If you tried to farm near the banks of the Nile, you would have been out of luck. Your farm plot would have disappeared under the waters for three months every year.[1363]

Then the Egyptians figured out how to tame the disaster. They did it by inventing a new niche, a new infrastructure of habit, a new lifestyle. They established towns on heights that the raging river could not reach. Heights from which they could walk down to their plots of land. They waited for the floods eagerly. Why? Because the waters of the Nile picked up rich, black soil[1364] in a 4,000 mile trip downstream from the African Great Lakes region where cichlid fish live and dropped those nutrient-rich muds on the river valley of Egypt's land, land that would remain exposed to the sun long enough to let you plant and harvest a crop.[1365] Despite being fringed by barren sands.

During the four months of floods, the Egyptians used their spare time to build pyramids. Then, when the floods receded, they went back to farming.[1366] With this new invention of a lifestyle, the flooding of the Nile was transformed from Egypt's destroyer to what ancient Greek historian Herodotus called Egypt's gift.[1367]

The moral of the story? Every disaster is an energy source waiting to be tamed. That includes the immense energy sources called tornadoes and hurricanes.

There's a tall tale from the 19th century that captures the spirit of the thing.[1368] Back in the 1800s in Texas, it is said that three cowboys were sitting on the rail fence around a horse corral. Sitting and bragging. Competing like male elephant seals. Over who could best take advantage of a new niche opened by the spread of cattle raising to the American West. One man on the fence described the orneriest horse ever seen by man and how he had tamed it. The second described a horse even meaner, and how he had slapped a saddle on its back and brought it to heel. The third cowboy, Pecos Bill, pointed at a tornado heading toward the ranch and said, "Watch this." Bill took his saddle, walked over to the tornado, slapped the saddle on its back, and jumped on.

The tornado bucked. The tornado kicked. The tornado tried to scrape Pecos Bill off by whopping him through the underbranches of trees. But no matter what the tornado did, Pecos Bill stayed on its back. Finally, the tornado was just plumb wore out. Bill stuck a bit between the whirlwind's teeth and with his reins told it when to go left and when to go right. Then the cowboy-to-top-all-cowboys rode the tornado over to the rail fence where his two companions were sitting slack-jawed and doffed his hat.

Pecos Bill had come out on top by doing the impossible.

We've done quite a few Pecos Bills. We've tamed fire with hearths, stoves, furnaces, combustion engines, and rockets. We've harnessed floods with watermills and hydroelectric dams. We've tamed gusts of wind with windmills, sails, and modern wind farms. That's three disasters down. More to go.

Now it's time to tame the whirlwind. It's time to harness the force that makes dust devils, tornadoes, and hurricanes. That force? Temperature and pressure differentials in the atmosphere. Opposites joined at the hip. Polymath Steve Nixon, who co-conceived an animation on space solar power satellites that won prize money in a contest run by the National Space Society, has a proposal to do just that.

Nixon wants to build what he's called megachimneys. Giant upside-down funnels rooted to the ground that stretch to the skies six miles high. Wind chimneys that set up a massive current of air. Air propelled by the huge temperature difference between the atmosphere at ground level and the air high in the troposphere where the thermometer can drop by 102 degrees Fahrenheit.[1369] Megachimneys that would also harvest the difference between air pressure at ground level and the 13% thinner air six miles above.

Megachimneys whose internal air currents could reach 200 miles per hour in their sprint to the heavens.

Yes, Nixon's megachimneys would tap the forces that generate spirals in the sky. They would tame the disasters of dancing wind. They would do what in the Bible[1370] is only a figure of speech, they would reap the whirlwind. To generate energy. To extract pure water. And to cool the atmosphere.

Right now, these megachimneys are hidden treasures. They are ideas Nixon is promoting to small groups online. But his chimneys may be harbingers of what's to come. They may hint at technologies that could pull off a real Pecos Bill. Using climate stabilization technologies to generate electricity, water, and cooling.

Remember, every catastrophe is an energy source waiting to be tamed. New niches are not found, they are invented.

***

One more note. If we really want the climate of the earth to be the way it was before the Industrial Revolution, then let's make it that way. But let's admit something. That is not returning the planet to its natural state. It is locking this globe into an anthropogenic state. A human-imposed state. A totally unnatural state.

But guess what? Nature loves the unnatural. Nature loves those who oppose her most.

CHAPTER 50

ARE HUMANS BAD FOR THE EARTH?

"Today the number of species is the largest in the history of life."
Pablo A. Tedesco,[1371] Institut de recherche pour le développement,
France,[1372] in the journal *Conservation Biology*

Are humans earth's curse, as many climate thinkers claim? Is humankind a cancer destroying the planet? Let's put that differently. Have we humans decreased the biosphere of this planet? Or have we added to it?

Have we upgraded or degraded the earth?

First off, we have performed a function for the biosphere that nature is normally forced to rely on bacteria to achieve. As you know, at this minute bacterial colonies[1373] of chemolithoautotrophs—stone eaters—twelve miles[1374] below your feet and mine are turning raw rock[1375] into food and fuel for life. That's called primary production.[1376] You may vaguely remember primary production from an early chapter on the starting days of life, when bacteria were eating volcanic glass and rock at the bottom of the seas.

Primary production is not doing things the easy way. It's not just feasting on your fellow living things. It's not just supping on microbes, plants, and animals. It's the hard part of life's ultimate mandate: kidnap, seduce, and recruit as many truly dead atoms as possible into the enterprise of life. Increase the total atomic weight of biomass. Wherever we've dug into the earth's crust, for example, bacteria are turning raw rock and peculiar chemicals into the stuff of life. That's doing life's job the hard way. And not complaining. That's primary production.

It turns out that, like bacteria, humans have made amazing contributions to primary production.

However there has been a theft of the term "primary production." A clique of environmental scientists—including *The Population Bomb*'s Paul Ehrlich[1377]—has invented a whole new way of twisting the phrase "primary production." They've ignored the contribution of the bacteria that yank dead atoms of sulphur, iron, nitrates,[1378] and methane into the life process. In the view of these environmental scientists, primary production is restricted to one thing and one thing only: photosynthesis. The process by which plants grab photons of light and jam them into the engines of life. And these researchers have invented a further term: the "Human Appropriation of Net Primary Production," HANPP.[1379] A term that accuses modern industrial humanity of thievery.

For example, Fridolin Krausmann of the Institute of Social Ecology in Vienna and seven colleagues contend that the human appropriation of net primary production on Earth has doubled from 1910 to 2005.[1380] With the word "appropriation," they mean that we humans have stolen the net primary productivity of the planet. They imply that we've yanked that productivity away from nature like a purse snatcher grabbing a Vuitton bag, running away with it, finding a safe alley, then pawing through the contents and putting the money in his pocket. In other words, these researchers regard the relationship between humans and nature as a robbery, a grand theft, a zero-sum game. But have we stolen from nature? Or have we added to the planet's greenery?

First, a question. Is a blue whale accused of stealing primary production when it gulps sixteen tons a day of plankton, krill, and entire schools of fish?[1381] Is a rabbit accused of stealing primary production when it eats the leaves of a young lettuce, a lettuce valiantly attempting to grow?[1382] Is a pride of lions accused of stealing primary production when it downs a gazelle? No. "Appropriation" is an accusation made solely against humans.

But the accusation is wrong. Unlike whales, rabbits and lions, we give back what we get. And more. We sow the seeds of the plants we eat. We go to great lengths to give those seeds everything they need to thrive. We reengineer those seeds so the number of their species that can grow on an acre of land increases. Dramatically. For example, American agricultural productivity per acre roughly tripled from 1948 to 2019.[1383] Yes, we nearly tripled the productivity of our favored species. And we did it smack dab in the middle of the period of time in which we are told that we stole more and more of plants' primary productivity.

Meanwhile we feed the farm animals we, unfortunately, also feast on. We protect these plants and animals from pathogens and predators. We help wheat, barley, corn, pigs, cows, and chickens fulfill the most basic imperative built into their biology: to have sex, to spread their genes, and to become as numerous as "the stars of heaven."[1384] Writes Vox, the world currently has "19.6 billion chickens, 1.4 billion cattle, and 980 million pigs."[1385] Yes, these animals sometimes live in conditions that need massive improvement. Massive. And a small army of animal rights activists like Mary Temple Grandin, England's Farm Animal Welfare Council, America's Humane Farm Animal Care, and England's Compassion in World Farming are working hard to improve those conditions. What's more, there's a whole new field of research: farm animal cognition.[1386] To get a handle on how farm animals feel. But 78[1387] billion animals' number would be far smaller if it weren't for human beings.

In fact, I suspect that we've enhanced the planet's net primary productivity—both its photosynthesis and its ability to snab dead atoms into the life process. We've done it by giving the biosphere new tools. How?

Eleven thousand years ago, plants lured us into the Agricultural Revolution. Much as they had recruited insects to carry their pollen. Yes, plants seduced us into the agricultural contract. And it turns out that plants made a wise move. The Agricultural Revolution rocketed the rate at which between 250 and 35,000[1388] species of plants could kidnap, seduce, and recruit raw materials into the grand enterprise of life.[1389] Plants hooked us on the fact that by serving them in the most effective way possible, we could raise the most food per acre for ourselves. So we dug holes for their seeds,[1390] broke up the ground to make life easier for their roots,[1391] put every sort of nutrient we could think of into the soil, from ash and bone[1392] to our feces and the feces of our farm animals,[1393] and pampered the plants with vital minerals and water. What's more, we invented a panoply of plant-servicing technologies from digging sticks, stone hoes, irrigation canals, plows, tractors, dams, and crop rotation, to chemical fertilizers.

Did the plants that managed to work out a deal with us benefit from recruiting us? You bet. Wild Emmer grass, the ancestor of wheat,[1394] for example, was limited to the Middle East.[1395] Its descendant, domesticated wheat, wheat tended by humans, on the other hand, is cultivated in China, India, Russia, Ukraine, and in the United States. That's an expansion to 12.7 times more territory. A massive increase. Thanks to wheat's recruitment of human beings as their servants: their transporters, their breeders, and their caretakers. But that's not all.

Cows, pigs, and chickens followed the example of wheat, trading the edibility of their flesh for an enhanced crack at manic mass production and a privileged place on the evolutionary ladder. With the tailor-made luxuries provided by the animals' employees, us humans. Remember, we nurture 78 billion domesticated animals at a time.[1396] And we cultivate over 250 species of plants.[1397]

But have we enslaved these living things? Or have they enslaved us?

In roughly 2,000 BC, [1398] 12,500 feet above sea level, in the Andes mountains of Peru and Bolivia, higher than the peak of Mount Moran in Wyoming, the pre-Tiwanaku people pioneered a new form of agriculture. Their environment, living around South

America's second largest lake, the Andean Lake Titicaca, was harsh on plants. It was a pampas, a low, flat, treeless grassland prone to periodic flooding. If the floods didn't get you, the frost at night did. The pampas was a plant killer.

Then there were the ravages of the wet and dry seasons. The floods came from the melting glaciers in the nearby mountain sierras, from the 27 rivers[1399] that flowed into Lake Titicaca, and from the single river that flowed out of the lake. Then there was the downpour of the rainy season—roughly two feet of water from the clouds above.[1400] Balanced by 3,000 hours of annual sunshine.[1401] To harness the disasters of the floods and the rain, the pre-Tiwanaku people dug parallel canals ten to thirty feet apart. They heaped the soil they'd excavated from these trenches on to the strips of land between the canals. And they formed raised beds on which plants could grow. Raised beds whose dark soil would absorb the rays of the sun during the day and continue to warm the plants during the murderous frost of night. Raised beds that one of the leading researchers on the topic, University of Pennsylvania professor of Anthropology and curator of the American Section of the Penn Museum, Clark Erickson, says improved drainage and maximized soil fertility.

In other words, the pre-Tiwanaku used their ingenuity to create a luxury resort for their favorite plants. Those plants were "cañihua: an Andean grain crop related to our weed lambsquarters" and "high in protein."[1402] And potatoes, a crop that ancient Native Americans in the Lake Titicaca region had created by selective breeding over 6,000 years earlier.[1403] The pre-Tiwanaku's paradise for potatoes and cañihua also turned out to be a wonderland for fish, domesticated alpaca, llamas, guinea pigs, and a wide variety of aquatic birds. Did the pre-Tiwanaku, with their raised-field agriculture, diminish nature's productivity? Or did they add to it?

New research shows that this disruption of the natural environment, raised-field agriculture, underlay civilizations as far away as the Amazon, where humans created landscapes we mistakenly believe were untouched, landscapes we worship as "the lungs of the planet."[1404] Human-made landscapes.[1405] Did the ancient South Americans, seeking new lands for their masters, their plants, destroy nature? Or did they help create it? Did humans lay the base for the rainforests that we are urged to preserve?[1406] Says the University of Oregon's Thomas Lee,[1407] these ancients "managed the landscape for natural resources while raising local biodiversity." And while creating the "nature" we bow down to today.

Meanwhile, using humans, there is a good chance that domesticated wheat, barley, rice, corn, broccoli, peas, potatoes, tomatoes, bok choy, choy sum, and carrots have outproduced every one of their wild cousins on earth. Why? We have custom-tailored their environments to precisely suit their tastes. And we've done that with 4.62 billion acres of cropland, land that we've turned into a plant paradise.[1408]

Back to the question, have we enslaved plants or have they enslaved us? Or is it a two-way street, a win-win, a symbiosis? Remember, 11,000 years ago, when plants seduced us the way that they had previously seduced insects, the deal they offered was simple. Remake wilderness into a utopia for us short-in-stature plant species. Free us from the light monopoly of the trees. Expand the land available to us by inventing irrigation systems and bringing water to intolerably dry spots. Enrich the soil. Put nutrients like human and animal waste into it. Shield us from frost. Keep out our plant competitors, weeds. And protect us from the animals who love to eat us—like locusts, pigs, rabbits, and deer.[1409] In exchange, we will feed you.

How did the deal work out for the plants? Human-managed areas today are 4.5 times more productive than wild areas.[1410] They are 4.5 times more heavenly for plants. Part of this productivity is because we have taken over the richest lands. But only part. In just the 53 years from 1961 to 2014, we humans increased "global cereal production"—the production of grains like wheat, oats, and corn—280 percent.[1411] But guess what? We nearly tripled our grain production using only 30% more land.[1412] And from 1961 to 2014, we doubled the number of humans we could feed with an acre of soil.[1413] That is not just because we have appropriated the richest lands. It is because we have made the richest lands richer.

The bottom line for plants has been huge. Of all the families of life on earth, plants dominate. They make up 80% of earth's biomass.[1414] By comparison, the total weight of all us humans on the planet is ridiculously small... roughly that of termites.[1415] Plants outweigh us by 7,500 to 1.[1416]

What's more, roughly ten thousand years ago, roughly forty species[1417] of animals chose to follow the example of the plants and make a deal with humankind. The result? Since pre-human days, "the total mass of mammals" has "increased approximately fourfold." That's according to a definitive 2018 study of "The

Biomass Distribution on Earth" in the Proceedings of the National Academy of Sciences.[1418]

Bacteria, too, have worked out deals with humans. There are 39 trillion of them[1419] in your gut right now. You are their transportation and gourmet-food provider. When they are hungry for a treat, you go to the grocery store and pick out the chocolate eclairs[1420] they want to eat. You can't actually digest the chocolate of those eclairs. But chocolate is gourmet fare for your bacteria. You are also your bacteria's crushing, mulching, and chewing machine. In exchange for your services, your bacteria digest[1421] your chocolate eclairs for you and manufacture vitamins like thiamine, folate, biotin, riboflavin, and pantothenic acid. [1422] Not to mention half of your daily vitamin K.

Has the bacterial deal with men and women paid off the way that the plant-human contract has? At this minute, there are 900 thousand billion billion bacteria living in human beings. One hundred sextillion. That's a nine with twenty-three zeroes after it. Bacteria are the number-two form of creatures on earth, making up 13% of the planet's biomass.[1423] So between gut bacteria, domesticated plants, and domesticated animals, a hefty percentage of earth's biomass benefits from the services of us human beings. A substantial percentage of earth's biomass has turned us into a niche and uses us to thrive.

But that's not all. Remember the sort of primary production that chemolithoautotrophs and sulfur eaters specialized in three billion years ago? Kidnapping, seducing, and recruiting truly dead atoms into the grand enterprise of life? Boosting the number of atoms in the earth's biomass? In 1913,[1424] we finally joined that exclusive club. The club of primary producers. We made the ultimate move to kidnap, seduce, and recruit dead atoms into the megaproject of life.

How?

***

Life desperately needs nitrogen to make amino acids, proteins, nucleotides, DNA, and RNA.[1425] And there is a huge amount of nitrogen in a convenient location, in the air we breathe, in the air over our heads, in the earth's atmosphere. In fact, there are 4,000 trillion tons of nitrogen above us.[1426] So why in past centuries were plants desperate for nitrogen?

There's a problem. The nitrogen in the air is locked in molecules, molecules in which two nitrogen atoms hug each other tight. Two nitrogen atoms glued together by nearly unbreakable triple bonds.[1427] But plants can only use nitrogen atoms that are single. And most plants have no machinery to break those nitrogen couples apart and make them usable. Most plants have no machinery to turn nitrogen pairs into single, solitary nitrogen atoms. Up until now, many plants have relied on nodules—blister and tumor-like growths on their roots—to produce nitrogen for them. What was in those growths? Bacterial partners, rhizobia, capable of primary production, capable of "fixing" the nitrogen couples—capable of breaking nitrogen couples' bonds and turning nitrogen molecules into single nitrogen atoms. But that has not been enough.

So we humans have pitched in, working our tails off to find single-atom nitrogen, usually in the form of the stuff we call fertilizer. In other words, our plant masters long ago enlisted us to help solve their nitrogen problem. In fact, in the 1800s, plants enlisted two of the top brains on the planet to solve their nitrogen conundrum. One of those geniuses was among the greatest scientific intellects of the 19th century. He was the younger brother of Prussia's Wilhelm von Humboldt, the younger brother of the man who was about to invent the modern university. His name was Alexander von Humboldt. And, as you know, he had made himself an expert in all the sciences of his time. Yes, all of them. What's more, he was sewing the individual patches of scientific knowledge into a quilt, into a single grand vision, an omnological vision. In fact, as you know, his biographer, Andrea Wulf, credits Alexander von Humboldt with "*The Invention of Nature.*" Wulf believes that van Humboldt gave us the modern view of a vastly interconnected web of life. Alexander von Humboldt was also the man who did more than any other to put the idea of a scientific expedition on the map. And he may have followed his curiosities more relentlessly than any other scientific thinker in history.

So relentlessly that, as you know, from 1799 to 1804 Alexander von Humboldt sailed off to explore South America, 32 years before Charles Darwin's Voyage of the Beagle. Von Humboldt was intent on doing science. So intent that he made his servants carry the 75 instruments[1428] in wooden cases lined with velvet that you've seen before. These were the finest cutting-edge instruments from Europe's most advanced instrument-makers. "Various telescopes, a sextant, a quadrant, a dipping needle, compasses, a pendulum, barometers, several thermometers, two electrometers, a microscope, a rain gauge and a cyanometer—to measure the blueness of the sky."[1429] As you know, in June 1802 von Humboldt climbed what was considered to be the highest mountain on earth, the 20,700-foot Chimborazo in the Ecuadorian Andes, thus showing that humans could go to heights some believed impossible. Stopping over and over again to make observations and to take samples. While he broke nature's most basic law—gravity. The only disappointment was that he didn't quite make it to the top.

When Alexander Von Humboldt returned to Europe on August 1, 1804,[1430] he had not sated his curiosity. Twenty-five years later, in 1829 he would travel to the tsar's court in Russia, then to the Russian border with China.[1431]

These adventures made Alexander von Humboldt world-famous. So famous that, as you know, to this day 1,600 locations around the world are named after him.[1432] And 400 species of plants and animals. Not to mention two asteroids, "4877 Humboldt"[1433] and 54 Alexandra.[1434]

Alexander Von Humboldt made one discovery after another in his five-year South American adventure. For example, he brought the attention of Europe to the Humboldt Current, "the cold, plankton and fish-rich... current, which feeds large populations of seabirds."[1435] Then there's the Humboldt Penguin, which lives, you guessed it, on the Humboldt Current.[1436] Not to mention the Humboldt hog-nosed South American skunk. The Humboldt big-eared South American brown bat.[1437] Humboldt's North American flying squirrel. Humboldt's Amazonian squirrel monkey. The Humboldt orchid.[1438] The Humboldt lily.[1439] And the "fearsome Humboldt squid,"[1440] another habitué of, yes, the Humboldt current.[1441] Repeats Germany's Humboldt Society, "almost 300 plants and more than 100 animals [were] named after Humboldt." And Humboldt's fame was spread in books about him, over 5,000 of them.[1442]

But Humboldt's most important discovery is one so seemingly mundane that he doesn't even mention it in his seven-volume[1443] account[1444] of his adventures in the Western Hemisphere. Specks of land off the coast of South America had been visited by birds for thousands of years—cormorants, boobies, and pelicans.[1445] Not to mention penguins. The results were islands of guano, small mountains of bird, penguin, and bat feces, bird shit, bulging above the sea. A shit that just so happened to be rich, rich, rich in single-atom nitrogen. The very substance the plant world cannot live without.

When he came home to Prussia, Humboldt brought back a sample of compacted bird, bat, and penguin feces.[1446] The feces were called guano. The Inca civilization of Peru[1447] had known for 1,600 years that guano was a powerful fertilizer.[1448] According to Garcilaso de la Vega in 1609, the Incas used "no other manure but the dung of sea birds,"[1449] The Spanish conquistadors had absorbed that knowledge and had used guano on their South American lands, but had never bothered to tell their European overlords about it.

When he was in Lima, Peru,[1450] Humboldt got his hands on a sample, "The first small lump of guano...from the Chincha Islands."[1451] When von Humboldt sailed home to Europe, he sent samples to two French chemistry superstars[1452] and to the English creator of the field of electrochemistry and future president of the Royal Society, Sir Humphry Davy. Which means that plants were harnessing some of the greatest minds of the nineteenth century to service their needs.

Next "the Scottish military engineer and 'experimental agriculturist' Alexander Beatson," the Lieutenant General Governor of the distant island of St. Helena, tested guano on the lawn of his St. Helena governor's mansion. And soon that patch of land was "covered with the most exuberant grass that can be imagined."[1453] In 1809, Beatson performed experiments in the mansion's gardens, tests on potatoes, pitting guano against more traditional European fertilizers—horse manure, pig manure, and human excreta. Guano outperformed the competition. Then Beatson wrote a book on the topic, and the book made a dent.[1454]

Beatson had a local supply of guano near his island, St. Helena, 1,210 miles from the European mainland in the southern Atlantic Ocean. But for everyone else, guano was expensive. Importing guano to Europe from Peru by sailing ship took at least three months. The lag time between placing your order and receiving your guano was eight months.[1455] So European landowners and farmers weren't interested.

But remember, plants—the dominant kingdom[1456] of life on this planet—had motivated an international team of scientific superstars to explore and to carry out studies on their behalf. They had harnessed a team of geniuses to fetch them nitrogen. Not to mention phosphorous, calcium, and potassium. The vital ingredients in guano. With one-atom nitrogen on top of the plants' shopping list.

Yet the appetite of European plants for guano would go unfilled for 37 more years. A transportation revolution would change that. Von Humboldt had traveled to South America on a sailing ship. Just a few years after Von Humboldt's travels, in 1807, Robert Fulton had done something that the celebrity author Sir Walter Scott had said was impossible. He had put a steam engine on a mode of transportation, a ship. In the late 1830s, steamships would produce a massive acceleration of global trade.[1457] They would produce what "one of the world's foremost thinkers on development history,"[1458] Vaclav Smil, calls an utter "transformation of world food production."[1459]

In 1839, 35 years after Von Humboldt's South American adventure ended, a Liverpool merchant, William Myers, used steam-powered cargo ships to transport 30 bags of guano to English shores. Soon massive amounts of bird and bat guano—bird and bat feces—from Peru, Bolivia, and Chile were imported from South America to Europe so that Europe's farmers could double or triple[1460] the productivity of their cultivated lands. The United Kingdom alone would import over two million tons.[1461] All to serve the hungers of plants. But in the late 1800s, the motherlodes of bird and bat guano in Peru would begin to run out. The situation would be so serious that The Farmer's Magazine of 1857 would call it "The Guano Crisis."[1462] The cravings of plants were in danger of going unmet.

Then in the dawning years of the 1900s, two German scientists, Fritz Haber and Carl Bosch,[1463] invented an industrial process to break apart the bonds of resisting nitrogen pairs from the air. In other words, Haber and Bosch invented a factory process to kidnap molecules of two nitrogen atoms from the atmosphere,[1464] to tear these atoms away from each other, and to make the individual nitrogen atoms available to the ravenous enterprise of life.

Who had made us carry out this act? Plants. The plants who had seduced us into agriculture. Now those plants motivated us to invent machines that could feed them vital atoms, atoms that had been all too scarce since the origin of land plants 470 million years ago—the atoms of nitrogen in the atmosphere. This was primary production at its best. Pulled off by the "cancer on the planet," humans.

In the process, plants had prodded us into increasing the earth's biomass. Dramatically. As Thomas Hager writes in his book on Haber and Bosch, *The Alchemy of Air,*

> Within a decade [of the Haber-Bosch process being commercialized], the artificial fertilizers it enabled had given European agriculture its greatest boost since the introduction of crop rotation and fallowing in the medieval era, and had provided the necessary impetus to make the great agricultural boom of the late 19th and early 20th centuries possible.[1465]

But when it came to the human contribution to plants, there was more. In 1967, when butterfly specialist Paul Ehrlich was on the radio lecturing the Commonwealth Club of California that we were about to run out of food and hundreds of millions would die, something called the Green Revolution had already been underway for twenty three years.[1466] The Green Revolution had already developed new high-yield plants like dwarf wheat and dwarf rice,[1467] plants that reduced their investment in tall stems and put what they saved into producing two or three times more grain. Plants that didn't topple over in the wind when their grains got too heavy. In addition, the Green Revolution had built dams, improved irrigation, called for pesticides and for more intensive use of fertilizers, and had taken advantage of tractors and harvesting machines.[1468] All to build a paradise for plants.

From the time Paul Ehrlich delivered his radio lecture to the year 2000, the Green Revolution would triple the world's agricultural productivity. Says the founding director of the Tata-Cornell Institute, Prabhu L. Pingali, between 1962 and 2012, "Although populations had more than doubled, the production of cereal crops tripled…, with," what you've already seen, "only a 30% increase in land area cultivated."[1469] In other words, the amount of biomass we could nurture on an acre of land grew explosively. As agricultural researcher William L. Cavert put it, "In the 1800s each farmer grew enough food each year to feed three to five people. By 1995, each farmer was feeding 128 people per year."[1470] What's more, he added, "In the 1800s, 90 percent of the population lived on farms; today it is around one percent." Instead of a population bomb killing millions, the Green Revolution had pulled one billion people out of poverty.[1471] By motivating us humans to feed the appetites of plants.

No, we did not steal primary productivity. We goosed it to radical new heights. We served our masters, plants. We gave them super-powers.

***

At the same time, we invented the environmental movement and a demand to plant more trees. The development of nitrogen fertilizers, mechanized farm equipment, and a host of other human improvements allowed us to feed more than six times as many people as we had in 1800.[1472] Despite that radical boost, between 2000 and 2020, says the journal *Science*, "the amount of forest increased by 1.3 million square kilometers."[1473] That's an increase of half a million square miles. An area that, as you know, is twice the size of France. At least eleven nations dramatically boosted the total square-footage of their forests.[1474] Those countries included Ukraine, Romania, Poland, Lithuania, Bulgaria, Spain, Algeria, Belarus, and the big ones, China, India, and Russia. Which means that in Europe and Asia, trees were taking advantage of tapped out farmland and turning it into forest. Using as their servants us human beings.

Meanwhile, just one human program planted over 13.96 billion[1475] trees in 193 countries between 2006 and 2020. Yes, 13.96 billion. That faithful footman of the tree was the United Nations Environment Programme's Tree Billion Campaign.[1476] A program that upped its ambitions in 2018 and became the Trillion Tree Campaign.[1477] Then there was The Bonn Challenge, launched in 2011, which "aims to restore 350 million hectares of degraded and deforested land worldwide by 2030."[1478] And The African Forest Landscape Restoration Initiative (AFR100), launched in 2015, which "aims to restore 100 million hectares of degraded and deforested land in Africa by 2030."[1479]

The result? When it came to gardening and greening the planet, plants have turned us into their ultimate servants.

***

Yes, we have changed the biosystems of the earth. But have those changes been for the better or the worse?

Have we humans increased or decreased the number of atoms and molecules kidnapped, seduced, and recruited into the grand scheme of life?

Have we humans increased or decreased the GAL, the planet's Gross Amount of Life? Even more important, have we increased or decreased the GAS, the Gross Amount of Sentience, the gross amount of awareness, cognition, and information, the gross amount of feeling, emotion, consciousness, and knowledge on this earth?

Have we added to or subtracted from what Carl Sagan saw as one of the greatest mandates in this universe, making the cosmos aware of herself?

In other words, have we lived out nature's mandate in ways that climate apocalyptics do not see? Have we "appropriated" primary production? Have we stolen from it, or have we added to it? Have we increased or decreased the number of tricks and tools that nature uses to make life thrive? Have we increased or decreased the ways in which nature can innovate? Have we increased or decreased her processing power? Have we served or sinned against the First Law of Flamboyance? Have we increased or decreased nature's gross capacity for supersized surprise?

CHAPTER 51

THE MYSTERY OF WISPS:
SENTIENCE

Environmental scientists believe that since the dawn of humans, the amount of biomass on this planet has plummeted by between 4.4 percent[1480] and 50 percent.[1481] But their measure of biomass is the stuff left in microorganisms, plants, or animals that were once alive but are now dead and dried. To weigh biomass, says the book *Techniques in Bioproductivity and Photosynthesis*, first you harvest the thing whose biomass you want to measure. In other words, you kill it. Then you dry it "to constant weight at 80 degrees C in a forced draught oven" and you allow it "to cool in a desiccator."[1482] That was back in 1986. Today to measure biomass you use advanced sensors and intense mathematics.[1483] But it all comes down to the same thing: dry weight.[1484] Is dead, dry, formerly-living mass really the measure of what nature or humans have achieved? Does life—actual living—contribute anything? If so, what? And have we humans contributed, or have we merely snatched, grabbed, and destroyed?

As you know, for its first thirteen billion years, this cosmos was massively materialistic. Galaxies, black holes, stars, planets, and moons kidnapped, seduced, and recruited mobs of atoms, blobs of stuff, gravity balls, and used them to invent larger amazements—like the planet-sized flares on the surface of stars or the spiral arms of galaxies.

But on one roiling catastrophe of a planet roughly 3.85 billion years ago[1485] something new entered the business of kidnap, seduce, and recruit. It was life. And life grew from less than an ounce to 550 gigatons of carbon in a long, slow process, a process that would include only one three-and-a-half-billionth of this planet's mass by modern times. Gravity balls had been in the business of kidnapping their smaller competitors and swallowing them whole. But life added something new to this matter-obsessed process. Just the way a massive herd of gravity balls produced something bigger than the sum of its parts—the intricate spirals of a galaxy or the five thousand light-year[1486] beacon of a black hole's flare—this new intruder on the evolutionary stage, life, made something that could not be measured in pounds, kilograms, or tons. What was it?

Sentience. And what in the world is sentience? We don't entirely know. Our ecological sciences have barely acknowledged its presence. But it was there from the beginning of the first communities of cells and DNA nearly four billion years ago.[1487] Remember the first form of life on earth was bacteria. Bacteria are your great great foremothers. And your bacterial ancestors were sentient.[1488] So were mine. How do we know?

In 1998, I started a group called The International Paleopsychology Project, an online gathering of scientists from all over the world, scientists whose work covered a vast variety of topics, from evolutionary biology, microbiology, anthropology, neuroscience, and psychology to physics. Our mandate was to "trace the evolution" of sentience from the Big Bang to you reading this page. My most exciting co-thinker was Eshel Ben-Jacob,[1489] head of the condensed matter physics department at the University of Tel Aviv. Ben-Jacob was also head of the Israel Physical Society,[1490] Israel's society of physicists. In other words, he was Israel's leading physicist. But he had done something strange.

In 1998, Ben-Jacob came to visit me in my Park Slope, Brooklyn, brownstone, despite the fact that I'd been bedridden with ME/CFS for ten years. Sitting with Eshel in a chair at the foot of my bed, his wife, Michal, asked if I had a copy of the latest *Scientific American*. Yes, I said, and took it out from under my right thigh where it was turned to the table of contents so I could circle the articles I wanted to read. Michal asked me to look at the cover. So I turned the pages back and closed the magazine. And there, taking up the whole front cover, was a gorgeous, full-color picture of one of Eshel's bacterial colonies.[1491] Why? Why would Israel's foremost physicist be growing colonies of bacteria?

When Eshel Ben-Jacob first saw the strange patterns that bacterial colonies make in a petri dish, he realized that they were fractal patterns. Patterns described by mathematics. Patterns that you can see when you crack a rock in half and polish the cleaved surface. The sort of patterns so startling and beautiful that these lapidary rocks are sold for up to $170.

Ben-Jacob was an omnologist. He refused to be caged by disciplinary boundaries. He reasoned that if the same sort of pattern that showed up in rocks showed up in a petri dish housing a bacterial colony, it was possible that the bacterial pattern had nothing to do with life's eccentricities. It was possible that the bacterial pattern was created by the same sort of forces that made these patterns in rocks. That's why Ben-Jacob, a consummate physicist, started something physicists don't do—growing bacterial colonies. And he came to a startling conclusion. The patterns he saw in his bacterial colonies were not the product of abiotic forces—the forces that govern lifeless lumps. They were the product of life. And of something more.

Wrote Ben-Jacob:

> Bacteria, the first and most fundamental of all organisms, lead rich social lives in complex, hierarchical communities. Collectively, they gather information from the environment, learn from past experience, and make decisions.

In other words, in their own primitive way, bacteria are sentient.

How do these microorganisms pull this off? Writes Ben-Jacob, the wee beasties have a rich chemical language.[1492] And they are in constant communication, swapping gossip. What's more,

> Bacteria do not store genetically all the information required to respond efficiently to all possible environmental conditions. Instead, to solve new encountered problems (challenges) posed by the environment, they first assess the problem via collective sensing, then recall stored information of past experience, and finally execute distributed information processing in networks ranging from a million to a billion bacteria in the colony, thus turning the colony into a super-brain.[1493]

An ecological researcher would measure the biomass of a petri dish of bacteria by killing it, drying it out, and weighing it. A very reductionist thing to do. That measure of dead stuff would miss all that life is. And all that life adds to the cosmos. Why? Because two of the most astonishing contents of that petri dish weigh nothing and have no constant set of atoms as their constituents. Those two amazements are life. And sentience.

Sentience? In bacteria? Surely, Mr. Bloom, you must be kidding. But I am not kidding.

Ben-Jacob says that bacteria extract information from what's around them, consult their memory of past experience, then make decisions. They act as a collective intelligence, a group learning machine, "a super-brain." They perceive, they store memories, they solve problems, and, like Hamlet, they make choices. They can literally solve puzzles.[1494] Says Ben-Jacob, "they solve optimization problems that are beyond what human beings can solve."[1495]

Ben-Jacob created a computer simulation of his social bacterial colonies and showed in theory that the micro-beasts could collaborate in a group mind to solve a maze and to find their way through the deceptive passages of that maze to food.[1496]

Yes, there is more than mere dry mass in your petri dish. Bacterial colonies have a collective intelligence.

And they store knowledge in two forms. In their super-brain's group memory.[1497] And in the memory storage of their genome. A molecular ring that contains all the most basic tricks that bacteria have learned since they first appeared in pools or at the bottom of the seas over 3.5 billion years ago.[1498]

So there is something in your petri dish of bacteria that you can't see. A something that, to repeat, weighs nothing. It's perception, memory, and the ability to choose between options. And sometimes to create options you never imagined existed. It's a primitive precursor of mind. It's sentience.

As life spread from a teaspoonful to 550 gigatons of dry mass, how much did that sentience grow? We do not know. It hasn't been measured.

Then how much more did sentience grow when humans came along? How much did the invention of stone tool making, the invention of language, the invention of body-and-face paint,[1499] the invention of painting on cave walls, the invention of the first list of prime numbers on a baboon bone,[1500] not to mention the invention of writing, and the invention of computers, how much did all that add to the planet? How much did 104 thousand museums,[1501] roughly 15 billion works of art,[1502] the creation of over 3,000 symphonies,[1503] 80 million songs,[1504] 100 million poems, far more than over 30,000 cultures,[1505] and 31,000 languages contribute?[1506] How much did over 30 million new inventions[1507] add to the tool kit of nature?

How much did these add not just to the biosphere, but to what legendary Catholic Jesuit priest, paleontologist, and philosopher Pierre Teilhard de Chardin, as you know, called the noosphere.[1508] The planet's sphere of knowledge. Not to mention the planet's sphere of passion, pain, joy, and wisdom?

We have one measure of sentience over the last 80 years. Alas, it's a reductionist measure. It's the amount of information in computer storage. And that figure goes from 2,560 bits of memory—two pages worth—in the British Manchester Mark I computer of the 1940s, to an estimated 16 zettabytes in 2025.[1509] The International Data Corporation is optimistic about the future of that information. It predicted in 2018 that global information storage capacity would hit 175 zettabytes by 2025.[1510] A zettabyte is one billion trillion bytes, 1,000,000,000,000,000,000,000 bytes. Just one zettabyte is the contents of two billion laptops. Or 4,000,000,000,000,000,000 pages. A pile of paper 0.04 light years thick.[1511]

But that measure of information in the biosphere is only a sliver of the sentience on this planet. It leaves out the ability of plants to sense a danger and to signal others in their community to go into defensive mode. It leaves out the pains and pleasures of cows, pigs, and chickens. And it leaves out the aspirations, insecurities, passions, and knowledge of people like you and me. It leaves out the mass moods and mass perceptions of groups of humans that show up in simple measures like a stock market chart.

And it leaves out another process, dematerialization. The increasing ability to create a massive amount of food for sentience, a massive store of knowledge and entertainment, while decreasing the use of material stuff. In 400 BC, all the written material in the world used a total of less than one ton of clay, papyrus, and parchment. In 2023, there were roughly 158 million books weighing a total of over 128 million pounds. That's the weight of over 28,500 elephant seals. Then came dematerialization.

In 1945, the first American electronic digital computer, ENIAC, weighed 27 tons. It had 320 bits of memory.[1512] Today, a computer with eight trillion bits of memory fits in your pocket. What's more, it gives you access to most of the world's written knowledge. Its weight? Not 27 tons. Six ounces. It's your smartphone. That's de-materialization. But the real power of that computer in your pocket can't be measured in pounds or tons. Your phone is a portal to a vast realm, a realm that is rapidly expanding, cyberspace. A place that is not made primarily of unchanging material particles. A place made by and for sentience.

That is massive growth. Growth that would not appear in dry weight. And growth that is not taken into account when eco-apocalyptics tell us we are destroying the planet.

***

Are we really savaging this earth? No, I'm sorry, we are growing the planet. And one of our ways of growing this earth is to give birth to fields like ecology that spot our mistakes and force us to correct them. Fields that force us to protect wild lands and to defend endangered species. Saving the planet is a radically new concept on this earth. A radically new concept in this universe. A radically new invention. And a radically new way to steer the forward thrust of the cosmos.

What's more, the concept of saving the earth is one of the unique contributions of guess what else? Western Civilization. Capitalist civilization. Not to mention a contribution of human beings.

Remember, the total system of life, a system that includes you and me, has gone from a tenth of an ounce to more than half a trillion tons in roughly 3.5 billion years.[1513] And life has a long way to go. Life and growth are synonymous.

But there is something else just as important. Something studiously ignored. Yes, we have increased the noosphere, the sphere of knowledge and self-awareness. And I suspect we have increased the GAL, the gross amount of life. But we have also increased the GAS. The gross amount of spirit. Even though we in science do not have a clue to what spirit is. And even though we in science ignore it.

I saw human spirit at its strongest in Buenos Aires in 1981[1514] in Argentina. As you know, in 1968 I left normal science and did a Darwin. Or a von Humboldt. I went on a twenty-year scientific expedition into a territory I knew nothing about: popular culture. Music and film. This sort of thing has a name: participant-observer science.[1515] In the subtitle of my first book, *The Lucifer Principle*, I would call it "a scientific expedition into the forces of history." Why take such a violent plunge? As you know, I was on a hunt for the dark underbelly where new myths and movements are made. And I was finding it.

I would eventually work to build or sustain the careers of people like Michael Jackson, Prince, Bob Marley, Bette Midler, Billy Joel, Billy Idol, Paul Simon, Peter Gabriel, David Byrne, Kiss, AC/DC, Aerosmith, Run DMC, Grandmaster Flash and the Furious Five, Joan Jett, and more. I would have the privilege of bringing souls to life. Including the souls of entire subcultures. A rare experience for a science person.

In fact, I would explain to my clients that music was not an exchange of plastic, downloads, or cash. Music was an exchange of soul. Raw, unadulterated soul. What the heck did I mean?

Flash back to 1981. I was asked to bring ten of America's most influential music writers to Argentina to see the first major rock tour in South American history.[1516] A tour by Queen. On February 28, 1981,[1517] something astounding happened. I was seated in one of the very last seats in an enormous soccer stadium in Buenos Aires. The distance to the stage was so great that when Queen stepped out to perform, the band members looked the size of periods and commas. It was impossible to see their faces, much less their body language.

But half way through the show, when the band broke into "We Are the Champions," something amazing happened. It was clear that singer Freddie Mercury and his crew were lifting the spirits of the audience members in a way that was transcendent. Ecstatic. How could you tell? Every one of the 54,000[1518] concert attendees rose to their feet with a BIC cigarette lighter in hand, holding the flame as high as they could reach. And every member of that audience sang along to the song. Word for word. Despite the fact that their native language was not English. It was Spanish.

That moment suddenly revealed the meaning of the word "anthem." An anthem is a song deliberately written to lift the human spirit to the sky. And Queen was the master of anthems. The crowd on its feet lifting BICs to the heavens was the living incarnation of what seminal sociologist Emil Durkheim in 1912 called "collective effervescence."[1519]

Which leads back to the question of whether the human spirit is just a figment of woohoo imagination or is something real. Any general worth his salt will tell you that the spirit of your troops is often what wins a battle. Or, as one epigram puts it, "It's not the metal in the weapons that counts. It's the mettle in the men." So human spirit can have a powerful impact on the material world. Just as it has a powerful impact in the world of elephant seal showdowns. In other words, spirit is as real in the enterprise of life as a dead, dried plant. Which is why it counts to question how much humans have increased the GAS—the Gross Amount of Spirit. The Gross Amount of Sentience. The Gross Amount of Soul.

***

Which does nature really demand of us?

- Rolling things backward—the strategy of the apocalypse? or
- Growth and invention, the process that many claim has brought catastrophe?

Let's reframe the question. Who has loved earth more and increased the gross amount of living matter the most on this planet, eco-apocalyptics or those who have made farms, cities, and industrially-produced nitrogen? Who has increased the green more, Greenpeace or the makers of the Green Revolution? The answer is that both have made a contribution. Both have added to the number of new doors opened to the grand enterprise of life. And each one needs the other. Yes, Greenpeace needs the industrialists, the capitalists, and the agriculturalists. And the industrialists, capitalists, and agriculturists need Greenpeace. But the eco-apocalyptics? Well, they are a different story.

And their biggest problem is their failure to understand something. Nature. And the ephemeral, unmeasurable, but beyond-value thing called life.

CHAPTER 52

GUILT OR AUDACITY? WHICH DOES NATURE DEMAND FROM YOU?

There is soul-lift in this truer tale of nature. Far more lift for the soul than in the false story of a nature that prizes an unchanging "equilibrium," a nature that prizes standing still. Far more elevation than in the picture of a cosmos of entropy, a cosmos continually falling apart. Not to mention a cosmos parsimoniously insisting on the shortest path between two points, a cosmos obsessed with thrift.

There is a simple fact. Nature is a maw of darkness, death, and pain. She is a maw of manic mass production and manic mass destruction. A maw of materialism, consumerism, and waste. But she is also a maker of ecstasies, exhilarations, soaring new planes of reality, and celebrations of the impossible and the strange.

The story of nature and the story of life is not a tale of plants, animals, and humans living in harmony with their environment. It is not a tale of a blissful balance. It is not the story of a warm and loving equilibrium.

And it is not the story of a green paradise, a garden of Eden. A garden trashed by the monstrous activities of humankind. The real story of life is the very opposite. It's the story of greening a planet of disaster and strife.

It's a tale of transformation. It's the tale of the uplift of a toxic planet, a poison pill of stone. It's the saga of how the first teaspoon of life and its children overcame 142 mass extinctions. It's the story of the savagery of non-stop climate change. But it's the story of more than a mere struggle for survival.

It's the story of how life harnessed disaster. It's the story of how life poisoned the atmosphere. Then how life invented a way to turn that poison into a power source: oxygen. It's the story of how life wrenched, wounded, and ate its environment of stone to create something radically unnatural—nature.

And it's the story of a battle for flamboyance, excess, and splendor. It's the story of a race for invention. A battle for self-upgrade. It's the story of audacity. It's the tale of how disrespectful macromolecules milked manna from Armageddon on this toxic ball of stone.

It's the tale of how an upstart macromolecular team upped the GAL, the gross amount of life, the gross amount of living matter, on this planet. Not to mention the GAS. The gross amount of spirit. The gross amount of sentience. The gross amount of soul.

Which means that the immediate challenge of humanity is not the one that you're being told. The story of life is not a warning to stop growth before it destroys us. It's the story of a challenge that dictates the very opposite: grow as much as you can. Grow as cleverly as you can. Grow as inventively as you can. Growth is the only way that you and the life around you can beat the odds. More important, growth is the only way to achieve two imperatives that nature has prodded us to invent: justice and peace.

Are you troubled by plastic waste? Turn it into a resource base. As bacteria have done.[1520] Are you disturbed by greenhouse gases? Harvest solar energy in space and end the burning of fossil fuels. While you're at it, capture the carbon of greenhouse gases to make miracles like carbon fiber and graphene. Just as plants have captured carbon to mass produce stems, trunks, and leaves.

The real story of nature means that it's time to abandon the end-of-the-world dictated by entropy. It's time to toss out the dystopia of a universe constantly falling apart. It's time to jettison the sugar cube whizzling away in a glass of water. It's time to abandon heat death.

And it's time to replace the Second Law of Thermodynamics with the First Law of Flamboyance. The Second Law of Thermodynamics says that entropy is always on the increase, that all things tend toward a random chaos. That this cosmos is constantly falling apart.

But the First Law of Flamboyance says that all things tend to astonishment. All things tend to fall together. All things tend to invent. All things tend to breakthrough. All things tend to flaunt, flash, and produce material miracles. If you'll excuse my use of the following verb: all things tend to bloom.

Is this statement mathematizable? Probably not. Which means we need a new math. One that starts with the rules of flamboyance, the rules of emergent properties, then works its way downward from there. A math that can explain the cosmos' gobsmacking creativity.

And it's time to accept the fact that this is a cosmos of what 19th century scientific thinker and omnologist Herbert Spencer called progress.

***

Twentieth-century experts like Harvard evolutionary biologist Stephen Jay Gould did everything they could to prove the concept of progress wrong. But this is a universe that has progressed

- from a nothing to a Big Bang,
- from a Big Bang to the birth of quarks,
- from quarks to protons and neutrons,
- from protons and neutrons to atoms,
- from atoms to galaxies, planets, and molecules,
- And from molecules to life...to you and me.

Nature's obsession with progress is real.

Yes, this cosmos is a search engine feeling out her potential, constantly reaching for her next big step up. And you and I are the fingertips of that probing. We are the antennae of a universe feeling out her next moves in possibility space. We are neurons in the cosmos' brain, processors in a cosmic learning machine.

Nature will progress to higher levels with us or without us. But I hope that we continue to be what nature demands—the biggest progress makers this cosmos has ever seen.

DOES NATURE LOOK DOWN OR UP?

Nature aspires to new heights. The proof? Nature is the mother of four great space programs:

- the upward lunge of the first bacteria to escape the sea and live on the land,
- the outrageous arrogance of the first plants to reach to the skies by erecting towers two inches high,
- the gumption of the first trees to defy gravity by pumping 11,000 gallons of water seven stories high, and generating up to 200,000 leaves apiece to hog the light from the heavens above, and
- the equally astonishing fuck-yous to gravity of the insects and dinosaurs who left the ground and took to the skies.

But the real proof of nature's audacity is in the longest-path-possible, most anti-entropic phenomenon this cosmos has ever invented—the wild excess we call sex.

***

Nature's real nature implies that we have a job ahead of us.

We are not the only creatures to master research and development. Bacteria do that at a speed that constantly threatens to outpace us.

And bacteria know something we don't. We are certain that we have run out of resources. But, as you know, for every ounce of biomass on this planet there are three and a half billion ounces of dead stuff waiting to be kidnapped, seduced, and recruited into the grand enterprise of life.

We may be blind to this bulging abundance of resources, but bacteria twelve miles beneath your feet and mine are not. They are turning raw rock to food even as we speak. They are carrying out life's mandate: kidnap, seduce, and recruit as many dead atoms as possible into the extraordinary enterprise of life.

Remember, bacteria do not live in harmony with nature. Bacteria are obstreperous. Bacteria are uppity. Bacteria do not take nature lying down. Bacteria are not mere survivors. Bacteria are doom-riders and catastrophe-tamers. Bacteria take nature apart and put her back together in whole new ways. The way bacteria desecrated the purity of stone and made three feet of mud at the bottom of the sea. Which makes bacteria nature's servants in the invention game.

Bacteria may sometimes be smarter than we are, but we are the only species capable of taking life beyond the gravity well. We are the only species capable of taking ecosystems to the heavens and beyond. Remember, nature favors those who oppose her most. Nature loves those who defy gravity. Nature favors migration. Nature favors those who grow her. Nature adores those who expand the bounds of life.

Which means that nature has given us a job: to garden the solar system and to green the galaxy. To bring space to life by bringing life to space. And to lift life, sentience, and soul beyond the skies.

***

Once upon a time, nature used materialism, consumerism, waste, vain display, and sex to invent you and me. Now what inventions will we give her in return?

ENDNOTES

1   J. Heitman, "Evolution Of Sexual Reproduction: A View From The Fungal Kingdom Supports An Evolutionary Epoch With Sex Before Sexes," *Fungal Biology Reviews* 29, no. 3–4 (1 December 2015): 108–117, https://doi.org/10.1016/j.fbr.2015.08.002 https://www.sciencedirect.com/science/article/pii/S1749461315000391

2   Alistair Cameron Crombie, *The History of Science from Augustine to Galileo*, Dover reprint of 1959 edition, 32.

3   Philip E.B. Jourdain, "The Nature and Validity of the Principle of Least Action," *The Monist* 23, no. 1913: 277–293.

4   P.L. Maupertuis (1748), "Mémoires de l'académie royale des sciences," Paris: Académie Royale des Sciences. Quoted in Pouchard, Michel, and Antoine Villesuzanne, "Are Superconductivity Mechanisms A Matter For Chemists?," *Condensed Matter* 5, no. 4 (2020): 67.

5   Pierre Louis de Maupertuis, *Essais De Cosmologie* (1750) (New Mexico: Legare Street Press, 2023).

6   Scottish Scientist Hall of Fame, "Lord Kelvin," https://digital.nls.uk/scientists/biographies/lord-kelvin/

7   Crosbie Smith. North British network, (active from approximately 1845 to 1890), *Oxford Dictionary of National Biography*, 24 May 2007.

8   William Thomson (Lord Kelvin), "On a Universal Tendency in Nature to the Dissipation of Mechanical Energy," *Proceedings of the Royal Society of Edinburgh*, 1857, https://www.cambridge.org/core/journals/proceedings-of-the-royal-society-of-edinburgh/article/2-on-a-universal-tendency-in-nature-to-the-dissipation-of-mechanical-energy/862309E0AF0924FA7C0AA7FA24B74F6F. Hermann von Helmholtz, Über die Erhaltung der Kraft (*On the Conservation of Force*), 1847. Hermann Helmholtz, "On the Conservation of Force; A Physical Memoir," in *Scientific Memoirs, selected from the transactions of foreign academies of science, and from foreign journals, Natural Philosophy*, edited by John Tyndall and William Francis (London: Taylor & Francis, 1853), 114–162. Crosbie Smith, M. Norton Wise, *Energy and Empire: A Biographical Study of Lord Kelvin* (Cambridge University Press, 1989), 500. David Cahan, "Helmholtz and the British Scientific Elite: From Force Conservation To Energy Conservation," *Notes And Records: The Royal Society Journal Of The History*

Of Science, 16 November 2011, https://doi.org/10.1098/rsnr.2011.0044 A. C. Crombie, "Helmholtz," *Scientific American* 198, no. 3 (March 1958): 94–103, https://www.jstor.org/stable/10.2307/24940945. S.S Thipse, *Advanced Thermodynamics*, Alpha Science International Limited, 2013. "Heat Death of the Universe: Origins of the Idea," Chemeurope.com, https://www.chemeurope.com/en/encyclopedia/Heat_death_of_the_universe.html#Origins_of_the_idea

9   A.C. Crombie, "Helmholtz," *Scientific American*, https://www.jstor.org/stable/10.2307/24940945.

10   "Shortly after the events we have described, Thomson [Lord Kelvin] published 'On a Universal Tendency in Nature to the Dissipation of Mechanical Energy' [13] in 1852 (where in the introductory passage, Thomson again invokes Carnot as the primary source of thermodynamic thinking). This paper begins a discussion of startling implication; that the universe is irrevocably heading towards a state of total dispersion, which was to be known by the 1860s as 'the heat death of the universe.'" Michael W. Collins and Richard C. Dougal, *Kelvin, Thermodynamics and the Natural World* (WIT Press, 2016), 266.

11   "Biodiversity," Galapagos Conservancy, https://www.galapagos.org/about_galapagos/biodiversity.

12   Rosemary B. Grant and Peter R. Grant, "What Darwin's Finches Can Teach Us about the Evolutionary Origin and Regulation of Biodiversity," *BioScience* 53, no. 10 (October 2003), 965–975, https://doi.org/10.1641/0006-3568(2003)053[0965:WDFCTU]2.0.CO;2. Grant, Peter R., and B. Rosemary Grant, *How and Why Species Multiply: The Radiation of Darwin's Finches* (Princeton University Press, 2007).

13   Akie Sato, Herbert Tichy, Colm O'hUigin, Peter R. Grant, B. Rosemary Grant, Jan Klein, "On the Origin of Darwin's Finches," *Molecular Biology and Evolution* 18, no. 3 (March 2001): 299–311, https://doi.org/10.1093/oxfordjournals.molbev.a003806.

14   Darwin's finches came either from the Caribbean or from South America according to Erik R. Funk, Kevin J. Burns, "Biogeographic Origins of Darwin's Finches (Thraupidae: Coerebinae)," *The Auk* 135, no. 3 (1 July 2018): 561–571, https://doi.org/10.1642/AUK-17-215.1.

15   S. Carvajal-Endara, A.P. Hendry, N.C. Emery, C.P. Neu, D. Carmona, K.M. Gotanda, T.J. Davies, A.J. Chaves, and M.T.J. Johnson, "The Ecology And Evolution Of Seed Predation By Darwin's Finches On Tribulus Cistoides On The Galápagos Islands," *Ecological Monographs* 90, no. 1 (2020).

16   Mehmet Akif Sarıkaya, Marek Zreda, and Attila Çiner, "Glaciations And Paleoclimate Of Mount Erciyes, Central Turkey, Since The Last Glacial Maximum, Inferred From 36cl Cosmogenic Dating And Glacier Modeling," *Quaternary Science Reviews* 28, no. 23-24 (2009): 2326–2341.

17   Rosemary B. Grant, Peter R. Grant, "What Darwin's Finches Can Teach Us about the Evolutionary Origin and Regulation of Biodiversity," https://doi.org/10.1641/0006-3568(2003)053[0965:WDFCTU]2.0.CO;2.

18   Jimenez, Raul, et al., "The Local and Distant Universe: Stellar Ages and H0," *Journal of Cosmology and Astroparticle Physics* 3, no. 28 (March 2019), https://iopscience.iop.org/article/10.1088/1475-7516/2019/03/043.

19   Quoted in the work of one of de Maupertuis' friends, mathematician Jean Le Rond d'Alembert, *Essai sur les éléments de philosophie* (Belgium, Olms, 1965, originally published 1759).

20   Reid, Daniel, "What If Nature Had Been Thrifty?," *The American Scholar* 76, no. 1 (2007): 147–49, http://www.jstor.org/stable/41221664.

21   Pierre-Louis Moreau De Maupertuis, "Accord de differentes loix de la nature qui avoient jusqu'ici paru incompatibles," *Histoire de l'academie royale des sciences et des belles-lettres de Berlin, 1744* (1748): 417–426, read to the French Academy on 15 April 1744. Philip E.B. Jourdain, "Maupertuis And The Principle Of Least Action," *The Monist* 22, no. 3 (July, 1912), http://www.jstor.org/stable/27900387.

22   Jean Le Rond d'Alembert, *Essai sur les éléments de philosophie* (1759) (Belgium, Olms, 1965).

23   Hofmann, Wilhelm, Hiroki Kotabe, and Maike Luhmann, "The spoiled pleasure of giving in to temptation," *Motivation and Emotion 37* (2013), pages 733-742. Though this quote is frequently attributed to Lord Chesterfield, Quoteresearcher says that researchers have not been able to find it in Chesterfield's body of work. Quoteresearch, "The Pleasure Is Momentary, the Position Is Ridiculous, the Expense Is Damnable," April 16, 2017, https://quoteinvestigator.com/2017/04/16/pleasure/

24   Housman published "When I Was One And Twenty" in 1896, Richard Perceval Graves, *A. E. Housman: The Scholar-Poet* (London: Faber & Faber, 2014).

25   Alain de Botton, *How to Think More About Sex (The School of Life)* (Picador, 2012), Kindle Locations 1167–1168.

26   Zoltán Szabadka and Vince Grolmusz, "High Throughput Processing of the Structural Information of the Protein Data Bank," *Journal of Molecular Graphics and Modelling* 25, no. 6 (March 2007): 831–836, https://www.sciencedirect.com/science/article/abs/pii/S1093326306001124.

27   J. William Schopf, et al., "Sims Analyses Of The Oldest Known Assemblage Of Microfossils Document Their Taxon-Correlated Carbon Isotope Compositions," *Proceedings of the National Academy of Sciences* 115, no. 1 (18 December 2017): 53–58, www.pnas.org/cgi/doi/10.1073/pnas.1718063115. Peter Ward and Joe Kirschvink, *A New History of Life: The Radical New Discoveries*

About the Origins and Evolution of Life on Earth (Bloomsbury, 2015), 45. Christian de Duve, *Life Evolving: Molecules, Mind, and Meaning* (Oxford University Press, 2002). Roy A. Black, Matthew C. Blosser, Benjamin L. Stottrup, Ravi Tavakley, David W. Deamer, and Sarah L. Keller, "Nucleobases Bind To And Stabilize Aggregates Of A Prebiotic Amphiphile, Providing A Viable Mechanism For The Emergence Of Protocells," *Proceedings of the National Academy of Sciences* 110, no. 33 (2013): 13272–13276, https://www.pnas.org/doi/abs/10.1073/pnas.1300963110. Sandra Hines, "Natural Affinities—Unrecognized Until Now—May Have Set Stage For Life To Ignite," press release, University of Washington, https://www.washington.edu/news/2013/07/29/natural-affinities-unrecognized-until-now-may-have-set-stage-for-life-to-ignite.

28 M. Homann, P. Sansjofre, M. Van Zuilen, et al., "Microbial Life And Biogeochemical Cycling On Land 3,220 Million Years Ago," *Nature Geoscience* 11, no. 23 (July 2018): 665–671, https://doi.org/10.1038/s41561-018-0190-9.

29 J. Olejarz, Y. Iwasa, A.H. Knoll, et al., "The Great Oxygenation Event As A Consequence Of Ecological Dynamics Modulated By Planetary Change," *Nature Communications* 12 (28 June 2021), https://doi.org/10.1038/s41467-021-23286-7.

30 "Ancient Origins Of Multicellular Life," *Nature* 533, no. 441 (25 May 2016), https://doi.org/10.1038/533441b. The more traditional date for the start of multicellular life is 600 million years ago. See John Staughton, "How Long Did It Take For Multicellular Life To Evolve From Unicellular Life?" *ScienceABC*, 6 Jan 2022, https://www.scienceabc.com/pure-sciences/how-long-did-it-take-for-multicellular-life-to-evolve-from-unicellular-life.html.

31 Eshel ben-Jacob, sitting at the foot of my bed in 1998, held up his open palm, pointed to it, and declared that the number of bacteria occupying a space the size of that palm would be seven trillion. Eshel Ben-Jacob, personal communication, October, 1998. To back up Ben-Jacob's figure of trillions, see Bridget Mintz Testa, "Down To The Atom," *Mechanical Engineering* 131, no. 2 (2009): 38–42, and: "Regina Bailey,Phases of the Bacterial Growth Curve," *ThoughtCo*, 19 September 2018, https://www.thoughtco.com/bacterial-growth-curve-phases-4172692.

32 Jun–ichi Wakita, et al., "Experimental Investigation On The Formation Of Dense-Branching-Morphology-Like Colonies In Bacteria," *Journal of the Physical Society of Japan* 67, no. 10 (1998): 3630–3636. Eshel ben-Jacob, "The Artistry of Microbes: Shaped to Survive," *Scientific American*, October 1998. A. Be'er, S.K. Strain, R.A. Hernández, E. Ben-Jacob, E.L. Florin, "Periodic Reversals In *Paenibacillus dendritiformis* Swarming," *Journal of Bacteriology* 195, no. 12 (June 2013): 2709–17, https://pubmed.ncbi.nlm.nih.gov/23603739/ J. Yan, H. Monaco, J.B. Xavier, "The Ultimate Guide to Bacterial Swarming: An Experimental Model to Study the Evolution of Cooperative Behavior," *Annual Review of Microbiology* 73 (8 September 2019): 293–312, https://doi.org/10.1146/annurev-micro-020518-120033, https://www.annualreviews.org/

doi/abs/10.1146/annurev-micro-020518-120033. A. Be'er, G. Ariel, "A Statistical Physics View of Swarming Bacteria," *Movement Ecology* 7, no. 9 (2019), https://doi.org/10.1186/s40462-019-0153-9. R.M. Harshey, J.D. Partridge, "Shelter In A Swarm," *Journal of Molecular Biology* 427, no. 23 (20 November 2015): 3683–94, https://doi.org/10.1016/j.jmb.2015.07.025. E. Ben-Jacob. Personal Communication, April 1996–December 1999. E. Ben-Jacob, H. Shmueli, O. Shochet, A. Tenenbaum, "Adaptive Self-Organization During Growth of Bacterial Colonies," *Physica A.* (15 September 1992): 378. E. Ben-Jacob, "From Snowflake Formation to the Growth of Bacterial Colonies. Part I. Diffusive Patterning in Azoic Systems," *Contemporary Physics* (September 1993): 247. E. Ben-Jacob, A. Tenenbaum, O. Shochet, "Holotransformations Of Bacterial Colonies And Genome Cybernetics," *Physica A* (January 1994). E. Ben-Jacob, A. Tenenbaum, O. Shochet, I. Cohen, A. Czirók and T. Vicsek, "Generic Modeling of Cooperative Growth Patterns in Bacterial Colonies," *Nature* 368 (1994): 46–49. E. Ben-Jacob, A. Tenenbaum, O. Shochet, I. Cohen, A. Czirók and T. Vicsek, "Communication, Regulation and Control During Complex Patterning of Bacterial Colonies," *Fractals* 2, no. 1 (1994): 14–44. E. Ben-Jacob, A. Tenenbaum, O. Shochet, I. Cohen, A. Czirók and T. Vicsek, "Cooperative Strategies in Formation of Complex Bacterial Patterns," *Fractals* 3, no. 4 (1995): 849–868. E. Ben-Jacob; I. Cohen: D.L. Gutnick, "Chemomodulation Of Cellular Movement, Collective Formation Of Vortices By Swarming Bacteria, And Colonial Development," *Physica A: Statistical and Theoretical Physics* (15 April 1997): 181–197. E. Ben-Jacob, "From Snowflake Formation to the Growth of Bacterial Colonies. Part II. Cooperative Formation of Complex Colonial Patterns," *Contemporary Physics* (May 1997): 205–241. Eshel Ben-Jacob, "Bacterial Wisdom, Gödel's Theorem And Creative Genomic Webs," *Physica A* 248 (1998): 57–76. Eshel Ben Jacob, Israela Becker, Yoash Shapira, and Herbert Levine, "Bacterial Linguistic Communication and Social Intelligence," *Trends in Microbiology* 12, no. 8 (1 August 2004): 366–372, https://doi.org/10.1016/j.tim.2004.06.006 and https://www.sciencedirect.com/science/article/pii/S0966842X04001386. Eshel Ben-Jacob, "Learning from Bacteria about Social Networks," talk at Google Headquarters, Mountain View, CA, 5 June 2015, M|M, Micro Minds Forum, https://microbes-mind.net/ben-jacob.

33 Lanyun Miao, Andrew H. Knoll, Yuangao Qu, and Maoyan Zhu, "1.63-Billion-Year-Old Multicellular Eukaryotes From The Chuanlinggou Formation In North China," *Science Advances* 10, no. 4 (24 Jan 2024), https://doi.org/10.1126/sciadv.adk3208. S. Hedges, S. Blair, et al., "A Molecular Timescale Of Eukaryote Evolution And The Rise Of Complex Multicellular Life," *BMC Evolutionary Biology* 4, no. 1 (2004): 1–9. L.A. Riedman, S.M. Porter, M.A. Lechte, A. dos Santos, and G.P Halverson, "Early Eukaryotic Microfossils Of The Late Palaeoproterozoic Limbunya Group, Birrindudu Basin, Northern Australia," *Papers in Palaeontology* 9, no. 6 (2023), https://onlinelibrary.wiley.com/

doi/full/10.1002/spp2.1538. S. Bengtson, T. Sallstedt, V. Belivanova, and M. Whitehouse, "Three-Dimensional Preservation Of Cellular And Subcellular Structures Suggests 1.6 Billion-Year-Old Crown-Group Red Algae," *PLoS Biology* 15, nol. 3 (2017), https://journals.plos.org/plosbiology/article?id=10.1371/journal.pbio.2000735.

34   Daniel S. Heckman, et al., "Molecular Evidence for the Early Colonization of Land by Fungi and Plants," *Science* 293, no. 10 (August 2001): 1129–1133, https://doi.org/10.1126/science.1061457. Barbara A. Ambrose, Michael D. Purugganan (eds.), "The Evolution of Plant Form," *Annual Plant Reviews* 45 (November 2012): xv.

35   James E. Stembridge, "Root Wedging and Rock Disintegration," *The Geographical Bulletin* 6 (1973): 33. J.A. Raven, D. Edwards, "Roots: Evolutionary Origins and Biogeochemical Significance," *Journal of Experimental Botany* 52, 1 (March 2001): 381–401, https://doi.org/10.1093/jexbot/52.suppl_1.381. C. Klappa, "Rhizoliths in Terrestrial Carbonates: Classification, Recognition, Genesis and Significance," *Sedimentology* 27 (December 1980): 613–629, https://doi.org/10.1111/J.1365-3091.1980.TB01651.X. B. Wild, R. Gerrits, and S. Bonneville, "The Contribution Of Living Organisms To Rock Weathering In The Critical Zone," *npj Materials Degradation* 6, no. 98 (20 December 2022), https://doi.org/10.1038/s41529-022-00312-7. "How Do the Roots of Plants Affect Rocks?" *Short-Fact*, 7 January 2020, https://short-fact.com/how-do-the-roots-of-plants-affect-rocks.

36   Luke Parry, "Discovering the Earthworm's Half a Billion Year Old Cousin," Blog, St. Edmund Hall, University of Oxford (16 June 2020), https://www.seh.ox.ac.uk/blog/discovering-the-earthworms-half-a-billion-year-old-cousin.

37   "It may be doubted whether there are many other animals which have played so important a part in the history of the world, as have these lowly organised creatures." Charles Darwin, *The Formation of Vegetable Mould, Through the Action of Worms, With Observations on Their Habits* (John Murray, 1881), 313. Also quoted in Amy Stewart, *The Earth Moved: On the Remarkable Achievements of Earthworms* (Algonquin Books, 2004), 1.

38   Charles Darwin, *The Formation of Vegetable Mould Through the Action of Worms*, Project Gutenberg, 1 Oct. 2000, www.gutenberg.org/ebooks/2355 Originally published 1881.

39   Kevin Laland and John Odling-Smee, "Life's Little Builders," *New Scientist*, November 15, 2003.

40   Clive A. Edwards (ed.), *Earthworm Ecology* (Boca Raton, Florida: CRC Press, 2004).

41   P.G. Eriksson, Wladyslaw Altermann, D.R. Nelson, W.U. Mueller, O. Catuneanu, Octavian Catuneanu (eds.), *The Precambrian Earth: Tempos and Events* (Elsevier Science, 2004), 361.

42   Ibid., 496–497.

43   Michael R. Rampino , Ken Caldeira & Yuhong Zhu, "A 27.5-My Underlying Periodicity Detected In Extinction Episodes Of Non-Marine Tetrapods," *Historical Biology* (2020), https://doi.org/10.1080/08912963 .2020.1849178. New York University, "Mass Extinctions Of Land-Dwelling Animals Occur In 27-Million-Year Cycle: Researchers Find That Timing Of Mass Extinctions Lines Up With Asteroid Impacts And Massive Volcanic Eruptions," *ScienceDaily*, https://www.sciencedaily.com/releases/2020/12/201211083113. htm

44   Rana Ezzeddine, et al. "Evidence for an Aspherical Population III Supernova Explosion Inferred from the Hyper-metal-poor Star HE 1327–2326," *The Astrophysical Journal* 876, no. 2 (2019): 97. Jennifer Chu, "Explosions of Universe's First Stars Spewed Powerful Jets," *MIT News*, 8 May 2019, https:// news.mit.edu/2019/universe-first-stars-jets-0508.

45   A. Soderberg, et. al., "An Extremely Luminous X-ray Outburst at the Birth of a Supernova," *Nature* 453 (2008): 469–474, https://doi.org/10.1038/ nature06997.

46   There is entire field and a movement called "degrowth economics." See: "What is degrowth," https://degrowth.info/degrowth.

47   National Geographic Society, "The Development of Agriculture," *National Geographic*, https://education.nationalgeographic.org/resource/development-agriculture.

48   John Bongaarts, "How Long Will We Live?" *Population and Development Review* 32, no. 4 (2006), 605–28, http://www.jstor.org/stable/20058921.

49   Steven Pinker, *The Better Angels Of Our Nature: Why Violence Has Declined* (New York: Penguin, 2012). Max Roser, "Data Review: Ethnographic And Archaeological Evidence On Violent Deaths: What Quantitative Data Is There About Violent Deaths In Non-State Societies?" published online at OurWorldInData.org, 2013, https://ourworldindata.org/ethnographic-and-archaeological-evidence-on-violent-deaths. Lawrence H. Keeley, *War Before Civilization: The Myth of the Peaceful Savage* (Oxford University Press, 1996). Ferris Jabr, "Steven Pinker: Humans Are Less Violent Than Ever," *New Scientist*, 12 October 2011, https://www.newscientist.com/article/mg21228340-100-steven-pinker-humans-are-less-violent-than-ever. Howard Bloom, *The Lucifer Principle: A Scientific Expedition Into the Forces of History* (New York: Atlantic Monthly Press, 1995).

50   Stanley Lebergott, "Prices and Wages by Decade: 1850–1859," in *Trends in the American Economy in the Nineteenth Century* (Princeton University Press, 1960), 449–500. https://www.nber.org/system/files/chapters/c2486/ c2486.pdf.*Wage Trends, 1800-1900*, Libraries University of Missouri, https:// libraryguides.missouri.edu/pricesandwages/1850-1859. Robert Whaples,

Wake Forest University, "Hours of Work in U.S. History," Economic History Association, https://eh.net/encyclopedia/hours-of-work-in-u-s-history. Phil Hyde, "History of the American Workweek, Timesizing not downsizing," https://www.timesizing.com/history-of-the-american-workweek.

51   "All Facts about IQ 131," *Sociosite, Social Science Information System*, University of Amsterdam, https://www.sociosite.net/iq-scores/131.

52   L.H. Trahan, K.K. Stuebing, J.M. Fletcher, and M. Hiscock, "The Flynn Effect: A Meta-Analysis," *Psychological Bulletin* 140, no. 5 (2014): 1332–1360, https://doi.org/10.1037/a0037173. Steven Johnson, *Everything Bad Is Good for You: How Today's Popular Culture Is Actually Making Us Smarter* (New York: Riverhead, 2005).

53   Elizabeth Pennisi, "Reforestation Means More Than Just Planting Trees: Scientists Are Figuring Out The Best Strategies To Regrow Lost Forests," *Science,* 22 November 2022, https://www.science.org/content/article/reforestation-means-just-planting-trees.

54   There were as few as one million and as many as ten million humans on earth 10,000 years ago. United States Census Bureau, Historical Estimates of World Population, https://www.census.gov/data/tables/time-series/demo/international-programs/historical-est-worldpop.html. Our World in Data believes the number of humans on the planet 12,000 years ago was 4 million. Max Roser and Hannah Ritchie (2023), "How Has World Population Growth Changed Over Time?" OurWorldInData.org, https://ourworldindata.org/population-growth-over-time. Columbia University's Dickson Despommier calculates the human population before the agricultural revolution at a mere million. D. Despommier, "Farming Up The City: The Rise Of Urban Vertical Farms," *Trends in Biotechnology* 31, no. 7 (2014): 388–9, https://doi.org/10.1016/j.tibtech.2013.03.008.

55   For the widely accepted date of 12,000 years ago, see sources like Graeme Barker, *The Agricultural Revolution in Prehistory: Why Did Foragers Become Farmers?* (Oxford University Press, 2006). For a South East Asian agricultural revolution 13,000 years ago, see Wilhelm G. Solheim II, "An Earlier Agricultural Revolution," *Scientific American* 226, no. 4 (April 1972): 34–41, https://www.jstor.org/stable/24927314.

56   Petroc Taylor,"Volume Of Data/Information Created, Captured, Copied, And Consumed Worldwide From 2010 To 2020, With Forecasts From 2021 To 2025," *Statista*, 16 November 2023, https://www.statista.com/statistics/871513/worldwide-data-created, quoted in Fabio Duarte, "Amount of Data Created Daily (2023)," *Exploding Topics*, 3 April 2023, https://explodingtopics.com/blog/data-generated-per-day.

57   Derek Gatherer, "What Really Killed Prince Albert?," *The Conversation*, 4 January 2018, https://theconversation.com/what-really-killed-prince-albert-85939.

58   CDC Centers for Disease Control and Prevention, "Typhoid Fever and Paratyphoid Fever," https://www.cdc.gov/typhoid-fever/symptoms.html.

59   Helen Rappaport, *A Magnificent Obsession: Victoria, Albert, and the Death That Changed the British Monarchy* (St. Martin's Publishing Group, 2012).

60   Howard Bloom, *The Genius of the Beast: A Radical Re-Vision of Capitalism* (Prometheus, 2009), https://www.amazon.com/Genius-Beast-Radical-Re-Vision-Capitalism/dp/1591027543.

61   Arthur Woods, "The Greater Earth System," https://greater.earth/GEO_DOCS/the_greater_earth_system.php.

62   Turan Kayaoglu, *The Organization of Islamic Cooperation: Politics, Problems, and Potential* (Routledge, 2015).

63   "Criminal Appeals No. 210 and 211 of 2015, (Against the judgment dated 09.03.2015 passed by the Islamabad High Court, Islamabad in Criminal Appeal No. 90 of 2011 and Capital Sentence Reference No. 01 of 2011)," 8. For different accounts of the number of bullets fired to kill the governor, Salmann Taseer, see: "Salmaan Taseer assassinated," *The Express Tribune*, Pakistan, 4 January 2011, http://tribune.com.pk/story/98988/salman-taseer-attacked-in-islamabad. And Shehrbano Taseer, "'27 bullets fired at my father, Salmaan Taseer'," *The New York Times*, 10 January 2011, in NDTV India, http://www.ndtv.com/world-news/27-bullets-fired-at-my-father-salmaan-taseer-444544.

64   Declan Walsh, "Salmaan Taseer, Aasia Bibi and Pakistan's struggle with extremism," *The Guardian*, 8 January 2011, http://www.theguardian.com/world/2011/jan/08/salmaan-taseer-blasphemy-pakistan-bibi. For the full story, see my book, Howard Bloom, *The Muhammad Code* (Feral House, 2016).

65   "List of Buildings in Dubai," *Wikipedia*, https://en.wikipedia.org/wiki/List_of_buildings_in_Dubai. "List of Tallest Buildings in the United Arab Emirates," *Wikipedia*, https://en.wikipedia.org/wiki/List_of_tallest_buildings_in_the_United_Arab_Emirates.

66   "List of Buildings in Dubai," https://en.wikipedia.org/wiki/List_of_buildings_in_Dubai.

67   The weight of the earth is 5.972×10^{21} tons. J. Henry, M. Yurukcu, and G. Nnanna, "How Fast Is the Distance between Earth and the Sun Changing in the Solar System?" Preprints.org, 2021, https://doi.org/10.20944/preprints202109.0508.v2. The biomass of living organisms on the planet is 545.8 gigatons of carbon. Without being reduced to mere carbon, there are 1.7 trillion tons (1.7 million gigatons) of living biomass on earth today, per M.F. Gaybullaeva, "The Role Of Biomass In Saving Natural Resources," *The American*

Journal of Horticulture and Floriculture Research 3, no. 2 (2021): 1–6, https://doi.org/10.37547/tajhfr/Volume03Issue02-01 This means that for every ounce of living biomass there are over 3.5 billion ounces of non-living matter.

68   "What Is the Volume of Earth?" *Physlink.com*, https://www.physlink.com/education/askexperts/ae419.cfm.

69   ScienceDirect, "Chemolithoautotroph," https://www.sciencedirect.com/topics/biochemistry-genetics-and-molecular-biology/chemolithoautotroph.

70   Katrina J. Edwards, Keir Becker, and Frederick Colwell, "The Deep, Dark Energy Biosphere: Intraterrestrial Life on Earth," *Annual Review of Earth and Planetary Sciences* (May 2012): 551–568. Mark A. Lever, Olivier Rouxel, Jeffrey C. Alt, et al, "Evidence for Microbial Carbon and Sulfur Cycling in Deeply Buried Ridge Flank Basalt," *Science* (15 March 2013): 1305–1308.

71   K. Pedersen, "Exploration of Deep Intraterrestrial Microbial Life: Current Perspectives," *FEMS Microbiology Letters* 185, no. 1 (April 2000): 9–16, https://doi.org/10.1111/j.1574-6968.2000.tb09033.x.

72   Karen G. Lloyd, Peter Barry, David Terry Fine, "Mysterious Microbes in Earth's Crust Might Help with the Climate Crisis," *Scientific American*, 29 March 2023, https://www.scientificamerican.com/video/mysterious-microbes-in-earths-crust-might-help-with-the-climate-crisis. C. Escudero, M. Vera, M. Oggerin, R. Amils, "Active Microbial Biofilms In Deep Poor Porous Continental Subsurface Rocks," *Scientific Reports* 8, no. 1 (24 January 2018): 1538, https://doi.org/10.1038/s41598-018-19903-z.

73   Jeffery DelViscio, "Meet the Magnificent Microbes of the Deep Unknown," *Science Quickly*, *Scientific American* podcast, 10 April 2023, https://www.scientificamerican.com/podcast/episode/meet-the-magnificent-microbes-of-the-deep-unknown.

74   Stephanie A. Napieralski, Heather L. Buss, Susan L. Brantley, Seungyeol Lee, Huifang Xu, and Eric E. Roden, "Microbial Chemolithotrophy Mediates Oxidative Weathering Of Granitic Bedrock," *Proceedings of the National Academy of Sciences* 116, no. 52 (16 December 2019), 26394–26401, https://www.pnas.org/doi/abs/10.1073/pnas.1909970117. Madison Payne, "Microbes That Eat Iron Cause Of Bedrock Erosion," *Sciworthy: The Encyclopedia of Science's Frontier*, 6 January 2022, https://sciworthy.com/microbes-that-eat-iron-are-a-cause-of-bedrock-erosion.

75   "Maxwell's Equations," International Society for Optics and Photonics, excerpted from David W. Ball, *Field Guide to Spectroscopy* (Bellingham, WA: SPIE Press, 2006), https://spie.org/publications/fg08_p07-08_maxwells_equation.

76   Steven Weinberg, *Dreams of a Final Theory: The Scientist's Search for the Ultimate Laws of Nature* (New York: Vintage Books, 2011).

77 Michio Kaku, *The God Equation: The Quest for a Theory of Everything*, Anchor, 2021.

78 T. Tashiro, A. Ishida, M. Hori, et al., "Early Trace Of Life From 3.5 Ga Sedimentary Rocks In Labrador, Canada," *Nature* 549 (2017): 516–518, https://doi.org/10.1038/nature24019. NASA, "Life on Earth Began at Least 3.85 Billion Years Ago, 400 Million Years Earlier than Previously Thought, Scientists Say," NASA News Release 96-230, 6 November 1996, https://www3.nasa.gov/home/hqnews/1996/96-230.txt.

79 IMDb, David Van Taylor, https://www.imdb.com/name/nm0888083.

80 For the story of the Park School's founding, see the book by the founder, Mary Hammett Lewis: M.H. Lewis, *An Adventure with Children* (Macmillan: 1929).

81 Stephen G. Brush, "Prediction and Theory Evaluation: Cosmic Microwaves and the Revival of the Big Bang," *Perspectives on Science* 1, no. 4 (1993): 565–602.

82 William Henry Overall, *Some Account of the Worshipful Company of Clockmakers of the City of London* (Blades, East & Blades, 1881),23.

83 "History of the Royal Society," The Royal Society, https://royalsociety.org/about-us/history.

84 Andrew Marvell, *Oxford Dictionary of National Biography*, https://www.oxforddnb.com/display/10.1093/ref:odnb/9780198614128.001.0001/odnb-9780198614128-e-18242;jsessionid=B4F3E488B8E2089A393F8C0B9B0D3FC9.

85 *Oxford English Dictionary*, "rate (n.1), sense I.3," December 2023, https://doi.org/10.1093/OED/5175542771.

86 Niels Steensgaard, "Freight Costs in the English East India Trade 1601–1657," *Scandinavian Economic History Review* 13, no. 2 (1965): 143–162, https://doi.org/10.1080/03585522.1965.10414367.

87 L.O. Petram, *The World's First Stock Exchange: How the Amsterdam Market for Dutch East India Company Shares Became a Modern Securities Market, 1602–1700* (thesis, University of Amsterdam, 2011), 12, https://pure.uva.nl/ws/files/1427391/85961_thesis.pdf.

88 Jean Sutton, *Lords of the East: The East India Company and Its Ships (1600-1874)* (London: Conway Maritime Press, 2000), 19. "East India Company," *Encyclopedia Britannica*, 29 December 2023, https://www.britannica.com/money/topic/East-India-Company.

89 Markman Ellis, *The Coffee House: A Cultural History* (Phoenix, 2004).

90 William Dalrymple, "The East India Company: The Original Corporate Raiders," *The Guardian*, 4 March 2015, https://www.theguardian.com/world/2015/mar/04/east-india-company-original-corporate-raiders. "Timur Ruby," *Wikipedia*, https://en.wikipedia.org/wiki/Timur_ruby.

91   McKie, Douglas, "The Origins and Foundation of the Royal Society of London," *Notes And Records: The Royal Society Journal Of The History Of Science* 15, no. 1 (1960): 1–37.

92   George Gamow, *My World Line: An Informal Autobiography* (New York: Viking, 1970). Artur D. Chernin, "George Gamow and the Big Bang," *Space Science Reviews* 74, nos. 3-4 (1995): 447–454.

93   Howard Bloom, "A Toroidal Model of the Cosmos: The Big Bagel," *Journal of Space Philosophy* (Spring 2013).

94   Joshua A. Frieman, Michael S. Turner, and Dragan Huterer, "Dark Energy And The Accelerating Universe," *Annual Review of Astronomy and Astrophysics* 46 (2008): 385–432, https://www.annualreviews.org/doi/abs/10.1146/annurev. astro.46.060407.145243.

95   Con Edison, "The New York Steam System," https://www.coned.com/en/ commercial-industrial/steam/why-steam.

96   Now known as the University at Buffalo, and now part of the State University of New York, https://www.buffalo.edu.

97   George Gamow, *The Creation of the Universe* (New York: Viking, 1952), 43. Fred Hoyle, *A Life in Science* (New York: Cambridge University Press, 2011), 88. Graham Farmelo, *The Strangest Man: The Hidden Life Of Paul Dirac, Mystic Of The Atom* (Basic Books, 2011), 151. Simon Mitton and Jane Wilson, "Several Lives and More," review of George Gamow's *My World Line: An Informal Autobiography*, in *Bulletin of the Atomic Scientists* (February 1971), 47. David Michael Harland, *The Big Bang: A View From The 21st Century* (London: Springer, 2003), 143.

98   "Current global population sizes" are "about 1.5 to 2 million whales" says renowned whale researcher Hal Whitehead. See Hal Whitehead, "Estimates Of The Current Global Population Size And Historical Trajectory For Sperm Whales," *Marine Ecology Progress* Series 242 (2002): 295–304, https://www. int-res.com/articles/meps2002/242/m242p295.pdf. See also: The International Whaling Commission, "Population Estimates: The IWC's Most Recent Abundance Estimates," https://iwc.int/about-whales/estimate.

99   Paul Ehrlich and Anne Ehrlich, *The End of Affluence* (Ballantine Books, 1974).

100   Paul Sabin, *The Bet* (Yale University Press, 2013), 96–99. Copyright Yale University Press, 2013. By permission of Yale University Press.

101   The population of earth in 1972 was 3.844 billion. See Database, "Earth Population by Country, 1972," https://database.earth/population/ by-country/1972. The earth's population in 2021 was 7.909 billion. "World 2021," *Population Pyramids of the World from 1950 to 2100*, https://www. populationpyramid.net/world/2021.

102   Director John P. Holdren, Office of Science and Technology Policy, The White House, under President Barack Obama, https://obamawhitehouse.archives.gov/administration/eop/ostp/about/leadershipstaff/director.

103   Our World in Data, "Prevalence Of Undernourishment In Developing Countries, 1970 To2015, Derived From Food And Agriculture Organization Of The United Nations," https://ourworldindata.org/grapher/prevalence-of-undernourishment-in-developing-countries-since-1970.

104   Three years longer lived, according to the U.S. Census Bureau: "The largest gains in life expectancy occurred between 1970 and 1980—an increase of about three years from 70.8 to 73.7 years." In Lauren Medina, Shannon Sabo, and Jonathan Vespa, "Living Longer: Historical and Projected Life Expectancy in the United States, 1960 to 2060—Population Estimates and Projections," *Current Population Reports*, February 2020, https://www.census.gov/content/dam/Census/library/publications/2020/demo/p25-1145.pdf.

105   According to the Brookings Institution, in 1970 there were 2.2 billion in poverty, 44% of the global population. In 1990, there were 1.9 billion in poverty, 36% of global population. H. Kharas, and M. Dooley, "The Evolution Of Global Poverty, 1990-2030. Brookings Global Working Paper, 166," Brookings Institution, https://www.brookings.edu/wp-content/uploads/2022/02/Evolution-of-global-poverty.pdf. Sevil Omer, "Global Poverty: Facts, FAQs, and How to Help," World Vision, https://www.worldvision.org/sponsorship-news-stories/global-poverty-facts. According to the World Bank, 9.2% of the global population lived in poverty in 2022. "Economic Poverty Factsheet 2023," *Economic Poverty Trends: Global, Regional and National, Development Initiatives*, February 2023, https://docs.google.com/document/d/1F44ftpateXZuBf_pxTSh4V7AwW2c4fqZXMgaEZ_yPR4/edit, https://devinit.org/resources/poverty-trends-global-regional-and-national.

106   Paul Ehrlich, *The Population Bomb* (Ballantine Books, 1970).

107   Erasmus Darwin's three chronological histories of the cosmos from a big explosion to the humans of his age were outlined in *The Botanic Garden* (1792), *Zoonomia* (1794–96), and *The Temple of Nature* (1803). See also: J.P. Daly, "The Botanic Universe: Generative Nature and Erasmus Darwin's Cosmic Transformism," *Republics of Letters* 6, no. 1 (2018), https://arcade.stanford.edu/rofl/botanic-universe-generative-nature-and-erasmus-darwins-cosmic-transformism. E.G. Hernández-Avilez, R. Ruiz-Gutiérrez, "From One Darwin to Another: Charles Darwin's Annotations to Erasmus Darwin's 'The Temple of Nature,'" *Humanities and Social Sciences Communications* 10, no. 143 (2023), https://doi.org/10.1057/s41599-023-01616-y.

108   Paul Sabin, *The Bet* (Yale University Press, Kindle Edition), 26–27. For the original Ehrlich essay, "Eco-catastrophe," see Glen Gaviglio and David E. Raye, *Society As It Is: A Reader* (MacMillan, 1976), 361.

109   Paul Ehrlich, Anne Ehrlich, and Gretchen C. Daily, *The Stork and the Plow*, (New Haven: Yale University Press, 1995).

110   Wahlberg, Niklas, Christopher W. Wheat, and Carlos Peña, "Timing And Patterns In The Taxonomic Diversification Of Lepidoptera (Butterflies And Moths)," *PLOS One* 8, no. 11 (2013), https://journals.plos.org/plosone/article?id=10.1371/journal.pone.0080875. A. Kawahara, et al., "Phylogenomics Reveals The Evolutionary Timing And Pattern Of Butterflies And Moths," *Proceedings of the National Academy of Sciences of the United States of America* 116 (2019): 22657–22663, https://doi.org/10.1073/pnas.1907847116

111   United States Department of Agriculture, "Monarch Butterfly Migration and Overwintering," U.S. Forest Service, https://www.fs.usda.gov/wildflowers/pollinators/Monarch_Butterfly/migration/index.shtml.

112   R.H. Nordlander and J.S. Edwards, "Morphology Of The Larval And Adult Brains Of The Monarch Butterfly, *Danaus plexippus plexippus, L*," *Journal of Morphology* 126 (1968): 67–93, https://doi.org/10.1002/jmor.1051260105.

113   Schuyler Null and Wilson Center, "What Paul Ehrlich Missed (and Still Does): The Population Challenge Is About Rights," *NewSecurityBeat*, the blog of the Environmental Change and Security Program, 3 June 2015, https://www.newsecuritybeat.org/2015/06/paul-ehrlich-missed-and-does-population-challenge-rights.

114   For a chart of the predictions of the coming of The Kingdom of God, see: "Predictions and Claims for the Second Coming," *Wikipedia*, https://en.wikipedia.org/wiki/Predictions_and_claims_for_the_Second_Coming.

115   "Religion by Country 2024," *World Population Review*, https://worldpopulationreview.com/country-rankings/religion-by-country.

116   "William Miller," *Encyclopedia Britannica*, 16 Dec. 2023, https://www.britannica.com/biography/William-Miller.

117   Vermont Historical Society, "The End Of The World Was Almost Today In 1843 And 1844: The Failed Prophesies Of The Millerites," *Vermont Digital Newspapers*, 21 March 2015, http://library.uvm.edu/vtnp/?p=2765.

118   J. Hewitson, "'To Despair at the Tedious Delay of the Final Conflagration': Hawthorne's Use of the Figure of William Miller," *ESQ, A Journal of the American Renaissance* 53 (2007): 111–89, https://doi.org/10.1353/esq.2007.0003.

119   New England Historical Society, "William Miller Convinced Thousands Of Millerites The End Was Near," 2023, https://newenglandhistoricalsociety.com/william-miller-convinced-thousands-millerites-world-end.

120   David Kramer, "The Terquasquicentennial Of The Day Of Wrath And The Great Disappointment Atop Cobb's Hill. Are The Ascension Robes A Myth?" *Talker of the Town*, 18 Oct 2019, https://talkerofthetown.com/2019/10/18/

remembering-the-terquasquicentennial-of-the-day-of-wrath-or-the-great-disappointment-atop-cobbs-hill-are-the-ascension-robes-a-myth. Ruth Alden Doan, "The Miller Heresy, Millennialism And American Culture (Religion, Antebellum)," The University of North Carolina at Chapel Hill, 1984, https://www.proquest.com/openview/476aa84c3750b2954870071b8c16d7b7/1. "Cobb's Hill–Rochester, New York," *Atlas Obscura*, 5 May 2023, https://www.atlasobscura.com/places/cobbs-hill-rochester-new-york.

121   K. Norwood, "The End Of The World Was Almost Today In 1843 And 1844: The Failed Prophesies Of The Millerites," Vermont Digital Newspaper Project (VTDNP), 21 March 2015, http://library.uvm.edu/vtnp/?p=2765.

122   Jessica Taylor, "All Your Questions About Seventh-Day Adventism And Ben Carson Answered," *NPR*, 27 October 2015, https://www.npr.org/sections/itsallpolitics/2015/10/27/452314794/all-your-questions-about-seventh-day-adventism-and-ben-carson-answered.

123   C.D. Navarrete, R. Kurzban, D.M. Fessler, and L.A. Kirkpatrick, "Anxiety and Intergroup Bias: Terror Management or Coalitional Psychology?" *Group Processes & Intergroup Relations* 7, no. 4 (2004): 370–397, https://journals.sagepub.com/doi/abs/10.1177/1368430204046144.

124   Donella H. Meadows, Dennis L. Meadows, Jorgen Randers, and William W. Behrens III, *The Limits to Growth; a Report for the Club of Rome's Project on the Predicament of Mankind*, New York: Universe Books, 1972.

125   Matthew Simmons, "Revisiting The Limits to Growth: Could The Club of Rome Have Been Correct, After All? (Part One)," *Resilience*, 29 September 2000, originally published by GreatChange.org, https://www.resilience.org/stories/2000-09-29/revisiting-limits-growth-could-club-rome-have-been-correct-after-all-part-one.

126   Bjorn Lomborg, "The Limits to Panic," *Project Syndicate*, 17 June 2013, https://www.project-syndicate.org/commentary/economic-growth-and-its-critics-by-bj-rn-lomborg.

127   Donella H. Meadows, Dennis L. Meadows, Jorgen Randers, and William W. Behrens III, *The Limits to Growth: A Report for the Club of Rome's Project on the Predicament of Mankind* (New York: Universe Books, 1972), 19, 51–55, 62–63.

128   "Almagest," Wikipedia, https://en.wikipedia.org/wiki/Almagest.

129   The Smithsonian says it is the largest museum, education, and research complex in the world. "Welcome," Smithsonian, https://www.si.edu.

130   Smithsonian Museum, "Earth Optimism x Folklife: Inspiring Conservation Communities," Smithsonian Folklife Festival, https://festival.si.edu/2022/earth-optimism.

131   Howard Bloom, "Climate Change Is Nature's Way: It's Our Good Luck One Of Earth's Many Ice Ages Ended 12,000 Years Ago," *Wall Street Journal*, 17 December 2009, https://www.wsj.com/articles/SB1000142405274870454100457459998193618834?page=1.

132   J. Lovelock, M. Whitfield, "Life Span of the Biosphere," *Nature* 296, no. 1 (April 1982): 561–563, https://doi.org/10.1038/296561a0. G.C. Reid, "Solar Variability and the Earth's Climate: Introduction and Overview," *Space Science Reviews* 94 (November 2000): 1–11, https://doi.org/10.1023/A:1026797127105. NASA, "Total Solar Irradiance: The Sun Also Changes, Understanding Earth's Energy Balance," NASA.

133   Technically these instant global warmings followed by long, slow global coolings are called "Dansgaard-Oeschger" events. In the 120,000 years since the end of the Eemian interglacial, these instant global warmings have occurred roughly every 1,500 years. Stefan Rahmstorf, "Ocean Circulation and Climate During the Past 120,000 Years," *Nature* 419 (12 September 2002): 207–214, https://doi.org/10.1038/nature01090. "Eemian: The Eemian Is The Interglacial Period Before The Inception Of The Last Glacial Cycle, From 130 To 115 Kyr Ago," *Quaternary Science Reviews*, 2017. "Eemian," *ScienceDirect*, https://www.sciencedirect.com/topics/earth-and-planetary-sciences/eemian. Paolo Stocchi Bas de Boer, Pippa L. Whitehouse, and Roderik S.W. van de Wal, "Current State And Future Perspectives On Coupled Ice-Sheet – Sea-Level Modelling," *Quaternary Science Reviews* 169 (2017): 13–28, https://doi.org/10.1016/j.quascirev.2017.05.013. Don Easterbrook, "Dansgaard-Oeschger event," *Encyclopedia Britannica*, 26 September 2019, https://www.britannica.com/science/Dansgaard-Oeschger-event.

134   Simone Ulmer, "An Ice Age :asting 115,000 Years in Two Minutes," Swiss National Supercomputing Centre, 6 November 2018, https://phys.org/news/2018-11-ice-age-years-minutes.html. "When Have Ice Ages Occurred? Explore The Ice Age Midwest.....Plants And Animals Of The Pleistocene," Illinois State Museum, 2015, https://iceage.museum.state.il.us/content/when-have-ice-ages-occurred.

135   Marion Prévost, Iris Groman-Yaroslavski, Kathryn M. Crater Gershtein, José-Miguel Tejero and Yossi Zaidner, "Early Evidence For Symbolic Behavior In The Levantine Middle Paleolithic: A 120 Ka Old Engraved Aurochs Bone Shaft From The Open-Air Site Of Nesher Ramla, Israel," *Quaternary International* 624 (20 January 2021): 80–93, http://doi.org/10.1016/j.quaint.2021.01.002. "Timeline of Prehistory," *Wikipedia*, 21 July 2023, https://en.wikipedia.org/wiki/Timeline_of_prehistory#cite_note-12.

136   Dirk L. Hoffmann, Diego E. Angelucci, Valentín Villaverde, Josefina Zapata, and João Zilhão, "Symbolic Use Of Marine Shells And Mineral Pigments By Iberian Neandertals 115,000 Years Ago," *Science Advances* 4, no. 2 (1 February 2018), https://doi.org/10.1126/sciadv.aar5255.

137   Howard Bloom, "Instant Evolution. The Influence Of The City On Human Genes: A Speculative Case," *New Ideas in Psychology* 19, no. 3 (December 2001): 203–220, https://doi.org/10.1016/S0732-118X(01)00004-6.

138   CERN, "The Early Universe, The Big Bang," https://home.cern/science/physics/early-universe.

139   Mark A. Hausman and Jeremiah P. Ostriker, "Galactic Cannibalism. Iii-The Morphological Evolution Of Galaxies And Clusters," *Astrophysical Journal* 224 (1 September 1978): 320–336. Mark A. Hausman and Jeremiah P. Ostriker, "Galactic Cannibalism. Iii-The Morphological Evolution Of Galaxies And Clusters, part 2," *The Astrophysical Journal* 224 (1978): 320–336, https://www.osti.gov/biblio/6879018.

140   385 trillion terawatts per second is the energy of 4.29 billion kg of matter, *WolframAlpha*, https://www.wolframalpha.com/input?i=energy+of+4.29+billion+kg+of+matter. Thanks to Steve Nixon for translating this from math into English. See also: Ali Sundermier, "The Sun Will Destroy Earth A Lot Sooner Than You Might Think," *Business Insider*, 18 September 2016, https://www.yahoo.com/news/sun-destroy-earth-lot-sooner-180000174.html.

141   J. H. Oort, "Superclusters," *Annual Review of Astronomy and Astrophysics* 21, no. 1 (September 1981): 373–428, https://www.annualreviews.org/doi/abs/10.1146/annurev.aa.21.090183.002105.

142   Albert Einstein, *The Meaning of Relativity: Including the Generalization of Gravitation Theory: Understanding the Foundations of Einstein's Revolutionary Theory* (Prabhat Prakashan, 1953). John Archibald Wheeler, *Geons, Black Holes, and Quantum Foam* (W. W. Norton & Company, 1998), 235.

143   "About How Many Stars Are in Space?" UCSB ScienceLine, University of South Carolina Beaufort, Question Date: 19 February 2013, http://scienceline.ucsb.edu/getkey.php?key=3775.

144   Figure based on the James Webb Space Telescope's observations of the earliest galaxies a mere 230 million years after the Big Bang. Fulvio Melia, "The Cosmic Timeline Implied By The Jwst High-Redshift Galaxies," *Monthly Notices of the Royal Astronomical Society: Letters* 521, no. 1 (May 2023): L85–L89, https://doi.org/10.1093/mnrasl/slad025. Ke-Jung Chen, Alexander Heger, Stan Woosley, et al., "Pair Instability Supernovae of Very Massive Population III Stars," *The Astrophysical Journal* 792, no. 1 (1 September 2014): https://iopscience.iop.org/article/10.1088/0004-637X/792/1/44/meta. Richard B. Larson and Volker Bromm, "The First Stars in the Universe: Exceptionally Massive And Bright, The Earliest Stars Changed The Course Of Cosmic History," *Scientific American*, 19 January 2009, https://www.scientificamerican.com/article/the-first-stars-in-the-un.

145   Salpeter, E. E., "The Other–The Effect Of Dying Stars (More Specifically Of Planetary Nebulae And Of Supernovae)," *Fundamental Theories in Physics 5* (2013): 223.

146   "The Life and Death of Stars," *Let's Talk Science: Stem Explained*, 23 July 2019, https://letstalkscience.ca/educational-resources/stem-in-context/life-and-death-stars.

147   Carolyn Ruth, "Where Do Chemical Elements Come from?" *ChemMatters*, October 2009, American Chemical Society, https://www.acs.org/content/dam/acsorg/education/resources/highschool/chemmatters/articlesbytopic/nuclearchemistry/chemmatters-oct2009-origin-chem-elem.pdf.

148   J. Johnson, "Populating the Periodic Table: Nucleosynthesis of the Elements," *Science* 363, no. 1 (February 2019): 474–478, https://doi.org/10.1126/science.aau9540. T. Weaver and S. Woosley, "Evolution And Explosion Of Massive Stars," *Reviews of Modern Physics* 74, no. 4 (1978): 1015–1071, https://doi.org/10.1111/j.1749-6632.1980.tb15942.x.

149   Isabella Ellinger and Adolf Ellinger, "Smallest Unit of Life: Cell Biology," *Comparative Medicine: Anatomy and Physiology* (2014): 19–33.

150   Mark Fischetti and Jen Christiansen, "Our Bodies Replace Billions of Cells Every Day: Blood and the Gut Dominate Cell Turnover," *Scientific American*, 1 April 2021, https://www.scientificamerican.com/article/our-bodies-replace-billions-of-cells-every-day.

151   "New England Patriots Team History," The Pro Football Hall of Fame, https://www.profootballhof.com/teams/new-england-patriots/team-history.

152   Gregor Tanner, Klaus Richter, and Jan-Michael Rost, "The Theory Of Two-Electron Atoms: Between Ground State And Complete Fragmentation," *Reviews of Modern Physics* 72, no. 2 (2000): 497, https://journals.aps.org/rmp/abstract/10.1103/RevModPhys.72.497.

153   W. Hillebrandt and K. Langanke, "Future Articles: Challenges in Nuclear Astrophysics," *Nuclear Physics News* 15 (2005): 21–31, https://doi.org/10.1080/10506890500454683. Laura Geggel, "Why Is Hydrogen the Most Common Element in the Universe?" *LiveScience*, 1 April 2017, https://www.livescience.com/58498-why-is-hydrogen-the-most-common-element.html.

154   C. Long, T. Pappas, K. Southerland, and C. Shortell, "An Analysis Of The Vascular Injuries And Attempted Resuscitation Surrounding The Assassination of Martin Luther King Jr.," *Journal of vascular surgery* 70, no. 5 (2019): 1652–1657, https://doi.org/10.1016/j.jvs.2019.06.203. Rob Hansen, "Nas: I'm Jesse Jackson on the Balcony when King got killed," *Revolt*, 14 August 2018, https://www.revolt.tv/article/2018-06-14/32177/nas-im-jesse-jackson-on-the-balcony-when-king-got-killed.

155   "Martin Luther King Jr. Assassination," *History.com*, 15 December 2023, https://www.history.com/topics/black-history/martin-luther-king-jr-assassination.

156   Yella Hewings-Martin, Ph.D, "Cell Death: Is Our Health at Risk?" *Medical News Today*, 11 August 2017, https://www.medicalnewstoday.com/articles/318927.

157   Stephan Wilkinson, "How The B-25 Became The Ultimate Strafer Of World War II," *History Net*, 28 April 2020, https://www.historynet.com/the-mighty-mitchell-how-b-25s-became-one-of-the-most-essential-aircraft-in-wwii.

158   Joseph Heller, *Catch-22, A Novel* (Simon & Schuster, 1999), 402, 404. First published 1961.

159   Ibid., 56, 214, 320, 402.

160   K. Moran, J. Backman, H. Brinkhuis, et al., "The Cenozoic Palaeoenvironment of the Arctic Ocean," *Nature* 441 (1 June 2006): 601-605, https://doi.org/10.1038/nature04800. Andrew J. Christ et al., "Deglaciation of Northwestern Greenland during Marine Isotope Stage 11," *Science* 381 (20 July 2023): 330–335, https://doi.org/10.1126/science.ade4248. Alex Kirby, "North Pole 'Was Once Subtropical'," BBC News Online, 7 September 2004, http://news.bbc.co.uk/2/hi/science/nature/3631764.stm.

161   James Zachos et al., "Trends, Rhythms, and Aberrations in Global Climate 65 Ma to Present," *Science*, 27 April 2001, 686–93, https://doi.org/10.1126/science.1059412. Senckenberg Research Institute and Natural History Museum, "Tropical Climate In The Antarctic: Palm Trees Once Thrived On Today's Icy Coasts 52 Million Years Ago," *ScienceDaily*, 1 August 2012, http://www.sciencedaily.com/releases/2012/08/120801132339.htm.

162   Claude J. Allgre and Stephen H, Schneider, "Evolution of Earth," *Scientific American*, July 2005, https://www.scientificamerican.com/article/evolution-of-earth.

163   Charles Herbert Langmuir, Wallace Smith Broecker, *How to Build a Habitable Planet: The Story of Earth from the Big Bang to Human Kind* (Princeton University Press, 2012).

164   Anirudh Chiti, Anna Frebel, Joshua D. Simon, Denis Erkal, Laura J. Chang, Lina Necib, Alexander P. Ji, Helmut Jerjen, Dongwon Kim, and John E. Norris, "An Extended Halo Around an Ancient Dwarf Galaxy," *Nature Astronomy* 5, no. 4 (2021): 392–400, https://www.nature.com/articles/s41550-020-01285-w. Sophie Lewis, "Astronomers Find Origins Of 'Galactic Cannibalism' With Discovery Of Ancient Dark Matter Halo," CBS News, 2 February 2021, https://www.cbsnews.com/news/astronomers-discover-galactic-cannibalism-ancient-dark-matter-halo-dwarf-galaxy.

165   B.A. Cohen et al., "Support for the Lunar Cataclysm Hypothesis from Lunar Meteorite Impact Melt Ages," *Science* 290 (1 Dec 2000): 1754–1756, https://doi.org/10.1126/science.290.5497.1754.

166   Vaclav Smil, *The Earth's Biosphere: Evolution, Dynamics, and Change,* (Cambridge, MA: MIT Press, 2003), 47.

167   Andrew Glikson, et al, "A New ~3.46 Ga Asteroid Impact Ejecta Unit At Marble Bar, Pilbara Craton, Western Australia: A Petrological, Microprobe And Laser Ablation Icpms Study," *Precambrian Research* 279 (2016): 103–122, https://www.sciencedirect.com/science/article/abs/pii/S0301926816300511. Adam Mann, "Bashing Holes in the Tale of Earth's Troubled Youth," *Nature* (24 January 2018), https://www.nature.com/articles/d41586-018-01074-6. Universe Today, "Massive Asteroid Hit Earth 3.5 Billion Years Ago, Dwarfing One That Killed The Dinosaurs," *Futurism*, 25 May 2016, https://futurism.com/30-km-wide-asteroid-impacted-australia-3-4-billion-years-ago.

168   R.M. Hazen, *The Story of Earth: The First 4.5 Billion Years, from Stardust to Living Planet* (Penguin Publishing Group, 2013).

169   Even today, says Idaho State University physicist Martin Hackworth, "an area like the Pinnacles National Park can have high temperatures of 38°C (100 °F) during a summer day, and then have lows of 5–10 °C (41≠50 °F)." That's a difference of up to 59 degrees Fahrenheit. But on the early earth, the difference between night and daytime temperature was so extreme that it cracked rocks. M. Hackworth "Weather & Climate," course notes, archived 12 October 2008 at the Wayback Machine, https://www.scribd.com/document/623191995/Diurnal-Air-Temperature-Variation. For the daily temperature change of the early earth cracking rocks, see David W. Schwartzman, *Life, Temperature, and the Earth* (Columbia University Press, 1999), and James F. Petersen, et al., *Physical Geography* (Cengage Learning, 2016). See also David Moore, *Fungal Biology in the Origin and Emergence of Life* (Cambridge University Press, 2013), 91.

170   A. Berger, ed., "Milankovitch and Climate: Understanding the Response to Astronomical Forcing," Springer Science & Business Media, 21 November 2013.

171   Alan Buis, NASA's Jet Propulsion Laboratory, "Milankovitch (Orbital) Cycles and Their Role in Earth's Climate," NASA News, 27 February 2020, https://climate.nasa.gov/news/2948/milankovitch-orbital-cycles-and-their-role-in-earths-climate.

172   C.J. Campisano, "Milankovitch Cycles, Paleoclimatic Change, and Hominin Evolution," *Nature Education Knowledge* 4, no. 3 (2012): 5, https://www.nature.com/scitable/knowledge/library/milankovitch-cycles-paleoclimatic-change-and-hominin-evolution-68244581.

173   David M. Raup and J. John Sepkoski, "Periodicity of Extinctions in the Geologic Past," *Proceedings of the National Academy of Sciences* (1 February 1984), 801–805.

174   Ibid.

175   Sami K. Solanki, "Solar Variability and Climate Change: Is There a Link?" *Astronomy & Geophysics* 43, no 5 (2002): 5, 9–5, 13.

176   N. Weiss and M. Thompson, "The Solar Dynamo," *Space Science Reviews* 144 (1 October 2008): 53–66, https://doi.org/10.1007/S11214-008-9435-Z.

177   NASA Science, "The Inconstant Sun," NASA News, 2003, https://www.spacedaily.com/news/climate-03a.html.

178   Glenn Elert, "Period of the Sun's Orbit around the Galaxy (Cosmic Year)," in Mark Morris (ed.), *The Physics Factbook*, http://hypertextbook.com/facts/2002/StacyLeong.shtml.

179   C.L. Kirkland, et al., "Did Transit Through The Galactic Spiral Arms Seed Crust Production On The Early Earth?" *Geology* 50, no. 11 (23 August 2022): 1312–1317, https://doi.org/10.1130/G50513.1. Michael Brown, et al., "Giant Impacts And The Origin And Evolution Of Archean Cratons," Goldschmidt 2023 Conference. Laura Fattaruso, "Movement Of The Solar System Through The Milky Way's Galactic Spiral Arms Helped Form Earth's First Continents," *ScienMag* (24 August 2022). Shannon Hal, "The Milky Way's Spiral Arms May Have Carved Earth's Continents: A Controversial New Theory Suggests The Milky Way Galaxy's Arms Sent Comets Hurtling Toward Early Earth, Where Impacts Built New Continental Crust," *Scientific American*, 30 September 2022, https://www.scientificamerican.com/article/the-milky-ways-spiral-arms-may-have-carved-earths-continents.

180   K.S. Carslaw, et al., "Cosmic Rays, Clouds, and Climate," *Science* 298 (29 November 2002): 1732–1737, https://doi.org/10.1126/science.1076964. Lev I. Dorman, "Space Weather and Cosmic Ray Effects," in Trevor M. Letcher editor, *Climate Change*, Third Edition (Elsevier, 2021), 711-768, https://doi.org/10.1016/B978-0-12-821575-3.00033-5, https://www.sciencedirect.com/science/article/pii/B9780128215753000335.

181   P. C. Frisch, "The Journey of the Sun," *arXiv.org Astronomy*, 1997, https://arxiv.org/abs/astro-ph/9705231. Dominik Koll, et al., "Interstellar 60Fe in Antarctica," *Physical Review Letters* 123, no. 7 (August 2019). Ginger Pinholster, "Someday, Cosmic Cloud Might Burst Earth's 'Breathing Bubble'," *University of Delaware Messenger* 7, no. 4 (1998), https://www1.udel.edu/PR/Messenger/98/4/someday.html.

182   J. Rojas, J. Duprat, C. Engrand, E. Dartois, L. Delauche, M. Godard, M. Gounelle, J.D. Carrillo-Sánchez, P. Pokorný and J.M.C. Plane, "The Micrometeorite Flux At Dome C (Antarctica), Monitoring The Accretion Of Extraterrestrial Dust On Earth," *Earth & Planetary Science Letters* (15 April 2021), https://doi.org/10.1016/j.epsl.2021.116794. "More Than 5,000 Tons Of Extraterrestrial Dust Fall To Earth Each Year," press release, CNRS Foundation (National Centre de la Recherche Scientifique), 8 April 2021, https://www.

cnrs.fr/en/more-5000-tons-extraterrestrial-dust-fall-earth-each-year Daisy Dobrijevic, and "5,200 Tons Of Space Dust Falls On Earth Each Year, Study Finds," *Space.com*, 22 April 2021, https://www.space.com/extraterrestrial-dust-falls-on-earth.

183   K. Parley, D. Patterson, "A 100-Kyr Periodicity In The Flux Of Extraterrestrial 3he To The Sea Floor," *Nature* 378 (7 December 1995): 600–603, https://doi.org/10.1038/378600a0 https://www.nature.com/articles/378600a0. V.I. Ermakov, V.P. Okhlopkov and Y.I. Stozhkov, "Influence Of Cosmic Rays And Cosmic Dust On The Atmosphere And Earth's Climate," *Bulletin of the Russian Academy of Science, Physics* 73 (8 April 2009): 416–418, https://doi.org/10.3103/S1062873809030411, https://link.springer.com/article/10.3103/S1062873809030411. Anders Cronholm and Birger Schmitz, "Extraterrestrial Chromite Distribution Across The Mid-Ordovician Puxi River Section, Central China: Evidence For A Global Major Spike In Flux Of L-Chondritic Matter," *Icarus* 208, no. 1 (2010): 36–48, https://doi.org/10.1016/j.icarus.2010.02.004, https://www.sciencedirect.com/science/article/pii/S0019103510000606. B. Peucker-Ehrenbrink, "Accretion Of Extraterrestrial Matter During The Last 80 Million Years And Its Effect On The Marine Osmium Isotope Record," *Geochimica et Cosmochimica Acta* 60, no. 17 (1996): 3187–3196, https://doi.org/10.1016/0016-7037(96)00161-5, https://www.sciencedirect.com/science/article/pii/0016703796001615. John M.C. Plane, "Cosmic Dust in the Earth's Atmosphere," *Royal Society of Chemistry, Chemical Society Review* 41 (7 June 2012): 6507-6518, https://doi.org/10.1039/C2CS35132C, https://pubs.rsc.org/en/content/articlehtml/2012/cs/c2cs35132c. Peucker-Ehrenbrink, Greg Ravizza, "The Effects Of Sampling Artifacts On Cosmic Dust Flux Estimates: A Reevaluation Of Nonvolatile Tracers (Os, Ir)," *Geochimica et Cosmochimica Acta* 64, no. 11 (2000): 1965–1970, https://doi.org/10.1016/S0016-7037(99)00429-9, https://www.sciencedirect.com/science/article/pii/S0016703799004299. C.S. Gardner, A.Z. Liu, D.R. Marsh, W. Feng, and J.M.C. Plane, "Inferring the Global Cosmic Dust Influx To The Earth's Atmosphere From Lidar Observations Of The Vertical Flux Of Mesospheric Na," *Journal of Geophysical Research: Space Physics* 119 (2014): 7870–7879, https://doi.org/10.1002/2014JA020383, https://agupubs.onlinelibrary.wiley.com/doi/full/10.1002/2014JA020383. Stephen J. Kortenkamp and Stanley F. Dermott, "Accretion of Interplanetary Dust Particles by the Earth," *Icarus* 135, no. 2 (1998): 469–495, https://doi.org/10.1006/icar.1998.5994, https://www.sciencedirect.com/science/article/pii/S0019103598959942. K.K. Turekian and M.P. Bacon, "Geochronometry of Marine Deposits," in Heinrich D. Holland, Karl K. Turekian (eds.), *Treatise on Geochemistry*, Second Edition (Elsevier, 2014), 335–353, https://doi.org/10.1016/B978-0-08-095975-7.00612-4, https://www.sciencedirect.com/science/article/pii/B9780080959757006124. John Plane, "CODITA—Cosmic Dust in the Terrestrial Atmosphere," *The Codita Project*, https://john-plane.

leeds.ac.uk/research/middle-upper-atmosphere/codita-cosmic-dust-in-the-terrestrial-atmosphere. Nancy Atkinson, "Getting a Handle on How Much Cosmic Dust Hits the Earth," *Universe Today*, 30 March 2012, https://www.universetoday.com/94392/getting-a-handle-on-how-much-cosmic-dust-hits-earth.

184 Vaclav Smil, *The Earth's Biosphere: Evolution, Dynamics, and Change* (Cambridge, MA: MIT Press, 2003), 93.

185 Michael Schirber, "'Snowball Earth' Might Have Been Slushy," *Astrobiology Magazine*, NASA's Goddard Institute for Space Studies (GISS), August 2015, https://www.giss.nasa.gov/research/features/201508_slushball.

186 Hidefumi Imura, *Environmental Systems Studies: A Macroscope for Understanding and Operating Spaceship Earth* (New York: Springer, 2013), 30. W.B. Harland, "Origins and Assessment of Snowball Earth Hypotheses," *Geological Magazine* 144, no. 4 (July 2007): 633–642. Robert E. Kopp, Joseph L. Kirschvink, Isaac A. Hilburn, and Cody Z. Nash, "The Paleoproterozoic Snowball Earth: A Climate Disaster Triggered By The Evolution Of Oxygenic Photosynthesis," *Proceedings of the National Academy of Sciences of the United States of America* 102, no. 32 (2005): 11131–11136. Gabrielle Walker, *Snowball Earth: The Story of a Maverick Scientist and His Theory of the Global Catastrophe that Spawned Life as We Know It* (New York: Broadway Books, Random House, 2004). Vaclav Smil, *The Earth's Biosphere: Evolution, Dynamics, and Change* (Cambridge, MA: MIT Press, 2003), 89.

187 "Three episodes of snowball Earth that likely occurred around 2,200, 710 and 640 million years ago." Philippe Bertrand and Louis Legendre, *Earth, Our Living Planet* (Springer International Publishing, 2021, Kindle Edition), 590.

188 J. Lovelock and M. Whitfield, "Life Span of the Biosphere," *Nature* 296 (1 April 1982): 561–563, https://doi.org/10.1038/296561a0. G.C. Reid, "Solar Variability and the Earth's Climate: Introduction and Overview," *Space Science Reviews* 94 (November 2000): 1–11, https://doi.org/10.1023/A:1026797127105. "Total Solar Irradiance: The Sun Also Changes, Understanding Earth's Energy Balance," NASA, 28 November 2017.

189 Technically these instant global warmings followed by long, slow global coolings are called "Dansgaard-Oeschger" events. In the 120,000 years since the end of the Eemian interglacial, these instant global warmings have occurred roughly every 1,500 years. Stefan Rahmstorf, "Ocean Circulation and Climate During the Past 120,000 Years," *Nature* 419 (12 September 2002): 207–214, https://doi.org/10.1038/nature01090. Paolo Stocchi Bas de Boer, Pippa L. Whitehouse, and Roderik S.W. van de Wal, "Current State And Future Perspectives On Coupled Ice-Sheet – Sea-Level Modelling," *Quaternary Science Reviews* 169 (2017): 13–28, https://doi.org/10.1016/j.quascirev.2017.05.013. Don Easterbrook, "Dansgaard-Oeschger event," *Encyclopedia Britannica*, 26

September 2019, https://www.britannica.com/science/Dansgaard-Oeschger-event.

190   Simone Ulmer, "An Ice Age Lasting 115,000 Years In Two Minutes," *Centro Svizzero di Calcolo Scientifico*, Swiss National Supercomputing Centre, 2 November 2018, https://phys.org/news/2018-11-ice-age-years-minutes. html. "When Have Ice Ages Occurred? Explore The Ice Age Midwest.....Plants And Animals of the Pleistocene," Illinois State Museum, 2015, https://iceage. museum.state.il.us/content/when-have-ice-ages-occurred.

191   Alan J. Kaufman and Shuhai Xiao, "High CO_2 Levels In The Proterozoic Atmosphere Estimated From Analyses Of Individual Microfossils," *Nature* 425 (18 September 2003): 279–282, https://www.nature.com/articles/nature01902. The authors estimate the CO_2 level in the atmosphere at "between 10 and 200 times" what it is today. See also Virginia Tech, "Atmospheric Carbon Dioxide Greater 1.4 Billion Years Ago," *ScienceDaily*, 19 September 2003, www. sciencedaily.com/releases/2003/09/030918092804.htm.

192   Stefan Rahmstorf, "Ocean Circulation and Climate During the Past 120,000 years," *Nature* 419 (12 September 2002): 207–214, http://www.nature. com/nature/journal/v419/n6903/abs/nature01090.html. Andrew J. Weaver, Oleg A. Saenko, Peter U. Clark, Jerry X, "Mitrovica—Meltwater Pulse 1A from Antarctica as a Trigger of the Bølling-Allerød Warm Interval," *Science* 299, no. 5613 (14 March 2003): 1709–1713, https://www.science.org/doi/full/10.1126/ science.1081002. Bruno Voituriez, "The Gulf Stream," Unesco, 2006, https:// unesdoc.unesco.org/ark:/48223/pf0000148252.

193   J. Gottschalk, L. Skinner, J. Lippold, et al., "Biological and Physical Controls In The Southern Ocean On Past Millennial-Scale Atmospheric Co_2 Changes," *Nature Communications* 7 (2016), https://doi.org/10.1038/ ncomms11539. National Park Service, "Geologic Formations," Guadalupe Mountains National Park, Texas, 2017, https://www.nps.gov/gumo/learn/ nature/geologicformations.htm.

194   David Braun, "Asteroid Terminated Dinosaur Era in Days," *National Geographic*, Nationalgeographic.com, 4 March 2010.

195   Juan Siliezar, "The Cataclysm that Killed the Dinosaurs," *Harvard Gazette*, 15 February 2021, https://news.harvard.edu/gazette/story/2021/02/new-theory-behind-asteroid-that-killed-the-dinosaurs.

196   Extrapolated from the figure of mass extinctions every 26 million years in D.M. Raup and J.J. Sepkoski, Jr, "Periodicity of Extinctions in the Geologic Past," *Proceedings of the National Academy of Sciences* 81, no. 3 (1 February 1984): 801–805, https://www.ncbi.nlm.nih.gov/pmc/articles/PMC344925.

197   The one in thirty billion figure comes from biophysicist Harold Morowitz. Two of the biological greats of the 20th century, Ernst Mayr and Nobel Prize winner Jacques Monod, put the odds of life evolving here or anywhere at close

to zero. Astrophysicist Fred Hoyle, another legend of mid-20th century science, gave slightly better odds. He calculated the chance of life's evolution at 10-40,000. Which is so close to zero that it's ridiculous. See also: Vaclav Smil, *The Earth's Biosphere: Evolution, Dynamics, and Change* (Cambridge, MA: MIT Press, 2003), 60–62.

198   Richard Wrangham, *Catching Fire: How Cooking Made Us Human* (Profile Books, 2010), 85–86. Stephen J. Pyne, *The Pyrocene: How We Created an Age of Fire, and What Happens Next* (University of California Press, 2022), 21. The following article not only refers to the use of fire 1.5 million years ago, but is complete with what the specialists call "pot lids:" S. Hlubik, F. Berna, C. Feibel, D. Braun, and J.W. Harris, (2017), "Researching The Nature Of Fire At 1.5 Mya On The Site Of Fxjj20 Ab, Koobi Fora, Kenya, Using High-Resolution Spatial Analysis And Ftir Spectrometry," *Current Anthropology* 58, no. (16 February 1984): S243–S257. C.K. Brain, *The Hunters or the Hunted?: An Introduction to African Cave Taphonomy* (University of Chicago Press, 1983). J.A. Gowlett, "The Discovery Of Fire By Humans: A Long And Convoluted Process," *Philosophical Transactions of the Royal Society B: Biological Sciences* 371, no. 1696 (5 June 2016), https://royalsocietypublishing.org/doi/10.1098/rstb.2015.0164.

199   Jordi Serangeli, et al., "Schöningen: A Reference Site for the Middle Pleistocene," *Journal of Mediterranean Earth Sciences* 15 (2023). Sahir Pandey, "Fur Coats from 300,000 Years Ago: Earliest Evidence of Bear Skin Use," *Ancient Origins* (29 December 2022), https://www.ancient-origins.net/news-evolution-human-origins/bear-skin-0017722.

200   François Djindjian, "Identifying The Hunter-Gatherer Systems Behind Associated Mammoth Bone Beds And Mammoth Bone Dwellings," *Quaternary international* 359 (2015): 47–57, https://www.sciencedirect.com/science/article/pii/S1040618214004595. Brian Handwerk, "A Mysterious 25,000-Year-Old Structure Built of the Bones of 60 Mammoths," *Smithsonian Magazine*, 16 March 2020, https://www.smithsonianmag.com/science-nature/60-mammoths-house-russia-180974426.

201   Arne Næss and George Sessions, "Basic Principles of Deep Ecology," The Anarchist Library, 1984, https://theanarchistlibrary.org/library/arne-naess-and-george-sessions-basic-principles-of-deep-ecology. Also available as a download at: *The Trumpeter* 3, no. 4, https://trumpeter.athabascau.ca/index.php/trumpet/article/download/44/39/102.

202   Peter Victor, "Questioning Economic Growth," *Nature* (8 November 2010): 370, https://www.nature.com/articles/468370a.

203   Vaclav Smil, *The Earth's Biosphere: Evolution, Dynamics, and Change*, (Cambridge, MA: MIT Press, 2003), 49. James F. Kasting, "Theoretical Constraints On Oxygen And Carbon Dioxide Concentrations In The Precambrian

Atmosphere," *Precambrian Research* 34, nos. 3–4 (January 1987): 205–229, https://doi.org/10.1016/0301-9268(87)90001-5.

204   S. Harmand, J. Lewis, C. Feibel, et al., "3.3-Million-Year-Old Stone Tools from Lomekwi 3, West Turkana, Kenya," *Nature* 521 (20 May 2015): 310–315, https://doi.org/10.1038/nature14464.

205   Shun-ichiro Karato, "On the Origin of the Asthenosphere," *Earth and Planetary Science Letters* 321 (2012): 95–103.

206   V.P. Trubitsyn, "Principles of the tectonics of floating continents," *Izvestiya Physics of the Solid Earth* 36, no. 9 (2000): 708–741.

207   L. Chen, X. Wang, X. Liang, et al., "Subduction Tectonics Vs. Plume Tectonics—Discussion On Driving Forces For Plate Motion," *Science China Earth Sciences* 63 (2 January 2020): 315–328, https://doi.org/10.1007/s11430-019-9538-2. C. Mallard, N. Coltice, M. Seton, et al., "Subduction Controls The Distribution And Fragmentation Of Earth's Tectonic Plates," *Nature* 535 (2016): 140–143, https://doi.org/10.1038/nature17992, https://www.nature.com/articles/nature17992. US Geological Survey, "EarthWord—Subduction," https://www.usgs.gov/news/science-snippet/earthword-subduction.

208   US Geological Survey, "Subduction Zone Science, Amazing Events in Subduction Zones," 7 September 2020, https://www.usgs.gov/special-topics/subduction-zone-science/science/introduction-subduction-zones-amazing-events.

209   The sea of objects through which the early earth swam is called "the near earth swarm of the protoplanetary cloud planetesimals." Oleg G. Sorokhtin, G.V. Chilingarian, N.O. Sorokhtin, *Evolution of Earth and its Climate: Birth, Life and Death of Earth* (Burlington, MA: Elsevier, 2004), 65. Felix M. Gradstein, James G. Ogg, Alan G. Smith (eds.), *A Geologic Time Scale* (Cambridge University Press, 2005), 143. Committee on Planetary and Lunar Exploration, "Exploration of Near Earth Objects," Commission on Physical Sciences, Mathematics, and Applications, Space Studies Board, Division on Engineering and Physical Sciences, Washington, DC, National Research Council (1998), 1.

210   H. Sierks, P. Lamy, C. Barbieri, et al, "Images of Asteroid 21 Lutetia: A Remnant Planetesimal from the Early Solar System," *Science* (28 October 2011): 487–490.

211   David A. Clague and G. Brent Dalrymple, "Tectonics, Geochronology, and Origin of the Hawaiian-Emperor Volcanic Chain," in *A Natural History of the Hawaiian Islands: Selected Readings II*, edited by E. Alison Kay (Honolulu: University of Hawai'i Press, 1994), 5.

212   J. Foster, *Hawai'i Volcanoes National Park* (Arcadia Publishing, 2015), 70.

213   Ken Rubin and Rochelle Minicola, "Mauna Loa Eruption History," Hawaii Center for Volcanology, School of Ocean and Earth Science and Technology,

University of Hawai'i at Manoa, http://www.soest.hawaii.edu/GG/HCV/mloa-eruptions.html.

214   T. E. Zegers and A. Ocampo, "Vaalbara And Tectonic Effects Of A Mega Impact In The Early Archean 3470 Ma," in Third International Conference on Large Meteorite Impacts 143, no. 3, Nordlingen, Germany, Lunar and Planetary Institute, Houston, TX.

215   S.M. Reddy, David A.D. Evans, "Palaeoproterozoic Supercontinents And Global Evolution: Correlations From Core To Atmosphere," in *Palaeoproterozoic Supercontinents and Global Evolution*, edited by Steven Michael Reddy, Bath, UK: Geological Society, Special Publication 323 (2009): 8. Philip Kearey, Keith A. Klepeis, Frederick J. Vine, *Global Tectonics* (Oxford, UK: Wiley-Blackwell, 2009), 374.

216   K.C. Condie, "Supercontinents, Superplumes and Continental Growth: the Neoproterozoic Record," in *Proterozoic East Gondwana: Supercontinent Assembly and Breakup,* edited by Masaru Yoshida, Brian F. Windley, and Somnath Dasgupta (Bath, UK: Geological Society of London, 2003), 1–16.

217   Hubert Staudigel, Harald Furnes, Nicola McLoughlin, et al., "3.5 billion Years of Glass Bioalteration: Volcanic Rocks As A Basis For Microbial Life?" *Earth-Science Reviews* 89, nos. 3–4 (August 2008): 156–176, https://doi.org/10.1016/j.earscirev.2008.04.005.

218   Foundation for Applied Molecular Evolution, "Scientists Announce a Breakthrough in Determining Life's Origin on Earth—and Maybe Mars," *Phys. org*, 3 June 2022, https://phys.org/news/2022-06-scientists-breakthrough-life-earthand-mars.html.

219   Frances Westall, M.J. De Wit, J. Dann, et al., "Early Archaean Fossil Bacteria And Biofilms In Hydrothermally-Influenced Shallow Water Sediments. Barberton Greenstone Belt. South Africa," *Precambrian Research* (February 2001): 93–116.

220   Uri Moran, "Number of Cells in Colony," BioNumb3R5, https://bionumbers.hms.harvard.edu/bionumber.aspx?s=n&v=8&id=104458. Puay Yen Yap and Dieter Trau, "Direct E.Coli Cell Count At Od600," Tip Biosystems Pte Ltd, Singapore. Matthew Wiens, "What Do You Think A Colony Is Composed Of, A Few Bacteria Or Closer To Millions Of Bacteria? How Do You Know?" *Quora*, https://www.quora.com/What-do-you-think-a-colony-is-composed-of-a-few-bacteria-or-closer-to-millions-of-bacteria-How-do-you-know.

221   Jordi van Gestel and Martin A. Nowak, "Phenotypic Heterogeneity And The Evolution Of Bacterial Life Cycles," *PLOS Computational Biology* 12, no. 2 (2016), https://journals.plos.org/ploscompbiol/article?id=10.1371/journal.pcbi.1004764. Yuriy Pichugin, et al., "Fragmentation Modes And The Evolution Of Life Cycles," *PLoS computational biology* 13, no. 11 (2017), https://journals.plos.org/ploscompbiol/article?id=10.1371/journal.pcbi.1005860. Jφrgen

Henrichsen, "Bacterial Surface Translocation: A Survey And A Classification," *Bacteriological Reviews* 36, no. 4 (1972): 478–503.

222   N. Steinberg, I. Kolodkin-Gal, "The Matrix Reloaded: How Sensing The Extracellular Matrix Synchronizes Bacterial Communities," *Journal of Bacteriology* 197, no. 13 (3 June 2015): 2092–2103, https://doi.org/10.1128/JB.02516-14, https://journals.asm.org/doi/10.1128/jb.02516-14. Eshel ben-Jacob, personal communication, 1999.

223   John Odling-Smee, University of Oxford, "Extended Evolutionary Synthesis," https://extendedevolutionarysynthesis.com/person/john-odling-smee—this is Dr. Smee's University of Oxford webpage.

224   John Odling-Smee, et al., "Niche Construction Theory: A Practical Guide for Ecologists," *The Quarterly Review of Biology* (March 2013): 4–28. Kevin N. Laland and Michael J. O'Brien, "Cultural Niche Construction: An Introduction," *Biological Theory* 6 (31 July 2012): 191–202, https://link.springer.com/article/10.1007/s13752-012-0026-6.

225   J. Rodney Quayle and T. Ferenci, "Evolutionary Aspects of Autotrophy," *Microbiological Reviews* 42, no. 2 (1978): 251–273. Eric S. Boyd, et al., "Bioenergetic Constraints on the Origin of Autotrophic Metabolism," *Philosophical Transactions of the Royal Society A* 378, no. 2165 (6 January 2020), https://royalsocietypublishing.org/doi/full/10.1098/rsta.2019.0151. John A. Raven, "The Evolution Of Autotrophy In Relation To Phosphorus Requirement," *Journal of Experimental Botany* 64, no. 13 (2013): 4023–4046.

226   Francis Westall, "Early Life: Nature, Distribution and Evolution," in *Origins and Evolution of Life: An Astrobiological Perspective,* edited by Muriel Gargaud, Purificación López-García and Hervé Martin (Cambridge, UK: Cambridge University Press, 2011), 402.

227   Richard Cowen, *History of Life* (Oxford, UK: Blackwell Publishing, 2013).

228   Vargas, Madeline, et al., "Microbiological Evidence for Fe (III) Reduction on Early Earth," *Nature* 395, no. 6697 (1998): 65–67.

229   D.W.J. Bosence and PH. Bridges, "A Review Of The Origin And Evolution Of Carbonate Mud-Mounds," in C.L.V. Monty, *Carbonate Mud-Mounds: Their Origin and Evolution*, The International Association of Sedimentologists (Oxford, UK: Blackwell-Science, 1995), 8.

230   Javier Tamames, et al., "Bringing Gene Order into Bacterial Shape," *Trends in Genetics* 17, no. 3 (2001): 124–126.

231   Francis Westall, "Early Life: Nature, Distribution and Evolution," in *Origins and Evolution of Life: An Astrobiological Perspective,* edited by Muriel Gargaud, Purificación López-García and Hervé Martin (Cambridge, UK: Cambridge University Press, 2011), 402.

232   Susan L. Brantley, "Rock to Regolith," *Nature Geoscience* 3, no. 5 (2010): 305–306. M.C. Ngoma, et al., "Sub-Core Scale Characterization of Microbial Invasion Impact in Carbonates: Implications for Mechanical Alteration," ARMA [American Rock Mechanics Association] US Rock Mechanics/Geomechanics Symposium, ARMA, 2023.

233   I.D.A. Ribeiro, C.G. Volpiano, L.K. Vargas, C.E. Granada, B.B. Lisboa, and L.M.P. Passaglia, "Use of Mineral Weathering Bacteria to Enhance Nutrient Availability in Crops: A Review," *Frontiers in Plant Science* 11 (11 December 2020), https://doi.org/10.3389/fpls.2020.590774.

234   Jost Wingender, Thomas R. Neu, and Hans-Curt Flemming, "What Are Bacterial Extracellular Polymeric Substances?," in *Microbial Extracellular Polymeric Substances: Characterization, Structure and Function*, edited by Jost Wingender, Thomas R. Neu, Hans-Curt Flemming (Berlin: Springer-Verlag, 1999), 12.

235   Frances Westall, M.J. De Wit, J. Dann, et al., "Early Archaean Fossil Bacteria And Biofilms In Hydrothermally-Influenced Shallow Water Sediments, Barberton Greenstone Belt, South Africa," *Precambrian Research* (February, 2001), 93–116.

236   Eshel Ben-Jacob, Personal communication, 1998–2001.

237   Eshel Ben-Jacob, Israela Becker and Yoash Shapira, "Reflections on Biochemical Linguistics of Bacteria," School of Physics and Astronomy, The Raymond and Beverly Sackler Faculty of Exact Sciences, Tel-Aviv University, 2003. Eshel Ben Jacob and Yoash Shapira, "Meaning-Based Natural Intelligence Vs. Information-Based Artificial Intelligence," School of Physics and Astronomy, Raymond & Beverly Sackler Faculty of Exact Sciences, Tel Aviv University, 25–28, https://www.researchgate.net/profile/Yoash-Shapira/publication/224909824_Meaning-based_natural_intelligence_vs_information-based_artificial_intelligence/links/09e415135eba87f921000000/Meaning-based-natural-intelligence-vs-information-based-artificial-intelligence.pdf. Eshel Ben Jacob, Yoash Shapira, and Alfred I. Tauber, "Seeking the Foundations of Cognition in Bacteria: From Schrodinger's Negative Entropy to Latent Information," *Physica A* 359 (2006): 506, 512. Eshel Ben-Jacob, personal communication, 19 January 2004.

238   Irving Kett, *Engineered Concrete: Mix Design and Test Methods*, Second Edition (Boca Raton, FL: CRC Press 2010). 5–6.

239   Robert E. Riding, Stanley M. Awramik, eds., *Microbial Sediments* (Berlin: Springer-Verlag, 2000), 214. Francis Westall, "Early Life: Nature, Distribution and Evolution," in *Origins and Evolution of Life: An Astrobiological Perspective*, edited by Muriel Gargaud, Purificación López-García, Hervé Martin (Cambridge, UK: Cambridge University Press, 2011), 402.

240   Westall, "Early Life," 402. Lost City Research, "Biology," in *Lost City: One of the Most Extreme Environments on Earth* website, University of Washington, 2003, http://www.lostcity.washington.edu/story/Biology. David G. Davies, "Regulation of Matrix Polymer in Biofilm Formation and Dispersion," in *Microbial Extracellular Polymeric Substances: Characterization, Structure and Function*, edited by Jost Wingender, Thomas R. Neu and Hans-Curt Flemming (Berlin: Springer-Verlag, 1999), 104.

241   "Ethylene Polymerization," *Wikipedia*, https://en.wikipedia.org/wiki/Ethylene#Polymerization.

242   M. Fata Moradali and Bernd H. A. Rehm, "Bacterial Biopolymers: From Pathogenesis to Advanced Materials," *Nature Reviews Microbiology* 18 (28 January 2020), 195–210, https://doi.org/10.1038/s41579-019-0313-3.

243   H.C. Flemming, E.D. van Hullebusch, T.R. Neu, et al., "The Biofilm Matrix: Multitasking in a Shared Space," *Nature reviews: Microbiology* 21 (20 September 2022): 70–86, https://doi.org/10.1038/s41579-022-00791-0.

244   N. Billings, A. Birjiniuk, T.S. Samad, P.S. Doyle, and K. Ribbeck, "Material Properties Of Biofilms—A Review Of Methods For Understanding Permeability And Mechanics," *Reports on Progress in Physics* 78, no. 3 (2015), https://doi.org/10.1088/0034-4885/78/3/036601.

245   Carey D. Nadell, Joao B. Xavier, Kevin R. Foster, "The sociobiology of biofilms," *FEMS Microbiology Reviews*, Volume 33, Issue 1, January 2009, pages 206–224, https://doi.org/10.1111/j.1574-6976.2008.00150.x

246   Sivan Elias , Ehud Banin, "Multi-species biofilms: living with friendly neighbors," *FEMS Microbiology Reviews*, Volume 36, Issue 5, September 2012, pages 990–1004, https://doi.org/10.1111/j.1574-6976.2012.00325.x Xiao-Lin Chu, Quan-Guo Zhang, Angus Buckling, Meaghan Castledine, "Interspecific Niche Competition Increases. Morphological Diversity in Multi-Species Microbial Communities," *Frontiers in Microbiology*, 30, July 2021, Volume 12 – 2021, https://doi.org/10.3389/fmicb.2021.699190

247   Frances Westall, "Early life: nature, distribution and evolution," in Muriel Gargaud (Editor), Purificación López-Garcìa (Editor), Hervé Martin (Editor), *Origins and Evolution of Life: An Astrobiological Perspective* (2011), Cambridge University Press, page 402.

248   Wolgemuth, Charles, et al., "How Myxobacteria Glide," *Current Biology* 12, no. 5 (2002): 369–377. Peter Weiss, "Microbes Fire An Oozie: Slime Engines May Push Bacteria Along," *Science News* 161, no. 12 (2002): 180.

249   Jost Wingender, Thomas R. Neu, and Hans-Curt Flemming, "What Are Bacterial Extracellular Polymeric Substances?" in *Microbial Extracellular Polymeric Substances: Characterization, Structure and Function*, edited by Jost Wingender, Thomas R. Neu, and Hans-Curt Flemming (Berlin: Springer-Verlag, 1999), 7–9.

250 Westall, "Early life," 402.

251 Richard F. Meyer, Emil D. Attanasi, and Philip A. Freeman, *Heavy Oil and Natural Bitumen Resources in Geological Basins of the World*, U.S. Geological Survey, 2007. American Geosciences Institute, "What Are Tar Sands?," 2018, https://www.americangeosciences.org/critical-issues/faq/what-are-tar-sands.

252 Ibid.

253 Ibid.

254 Mark Shrope and John Pickrell, "Introduction: Mysteries of the Deep Sea," *New Scientist* (4 September 2006), https://www.newscientist.com/article/dn9967-introduction-mysteries-of-the-deep-sea.

255 Helen Matsos, "Earth's Hidden Biospheres," *Astrobiology Magazine*, NASA, https://astrobiology.nasa.gov/news/earths-hidden-biospheres.

256 Mark Shrope and John Pickrell, "Introduction: Mysteries of the Deep Sea," *New Scientist*, 4 September 2006, http://www.newscientist.com/article/dn9967#.Ud8nYoGTjeI.

257 Ibid.

258 Helen Matsos, "Lost City Life Methane-Powered," *Astrobiology Magazine*, 3 March 2005.

259 Ibid.

260 Lost City Research, "Biology," *Lost City: One of the Most Extreme Environments on Earth*, University of Washington, 2003, http://www.lostcity.washington.edu/story/Biology.

261 Ibid.

262 Ibid.

263 M.J. Russell, A.J. Hall, and W. Martin, "Serpentinization as a Source of Energy at the Origin of Life," *Geobiology* (December 2010): 355–371.

264 Alexis Templeton and the Rock Powered Life Lab, "Atlantis Massif," University of Colorado, Boulder, College of Engineering and Applied Science, 2016, https://www.colorado.edu/lab/rockpoweredlife/research/theme-i-field-investigations/atlantis-massif.

265 G. Proskurowski, M.D. Lilley, J.S. Seewald, G.L. Früh-Green, et al., "Abiogenic Hydrocarbon Production at Lost City Hydrothermal Field," *Science* 319, no. 5863 (2008): 604–607, https://doi.org/10.1126/science.1151194.

266 H.E. Elsaied, T. Hayashi, and T. Naganuma, "Molecular Analysis Of Deep-Sea Hydrothermal Vent Aerobic Methanotrophs By Targeting Genes Of 16s Rrna And Particulate Methane Monooxygenase," *Marine Biotechnology* (September-October 2004): 503–509.

267   Lost City Research, "Chemistry," *Lost City: One of the Most Extreme Environments on Earth,* University of Washington, 2003, http://www.lostcity. washington.edu/story/Chemistry.

268   Helen Matsos, "Lost City Life Methane-Powered," *Astrobiology Magazine,* posted: March 3, 2005.

269   Cox, M. and Battista, J., *"Deinococcus radiodurans*—The Consummate Survivor," *Nature Reviews, Microbiology* 3 (2005): 882–892, https://doi. org/10.1038/nrmicro1264

270   Mark Shrope and John Pickrell, "Introduction: Mysteries of the Deep Sea," *New Scientist,* 4 September 2006, https://www.newscientist.com/article/ dn9967-introduction-mysteries-of-the-deep-sea.

271   Ibid.

272   Helen Matsos, "Earth's Hidden Biospheres," Astrobiology Magazine, March 3, 2005.

273   J. Fang, C. Kato, G.M. Runko, et al., "Predominance Of Viable Spore-Forming Piezophilic Bacteria In High-Pressure Enrichment Cultures From ~1.5 To 2.4 Km-Deep Coal-Bearing Sediments Below The Ocean Floor," *Frontiers in Microbiology* 8, no. 137 (2017), https://www.ncbi.nlm.nih.gov/pmc/articles/ PMC5292414.

274   Westall, "Early Life," 402.

275   Ibid.

276   Ibid.

277   Aharon Oren, Joseph Seckbach (eds.), *Microbial Mats: Modern and Ancient Microorganisms in Stratified Systems* (Springer, 2010).

278   Drew Lohrera and Nicole Hancock, "Marine Soft Sediments: More Diversity Than Meets The Eye," *Water and Atmosphere* 12 (2004): 26–27, https://niwa. co.nz/sites/niwa.co.nz/files/import/attachments/sediment.pdf.

279   Ibid.

280   D.W.J. Bosence and PH. Bridges, "A Review Of The Origin And Evolution Of Carbonate Mud-Mounds," in C.L.V. Monty, *Carbonate Mud-Mounds: Their Origin and Evolution,* The International Association of Sedimentologists (Oxford, UK: Blackwell-Science, 1995): 8.

281   Ibid.

282   Dorion Sagan, *Cosmic Apprentice: Dispatches from the Edges of Science* (Minneapolis: University Of Minnesota Press, 2013), 117.

283   Rhys Taylor (Arecibo Observatory, Puerto Rico), "Galaxy Size Comparison Chart," first printed in *Ciel et Espace* (June 2013), https://astrorhysy.blogspot. com/2013/04/infographic-galaxy-size-comparison-chart.html.

284   Fred Hoyle, *The Intelligent Universe* (New York: Holt, Rinehart, and Winston, 1984).

285   Genesis 1:28, New American Standard Bible (1995).

286   Genesis 26:4, English Standard Version (2001).

287   I owe the enthusiastic use of the word "abundance" to Peter Diamandis. See Peter Diamandis and Steven Kotler, *Abundance: The Future is Better Than You Think* (Simon and Schuster, 2012).

288   K. Takai, M. Suzuki, S. Nakagawa, et al., "Sulfurimonas Paralvinellae Sp. Nov., A Novel Mesophilic, Hydrogen-And Sulfur-Oxidizing Chemolithoautotroph Within The Epsilonproteobacteria Isolated From A Deep-Sea Hydrothermal Vent Polychaete Nest, Reclassification Of Thiomicrospira Denitrificans As Sulfurimonas Denitrificans Comb. Nov. And Emended Description Of The Genus Sulfurimonas," *International Journal of Systematic and Evolutionary Microbiology* 56, no. 8 (2006): 1725–1733.

289   I. Enami, and I. Fukuda, "Mechanisms of The Acido-And Thermophily Of Cyanidium Caldarium Geitler Iii. Loss Of These Characteristics Due To Detergent Treatment," *Plant and Cell Physiology* 18, no. 3 (1977): 671–680.

290   Yanhe Ma, et al., "Halophiles 2010: Life in Saline Environments," *Applied and Environmental Microbiology* 76, no. 21 (2010): 6971–6981. Shiladitya DasSarma and Priya Arora, "Halophiles," *Encyclopedia of Life Sciences*, 2001. Trüper, Hans G., Jörg Severin, Axel Wohlfarth, Ewald Müller, and Erwin A. Galinski, "Halophily, Taxonomy, Phylogeny and Nomenclature," in F. Rodriguez-Valera (ed.), *General and Applied Aspects of Halophilic Microorganisms* (1991): 3–7, https://doi.org/10.1007/978-1-4615-3730-4_1.

291   R. Lloyd and Lydia D. Orr, "The Diuretic Response By Rainbow Trout To Sub-Lethal Concentrations Of Ammonia," *Water Research* 3, no. 5 (1969): 335–344.

292   Patrick Chain, Jane Lamerdin, Frank Larimer, Warren Regala, Victoria Lao, Miriam Land, Loren Hauser, Daniel Arp, et.al., "Complete Genome Sequence Of The Ammonia-Oxidizing Bacterium And Obligate Chemolithoautotroph Nitrosomonas Europaea," *Journal of Bacteriology* 185, no. 9 (2003).

293   More properly, feeding on ammonia is called "nitrifying." For example, A. Cabezas, P. Draper and C. Etchebehere, "Fluctuation of Microbial Activities After Influent Load Variations In A Full-Scale Sbr: Recovery Of The Biomass After Starvation," *Applied Microbiology and Biotechnology* 84 (2009): 1191–1202, https://doi.org/10.1007/s00253-009-2138-x.

294   Richard E. Michod and Aurora M. Nedelcu, "Cooperation and Conflict During The Unicellular–Multicellular And Prokaryotic–Eukaryotic Transitions," in Andrés Moya and Enrique Font (eds.), *Evolution: From Molecules to Ecosystems* (Oxford University Press, 2003), http://www.eebweb.arizona.

edu/faculty/michod/Downloads/Valencia%20chapter.pdf. Chang Yin, Xiaoping Fan, Hao Chen, et al., "Substrate Competition Contributes To The Niche Differentiation Between Ammonia-Oxidizing Bacteria And Archaea In Ammonium-Rich Alkaline Soils," 1 April 2020, https://doi.org/10.21203/rs.3.rs-20413/v1. Neil W. Blackstone, "An Evolutionary Framework for Understanding the Origin of Eukaryotes," *Biology* 5, no. 2 (27 April 2016), https://doi.org/10.3390/biology5020018.

295   James R. Brown and W. Ford Doolittle, "Archaea and the Prokaryote-to-Eukaryote Transition," *Microbiology and Molecular Biology Reviews* 61, no. 4 (1991): 456–502. Werner Schwemmler, "Symbiogenesis in Insects as a Model for Morphogenesis, Cell Differentiation, and Speciation," in Lynn Margulis, René Feste (eds.), *Symbiosis As a Source of Evolutionary Innovation: Speciation and Morphogenesism* (MIT Press, 1991), 194.

296   Elisa T. Granato, Thomas A. Meiller-Legrand, and Kevin R. Foster, "The Evolution and Ecology of Bacterial Warfare," *Current Biology* 29, no. 11 (2019): R521-R537, https://www.sciencedirect.com/science/article/pii/S0960982219304221. See-Yeun Ting, Dustin E. Bosch, Sarah M. Mangiameli, Matthew C. Radey, Shuo Huang, Young-Jun Park, Katherine A. Kelly, et al., "Bifunctional Immunity Proteins Protect Bacteria Against Ftsz-Targeting Adp-Ribosylating Toxins," *Cell* 175, no. 5 (2018): 1380–1392. "Bacteria have Familiar Weapon in War," *Technology Networks*, 19 October 2018, https://www.technologynetworks.com/immunology/news/bacteria-have-familiar-weapon-in-war-310786. Texas A&M, "Of Micro Combat: Study Looks At How Bacteria Wage War, Resist Occupation," Research@Texas A&M, https://research.tamu.edu/2016/01/12/of-micro-combat-study-looks-at-how-bacteria-wage-war-resist-occupation.

297   A trillion trillion trillion is roughly the number of all the organisms who have ever lived. Each was one of nature's bets. Each was one of nature's feelers into possibility space.

298   Chris Parsons, "A Moment That Changed Earth: Chicxulub Crater Research Explores How An Asteroid Impact Led To A Global Catastrophe And The Extinction Of The Dinosaurs," *Science Matters*, National Science Foundation, 15 June 2022, https://new.nsf.gov/science-matters/moment-changed-earth. Wood, Barry, "The Chicxulub File: Discovering the K-Pg Mass Extinction: A Four Decade Perspective," *Journal of Big History* 5, no. 1 (2022). Amos, Jonathan, "Dinosaur Asteroid Hit 'Worst Possible Place'," www.bbc.com (2017). James Rogers, "Asteroid That Wiped Out Dinosaurs Had Power Of 10 Billion Atomic Bombs: Study," *New York Post*, 10 September 2019, https://nypost.com/2019/09/10/asteroid-that-wiped-out-dinosaurs-had-power-of-10-billion-atomic-bombs-study.

299   Nir J. Shaviv, "The Spiral Structure Of The Milky Way, Cosmic Rays, And Ice Age Epochs On Earth," *New Astronomy* 8, no. 1 (2003): 39–77. P.C. Frisch,

"The Journey of the Sun," *arXiv.org Astronomy*, 1997, https://arxiv.org/abs/astro-ph/9705231.

300 ABC News, "This Is the Way the World Ends? Volcanoes Could Darken World," 6 June 2012, https://abcnews.go.com/Technology/end-world-super-volcanoes-form-quickly-destructive-asteroid/story?id=16508702.

301 Shuaiqi Guo, Corey A. Stevens, Tyler D. R. Vance, Luuk L. C. Olijve, and Peter L. Davies, "Structure Of A 1.5-Mda Adhesin That Binds Its Antarctic Bacterium To Diatoms And Ice," *Science Advances* 3, no. 8, https://www.science.org/doi/full/10.1126/sciadv.1701440.

302 Ines M. Hauner, Antoine Deblais, James K. Beattie, Hamid Kellay, and Daniel Bonn, "The Dynamic Surface Tension of Water," *The Journal of Physical Chemistry Letters* 8, no. 7 (2017): 1599–1603, https://pubs.acs.org/doi/full/10.1021/acs.jpclett.7b00267.

303 M. Homann, P. Sansjofre, M. Van Zuilen, et al., "Microbial Life And Biogeochemical Cycling On Land 3,220 Million Years Ago," *Nature Geoscience* 11 (2018): 665–671, https://doi.org/10.1038/s41561-018-0190-9. Gregory J. Retallack, Evelyn S. Krull, Glenn D. Thackray, Dula Parkinson, "Problematic Urn-Shaped Fossils from a Paleoproterozoic (2.2Ga) Paleosol in South Africa," *Precambrian Research* 235 (2013): 71, https://doi.org/10.1016/j.precamres.2013.05.015. University of Oregon, "Greening of the Earth Pushed Way Back in Time," *ScienceDaily*, 22 July 2013, www.sciencedaily.com/releases/2013/07/130722141548.htm.

304 Bacteria took to the land approximately 3.22 billion years ago. See: M. Homann, P. Sansjofre, M. Van Zuilen, et al., "Microbial Life and Biogeochemical Cycling on Land 3,220 million years Ago," *Nature Geoscience* 11 (23 July 2018): 665–671, https://doi.org/10.1038/s41561-018-0190-9.

305 Denis Wood, *Five Billion Years of Global Change: A History of the Land* (Guilford Press, 2004), 59–62.

306 Melanie Ghoul and Sara Mitri, "The Ecology and Evolution of Microbial Competition," *Trends in Microbiology* 24, no. 10 (October 2016): 833, http://dx.doi.org/10.1016/j.tim.2016.06.011.

307 George Y. Liu and Victor Nizet, "Color Me Bad: Microbial Pigments As Virulence Factors," *Trends in Microbiology* 17, no. 9 (2009): 406–413, https://www.cell.com/trends/microbiology/abstract/S0966-842X(09)00151-6.

308 W. Kardinaal, A. Edwin, et al., "Competition for Light Between Toxic And Nontoxic Strains Of The Harmful Cyanobacterium Microcystis," *Applied and Environmental Microbiology* 73, no. 9 (2007): 2939–2946.

309 Denis Wood, *Five Billion Years of Global Change: A History of the Land* (Guilford Press, 2004), 59–62.

310   M.P. Johnson, "Photosynthesis," *Essays in Biochemistry* 60, no. 3 (31 October 2016): 255–273, https://doi.org/10.1042/EBC20160016.

311   Wood, *Five Billion Years of Global Change*, 59–62.

312   Ibid.

313   Ibid.

314   Spadafora, Alessandra, et al., "Microbial Biomineralization Processes Forming Modern Ca:mg Carbonate Stromatolites," *Sedimentology* 57, no. 1 (2010): 27–40, https://onlinelibrary.wiley.com/doi/full/10.1111/j.1365-3091.2009.01083.x.

315   Richard K. Grosberg, Geerat J. Vermeij , Peter C. Wainwright, et al, "Biodiversity in Water and On Land," *Current Biology* 22, no. 21 (6 November 2012): R900–R903, https://www.sciencedirect.com/science/article/pii/S0960982212011529. Hannah Ritchie, "Oceans, Land, And Deep Subsurface: How Is Life Distributed Across Environments?" Published online at OurWorldInData.org (2019), https://ourworldindata.org/life-by-environment.

316   H. Wu, Y. Fang, J. Yu, and Z. Zhang, "The Quest For A Unified View Of Bacterial Land Colonization," *ISME Journal*, International Society for Microbial Ecology 8, no. 7 (July 2014): 1358–69, https://www.nature.com/articles/ismej2013247, https://www.ncbi.nlm.nih.gov/pmc/articles/PMC4069389. University of Tennessee at Knoxville, "Research Reveals Aquatic Bacteria More Recent Move To Land," *Phys.org*, 22 December 2011, https://phys.org/news/2011-12-reveals-aquatic-bacteria.html. Fabia U. Battistuzzi, Andreia Feijao, and S. Blair Hedges, "A Genomic Timescale Of Prokaryote Evolution: Insights Into The Origin Of Methanogenesis, Phototrophy, And The Colonization Of Land," *BMC Evolutionary Biology* 4 (2004): 1–14, https://bmcecolevol.biomedcentral.com/articles/10.1186/1471-2148-4-44.

317   Lynn Margulis and Dorion Sagan, *Microcosmos: Four Billion Years of Microbial Evolution* (University of California Press, 1997): 74.

318   R.E. Blankenship, "Early Evolution of Pphotosynthesis," *Plant Physiology* 154, no. 2 (October 2010): 434–8, https://doi.org/10.1104/pp.110.161687, https://academic.oup.com/plphys/article/154/2/434/6111499. David Biello, "The Origin of Oxygen in Earth's Atmosphere," *Scientific American*, 19 August 2009, https://www.scientificamerican.com/article/origin-of-oxygen-in-atmosphere. Robert Perkins, "How do Plants Make Oxygen? Ask Cyanobacteria," *CalTech*, 30 March 2017, https://www.caltech.edu/about/news/how-do-plants-make-oxygen-ask-cyanobacteria-54559.

319   David Chandler, "How Oxygen Gas Is Produced During Photosynthesis?" *Sciencing*, 5 April 2018, https://sciencing.com/oxygen-gas-produced-during-photosynthesis-6365699.html.

320   Eshel ben-Jacob, personal communication, October, 1998.

321   Song-Can Chen, Guo-Xin Sun, Yu Yan, Konstantinos T. Konstantinidis, Si-Yu Zhang, Ye Deng, Xiao-Min Li, et al., "The Great Oxidation Event Expanded The Genetic Repertoire Of Arsenic Metabolism And Cycling," *Proceedings of the National Academy of Sciences* 117, no. 19 (2020): 10414–10421, https://www.pnas.org/content/117/19/10414.

322   Phil Plait, "Poisoned Planet," *Slate*, 28 July 2014, https://slate.com/technology/2014/07/the-great-oxygenation-event-the-earths-first-mass-extinction.html.

323   R. Ligrone, "The Great Oxygenation Event," in: *Biological Innovations that Built the World* (Switzerland: Springer), https://doi.org/10.1007/978-3-030-16057-9_5.

324   Nick Lane, "Serial Endosymbiosis or Singular Event at the Origin of Eukaryotes?" *Journal of Theoretical Biology* 434 (7 December 2017): 58–67, https://www.sciencedirect.com/science/article/pii/S0022519317302011.

325   British Society for Cell Biology, LysosomeSection 4: "What Causes Mutations?—Mitochondrion—Much More Than an Energy Converter," https://bscb.org/learning-resources/softcell-e-learning/mitochondrion-much-more-than-an-energy-converter.

326   Aloysius G.M. Tielens, et al., "Mitochondria As We Don't Know Them," *Trends in Biochemical Sciences* 27, no. 11 (2002): 564–572, https://www.cell.com/trends/biochemical-sciences/abstract/S0968-0004(02)02193-X.

327   Baljit S. Khakh and Geoffrey Burnstock, "The Double Life of ATP," *Scientific American* 301, no1. 6 (2009): 84.

328   Phil Plait, "Poisoned Planet," *Slate*, 28 July 2014, https://slate.com/technology/2014/07/the-great-oxygenation-event-the-earths-first-mass-extinction.html.

329   Matthew R. Warke, Tommaso Di Rocco, Aubrey L. Zerkle, Aivo Lepland, Anthony R. Prave, Adam P. Martin, Yuichiro Ueno, Daniel J. Condon, and Mark W. Claire, "The Great Oxidation Event Preceded A Paleoproterozoic 'Snowball Earth'," *Proceedings of the National Academy of Sciences* 117, no. 24 (2020): 13314-13320, https://www.pnas.org/doi/abs/10.1073/pnas.2003090117. R.E. Kopp, J.L. Kirschvink, I.A. Hilburn, C.Z. Nash, "The Paleoproterozoic Snowball Earth: A Climate Disaster Triggered By The Evolution Of Oxygenic Photosynthesis," *Procedings of the National Academy of Sciences* 102 (2005): 11131–11136. Cavalier-Smith Thomas, "Cell Evolution and Earth History: Stasis and Revolution," *Philosophical Transactions of the Royal Society* (2006) B361969–1006, http://doi.org/10.1098/rstb.2006.1842.

.330   Denis Wood, *Five Billion Years of Global Change: A History of the Land* (Guilford Press, 2004), 59–62.

331 David M. Raup, "Biological Extinction in Earth History," *Science* 231, no. 4745 (1986): 1528–1533.

332 Patricia Sánchez Baracaldo and Tanai Cardona, "On the Origin Of Oxygenic Photosynthesis and Cyanobacteria," *New Phytologist* 225, no. 4 (2020): 1440–1446.

333 "Annual Plant Reviews, Volume 45," in *The Evolution of Plant Form*, edited by Barbara A. Ambrose, Michael D. Purugganan (Wiley-Blackwell, February 2013), xv. Elizabeth Pennisi, "Land Plants Arose Earlier Than Thought—And May Have Had A Bigger Impact On The Evolution Of Animals," *Science*, 19 February 2018, https://www.sciencemag.org/news/2018/02/land-plants-arose-earlier-thought-and-may-have-had-bigger-impact-evolution-animals.

334 The first biofilms appeared roughly 3.25 billion years ago. See: Hall-Stoodley, L., Costerton, J. & Stoodley, P., "Bacterial Biofilms: From The Natural Environment To Infectious Diseases," *Nature Reviews Microbiology* 2 (2004): 95–108, https://doi.org/10.1038/nrmicro821.

335 C.J. Donoghue Philip and M. Paul Smith (eds.), *Telling the Evolutionary Time: Molecular Clocks and the Fossil Record* (Boca Raton, Florida: CRC Press, 2003), 124.

336 T. Heulin, et al., "Bacterial Adaptation to Hot and Dry Deserts," in H. Stan-Lotter and S. Fendrihan (eds.), *Adaption of Microbial Life to Environmental Extremes* (Switzerland: Springer, 2017), https://doi.org/10.1007/978-3-319-48327-6_4. E. Laskowska and D. Kuczyńska-Wiśnik, "New Insight Into The Mechanisms Protecting Bacteria During Desiccation," *Current Genetics* 66 (26 September 2019): 313–318, https://doi.org/10.1007/s00294-019-01036-z. L.K. Vestby, T. Grønseth, R. Simm, L.L. Nesse, "Bacterial Biofilm and its Role in the Pathogenesis of Disease," *Antibiotics* 9, no. 2 (3 February 2020): 59, https://doi.org/10.3390/antibiotics9020059.

337 Plants took to the land 470 million years ago, according to A.J. Shaw, P. Szövényi, B. and Shaw, "Bryophyte Diversity And Evolution: Windows Into The Early Evolution Of Land Plants," *American Journal of Botany* 98 (1 March 2011): 352–369, https://doi.org/10.3732/ajb.1000316.

338 Patricia G. Gensel and Dianne Edwards (eds.), *Plants Invade the Land: Evolutionary and Environmental Perspectives* (Columbia University Press, 2001), 3.

339 Jiasong Fang, Chiaki Kato, Gabriella M. Runko, Yuichi Nogi, Tomoyuki Hori, Jiangtao Li, Yuki Morono, and Fumio Inagaki, "Predominance of Viable Spore-Forming Piezophilic Bacteria In High-Pressure Enrichment Cultures From~ 1.5 To 2.4 Km-Deep Coal-Bearing Sediments Below The Ocean Floor," *Frontiers in microbiology* 8 (2017): 137, https://doi.org/10.3389/fmicb.2017.00137.

340   Erik P. Sunde, Peter Setlow, Lars Hederstedt, and Bertil Halle, "The Physical State Of Water In Bacterial Spores," *Proceedings of the National Academy of Sciences, Biophysics And Computational Biology* 106, no. 46 (17 November 2009): 19334–19339, https://doi.org/10.1073/pnas.0908712106.

341   George Wong, "Spore Dispersal in Fungi, in Magical Mushrooms and Mystical Molds," *Botany* 135, University of Hawaii at Manoa, http://www.botany. hawaii.edu/faculty/wong/BOT135/Lect05_a.htm. In ferns, "spores are produced continually and are unlimited in number." See L.G. Hickok and T.R. Warne, "Laboratory Investigations with C-Fern™ (*Ceratopteris richardii*)," in *Tested Studies for Laboratory Teaching* 19, S. J. Karcher (ed.), *Proceedings of the 19th Workshop/Conference of the Association for Biology Laboratory Education (ABLE)* (1998), 147, https://www.ableweb.org/biologylabs/wp-content/uploads/ volumes/vol-19/10-hickok.pdf.

342   D. Cressey, "Fossils Rewrite History of Penetrative Sex," *Nature* (2014), https://doi.org/10.1038/nature.2014.16173. Terrence McCoy, "Scientists Discover the Awkward Origins of Sex," *Washington Post* (20 October 2014), https://www.washingtonpost.com/news/morning-mix/wp/2014/10/20/ scientists-discover-the-kinda-disgusting-origins-of-sex. D. Speijer "What Can We Infer About The Origin Of Sex In Early Eukaryotes?" *Philosophical Transactions of the Royal Society of London. Series B, Biological Sciences* 271, no. 1706 (19 Oct 2016), https://doi.org/10.1098/rstb.2015.0530. Linda A. Hufnagel, "The Cilioprotist Cytoskeleton, a Model for Understanding How Cell Architecture and Pattern Are Specified: Recent Discoveries from Ciliates and Comparable Model Systems," *Cytoskeleton* (21 September 2021): 251–295, https://doi. org/10.1007/978-1-0716-1661-1_13.

343   Sarah Otto, "Sexual Reproduction and the Evolution of Sex," SciTable, *Nature Education* 1, no. 1 (2008): 182, https://www.nature.com/scitable/top- icpage/sexual-reproduction-and-the-evolution-of-sex-824. A.C. Kemen, M.T. Agler and E. Kemen, , "Host–Microbe And Microbe–Microbe Interactions In The Evolution Of Obligate Plant Parasitism," *New Phytologist* 206 (26 January 2015): 1207–1228, https://doi.org/10.1111/nph.13284. Christie Wilcox, "Nonbinary Nature," *ScienceAdviser*, American Academy of Science (30 June 2023), https:// view.aaas.sciencepubs.org/?qs=356d987418aae1e1c01ef367ec5c1d397a7e- b4802abd6127a6bbbba88319b65f84aa951c7997a4393e37ebc5f6f3e6e444c8abc- c0695944f91873776c4adbf5a47953cae9051240ea26bb47a91563b6e.

344   Pierre Louis de Maupertuis: "Whenever any change occurs in nature, the quantity of action employed for this is always the smallest possible," Quoted in Jerome Fee, "Maupertuis, and the Principle of Least Action," *The Scientific Monthly* 52, no. 6 (June 1941): 496–503.

345   Mary Terrall, "Salon, Academy, and Boudoir: Generation and Desire in Maupertuis's Science of Life," *Isis* 87, no. 2 (1996): 217–229.

346   The phrases Charles Darwin used in his second volume of the *Origin of Species* were "Struggle for Existence" and "survival of favoured individuals and races." The phrases that Herbert Spencer suggested in a conversation with Darwin were "survival of the fittest" and "natural selection, or survival of the fittest." Cbarles Darwin, *The Origin of Species by Means of Natural Selection: or, the Preservation of Favored Races in the Struggle for Life*, Volume II (New York: Appleton, 1897). Originally published 1859.

347   L. Dolan, "Body Building On Land: Morphological Evolution Of Land Plants," *Current Opinion in Plant Biology* 12, no. 1 (2009): 4–8, https://doi.org/10.1016/j.pbi.2008.12.001.

348   M.Y. Galperin, S.L. Mekhedov, P. Puigbo, S. Smirnov, Y.I Wolf, D.J. Rigden, "Genomic Determinants Of Sporulation In Bacilli And Clostridia: Towards The Minimal Set Of Sporulation-Specific Genes," *Environmental Microbiology* 11 (14 Nov 2021): 2870–90, https://www.ncbi.nlm.nih.gov/pmc/articles/PMC3533761.

349   Terri D. Fisher, Zachary T. Moore, and Mary-Jo Pittenger, "Sex On The Brain? An Examination Of Frequency Of Sexual Cognitions As A Function Of Gender, Erotophilia, And Social Desirability," *Journal of Sex Research* 49, no. 1 (2012): 69–77. Pamela Paul, "When Thoughts Turn to Sex, or Not," *New York Times* (9 December 2011), https://www.nytimes.com/2011/12/11/fashion/sex-on-the-brain-studied.html.

350   Mark Atkinson, "Mapping The 100 Trillion Cells That Make Up Your Body," University of Florida News, 16 October 2018, https://news.ufl.edu/articles/2018/10/mapping-the-100-trillion-cells-that-make-up-your-body.html.

351   Obubu Maxwell, Ikediuwa Udoka Chinedu, and Anabike Charles Ifeanyi, "Numbers in Life: A Statistical Genetic Approach," *Scientific Review* 5, no. 7 (2019): 142–149. A bryophyte has over 14,000 protein-coding genes and god knows how many "junk" genes. Jian Zhang, Xin-Xing Fu, Rui-Qi Li, Xiang Zhao, Yang Liu, Ming-He Li, Arthur Zwaenepoel, et al., "The Hornwort Genome And Early Land Plant Evolution," *Nature Plants* 6, no. 2 (2020): 107–118, https://www.nature.com/articles/s41477-019-0588-4. It is assumed these days that the phrase "junk dna" was in error, and that many genes do things we simply don't yet understand.

352   With the exception of your blood cells, which eject their nuclei before they enter the blood stream. See: "No Nucleus? No Problem: Red Blood Cells and Platelets," *Visible Body* (2023), https://www.visiblebody.com/learn/biology/blood-cells/red-blood-cells-platelets.

353   Chial, Heidi, "DNA Sequencing Technologies Key to the Human Genome Project," *Nature Education* 1, no. 1 (2008), https://www.nature.com/scitable/topicpage/dna-sequencing-technologies-key-to-the-human-828. Corrinne E. Grover and Jonathan F. Wendel, "Recent Insights Into Mechanisms Of

Genome Size Change In Plants," *Journal of Botany* (30 May 2010), https://doi.org/10.1155/2010/382732.

354 Extrapolating from the following article, the number of atoms in a human genome is roughly 60 billion. Wim Hordijk, "Exploring The Origins Of Life With Autocatalytic Sets," *Research Outreach*, 25 February 2020, https://researchoutreach.org/articles/exploring-origins-life-autocatalytic-sets. However, extrapolating from the following three articles, the number of atoms in a single human genome is roughly 128 billion. J. Craig Venter, Hamilton O. Smith, Mark D. Adams, "The Sequence of the Human Genome," *Clinical Chemistry* 61, no. 9 (1 September 2015): 1207–1208, https://doi.org/10.1373/clinchem.2014.237016. T. Lencz and A. Darvasi, "Single Nucleotide Polymorphisms (SNPs)," *Reference Module in Life Sciences 2017* (Elsevier, 2016) https://doi.org/10.1016/B978-0-12-809633-8.07157-0. Robert A. Lue, "DNA Structure & Chemistry," in *Life Sciences 1A*, Harvard University, 24 September 2018, https://projects.iq.harvard.edu/files/lifesciences1abookv1/files/8_-_dna_replication_revised_9-24-2018.pdf.

355 François Serra, et al., "Restraint-Based Three-Dimensional Modeling Of Genomes And Genomic Domains," *FEBS letters [Federation of European Biochemical Societies]* 589, no. 20 (2015): 2987–2995, https://www.sciencedirect.com/science/article/pii/S0014579315003889.

356 Optical Microscopy Division of the National High Magnetic Field Laboratory, "Chromatin and Chromosomes," *Molecular Exressions: Cell Biology and Microscopy—Structure and Function of Cells & Viruses*, Optical Microscopy Division of the National High Magnetic Field Laboratory, a joint venture of The Florida State University, the University of Florida, and the Los Alamos National Laboratory, 19 June 2013, https://micro.magnet.fsu.edu/cells/nucleus/chromatin.html.

357 "Genome" | Learn Science at Scitable," *Nature Education*, https://www.nature.com/scitable/definition/genome-43. For slightly different figures on the number of base pairs in your genes, 3.05 billion base pairs, see Carl Zimmer, "Scientists Say They Have Finally Sequenced the Entire Human Genome," *The New York Times*, 23 July 2021, https://www.nytimes.com/2021/07/23/science/human-genome-complete.html. See also Sergey Nurk, Sergey Koren, Arang Rhie, Mikko Rautiainen, Andrey V. Bzikadze, Alla Mikheenko, Mitchell R. Vollger, et al., "The Complete Sequence of a Human Genome," *Science* 376, no. 6588 (2022): 44–53, https://www.science.org/doi/full/10.1126/science.abj6987.

358 Optical Microscopy Division of the National High Magnetic Field Laboratory, "Chromatin and Chromosomes," *Molecular Exressions: Cell Biology and Microscopy—Structure and Function of Cells & Viruses*, Optical Microscopy Division of the National High Magnetic Field Laboratory, a joint venture of The Florida State University, the University of Florida, and the Los Alamos National Laboratory, https://micro.magnet.fsu.edu/cells/nucleus/chromatin.html. And

Life Sciences 1A, "The Cell Nucleus," *National MagLab* (24 September 2018), https://micro.magnet.fsu.edu/cells/nucleus/nucleus.html

359   Timothy J. Richmond, "Predictable Packaging," *Nature* 442, no. 7104 (2006): 750–751, https://www.nature.com/articles/442750a.

360   R. Brown and B. Lemmon, "Spores Before Sporophytes: Hypothesizing The Origin Of Sporogenesis At The Algal-Plant Transition," *The New Phytologist*, 190, no. 4 (2011): 875–81, https://doi.org/10.1111/j.1469-8137.2011.03709.x. The seaweed site, "Information on Marine Algae," *Seaweed* (January 2022), https://www.seaweed.ie/algae/algae.php.

361   "When infected, the corn plant will have black galls of various sizes. A gall that is about 1 in 3 may contain approximately 25 billion spores! Multiply that by all of the galls that may be present in a single corn plant and we will literally have billions and billions of spores. *Ganoderma applanatum*, the Artist Fungus (Figure 1c-d), produces a perennial fruiting body, which may disperse 5.4 trillion spores over a six month period." W.H. Wong, "Symbiosis: Mycorrhizae and Lichens," University of Hawaii, http://www.botany.hawaii.edu/faculty/wong/BOT135/Lect05_a.htm.

362   "Neurons are estimated to express 40% of their total genes, some of which are also expressed in other types of cells," Bong-Kiun Kaang, "Genes and Neurons," in *Molecular Pain*, edited by Min Zhuo (Springer, 2007), 3°15.

363   *MedlinePlus* [Internet], Bethesda (MD): National Library of Medicine (US), "How Many Chromosomes Do People Have?," 24 June 2020, https://medlineplus.gov/genetics/understanding/basics/howmanychromosomes.

364   Elizabeth A. Kellogg and Jeffrey L. Bennetzen, "The Evolution Of Nuclear Genome Structure In Seed Plants," *American Journal of Botany* 91, no. 10 (2004): 1709–1725, https://bsapubs.onlinelibrary.wiley.com/doi/full/10.3732/ajb.91.10.1709.

365   "Meiosis," *Simple English Wikipedia, The Free Encyclopedia* (February 2024), http://simple.wikipedia.org/wiki/Meiosis.

366   Adèle L. Marston and Angelika Amon, "The Mitotic And Meiotic Cell Cycles, Meiosis: Cell-Cycle Controls Shuffle And Deal," *Nature Reviews Molecular Cell Biology* 6, no1. 818 (October 2005), https://www.nature.com/articles/nrm1759.pdf, https://www.nature.com/articles/nrm1759.

367   Helene Guillon and Bernard de Massy, "An Initiation Site For Meiotic Crossing-Over And Gene Conversion In The Mouse," *Nature Genetics* 32, no. 2 (2002): 296–299. Kris A. Wetterstrand, "Crossing Over," *National Human Genome Research Institute* (4 October 2022), https://www.genome.gov/genetics-glossary/Crossing-Over.

368   M. Bonora, A. Bononi, E. de Marchi, C. Giorgi, et al., "Molecular Mechanisms Of Cell Death: Central Implication Of Atp Synthase In

Mitochondrial Permeability Transition," *Oncogene* 34, no. 11 (12 Mar 2015): 1475–86, https://doi.org/10.1038/onc.2014.96. P.P. Khil and R.D. Camerini-Otero, "Variation in Patterns Of Human Meiotic Recombination." *Meiosis* 5 (2009): 117–127, https://www.ncbi.nlm.nih.gov/pmc/articles/PMC3105470.

369   National Human Genome Research Institute, "Human Genome Project Completion: Frequently Asked Questions," *Genome.gov*, 2010, https://www.genome.gov/human-genome-project/Completion-FAQ.

370   Elizabeth A. Kellogg and Jeffrey L. Bennetzen, "The Evolution Of Nuclear Genome Structure In Seed Plants," *American Journal Of Botany* 91, no. 10 (2004): 1709–1725, https://bsapubs.onlinelibrary.wiley.com/doi/full/10.3732/ajb.91.10.1709.

371   R.H. Ramírez-González, D. Borrill, S.A. Harrington, J. Brinton, L. Venturini, M. Davey et al, "The Transcriptional Landscape of Polyploid Wheat," *Science* 361, no. 6403 (2018), https://www.science.org/doi/full/10.1126/science.aar6089. Pallab Ghosh, "Wheat Gene Map To Help 'Feed The World'," BBC, 16 August 2018, https://www.bbc.com/news/science-environment-45173968.

372   There have been roughly 5 billion species on this planet so far. If each of these produced only a quadrillion organisms, that would be 5 septillion creatures who have ever lived. An underestimate. Philippe Bertrand, Louis Legendre, *Earth: Our Living Planet* (Springer International Publishing, Kindle Edition), 249.

373   Adèle L. Marston and Angelika Amon, "Erratum: Meiosis: Cell-cycle Controls Shuffle and Deal," *Nature reviews Molecular Cell Biology* 6, no. 10 (2005): 818. Frank Uhlmann, "Chromosome Cohesion And Segregation In Mitosis And Meiosis," *Current Opinion In Cell Biology* 13, no. 6 (2001): 754–761. Ke Zheng and P. Jeremy Wang, "Blockade of Pachytene Pirna Biogenesis Reveals A Novel Requirement For Maintaining Post-Meiotic Germline Genome Integrity," *PLoS Genetics* 8, no. 11 (2012), https://journals.plos.org/plosgenetics/article?id=10.1371/journal.pgen.1003038.

374   MedlinePlus, "What Is a Chromosome?," Bethesda (MD): National Library of Medicine (24 Jun 2020), https://medlineplus.gov/genetics/understanding/basics/chromosome.

375   Clift D, Schuh M, "Restarting Life: Fertilization And The Transition From Meiosis To Mitosis," *Nature Reviews Molecular Cell Biology* 14, no. 9 (September 2013): 549–62, https://doi.org/10.1038/nrm3643, https://www.nature.com/articles/nrm3643.

376   Ajeet Chaudhary, Rachele Tofanelli, Kay Schneitz, "Plant Reproduction: Shaping the Genome of Plants," *eLife* (6 February 2020), https://doi.org/10.7554/eLife.54874.

377   Didier G. Schaefer and Jean-Pierre Zrÿd, "The Moss *Physcomitrella patens*, Now and Then," *Plant Physiology* 127, no. 4 (December 2001): 1430–

1438, https://doi.org/10.1104/pp.010786. Kimihiro Terasawa, et al., "The Mitochondrial Genome of the Moss Physcomitrella patens Sheds New Light on Mitochondrial Evolution in Land Plants," *Molecular Biology and Evolution* 24, no. 3 (March 2007): 699–709, https://doi.org/10.1093/molbev/msl198. Jennifer L. Morris, Mark N. Puttick, James W. Clark, Dianne Edwards, Paul Kenrick, Silvia Pressel, Charles H. Wellman, Ziheng Yang, Harald Schneider, and Philip C.J. Donoghue, "The Timescale Of Early Land Plant Evolution," *Proceedings of the National Academy of Sciences* 115, no. 10 (2018): E2274-E2283, https://www.pnas.org/doi/abs/10.1073/pnas.1719588115. Katie Pavid, "Plant Life On Earth Is Much Older Than We Thought," Natural History Museum, London (February 2018), https://www.nhm.ac.uk/discover/news/2018/february/plant-life-on-earth-is-much-older-than-we-thought.html.

378   Micael Jonsson, Paul Kardol, Michael J. Gundale, Sheel Bansal, Marie-Charlotte Nilsson, Daniel B. Metcalfe, and David A. Wardle, "Direct And Indirect Drivers Of Moss Community Structure, Function, And Associated Microfauna Across A Successional Gradient," *Ecosystems* 18 (2015): 154–169, https://doi.org/10.1007/s10021-014-9819-8.

379   The total land surface area of Earth is about 57 million square miles. R. Gowtham, J. Daniel, S. Karthik, and R. Vignesh, "Revamping Digital Land Survey using GPS and Internet of Things (IoT)," 2018 International Conference on Soft-computing and Network Security (ICSNS) (2018): 1–9, https://doi.org/10.1109/ICSNS.2018.8573672. For an older figure of 52 million square miles, a figure from the 1930s, see "War and Populations," *Nature* 137 (1936), 1025–1025, https://doi.org/10.1038/1371025b0. Pianka, Eric R, "Land," University of Texas at Austin, http://www.zo.utexas.edu/courses/Thoc/land.html.

380   E. Ayres, R. van der Wal, M. Sommerkorn and R.D. Bardgett, "Direct Uptake Of Soil Nitrogen By Mosses," *Biology Letters* 2, no. 2 (22 June 2006): 286–8, https://royalsocietypublishing.org/doi/full/10.1098/rsbl.2006.0455. D. Wang, A. Xu, C. Elmerich, et al., "Biofilm Formation Enables Free-Living Nitrogen-Fixing Rhizobacteria To Fix Nitrogen Under Aerobic Conditions," *ISME Journal [The International Society for Microbial Ecology]* 11 (24 March 2017): 1602–1613, https://doi.org/10.1038/ismej.2017.30. Serena Rinaldo, Giorgio Giardina, Federico Mantoni, Alessio Paone, Francesca Cutruzzolà, "Beyond Nitrogen Metabolism: Nitric Oxide, Cyclic-Di-Gmp And Bacterial Biofilms," *FEMS Microbiology Letters [Federation of European Microbiological Societies]* 365, no. 6 (March 2018), https://doi.org/10.1093/femsle/fny029. Lee Billings, "Bacteria Got an Early Fix on Nitrogen," *Sixty Second Science*, Scientific American Podcast, 23 February 2015, https://www.scientificamerican.com/podcast/episode/bacteria-got-an-early-fix-on-nitrogen. According to this podcast's script, bacteria started fixing nitrogen 3.2 billion years ago. Nitrogen would prove a vital nutrient for later organisms, as you'll see later in this book.

381 J.M. Glime, 2017, "Life Cycles: Surviving Change," in J.M. Glime, *Bryophyte Ecology, Volume 1, Physiological Ecology*, Michigan Technological University and the International Association of Bryologists (9 April 2021), http://digitalcommons.mtu.edu/bryophyte-ecology.

382 Human eggs are produced at the rate of one per menstrual cycle. Human sperm are produced at the rate of 250-280 million per ejaculation. Plus, the egg is 10,000 times the size of the sperm. Khan Academy, "Egg Meets Sperm," https://www.khanacademy.org/test-prep/mcat/cells/embryology/a/egg-meets-sperm.

383 Martin John Ingrouille and Bill Eddi, *Plants: Evolution and Diversity* (Cambridge University Press, 2006), 137. For the large number of antheridia per plant, see this photo: http://www.anbg.gov.au/bryophyte/photos-captions/rosulabryum-billarderi-170.html, and the explanatory caption: "Individual antheridia and archegonia are microscopic but at times you can see where they are formed. In this photo of the moss Rosulabryum billardieri each yellow ball is a cluster of antheridia," http://www.anbg.gov.au/bryophyte/sexual-reproduction.html.

384 Naina Kumar and Amit Kant Singh, "The Anatomy, Movement, And Functions Of Human Sperm Tail: An Evolving Mystery," *Biology of Reproduction* 104, no. 3 (March 2021): 508–520, https://doi.org/10.1093/biolre/ioaa213. Kiyoshi Miki and David E. Clapham, "Rheotaxis Guides Mammalian Sperm," *Current Biology 23* (18 March 2013): 443–452, https://www.sciencedirect.com/science/article/pii/S0960982213001486. J. Gray and G.J. Hancock, "The Propulsion of Sea-Urchin Spermatozoa," *Journal of Experimental Biology* 32, no. 4 (1 December 1955): 802–814, https://doi.org/10.1242/jeb.32.4.802.

385 Ben Lovejoy, "Foxconn Moving Some Ipad Production Outside China For The First Time," *9to5Mac* (26 Nov 2020), https://9to5mac.com/2020/11/26/ipad-production-moving-outside-china.

386 Irene Bisang, Johan Ehrlén, and Lars Hedenäs, "Reproductive Effort And Costs Of Reproduction Do Not Explain Female Biased Sex Ratios In The Moss Pseudocalliergon Trifarium (Amblystegiaceae)," *American Journal of Botany* 93, no. 9 (2006): 1313–1319, https://doi.org/10.3732/ajb.93.9.1313. Johan Ehrlén, Irène Bisang, and Lars Hedenäs, "Costs of Sporophyte Production In The Moss, Dicranum Polysetum," *Plant Ecology* (August 2000): 207–217, https://link.springer.com/article/10.1023/A:1026531122302.

387 This protective vessel is called an archegonium. See: "Sexual Reproduction – bryophyte," ANBG (Australian National Herbarium of the Australian National Botanical Garden), https://www.anbg.gov.au/bryophyte/sexual-reproduction.html.

388 Rodrigo B. Singer and Samantha Koehler, "Notes On The Pollination Biology Of Notylia Nemorosa (Orchidaceae): Do Pollinators Necessarily Promote

Cross Pollination?" *Journal of Plant Research* 116 (2003): 19–25, http://www.ncbi.nlm.nih.gov/pubmed/12605296.

389 "Sexual Reproduction—Bryophyte," ANBG (Australian National Herbarium of the Australian National Botanical Garden), https://www.anbg.gov.au/bryophyte/sexual-reproduction.html.

390 Heino Lepp, "Reproduction & Dispersal: Sexual Reproduction," Australian National Botanic Gardens and Australian National Herbarium, November 2008, http://www.anbg.gov.au/bryophyte/sex-sperm-dispersal.html.

391 Mati Kahru, Juha-Markku Leppanen, and Ove Rud, "Cyanobacterial Blooms Cause Heating of the Sea Surface," *Marine Ecology Progress Series* (1993): 1–7, https://www.int-res.com/articles/meps/101/m101p001.pdf. E.A. Widder, "Bioluminescence in the Ocean: Origins of Biological, Chemical, and Ecological Diversity," *Science* 328 (2010): 704–708, https://doi.org/10.1126/science.1174269.

392 John Waterbury, "Little Things Matter A Lot," *Oceanus: The Journal of Our Ocean Planet*, Woods Hole Oceanographic Institution (11 March 2005), https://www.whoi.edu/oceanus/feature/little-things-matter-a-lot.

393 Ibid.

394 Jennifer Frazer, in the *Scientific American,* disagrees. She says that sperm are poor and clumsy swimmers. Her source is Robin Wall Kimmerer's book *Gathering Moss: A Natural and Cultural History of Mosses* (Oregon State University Press, 2003), 26. But other sources admire the ability of bryophyte sperm to go great distances. They point out that bryophyte sperm are sluggish when they don't have the water they need to travel and when they lack the come-hither scents of females to guide them. But they persist, remaining functional for up to 100 hours. When circumstances turn ripe for travel and romance, bryophyte sperm perk up. And they swim like experts. D. Haig "Living Together and Living Apart: The Sexual Lives of Bryophytes," *Philosophical Transactions of the Royal Society of London, B Biological Sciences* 371, no. 1706 (19 October 2016), https://doi.org/10.1098/rstb.2015.0535, https://www.ncbi.nlm.nih.gov/pmc/articles/PMC5031620. J.M. Glime, "Life Cycles: Surviving Change," in J.M. Glime, "Bryophyte Ecology, Volume 1. Physiological Ecology," *Journal of Plant Research* 130, no. 3 (2017): 455–464, https://digitalcommons.mtu.edu/bryophyte-ecology. L. Alvarez, "The Tailored Sperm Cell," *Journal of Plant Research* 130 (29 March 2017): 455–464, https://doi.org/10.1007/s10265-017-0936-2, https://link.springer.com/article/10.1007/s10265-017-0936-2. Todd N. Rosenstiel and Sarah M.Eppley, "Long-lived Sperm in the Geothermal Bryophyte *Pohlia nutans*," *Royal Society, Biology Letters* 5, no. 6 (2009): 857–860, https://royalsocietypublishing.org/doi/full/10.1098/rsbl.2009.0380. Brent D. Mishler and Steven P. Churchill, "Transition To A Land Flora: Phylogenetic Relationships Of The Green Algae And Bryophytes," *Cladistics* 1, no. 4 (1985):

305–328, https://onlinelibrary.wiley.com/doi/abs/10.1111/j.1096-0031.1985. tb00431.x. "Australian Bryophytes: Sporophyte development," Australian National Herbarium, Australian National Botanical Garden, https://www.anbg. gov.au/bryophyte/life-cycle-sporophyte-dev.html.

395   "Flagella and Cilia," *Plant Life* (4 April 2011), http://lifeofplant.blogspot. com/2011/04/flagella-and-cilia.html.

396   Martin John Ingrouille and Bill Eddi, *Plants: Evolution And Diversity* (Cambridge University Press, 2006), 137.

397   Australian National Botanic Gardens, "Liberation and dispersal of Sperm—Bryophyte," 2005, http://www.anbg.gov.au/bryophyte/sex-sperm-dispersal.html. Todd N. Rosenstiel and Sarah M. Eppley, "Long-lived Sperm In The Geothermal Bryophyte Pohlia Nutans," *Royal Society, Biology Letters* 5, no. 6 (2009): 857–860, https://royalsocietypublishing.org/doi/full/10.1098/ rsbl.2009.0380, https://royalsocietypublishing.org/doi/abs/10.1098/ rsbl.2009.0380.

398   See chart at Khan Academy, "Egg meets sperm," https://www. khanacademy.org/test-prep/mcat/cells/embryology/a/egg-meets-sperm.

399   Peter H. Raven, Ray F. Evert, and Susan E. Eichhorn, *Biology of Plants* (New York: W.H. Freeman, 2005), 349. Ingrouille, Martin John, and Bill Eddi, *Plants: Evolution and Diversity*, Cambridge University Press, 2006, pages 136-137.

400   Patricia G. Gensel and Dianne Edwards (eds.), *Plants Invade the Land: Evolutionary and Environmental Perspectives* (Columbia University Press, 2001), https://doi.org/10.7312/gens11160, https://www.degruyter.com/ document/doi/10.7312/gens11160/html.

401   "A single ragweed plant can produce billions of pollen grains in one season." Texas Parks and Wildlife Department, "Airborne Pollen," *Young Naturalist*, https://tpwd.texas.gov/publications/nonpwdpubs/young_naturalist/ plants/airborne_pollen. "Spruce, like birch, produced about 5.5 billion grains in a year. Cereal rye grass produced 4.25 million pollen grains per inflorescence." Stanley J. Szefler, Francisco A. Bonilla, and Cezmi A. Akdis, *Pediatric Allergy: Principles and Practice* (Elsevier, 2016), 186.

402   Kimitsune Ishizaki, Ryuichi Nishihama, Katsuyuki T. Yamato, and Takayuki Kohchi, "Molecular Genetic Tools and Techniques for *Marchantia polymorpha* Research," *Plant and Cell Physiology* 57, no. 2 (February 2016): 262–270, https://doi.org/10.1093/pcp/pcv097. Bruce E. Fleury, "Bryophytes, Ferns and Fern Allies," Tulane University. Jennifer Frazer, "How Mosses Have Sex in Spite of Their Swimming-Challenged Sperm," *Scientific American* (27 July 2012), https://blogs.scientificamerican.com/artful-amoeba/how-mosses-have-sex-in-spite-of-their-swimming-challenged-sperm.

403   T.R. Birkhead, A.P. Møller, and W.J. Sutherland, "Why Do Females Make It So Difficult For Males To Fertilize Their Eggs?," *Journal of Theoretical Biology* 161, no. 1 (1993): 51–60. Don. R. Levitan, "Do Sperm Really Compete And Do Eggs Ever Have A Choice? Adult Distribution And Gamete Mixing Influence Sexual Selection, Sexual Conflict, And The Evolution Of Gamete Recognition Proteins In The Sea," *The American Naturalist* 191, no. 1 (2018): 88–105. William V. Holt and Katrien J.W. van Look, "Concepts in Sperm Heterogeneity, Sperm Selection And Sperm Competition As Biological Foundations For Laboratory Tests Of Semen Quality," *Reproduction* 127, no. 5 (2004): 527–535. Mats Olsson, et al., "Sperm Selection by Females," *Nature* 383, no. 6601 (1996): 585, https://www.nature.com/articles/383585a0.

404   D. Clift and M. Schuh, "Restarting Life: Fertilization And The Transition From Meiosis To Mitosis," *Nature Reviews Molecular Cell Biology* 14, no. 9 (September 2013): 549–62, https://doi.org/10.1038/nrm3643. Bruce Alberts, Alexander Johnson, Julian Lewis, Martin Raff, Keith Roberts, and Peter Walter, "Meiosis and Fertilization," in *Molecular Biology of the Cell* (W.W. Norton & Co, 2002). G.M. Cooper, *The Cell: A Molecular Approach*, 2nd edition (Sunderland (MA): Sinauer Associates; 2000), https://www.ncbi.nlm.nih. gov/books/NBK9901. Bruce Alberts, Alexander Johnson, Julian Lewis, David Morgan, Martin Raff, Keith Roberts, and Peter Walter, "Genetic Recombination," *ScienceDirect* (Elsevier, 2015), https://www.sciencedirect.com/topics/medicine-and-dentistry/genetic-recombination. Suzanna Clancy, "Genetic Recombination," *Nature Education* 1, no. 1 (2008): 401, https://www.nature.com/scitable/ topicpage/genetic-recombination-514.

405   Andrew P. Hendry, Kiyoko M. Gotanda, and Erik I. Svensson. "Human Influences On Evolution, And The Ecological And Societal Consequences," *Philosophical Transactions of the Royal Society B* 372, no. 1712 (2017), https:// royalsocietypublishing.org/doi/10.1098/rstb.2016.0028. Isabel Alves, Armande Ang Houle, Julie G. Hussin, and Philip Awadalla, "The Impact Of Recombination On Human Mutation Load And Disease," *Philosophical Transactions of the Royal Society B: Biological Sciences* 372, no. 1736 (2017), https:// royalsocietypublishing.org/doi/10.1098/rstb.2016.0465.

406   Ron Milo and Rob Phillips, "How Many Cells Are in the Human Body?" in Ron Milo and Rob Phillips, *Cell biology By The Numbers* (Garland Science, 2015). Yella Hewings-Martin, "How Many Cells Are in the Human Body?" *Medical News Today* (2017).

407   Don R. Levitan, "Do Sperm Really Compete and Do Eggs Ever Have a Choice? Adult Distribution and Gamete Mixing Influence Sexual Selection, Sexual Conflict, and the Evolution of Gamete Recognition Proteins in the Sea, Conspecific sperm precedence (CSP)," *The American Naturalist* 191, no. 1 (2018), https://www.journals.uchicago.edu/doi/full/10.1086/694780. I.A. Brewis and C.H. Wong, "Gamete Recognition: Sperm Proteins That Interact With

The Egg Zona Pellucida," *Reviews of Reproduction* 4, no. 3 (September 1999): 135–42, https://doi.org/10.1530/ror.0.0040135, https://pubmed.ncbi.nlm.nih.gov/10521150. J.H. Nadeau, "Do Gametes Woo? Evidence for Their Nonrandom Union at Fertilization," *Genetics* 207, no. 2 (October 2017): 369–387, https://academic.oup.com/genetics/article/207/2/369/5930780. Carrie Arnold, "Choosy Eggs May Pick Sperm for Their Genes, Defying Mendel's Law," *Quanta Magazine* (15 November 2017), https://www.quantamagazine.org/choosy-eggs-may-pick-sperm-for-their-genes-defying-mendels-law-20171115. R.C. Firman, C. Gasparini, M.K. Manier, and T. Pizzari, "Postmating Female Control: 20 Years of Cryptic Female Choice," *Trends in Ecology & Evolution* 32, no. 5 (2017): 368–382, https://doi.org/10.1016/j.tree.2017.02.010, https://www.cell.com/trends/ecology-evolution/fulltext/S0169-5347%2817%2930046-0.

408 S. Pressel and J.G. Duckett, "Do Motile Spermatozoids Limit The Effectiveness Of Sexual Reproduction In Bryophytes? Not In The Liverwort Marchantia Polymorpha," *Journal of Sytematics Evolution* 57 (2019): 371–381, https://doi.org/10.1111/jse.12528, https://onlinelibrary.wiley.com/doi/10.1111/jse.12528.

409 M. Maschler, E. Solan, and S. Zamir, "Mixed Strategies," in Maschler, Solan, and Zamir, *Game Theory* (Cambridge University Press, 2013), 144–218, https://doi.org/10.1017/CBO9780511794216.006.

410 You plants may have made it to land earlier than 450 million years ago. See: Elizabeth Pennisi, "Land Plants Arose Earlier Than Thought—And May Have Had A Bigger Impact On The Evolution Of Animals: Older Birth Date Suggests Plants And Animals Arose In Parallel 500 Million Years Ago," *Science* (19 February 2018), https://www.science.org/content/article/land-plants-arose-earlier-thought-and-may-have-had-bigger-impact-evolution-animals.

411 Filippo Aureli, et al., "Fission-Fusion Dynamics: New Research Frameworks," *Current Anthropology* 49, no. 4 (2008): 627–654. I. Couzin and M. Laidre, "Fission-Fusion Populations," *Current Biology* 19 (2009): R633–R635, https://doi.org/10.1016/j.cub.2009.05.034.

412 Don R. Levitan, "Do Sperm Really Compete and Do Eggs Ever Have a Choice? Adult Distribution and Gamete Mixing Influence Sexual Selection, Sexual Conflict, and the Evolution of Gamete Recognition Proteins in the Sea, Conspecific sperm precedence (CSP)," *The American Naturalist* 191, no. 1 (2018), https://www.journals.uchicago.edu/doi/full/10.1086/694780. I.A. Brewis, C.H. Wong, "Gamete Recognition: Sperm Proteins That Interact With The Egg Zona Pellucida," *Reviews of Reproduction* 4, no. 3 (September 1999): 135–42, https://pubmed.ncbi.nlm.nih.gov/10521150, https://rep.bioscientifica.com/downloadpdf/view/journals/revreprod/4/3/135.pdf.

413   Stuart Kauffman and Lee Smolin, "A Possible Solution To The Problem Of Time In Quantum Cosmology," arXiv preprint (1997), https://arxiv.org/abs/gr-qc/9703026.

414   Stuart Kauffman, "The Adjacent Possible," *Edge* (9 November 2003), https://www.edge.org/conversation/stuart_a_kauffman-the-adjacent-possible. Stuart Kauffman, *A World Beyond Physics: The Emergence And Evolution Of Life* (Oxford University Press, 2019).

415   Colin Blakemore, Roger H.S. Carpenter, and Mark A. Georgeson, "Lateral Inhibition Between Orientation Detectors In The Human Visual System," *Nature* 228, no. 5266 (1970): pages 37–39, https://www.nature.com/articles/228037a0.

416   Charles Darwin, *On the Origin of Species by Means of Natural Selection, or Preservation of Favoured Races in the Struggle for Life* (London: John Murray, 1859). Charles Darwin, *The Origin of Species by Means of Natural Selection: or, the Preservation of Favored Races in the Struggle for Life*, Volume II (New York: Appleton, 1897).

417   Plants took to the land 470 million years ago, according to A.J. Shaw, P. Szövényi, B. Shaw, "Bryophyte Diversity And Evolution: Windows Into The Early Evolution Of Land Plants," *American Journal of Botany* 98 (2011): 352–369, https://doi.org/10.3732/ajb.1000316.

418   Carl Zimmer, "Dawn of the Leafy Age," The Loom (Zimmer's blog) (5 July 2004), http://www.corante.com/loom/archives/004766.html. See also: Australian National Botanic Gardens, "Reproduction and Dispersal—Bryophyte," http://www.anbg.gov.au/bryophyte/reproduction-dispersal.html. Bryophytes, liverworts, and mosses can reproduce sexually or asexually.

419   "Sexual Reproduction," *Wikipedia, The Free Encyclopedia* (17 February 2024), http://en.wikipedia.org/wiki/Sexual_reproduction#Bryophytes.

420   "Bryophytes have neither pollen nor flowers and rely on water to carry the male gametes (the sperm) to the female gametes (the eggs). The spore capsules are produced after the sperm have fertilized the eggs. Hence the spores are part of the sexual reproductive cycle." Australian National Botanic Gardens, "Reproduction and Dispersal—Bryophyte," 2019, http://www.anbg.gov.au/bryophyte/reproduction-dispersal.html.

421   Howard Bloom, "The Xerox Effect: On the Importance of Pre-Biotic Evolution," *PhysicaPlus, the online publication of the Israeli Physical Society*, 13 November 2018. Reprinted in *Psychology Today* (20 May 2020), https://cdn2.psychologytoday.com/assets/the_xerox_effect-text_with_illustrations_0520-02.pdf.

422   Carl Zimmer, "Dawn of the Leafy Age," The Loom, (5 July 2004), http://www.corante.com/loom/archives/004766.html.

423   Robert A. Gastaldo, Patricia G. Gensel, Ian J. Glasspool, Steven J. Hinds, Olivia A. King, Duncan McLean, Adrian F. Park, Matthew R. Stimson, and Timothy Stonesifer, "Enigmatic Fossil Plants With Three-Dimensional, Arborescent-Growth Architecture From The Earliest Carboniferous Of New Brunswick, Canada," *Current Biology* (2024), https://bio.unc.edu/wp-content/uploads/sites/353/2024/02/GastaldoGensel2024-Sanfordiacaulis1.pdf.

424   Geoffrey M. Cooper, *The Cell: A Molecular Approach* (Sunderland MA: Sinauer Associates, 2000), https://www.ncbi.nlm.nih.gov/books/NBK9905. Ron Milo and Rob Phillips, "How Large Are Chloroplasts?" *BioNumbers: Cell Biology by the Numbers*, http://book.bionumbers.org/how-large-are-chloroplasts.

425   Dennis Normile, "The World's First Trees Grew By Splitting Their Guts: Spectacularly Well Preserved Specimens Reveal Complex Structure," *Science* (23 October 2017), https://www.science.org/content/article/world-s-first-trees-grew-splitting-their-guts.

426   Matthew P. Nelsen, William A. DiMichele, Shanan E. Peters, and C. Kevin Boyce, "Delayed Fungal Evolution Did Not Cause The Paleozoic Peak In Coal Production," *Proceedings of the National Academy of Sciences* 113, no. 9 (2016): pages 2442–2447, https://www.pnas.org/doi/abs/10.1073/pnas.1517943113.

427   David J. Nowak, "Water & Forests: The Role Trees Play in Water Quality," US Department of Agriculture, Forest Service (12 April 2023), https://www.fs.usda.gov/Internet/FSE_DOCUMENTS/stelprdb5269813.pdf.

428   Ibid.

429   Amy Ellis Nutt, "Why Do Some Male Trees Turn Female?" *Washington Post* (27 February 2018), https://www.washingtonpost.com/news/speaking-of-science/wp/2018/02/27/why-do-some-male-trees-turn-female.

430   Tim J. Brodribb and Taylor S. Field, "Leaf Hydraulic Evolution Led A Surge In Leaf Photosynthetic Capacity During Early Angiosperm Diversification," *Ecology Letters* 13, no. 2 (2010): 175–183, https://onlinelibrary.wiley.com/doi/full/10.1111/j.1461-0248.2009.01410.x. "How Did Flowering Plants Evolve To Dominate Earth?" *ScienceDaily* (9 December 2009), https://www.sciencedaily.com/releases/2009/12/091201100221.htm.

431   Ibid.

432   Maddie Stone, "Human-Made Materials Now Equal Weight Of All Life On Earth," *National Geographic* (9 December 2020), https://www.nationalgeographic.com/environment/article/human-made-materials-now-equal-weight-of-all-life-on-earth.

433   Tim J. Brodribb and Taylor S. Field, "Leaf Hydraulic Evolution Led A Surge In Leaf Photosynthetic Capacity During Early Angiosperm Diversification," *Ecology Letters* 13, no. 2 (2010): 175–183. "How Did Flowering Plants

Evolve To Dominate Earth?" *ScienceDaily*, http://www.sciencedaily.com/releases/2009/12/091201100221.htm.

434 Ibid.

435 Sherwin Carlquist, "Xylem Heterochrony: An Unappreciated Key To Angiosperm Origin And Diversifications," *Botanical Journal of the Linnean Society* 161, no. 1 (2009): 26, https://academic.oup.com/botlinnean/article/161/1/26/2418353. Wiley-Blackwell, "Weeds That Reinvented Weediness: New Research Sheds Light On Origins And Success Of Flowering Plants," *ScienceDaily* (5 September 2009), http://www.sciencedaily.com/releases/2009/09/090903064929.htm.

436 Ibid.

437 Ibid.

438 Ibid.

439 Ibid.

440 Leaves first appeared 380 million years ago. See: Peter A. Ensiminger, "Leaf – Evolution," *Science Encyclopedia*, https://science.jrank.org/pages/3875/Leaf-Evolution.html.

441 Andrew B. Leslie, Carl Simpson, and Luke Mander, "Reproductive Innovations And Pulsed Rise In Plant Complexity," *Science* 373, no. 6561 (2021): 1368–1372, https://www.science.org/doi/full/10.1126/science.abi6984.

442 Ibid. Stanford University, "Plants Didn't Evolve Gradually—They Evolved Complexity in Two Dramatic Bursts 250-Million-Years Apart," September 20, 2021, https://scitechdaily.com/plants-didnt-evolve-gradually-they-evolved-complexity-in-two-dramatic-bursts-250-million-years-apart.

443 Khramov, Alexander V., Tatiana Foraponova, and Piotr Węgierek, "The Earliest Pollen-Loaded Insects from the Lower Permian of Russia," *Biology Letters* 19, no. 3 (2023), https://royalsocietypublishing.org/doi/full/10.1098/rsbl.2022.0523. Constance Holden, "Permian Pollen Eaters," *Science* (16 May 1997), https://www.science.org/content/article/permian-pollen-eaters.

444 Alexander V. Khramov, Sergey V. Naugolnykh, and Piotr Węgierek, "Possible Long-Proboscid Insect Pollinators From The Early Permian Of Russia," *Current Biology* 32, no. 17 (2022): 3815–3820.

445 Constance Holden, "Permian Pollen Eaters," *Science* (16 May 1997), https://www.science.org/content/article/permian-pollen-eaters.

446 J.L. Capinera, "Internal Anatomy of Insects," in J.L. Capinera (ed.), *Encyclopedia of Entomology* (Dordrecht: Springer, 2008), 2020–2024, https://doi.org/10.1007/978-1-4020-6359-6_1562.

447 Janice M. Glime and Irene Bisang, "Sexuality: Sexual Strategies," in *Bryophyte Ecology* (International Association of Bryologists, Michigan, USA, 2017), 3.

448 Stephen Blackmore, Alexandra H. Wortley, John J. Skvarla, and John R. Rowley, "Pollen Wall Development in Flowering Plants," *New Phytologist* 174, no. 3 (2007): 483–498, https://nph.onlinelibrary.wiley.com/doi/10.1111/j.1469-8137.2007.02060.x.

449 O. Shu, "On The First Appearance Of Some Gymnospermous Pollen And Gsdp Assemblages In The Sub-Angara, Euramerian And Cathaysia Provinces," *Journal of Palaeosciences* 45 (1996): 20–32, https://doi.org/10.54991/jop.1996.1215. Jan Muller, "Significance of Fossil Pollen for Angiosperm History," *Annals of the Missouri Botanical Garden* 71, no. 2 (1984): 419–43, https://doi.org/10.2307/2399033.

450 M.Y. Galperin, S.L. Mekhedov, P. Puigbo, S. Smirnov, Y.I. Wolf, D.J. Rigden, "Genomic Determinants Of Sporulation In Bacilli And Clostridia: Towards The Minimal Set Of Sporulation-Specific Genes," *Environmental Microbiology* 14, no. 11 (November 2012): 2870–90, https://doi.org/10.1111/j.1462-2920.2012.02841.x.

451 Zenkteler, Maciej, and Agnieszka Maria Bagniewska-Zadworna, "Distant in Vitro Pollination of Ovules," *Phytomorphology: An International Journal of Plant Morphology* 51 (2001), https://www.semanticscholar.org/paper/Distant-in-vitro-pollination-of-ovules-Zenkteler-Bagniewska-Zadworna/40ea8d61eaf4ce8a2231b575e539a047dca2a329.

452 "The stylar fluid contains sucrose, glucose and fructose as the main carbohydrates, as well as osmiophilic droplets and proteins." M.T.M. Willemse"Progamic Phase And Fertilization In *Gasteria verrucosa* (*Mill.*) *H. Duval*: Pollination Signals," *Sexual Plant Reproduction* 9 (1996): 348–352, https://doi.org/10.1007/BF02441954 https://link.springer.com/article/10.1007/BF02441954. See also: "Information Sheet 9: Parts of Flowers," Africanized Honey Bees on the Move, Africanized Honey Bee Education Project, The University of Arizona, https://cals.arizona.edu/pubs/insects/ahb/inf9.html.

453 S. Zhong, J. Zhang and L. Qu, "The Signals To Trigger The Initiation Of Ovule Enlargement Are From The Pollen Tubes: The Direct Evidence," *Journal of Integrative Plant Biology* 59, no. 9 (2017): 600–603, https://doi.org/10.1111/jipb.12577 https://onlinelibrary.wiley.com/doi/10.1111/jipb.12577.

454 Valayamghat Raghavan, "Pollen Nutritional Content And Digestibility For Animals," in Brian E.S. Gunning, Martin William Steer (eds.), *Plant Cell Biology: Structure and Function* (Sudbury, Massachusetts: Jones and Bartlett Publishers, 1996). I.H. Roulslon and J.H. Cane, "Molecular Embryology of Flowering Plants," in *Pollen and Pollination*, edited by Amots Dafni, Michael Hesse, Ettore Pacini (Vienna: Springer, 2000): 187, 210–211.

455 David L. Mulcahy and Gabriella Bergamini Mulcahy, "The Effects of Pollen Competition," *American Scientist* 75, no. 1 (1987): 44–50.

456   Conrad C. Labandeira, Jiri Kvacek, and Mikhail B. Mostovski, "Pollination Drops, Pollen, and Insect Pollination of *Mesozoic gymnosperms*," *Taxon* 56, no. 3 (August 2007), https://onlinelibrary.wiley.com/doi/abs/10.2307/25065852.

457   Teresa Friedrich Finnern, "Cycads and Ginkos," *LibreTexts, Biology*, https://bio.libretexts.org/Courses/Norco_College/BIO_5%3A_General_Botany_(Friedrich_Finnern)/21%3A_Seed_Plants/21.03%3A_Cycads_and_Ginkos.

458   Maria V. Tekleva, Valentin A. Krassilov, and Jiří Kvaček, "Pollen Genus Eucommiidites: Ultrastructure," *Acta Palaeobotanica* 46, no. 2 (2006): 137–155.

459   Labandeira, Kvacek, and Mostovski, "Pollination Drops."

460   T'ai H. Roulston and James H. Cane, "Pollen Nutritional Content and Digestibility for Animals," *Plant Systematics and Evolution* 222 (2000): 187–209, https://link.springer.com/article/10.1007/bf00984102.

461   Constance Holden, "Permian Pollen Eaters," *Science* (16 May 1997).

462   Your discovery appears to have taken place 95 million years ago. David Winship and Shusheng Hu, "Coevolution of Early Angiosperms And Their Pollinators: Evidence From Pollen," *Palaeontographica Abteilung B Band* 283 (4 October 2010): 103–135, https://www.researchgate.net/publication/281233393_Coevolution_of_early_angiosperms_and_their_pollinators_Evidence_from_pollen.

463   E. Nikinmaa, T. Hölttä, P. Hari, P. Kolari, A. Mäkelä, S. Sevanto and T. Vesala T., "Assimilate Transport In Phloem Sets Conditions For Leaf Gas Exchange," *Plant, Cell & Environment* 36, no. 3 (March 2013): 655–69, https://onlinelibrary.wiley.com/doi/full/10.1111/pce.12004. Karen Wilson, David Morrison (eds.), *Monocots: Systematics and Evolution* (CSIRO Publishing, 2000).

464   Ibid. Milton Clarence Gugler, "Sap Exudation Phenomena Of The Elm Tree," *HathiTrust Digital Library*, http://catalog.hathitrust.org/Record/005742806. H.K. Jensen, J.A. Savage, N.M. Holbrook, "Optimal Concentration For Sugar Transport In Plants," *Journal of the Royal Society Interface* 10, no. 83 (20 March 2013), https://royalsocietypublishing.org/doi/full/10.1098/rsif.2013.0055.

465   J.D. Rejón, F. Delalande, C. Schaeffer-Reiss, et al., "The Plant Stigma Exudate: A Biochemically Active Extracellular Environment For Pollen Germination?" *Plant Signaling and Behavior* 9, no. 4 (2014), https://www.tandfonline.com/doi/full/10.4161/psb.28274. J.D. Rejón, F. Delalande, C. Schaeffer-Reiss, et al., "Proteomics Profiling Reveals Novel Proteins And Functions Of The Plant Stigma Exudate," *Journal of Experimental Botany* (2013) https://academic.oup.com/jxb/article/64/18/5695/610064.

466   Tokushiro Takaso, "'Pollination Drop' Time at the Arnold Arboretum," *Arnoldia* 50, no. 2 (1990): 2–7, https://www.jstor.org/stable/42954377. Patrick von Aderkas, Natalie A. Prior, and Stefan A. Little, "The Evolution Of Sexual Fluids In Gymnosperms From Pollination Drops To Nectar," *Frontiers in Plant Science* 9 (2018): 1844, https://www.frontiersin.org/journals/plant-science/articles/10.3389/fpls.2018.01844/full.

467   M. Nepi, P. von Aderkas, R. Wagner, S. Mugnaini, A. Coulter, E. Pacini, "Nectar and Pollination Drops: How Different Are They?" *Annals of Botany* (August 2009): 205–19, https://academic.oup.com/aob/article/104/2/205/105502.

468   F.W. Martin and J.L. Brewbaker, "The Nature of the Stigmatic Exudate and its Role in Pollen Germination," in *Pollen*, edited by Butterworth-Heinemann (1971), 262–266, https://doi.org/10.1016/B978-0-408-70149-5.50034-8, https://www.sciencedirect.com/science/article/pii/B9780408701495500348.

469   Karen Wilson, David Morrison (eds.), *Monocots: Systematics and Evolution: Systematics and Evolution* (CSIRO Publishing, 2000).

470   Massimo Nepi, Patrick von Aderkas, and Ettore Pacini, "Sugary Exudates in Plant Pollination," in *Secretions and Exudates in Biological Systems*, edited by Jorge M. Vivanco and František Baluška (Springer Science & Business Media, 2012), 155–185.

471   Gretchen Vogel, "Early Start for Plant-Insect Dance," *Science*, 273 (16 August 1996): 872, https://www.science.org/doi/10.1126/science.273.5277.872.

472   "Nectar," GreekMythology.com (17 May 2015), https://www.greekmythology.com/Myths/Elements/Nectar/nectar.html.

473   Graves Robert, *The Greek Myths* (London: Penguin, 1980). Sean Kelly, "Ambrosia and Nectar: The Food and Drink of the Gods," *Classical Wisdom*, (27 October 2021), https://classicalwisdom.com/mythology/ambrosia-and-nectar-the-food-and-drink-of-the-gods.

474   Juan P. González-Varo, F. Javier Ortiz-Sánchez, and Montserrat Vilà, "Total Bee Dependence on One Flower Species Despite Available Congeners of Similar Floral Shape," *PLOS One* (22 September 2016), https://doi.org/10.1371/journal.pone.0163122.

475   Andrew B. Leslie, Carl Simpson, and Luke Mander, "Reproductive Innovations And Pulsed Rise In Plant Complexity," *Science* (17 September 2021), https://www.science.org/doi/full/10.1126/science.abi6984. "Plants Didn't Evolve Gradually—They Evolved Complexity in Two Dramatic Bursts 250-Million-Years Apart," *SciTechDaily* (20 September 2021), https://scitechdaily.com/plants-didnt-evolve-gradually-they-evolved-complexity-in-two-dramatic-bursts-250-million-years-apart. Peter A. Hochuli and Susanne Feist-Burkhardt, "Angiosperm-like Pollen and Afropollis from the Middle Triassic (Anisian) of the Germanic Basin (Northern Switzerland)," *Frontiers in Plant Science* 4

(2013), https://www.frontiersin.org/journals/plant-science/articles/10.3389/fpls.2013.00344/full. "New Fossils Push The Origin Of Flowering Plants Back By 100 Million Years To The Early Triassic," *Science Daily* (1 October 2013), https://www.sciencedaily.com/releases/2013/10/131001191811.htm.

476   Simon Wallace, Andrew Fleming, Charles H. Wellman and David J. Beerling, "Evolutionary Development Of The Plant And Spore Wall," *AoB Plants, Annals of Botany* (7 October 2011), https://doi.org/10.1093/aobpla/plr027, https://www.ncbi.nlm.nih.gov/pmc/articles/PMC3220415.

477   Alexander V. Khramov, Tatiana Foraponova, and Piotr Węgierek, "The Earliest Pollen-Loaded Insects from the Lower Permian of Russia," *The Royal Society: Biology Letters* (1 March 2023), https://doi.org/10.1098/rsbl.2022.0523. Will Sullivan, "Scientists Discover Oldest Known Fossils of Pollen-Carrying Insects," *Smithsonian Magazine* (6 March 2023), https://www.smithsonianmag.com/smart-news/scientists-discover-oldest-known-fossils-of-pollen-carrying-insects-180981721.

478   X. Wang and S. Zheng, "The Earliest Normal Flower From Liaoning Province, China," *Journal Of Integrative Plant Biology* 51, no. 8 (2009): 800–11, https://doi.org/10.1111/j.1744-7909.2009.00838.x.

479   Bloom, Howard, "Instant Evolution. The Influence Of The City On Human Genes: A Speculative Case," *New Ideas in Psychology* 19, no. 3 (2001): 203–220, https://www.sciencedirect.com/science/article/abs/pii/S0732118X01000046.

480   Susan M. Gaines, Geoffrey Eglinton, and Jurgen Rullkotter, *Echoes of Life: What Fossil Molecules Reveal About Earth History* (Oxford University Press, 2009).

481   C. Coiffard, B. Gomez, V. Daviero-Gomez, D. L. Dilcher, "Rise to Dominance Of Angiosperm Pioneers In European Cretaceous Environments," *Proceedings of the National Academy of Sciences* (2012), https://www.pnas.org/doi/full/10.1073/pnas.1218633110. Indiana University, "Emergence of Flowering Plants: New Light Shed On Darwin's 'Abominable Mystery'," *ScienceDaily* (6 December 2012), https://www.sciencedaily.com/releases/2012/12/121206094131.htm.

482   Chanderbali et al., "Transcriptional Signatures Of Ancient Floral Developmental Genetics In Avocado (*Persea americana; Lauraceae*)," *Proceedings of the National Academy of Sciences* (2009), https://www.pnas.org/doi/full/10.1073/pnas.0811476106, https://doi.org/10.1073/pnas.0811476106. "Insight Into Evolution Of First Flowers," *ScienceDaily* (19 May 2009), https://www.sciencedaily.com/releases/2009/05/090518172453.htm.

483   Michael J. Moore, Pamela S. Soltis, Charles D. Bell, J. Gordon Burleigh, and Douglas E. Soltis, "Phylogenetic Analysis Of 83 Plastid Genes Further Resolves The Early Diversification Of Eudicots," *Proceedings of the National Academy*

of Sciences (2010), https://www.pnas.org/doi/full/10.1073/pnas.0907801107. University of Florida, "DNA Sequencing Unlocks Evolutionary Origins, Relationships Among Flowering Plants," *ScienceDaily* (24 February 2010), https://www.sciencedaily.com/releases/2010/02/100223161831.htm.

484   E.A. Kellogg, "The Evolutionary History of Ehrhartoideae, Oryzeae, and Oryza. RICE 2, 1–14," *Springer Open* (30 January 2009), https://thericejournal.springeropen.com/articles/10.1007/s12284-009-9022-2. https://doi.org/10.1007/s12284-009-9022-2.

485   PBS, "First Flower," (17 April 2007), https://www.pbs.org/wgbh/nova/flower.

486   Michael Klesius, "The Big Bloom—How Flowering Plants Changed the World," *National Geographic* (20 September 2021), https://www.nationalgeographic.com/science/article/big-bloom.

487   Columba Stewart, "Evagrius Ponticus and the 'Eight Generic Logismoi'" (2005), College of Saint Benedict/Saint John's University, https://digitalcommons.csbsju.edu/sot_pubs/56.

488   Eileen Sweeney, "Aquinas on the Seven Deadly Sins: Tradition and Innovation," in *Sin in Medieval and Early Modern Culture: The Tradition of the Seven Deadly Sins* (York Medieval Press/Boydell and Brewer, 2012).

489   Giorgio Riello and Ulinka Rublack (eds.), *The Right to Dress: Sumptuary Laws in a Global Perspective, C.1200-1800* (Cambridge University Press, 2019). "Sumptuary law," *Wikipedia, The Free Encyclopedia*, https://en.wikipedia.org/wiki/Sumptuary_law.

490   Monica A. Geber, Todd E. Dawson, and Lynda F. Delph (eds.), *Gender and Sexual Dimorphism in Flowering Plants* (Springer Science & Business Media, 2012), 161. A.J. Richards, *Plant Breeding Systems* (Germany: Chapman & Hall, 1997), 145. Gustavo Brant Paterno, Carina Lima Silveira, Johannes Kollmann, Mark Westoby, and Carlos Roberto Fonseca, "The Maleness of Larger Angiosperm Flowers," *Proceedings of the National Academy of Sciences* 117, no. 20 (19 May 2020): 10921–10926, https://www.pnas.org/doi/abs/10.1073/pnas.1910631117, see chart, page 3.

491   Vincent P. Gutschick, *A Functional Biology of Crop Plants* (Springer Science & Business Media, 2012), 158. Candace Galen, "Why Do Flowers Vary? The Functional Ecology Of Variation In Flower Size And Form Within Natural Plant Populations," *Bioscience* 49, no. 8 (1 August 1999): 631–640, https://doi.org/10.2307/1313439, https://academic.oup.com/bioscience/article/49/8/631/254663.

492   Adam Smith, *An Inquiry Into The Nature And Causes Of The Wealth Of Nations* (Philadelphia: Thomas Dobson, 1776).

493   Herbert Spencer, *First Principles* (London: Williams and Northgate, 1867), 4, 91.

494   Jordi Bascompte, "Mutualism and Biodiversity," *Current Biology* 29, no. 11 (2019): R467–R470, https://www.cell.com/current-biology/pdf/S0960-9822(19)30390-2.pdf.

495   Richard J. Butler, et al., "Diversity Patterns Amongst Herbivorous Dinosaurs And Plants During The Cretaceous: Implications For Hypotheses Of Dinosaur/Angiosperm Co-Evolution," *Journal of Evolutionary Biology* 22, no. 3 (2009): 446–459. Fabien Génin, et al., "Co-Evolution Assists Geographic Dispersal: The Case Of Madagascar," *Biological Journal of the Linnean Society* 137, no. 2 (2022): 163–182, https://academic.oup.com/biolinnean/article/137/2/163/6672676.

496   X. Wang and S. Zheng, "The Earliest Normal Flower from Liaoning Province, China," *Journal of Integrative Plant Biology* 51, no. 8 (2009): 800–11, https://doi.org/10.1111/j.1744-7909.2009.00838.x, https://onlinelibrary.wiley.com/doi/10.1111/j.1744-7909.2009.00838.x. Mario Vallejo-Marin, "Revealed: The First Flower, 140-million Years Old, Looked Like a Magnolia," *Scientific American* (1 August 2017), https://www.scientificamerican.com/article/revealed-the-first-flower-140-million-years-old-looked-like-a-magnolia.

497   Stephen Jay Gould, Elisabeth S. Vrba, "Exaptation—A Missing Term in the Science of Form," *Paleobiology* 8, no. 1 (1982): 4–15, https://doi.org/10.1017/S0094837300004310, https://www.cambridge.org/core/journals/paleobiology/article/abs/exaptationa-missing-term-in-the-science-of-form/A672662BA208D220B9F9A06DE5D804B8.

498   R. Sablowski, Control of Patterning, Growth, And Differentiation By Floral Organ Identity Genes," *Journal of Experimental Botany* 66, no. 4 (2015): 1065–73, https://doi.org/10.1093/jxb/eru514, https://academic.oup.com/jxb/article/66/4/1065/593766.

499   Larry D. Noodén (ed.), *Senescence and Aging in Plants* (Elsevier, 2012), xxi.

500   Simon Geir Møller (ed.), *Plastids Vol. 13* (Boca Raton, FLA, CRC Press, 2005), 52.

501   "Which Cut Flowers Last The Longest?" *Floraly* (21 March 2019), https://www.floraly.com.au/blogs/news/longest-lasting-flowers.

502   H.J. Rogers, "From Models to Ornamentals: How Is Flower Senescence Regulated?" *Plant Molecular Biology* 82, no. 15 (September 2013): 563–574, https://doi.org/10.1007/s11103-012-9968-0. Y. Guo and S. Gan, "Leaf Senescence: Signals, Execution, and Regulation," *Current Topics in Developmental Biology* 71 (2005): 83–112, https://www.sciencedirect.com/science/article/pii/S0070215305710036. Cheng-Hung Yen and Chang-Hsien Yang, "Evidence for Programmed Cell Death During Leaf Senescence In Plants,"

Plant and Cell Physiology 39, no. 9 (1998): 922–927, https://academic.oup.com/pcp/article/39/9/922/1819124.

503   Michael Pollan, *The Botany of Desire: A Plant's Eye View of the World* (New York: Random House, 2001).

504   Silas Busck Mellor, et al., "Non-Photosynthetic Plastids As Hosts For Metabolic Engineering," *Essays in Biochemistry* 62, no. 1 (April 2018): 41–50, https://portlandpress.com/essaysbiochem/article-abstract/62/1/41/78517/Non-photosynthetic-plastids-as-hosts-for-metabolic.

505   Esther Iglich, "Flower Structure," in *Botany: The Appreciation of Plant Life* (6 September 2019), https://www.merlot.org/merlot/viewMaterial.htm?id=1118124.

506   Heather M. Whitney, Georgina Milne, Sean A. Rands, Silvia Vignolini, Cathie Martin, and Beverley J. Glover, "The Influence Of Pigmentation Patterning On Bumblebee Foraging From Flowers of *Antirrhinum majus*," *Naturwissenschaften* (19 February 2013), https://link.springer.com/article/10.1007/s00114-013-1020-y. "Bees Attracted To Contrasting Colors When Looking For Nectar," *ScienceDaily* (21 February 2013), https://www.sciencedaily.com/releases/2013/02/130221084707.htm.

507   Esther Iglich, "Flower Structure," in *Botany: The Appreciation of Plant Life*, https://www.merlot.org/merlot/viewMaterial.htm?id=1118124.

508   K. Salomo, J.F. Smith, et al., "The Emergence of Earliest Angiosperms may be Earlier than Fossil Evidence Indicates," Systematic Botany 42, no. 4 (December 2017): 607–619, https://doi.org/10.1600/036364417X696438, https://www.ingentaconnect.com/content/aspt/sb/2017/00000042/00000004/art00001;jsessionid=2njojod1a9qk1.x-ic-live-02.

509   Rigon, Riccardo, "How Many Leaves Has a Tree?" *About Hydrology* (22 November 2015), http://abouthydrology.blogspot.com/2015/11/how-many-leaves-has-tree.html.

510   M. Gagliano, V. Vyazovskiy, A. Borbély, et al., "Learning by Association in Plants," *Scientific Reports* 6, no. 38427 (2016), https://www.nature.com/articles/srep38427, https://doi.org/10.1038/srep38427. Frantisek Baluska, Monica Gagliano and Guenther Witzany (eds.), *Memory and Learning in Plants* (Springer Nature, 2018), https://link.springer.com/content/pdf/10.1007/978-3-319-75596-0.pdf.

511   Peter Bernhardt, *Wily Violets and Underground Orchids: Revelations of a Botanist* (University of Chicago Press, 2003), 64.

512   Sarah Zielinski, "Which Came First—The Plant Or Its Pollinator? The Evolution of the Orchid and the Orchid Bee," *Smithsonian Magazine* (23

September 2011), https://www.smithsonianmag.com/science-nature/the-evolution-of-the-orchid-and-the-orchid-bee-87336709.

513   Michael Klesius, "The Big Bloom—How Flowering Plants Changed the World," *National Geographic* (July 2002), https://www.nationalgeographic.com/science/article/big-bloom.

514   Andrew Groover and Quentin Cronk (eds.), *Comparative And Evolutionary Genomics Of Angiosperm Trees, Vol. 21* (New York: Springer, 2017). Quentin C.B. Cronk and Andrew T. Groover, "Introduction," in *Comparative Genomics of Angiosperm Trees: A New Era of Tree Biology* (Springer 2016), 1–11, https://link.springer.com/chapter/10.1007/7397_2016_33.

515   "State of the World's Plants," Royal Botanic Gardens, United Kingdom (2016), https://kew.iro.bl.uk/concern/reports/f931f1de-72c7-46b4-b57c-28eb417c53ec?locale=en. Shreya Dasgupta, "How Many Plant Species Are There In The World? Scientists Now Have An Answer," *Mongabay: News & Inspiration From Nature's Frontline* (12 May 2016), https://news.mongabay.com/2016/05/many-plants-world-scientists-may-now-answer. Rebecca Morelle, "Kew Report Makes New Tally For Number Of World's Plants," *BBC* (10 May 2016), https://www.bbc.com/news/science-environment-36230858. "Plants Didn't Evolve Gradually – They Evolved Complexity in Two Dramatic Bursts 250-Million-Years Apart," *SciTech Daily* (20 September 2021), https://scitechdaily.com/plants-didnt-evolve-gradually-they-evolved-complexity-in-two-dramatic-bursts-250-million-years-apart. "Plants Evolved Complexity In Two Bursts — With A 250-Million-Year Hiatus," *Science Daily* (16 September 2021), https://www.sciencedaily.com/releases/2021/09/210916142851.htm.

516   M.J. Benton, P. Wilf and H. Sauquet, "The Angiosperm Terrestrial Revolution and the Origins Of Modern Biodiversity," *New Phytologist* 233 (2022): 2017–2035, https://doi.org/10.1111/nph.17822, https://nph.onlinelibrary.wiley.com/doi/10.1111/nph.17822.

517   O. Eriksson, "Evolution Of Angiosperm Seed Disperser Mutualisms: The Timing Of Origins And Their Consequences For Coevolutionary Interactions Between Angiosperms And Frugivores," *Biological Reviews* 91 (2016): 168–186 (20 December 2014), https://onlinelibrary.wiley.com/doi/10.1111/brv.12164, https://doi.org/10.1111/brv.12164. Sussman, R.W. "Primate Origins And The Evolution Of Angiosperms," *American Journal of Primatology* 23 (1991): 209–223, https://doi.org/10.1002/ajp.1350230402. Dirk Ahrens, Julia Schwarzer, and Alfried P. Vogler, "The Evolution Of Scarab Beetles Tracks The Sequential Rise Of Angiosperms And Mammals," *Proceedings of the Royal Society B: Biological Sciences* 281, no. 1791 (22 September 2014), https://royalsocietypublishing.org/doi/full/10.1098/rspb.2014.1470.

518   M.J. Benton, P. Wilf and H. Sauquet, "The Angiosperm Terrestrial Revolution and the Origins Of Modern Biodiversity," *New Phytologist*

233 (2022): 2017–2035, https://doi.org/10.1111/nph.17822, https://nph. onlinelibrary.wiley.com/doi/10.1111/nph.17822.

519   Water Science School, "The Distribution Of Water On, In, And Above The Earth," *US Geological Survey*, https://www.usgs.gov/media/images/ distribution-water-and-above-earth.

520   Danielle Torrent Tucker, "Plants Evolved Complexity In Two Bursts – With A 250-Million-Year Hiatus," Stanford University, Doerr School of Sustainability, (18 September 2021), https://earth.stanford.edu/news/plants-evolved-complexity-two-bursts-250-million-year-hiatus.

521   Roger Highfield and Roger Pennell, "On the Scent of Designer Blooms," *The Telegraph* (10 July 2002), https://www.telegraph.co.uk/news/science/science-news/4768723/On-the-scent-of-designer-blooms.html.

522   N. Dudareva, A. Klempien, J.K. Muhlemann, and I. Kaplan, "Biosynthesis, Function And Metabolic Engineering Of Plant Volatile Organic Compounds," *New Phytologist* 198 (2013): 16–32, https://nph.onlinelibrary.wiley.com/ doi/10.1111/nph.12145, https://doi.org/10.1111/nph.12145.

523   Among other signaling mechanisms, plants use Volatile Organic Compounds, VOCs. See H. Ueda Y. Kikuta, and K. Matsuda, "Plant Communication: Mediated by Individual or Blended VOCs?" *Plant Signaling Behavior* 7, no. 2 (1 February 2012): 222–6, https://doi.org/10.4161/psb.18765, https://www.tandfonline.com/doi/full/10.4161/psb.18765.

524   Esther Iglich, "Flower Structure," in *Botany: The Appreciation of Plant Life* (2019), https://www.merlot.org/merlot/viewMaterial.htm?id=1118124.

525   Roger Highfield and Roger Pennell, "On the Scent of Designer Blooms," The Telegraph (10 July 2002), https://www.telegraph.co.uk/news/science/science-news/4768723/On-the-scent-of-designer-blooms.html.

526   William Agosta, *Thieves, Deceivers, and Killers: Tales of Chemistry in Nature* (Princeton University Press, 2009), 44.

527   Roger Highfield and Roger Pennell, "On the Scent of Designer Blooms," *The Telegraph* (10 July 2002), https://www.telegraph.co.uk/news/science/science-news/4768723/On-the-scent-of-designer-blooms.html.

528   Agosta, *Thieves, Deceivers, and Killers*, 44.

529   Karen Wilson and David Morrison (eds.), *Monocots: Systematics and Evolution* (CSIRO Publishing, 2000).

530   J.E. Meisel, R.S. Kaufmann, F. Pupulin, *Orchids of Tropical America: An Introduction and Guide* (Cornell University Press, 2015), 173.

531   For fads and fashions among pre-human creatures, in this case guppies, see Lee AlanDugatkin, "Interface Between Culturally Based Preferences And Genetic Preferences: Female Mate Choice In Poecilia Reticulate," *Proceedings of the*

National Academy of Sciences 93, no. 7 (1996): 2770–2773, https://www.pnas.org/doi/abs/10.1073/pnas.93.7.2770.

532   J. Ackerman, "Geographic and Seasonal Variation in Fragrance Choices and Preferences of Male Euglossine Bees," *Biotropica* 21, no. 340 (December 1989), https://doi.org/10.2307/2388284. N. Williams and C. Dodson, "Selective Attraction Of Male Euglossine Bees To Orchid Floral Fragrances And Its Importance In Long Distance Pollen Flow," *Evolution* 26, https://doi.org/10.1111/j.1558-5646.1972.tb00176.x, https://academic.oup.com/evolut/article/26/1/84/6867461. Ø.H., Opedal, A.A. Martins and E.L. Marjakangas, "A Database And Synthesis Of Euglossine Bee Assemblages Collected At Fragrance Baits," *Apidologie* 51 (2020): 519–530, https://doi.org/10.1007/s13592-020-00739-4, https://link.springer.com/article/10.1007/s13592-020-00739-4.

533   T. Eltz, M. Ayasse and K. Lunau, "Species-Specific Antennal Responses to Tibial Fragrances by Male Orchid Bees," *Journal of Chemical Ecology* 32 (2006): 71–79, https://doi.org/10.1007/s10886-006-9352-0, https://link.springer.com/article/10.1007/s10886-006-9352-0. T. Eltz, D. Roubik and K. Lunau, "Experience-Dependent Choices Ensure Species-Specific Fragrance Accumulation In Male Orchid Bees," *Behavioral Ecology and Sociobiology* 59 (2005): 460, https://doi.org/10.1007/S00265-005-0093-9, https://link.springer.com/article/10.1007/s00265-005-0093-9. S. Cappellari and B. Harter-Marques, "First Report of Scent Collection by Male Orchid Bees (Hymenoptera: Apidae: Euglossini) from Terrestrial Mushrooms," *Journal of the Kansas Entomological Society* 83,no. 3 (2010): 264–266, https://doi.org/10.2317/JKES0911.16.1, https://bioone.org/journals/journal-of-the-kansas-entomological-society/volume-83/issue-3/JKES0911.16.1/First-Report-of-Scent-Collection-by-Male-Orchid-Bees-Hymenoptera/10.2317/JKES0911.16.1.short.

534   T. Pokorny, M. Hannibal, J.J.G. Quezada-Euan, et al., "Acquisition of Species-Specific Perfume Blends: Influence Of Habitat-Dependent Compound Availability On Odour Choices Of Male Orchid Bees (*Euglossa spp.*)," *Oecologia* 172 (2013): 417–425, https://doi.org/10.1007/s00442-013-2620-0, https://link.springer.com/article/10.1007/s00442-013-2620-0.

535   K. Brandt, S. Dötterl, et al., "Unraveling the Olfactory Biases of Male Euglossine Bees: Species-Specific Antennal Responses and Their Evolutionary Significance for Perfume Flowers," *Frontiers in Ecology and Evolution, Section on Behavioral and Evolutionary Ecology* 9, no. 15 (October 2021), https://doi.org/10.3389/fevo.2021.727471, https://www.frontiersin.org/articles/10.3389/fevo.2021.727471/full.

536   Adrian Prisca, "The 8 Most Expensive Perfumes in the World," *Luxatic*, 15 November 2021, https://luxatic.com/the-8-most-expensive-perfumes-in-the-world.

537   Sara Banayan, "Most Expensive Perfumes," Forbes February 3, 2006, https://www.forbes.com/2006/02/03/most-expensive-perfumes-cx_sb_0203fashion3_ls.html?sh=23ab73ab70a0

538   To paraphrase Matthew 22:14, "Many are called, but few are chosen." A rule that appears over and over again in female mate choice.

539   Eltz T., Zimmermann Y., Pfeiffer C., Pech J.R., Twele R., Francke W., Quezeda-Euan J.J.G., Lunau K., "An olfactory shift is associated with male perfume differentiation and sibling species divergence in orchid bees," Current Biology, 2008; 18, pages 1844-1848. Duncan E. Jackson, "Sympatric Speciation: Perfume Preferences of Orchid Bee Lineages," Current Biology, Volume 18, Issue 23, December 09, 2008, pages R1092-R1093, https://doi.org/10.1016/j.cub.2008.10.023

https://www.cell.com/current-biology/fulltext/S0960-9822(08)01354-7

540   Alcock, John, "Warty Hammer Orchids, Adaptations, and Darwin", a chapter in An Enthusiasm for Orchids: Sex and Deception in Plant Evolution, New York, 2006, online edition, Oxford Academic, 1 Sept. 2007, https://doi.org/10.1093/acprof:oso/9780195182743.003.0001

PBS, Mimicry: The Orchid and the Bee, https://www.pbs.org/wgbh/evolution/library/01/1/l_011_02.html

541   Bohman, Björn, Jeffares, Lynne, Flematti, Gavin, Byrne, Lindsay T. et al, "Discovery of Tetrasubstituted Pyrazines As Semiochemicals in a Sexually Deceptive Orchid," Journal of Natural Products, Vol. 75, issue 9, American Chemical Society, September 28, 2012, https://pubs.acs.org/doi/10.1021/np300388y

542   W. Whitten, A. Young, and D. Stern, "Nonfloral Sources of Chemicals That Attract Male Euglossine Bees (Apidae: Euglossini)," Journal of Chemical Ecology 20 (1993): 821–822, https://doi.org/10.1007/BF02059617, https://link.springer.com/article/10.1007/BF02059617.

543   Jonas Henske et al., "Function of Environment-Derived Male Perfumes in Orchid Bees," Current Biology (2023), https://doi.org/10.1016/j.cub.2023.03.060, https://www.cell.com/current-biology/abstract/S0960-9822(23)00380-9.

544   Simone Caroline Cappellari and Birgit Harter-Marques, "First Report of Scent Collection by Male Orchid Bees (Hymenoptera: Apidae: Euglossini) from Terrestrial Mushrooms," Journal of the Kansas Entomological Society 83, no. 3 (July 2010): 264–266, https://www.researchgate.net/publication/256244107_First_Report_of_Scent_Collection_by_Male_Orchid_Bees_Hymenoptera_Apidae_Euglossini_from_Terrestrial_Mushrooms. W. Mark Whitten, Allen M. Young, and David L. Stern, "Nonfloral Sources of Chemicals That Attract Male Euglossine Bees (Apidae: Euglossini)," Journal of Chemical Ecology 19 (1993): 3017–3027, https://link.springer.com/article/10.1007/BF00980599.

545 Jonas Henske et al., "Function of Environment-Derived Male Perfumes in Orchid Bees," *Current Biology* (April 12, 2023), https://www.cell.com/current-biology/abstract/S0960-9822(23)00380-9. Ruhr-Universität-Bochum, "Why Orchid Bees Concoct Their Own Fragrance," *Science Daily*, April 13, 2023, https://www.sciencedaily.com/releases/2023/04/230413154438.htm.

546 Nikolaas Tinbergen, The Study of Instinct (2020; Pygmalion Press, an imprint of Plunkett Lake Press, originally published 1951). Deidre Barrett, "Supernormal Stimuli," *Encyclopedia of Evolutionary Psychological Science* (1 January 2016), https://link.springer.com/referenceworkentry/10.1007/978-3-319-16999-6_94-1. V. Barnett, "Supernormal Stimuli (Konrad Lorenz)," in *Encyclopedia of Evolutionary Psychological Science*, ed. T.K. Shackelford and V.A. Weekes-Shackelford (Cham, Switzerland: Springer, first online January 2021), https://link.springer.com/referenceworkentry/10.1007/978-3-319-19650-3_3573, https://doi.org/10.1007/978-3-319-19650-3_3573. Deirdre Barrett, *Supernormal Stimuli: How Primal Urges Overran Their Evolutionary Purpose* (New York: WW Norton & Company, 2010).

547 Andrew B. Leslie, Carl Simpson, and Luke Mander, "Reproductive Innovations and Pulsed Rise in Plant Complexity," Science (2021), https://www.science.org/doi/10.1126/science.abi6984.

548 Kevin Thiele, "Thynnids Are Seriously Sexy Wasps," Taxonomy Australia (22 February 2019), https://www.taxonomyaustralia.org.au/post/thynnids-are-seriously-sexy-wasps.

549 C.C. McDonald, J. Podesta, C.C. Fortuin, and K.J. Gandhi, "Expanded Range of Eight Orchid Bee Species (Hymenoptera, Apidae, Euglossini) in Costa Rica," *Biodiversity Data Journal* (28 July 2022), https://bdj.pensoft.net/article/81220.

550 D. Wcisło, G. Vargas, K. Ihle, and W. Wcisło, "Nest Construction Behavior by the Orchid Bee Euglossa Hyacinthine," *Journal of Hymenoptera Research* 29 (2012): 15–20, https://doi.org/10.3897/JHR.29.4067, https://jhr.pensoft.net/articles.php?id=1615.

551 Charles Darwin, *The Various Contrivances by Which Orchids are Fertilised by Insects* (London: John Murray, 1862).

552 Rodrigo B. Singer and Samantha Koehler, "Notes on the Pollination Biology of Notylia nemorosa (Orchidaceae): Do Pollinators Necessarily Promote Cross Pollination?," *Journal of Plant Research* 116 (2003): 19–25, https://link.springer.com/content/pdf/10.1007/s10265-002-0064-4.pdf, http://www.ncbi.nlm.nih.gov/pubmed/12605296, https://pubmed.ncbi.nlm.nih.gov/12605296.

553 Conrad C. Labandeira, "Insect Mouthparts: Ascertaining the Paleobiology of Insect Feeding Strategies," *Annual Review of Ecology and Systematics* 28, no.

1 (1997): 153–193, https://www.annualreviews.org/content/journals/10.1146/annurev.ecolsys.28.1.153.

554   Constance Holden, "Permian Pollen Eaters," *Science*, (16 May 1997), https://www.science.org/content/article/permian-pollen-eaters.

555   T. Seeley and S. Buhrman, "Group Decision Making in Swarms of Honey Bees," *Behavioral Ecology and Sociobiology* 45 (January 1999): 19–31, https://doi.org/10.1007/s002650050536, https://link.springer.com/article/10.1007/s002650050536.

556   Teodoro Pittman, "How Far Do Bees Travel? (Scouting for Food, Mating & More)," *Misfit Animals*, https://misfitanimals.com/bees/how-far-do-bees-travel.

557   P. Visscher, "How Self-Organization Evolves," *Nature* 421 (20 February 2003): 799–800, https://doi.org/10.1038/421799a, https://www.nature.com/articles/421799a.

558   T. Inagaki, M. Irwin, M. Moieni, I. Jevtić, and N. Eisenberger, "A Pilot Study Examining Physical and Social Warmth: Higher (Non-Febrile) Oral Temperature Is Associated with Greater Feelings of Social Connection," *PLoS ONE* 11 (2016), https://doi.org/10.1371/journal.pone.0156873. J. Dabbs and J. Moorer, "Core Body Temperature and Social Arousal," *Personality and Social Psychology Bulletin* 1 (1975): 517–520, https://doi.org/10.1177/014616727500100312. Y. Jason Castro, "Bees Appear to Experience Moods," *Scientific American* (1 January 2012), https://www.scientificamerican.com/article/the-secret-inner-life-of-bees.

559   T. D. Seeley, *The Wisdom of the Hive: The Social Physiology of Honey Bee Colonies* (Harvard University Press, 2009).

560   Bee enthusiasm can be measured by the temperature of the bee. When a bee is excited about a new find of flowers, her temperature goes up. N. Sadler and J. Nieh, "Honey Bee Forager Thoracic Temperature Inside the Nest Is Tuned to Broad-Scale Differences in Recruitment Motivation," *Journal of Experimental Biology* 214 (2011): 469–475, https://doi.org/10.1242/jeb.049445. A. Stabentheiner, "Thermoregulation of Dancing Bees: Thoracic Temperature of Pollen and Nectar Foragers in Relation to Profitability of Foraging and Colony Need," *Journal of Insect Physiology* 47, no. 4-5 (2001): 385–392, https://doi.org/10.1016/S0022-1910(00)00132-3.

561   Karl von Frisch, translated by Leigh E. Chadwick, *The Dance Language and Orientation of Bees* (Cambridge, MA: The Belknap Press of Harvard University Press, 1967). Thomas D. Seeley, *Honeybee Ecology: A Study of Adaptation in Social Life* (Princeton, NJ: Princeton University Press, 1985). Thomas D. Seeley and Royce A. Levien, "A Colony of Mind: The Beehive As Thinking Machine," *The Sciences* (July/August 1987): 38–42. Thomas D. Seeley, *The Wisdom of the Hive: The Social Physiology of Honey Bee Colonies* (Cambridge, Massachusetts: Harvard University Press, 1995).

562   Steven N.S. Cheung, "The Fable of the Bees: An Economic Investigation," *The Journal of Law and Economics* 16, no. 1 (1973): 11–33.

563   Seeley, *Honeybee Ecology*. Seeley and Levien, "A Colony of Mind. Seeley, *The Wisdom of the Hive*.

564   Charles Robert Darwin to Asa Gray, April 3, 1860, letter, *The Darwin Project*, https://www.darwinproject.ac.uk/letter/?docId=letters/DCP-LETT-2743.xml.

565   P.G. Wodehouse, *Pigs Have Wings* (New York: Doubleday & Company, 1952).

566   C. Coiffard, B. Gomez, V. Daviero-Gomez, and D.L. Dilcher, "Rise to Dominance of Angiosperm Pioneers in European Cretaceous Environments," *Proceedings of the National Academy of Sciences*, 2012, https://doi.org/10.1073/pnas.1218633110, https://www.pnas.org/doi/full/10.1073/pnas.1218633110. Indiana University, "Emergence of Flowering Plants: New Light Shed on Darwin's 'Abominable Mystery,'" *ScienceDaily*, (6 December 2012), https://www.sciencedaily.com/releases/2012/12/121206094131.htm.

567   Ibid.

568   Ibid.

569   Ibid.

570   Frank Berendse and Marten Scheffer, "The Angiosperm Radiation Revisited, an Ecological Explanation for Darwin's 'Abominable Mystery,'" *Ecology Letters*, Published Online (2 July 2009), https://onlinelibrary.wiley.com/doi/full/10.1111/j.1461-0248.2009.01342.x. Wageningen University and Research Centre, "Darwin's Mystery of Appearance of Flowering Plants Explained," *ScienceDaily* (14 July 2009), http://www.sciencedaily.com/releases/2009/07/090713211621.htm.

571   Ibid.

572   Charles N. Miller, Jr., "Mesozoic Conifers," *Botanical Review* 43, no. 2 (April–June 1977), https://link.springer.com/article/10.1007/BF02860718. K. Denniston, "Ancient Trees," *Northwest Conifer Connections* (10 August 2014), http://nwconifers.blogspot.com/2014/08/ancient-trees.html.

573   Spencer C.H. Barrett, "The Evolution of Plant Sexual Diversity," *Nature Reviews Genetics* 3, no. 4 (2002): 274–284, https://www.nature.com/articles/nrg776.

574   Frank Berendse and Marten Scheffer, "The Angiosperm Radiation Revisited, an Ecological Explanation for Darwin's 'Abominable Mystery,'" *Ecology Letters*, Published Online (2 July 2009), https://onlinelibrary.wiley.com/doi/full/10.1111/j.1461-0248.2009.01342.x. Wageningen University and Research Centre, "Darwin's Mystery of Appearance of Flowering Plants Explained," *ScienceDaily* (14 July 2009), http://www.sciencedaily.com/

releases/2009/07/090713211621.htm. M.J. Benton, P. Wilf, and H. Sauquet, "The Angiosperm Terrestrial Revolution and the Origins of Modern Biodiversity," *New Phytologist* 233 (2022): 2017–2035, https://doi.org/10.1111/nph.17822, https://nph.onlinelibrary.wiley.com/doi/10.1111/nph.17822. Sakia Fields, "Diversification of Angiosperms During the Cretaceous Period," University of Nebraska, Lincoln, DigitalCommons@University of Nebraska – Lincoln, Environmental Studies Undergraduate Student Theses, Environmental Studies Program, 2021, https://digitalcommons.unl.edu/cgi/viewcontent.cgi?article=1296&context=envstudtheses.

575   Charles Darwin, Letter to J.D. Hooker, Down, 22 July [1879], in Charles Darwin, More Letters of Charles Darwin, Volume 2, 378, *Darwin Correspondence Project*, https://www.darwinproject.ac.uk/letter/DCP-LETT-12167.xml.

576   NWO (Netherlands Organization for Scientific Research), "Evolution Experiments with Flowers," *ScienceDaily* (30 December 2009), http://www.sciencedaily.com/releases/2009/10/091029150610.htm.

577   Frank Berendse and Marten Scheffer, "The Angiosperm Radiation Revisited, an Ecological Explanation for Darwin's 'Abominable Mystery,'" *Ecology Letters* (2 July 2009), https://onlinelibrary.wiley.com/doi/full/10.1111/j.1461-0248.2009.01342.x. Wageningen University and Research Centre, "Darwin's Mystery of Appearance of Flowering Plants Explained," Frank Berendse and Marten Scheffer, "The Angiosperm Radiation Revisited, an Ecological Explanation for Darwin's 'Abominable Mystery,'" *Ecology Letters* (2 July 2009), https://onlinelibrary.wiley.com/doi/full/10.1111/j.1461-0248.2009.01342.x. Wageningen University and Research Centre, "Darwin's Mystery of Appearance of Flowering Plants Explained," *ScienceDaily* (14 July 2009), http://www.sciencedaily.com/releases/2009/07/090713211621.htm., (14 July 2009), http://www.sciencedaily.com/releases/2009/07/090713211621.htm.

578   Ibid.

579   Ibid.

580   Wageningen University and Research Centre, "Darwin's Mystery of Appearance of Flowering Plants Explained," *ScienceDaily* (14 July 2009), http://www.sciencedaily.com/releases/2009/07/090713211621.htm.

581   Frank Berendse and Marten Scheffer, "The Angiosperm Radiation Revisited, an Ecological Explanation for Darwin's 'Abominable Mystery,'" *Ecology Letters* (2 July 2009), https://onlinelibrary.wiley.com/doi/full/10.1111/j.1461-0248.2009.01342.x.

582   Kevin A. Simonin and Adam B. Roddy, "Genome Downsizing, Physiological Novelty, and the Global Dominance of Flowering Plants," *PLoS Biology* 16, no. 1 (11 January 2018), https://journals.plos.org/plosbiology/article?id=10.1371/journal.pbio.2003706.

583   "Chicxulub, A Massive Asteroid That Hit Earth 65 Million Years Ago," *New Scientist*, https://www.newscientist.com/definition/chicxulub.

584   Jamie B. Thompson and Santiago Ramírez-Barahona, "No Phylogenetic Evidence for Angiosperm Mass Extinction at the Cretaceous–Palaeogene (K-Pg) Boundary," *Biology Letters* 19, no. 9 (2023), https://royalsocietypublishing. org/doi/full/10.1098/rsbl.2023.0314. StudyFinds, "Flower Power: Ancestors of Modern-Day Plants Survived Mass Extinction of the Dinosaurs," (22 September 2023), https://studyfinds.org/plants-survived-extinction. University of Bath, "Nature's Great Survivors: Flowering Plants Survived the Mass Extinction That Killed the Dinosaurs," *EurekAlert* (12 September 2023), https://www.eurekalert. org/news-releases/1001247.

585   Chanderbali et al., "Transcriptional Signatures of Ancient Floral Developmental Genetics in Avocado (Persea americana; Lauraceae)," *Proceedings of the National Academy of Sciences* (2009), https://doi. org/10.1073/pnas.0811476106, https://www.pnas.org/doi/full/10.1073/ pnas.0811476106. University of Florida, "Insight Into Evolution of First Flowers," *ScienceDaily* (19 May 2009), https://www.sciencedaily.com/ releases/2009/05/090518172453.htm.

586   Fabien L. Condamine, Eva B. Koppelhus, and Alexandre Antonelli, "The Rise of Angiosperms Pushed Conifers to Decline During Global Cooling," *Proceedings of the National Academy of Sciences* 117, no. 46 (2 November 2020): 28867–28875, https://www.pnas.org/doi/abs/10.1073/pnas.2005571117. H. John B. Birks, "Angiosperms Versus Gymnosperms in the Cretaceous," *Proceedings of the National Academy of Sciences* 117, no. 49 (13 November 2020): 30879–30881, https://doi.org/10.1073/pnas.2021186117.

587   Brian Tomasik, "How Many Wild Animals Are There?" (7 August 2019), Reducing Suffering.org, https://reducing-suffering.org/how-many-wild-animals-are-there. See also the figures cited from Brian Tomasik at: Ben Team, "Number of Animals on the Earth," Wild Sky Media, Mom.com, https://animals.mom. com/number-animals-earth-3994.html.

588   Lynn Margulis, *Symbiotic Planet: A New Look at Evolution* (Basic Books, 2008).

589   P. Brand, V. Larcher, A. Couto, J. Sandoz, and S. Ramírez, "Sexual Dimorphism in Visual and Olfactory Brain Centers in the Perfume-Collecting Orchid Bee Euglossa Dilemma (Hymenoptera, Apidae)," *The Journal of Comparative Neurology* 526 (2018): 2068–2077, https://doi.org/10.1002/ cne.24483.

590   E. Snively and J. Theodor, "Common Functional Correlates of Head-Strike Behavior in the Pachycephalosaur Stegoceras Validum (Ornithischia, Dinosauria) and Combative Artiodactyls," *PLoS ONE* 6 (2011), https://doi.org/10.1371/ journal.pone.0021422. Robert Sanders, "Did Dome-Headed Dinosaurs Sport

Bristly Headgear?" *Berkeley News* (23 May 2023), https://news.berkeley.edu/2023/05/23/did-dome-headed-dinosaurs-sport-bristly-headgear.

591   D. Cary Woodruff, Darren Naish, and Jamie Dunning, "Photoluminescent Visual Displays: An Additional Function of Integumentary Structures in Extinct Archosaurs?" *Historical Biology* (2020), https://doi.org/10.1080/08912963.2020.1731806, https://www.tandfonline.com/doi/full/10.1080/08912963.2020.1731806.

592   Pascal Godefroit et al., "A Jurassic Ornithischian Dinosaur from Siberia with Both Feathers and Scales," *Science* 345 (2014): 451–455, https://doi.org/10.1126/science.1253351, https://www.science.org/doi/10.1126/science.1253351. Michael Balter, "Earliest Dinosaurs May Have Sported Feathers: First Feathers May Have Been for Insulation or Sexual Display," *Science* (24 July 2014), https://www.science.org/content/article/earliest-dinosaurs-may-have-sported-feathers.

593   For a demonstration of fads and fashions in animal mate choice, see: Dugatkin Lee Alan and J. Godin Jean-Guy, "Reversal of Female Mate Choice by Copying in the Guppy (Poecilia Reticulata)," *Proceedings of the Royal Society of London B* 249 (1992): 179–184, http://doi.org/10.1098/rspb.1992.0101, https://royalsocietypublishing.org/doi/10.1098/rspb.1992.0101.

594   Justine Chow, "A Whiff of Bee Evolution," Student Voices, Scitable, *Nature Education* (30 July 2010), https://www.nature.com/scitable/blog/student-voices/a_whiff_of_bee_evolution. Stephen Buchmann, "Orchid Bees (The Euglossines)," US Forest Service, US Department of Agriculture (22 May 2015), https://www.fs.usda.gov/wildflowers/pollinators/pollinator-of-the-month/orchid_bees.shtml.

595   Stephen Buchmann, *Letters from the Hive: An Intimate History of Bees, Honey, and Humankind* (Banning, CA: Banning Repplier, 2006), 108.

596   Stephen Taber, III, "The Frequency of Multiple Mating of Queen Honey Bees," *Journal of Economic Entomology* 47, no. 6 (1 December 1954): 995–998, https://doi.org/10.1093/jee/47.6.995, https://academic.oup.com/jee/article-abstract/47/6/995/2205757. Christina Grozinger, "How Many Times Does a Queen Honey Bee Mate?" *Bee Health* (20 August 2019), https://bee-health.extension.org/how-many-times-does-a-queen-honey-bee-mate.

597   N.E. Gary, "Observations of Mating Behaviour in the Honeybee," *Journal of Apicultural Research* 2, no. 1 (1963): 3–13, https://doi.org/10.1080/00218839.1963.11100050, https://www.tandfonline.com/doi/abs/10.1080/00218839.1963.11100050. S. Hayashi and T. Satoh, "Sperm Maturation Process Occurs in the Seminal Vesicle Following Sperm Transition from Testis in Honey Bee Males," *Apidologie* 50 (2019): 369–378, https://doi.org/10.1007/s13592-019-00652-5, https://link.springer.com/article/10.1007/s13592-019-00652-5. Alex Walls, "Bees Are Explosively Ejaculating to Death. A Polystyrene Cover Could Help Stop

It," The University of British Columbia (22 February 2022), https://news.ubc.ca/2022/02/22/bees-are-explosively-ejaculating-to-death-a-polystyrene-cover-could-help-stop-it. Rachael Funnell, "If You Listen Carefully, You Can Hear a Bee Ejaculate," *IFLScience* (20 July 2023), https://www.iflscience.com/if-you-listen-carefully-you-can-hear-a-bee-ejaculate-69912.

598   Lovleen Marwaha, *The Polyandrous Queen Honey Bee: Biology and Apiculture* (Bentham Science Publishers, 2023), 241. Prudence Wood, "What Is a Honey Bee Drone Congregation Area?" *Bee Professor* (10 March 2023), https://beeprofessor.com/what-is-a-drone-congregation-area.

599   Arati Kumar-Rao, "No Honey, No Hives, But Solitary Bees Have Important Lives," *Mongabay* (8 March 2021), https://india.mongabay.com/2021/03/no-honey-no-hives-but-solitary-bees-have-important-lives.

600   Gustavo Brant Paterno, Carina Lima Silveira, Johannes Kollmann, Mark Westoby, and Carlos Roberto Fonseca, "The Maleness of Larger Angiosperm Flowers," *Proceedings of the National Academy of Sciences* 117, no. 20 (2020): 10921–10926. See chart, page 3. Thomas D. Seeley, *Honeybee Ecology: A Study of Adaptation in Social Life* (Princeton, NJ: Princeton University Press, 1985). Thomas D. Seeley and Royce A. Levien, "A Colony of Mind: The Beehive As Thinking Machine," The Sciences, (July/August 1987): 38–42. Thomas D. Seeley, *The Wisdom of the Hive: The Social Physiology of Honey Bee Colonies* (Cambridge, Massachusetts: Harvard University Press, 1995).

601   Chrys Voyiatzi and D.G. Voyiatzis, "In Vitro Shoot Proliferation Rate of Dieffenbachia Exotica Cultivar 'Marianna' As Affected by Cytokinins, the Number of Recultures and the Temperature," *Scientia Horticulturae* 40, no. 2 (1989): 163–169, https://doi.org/10.1016/0304-4238(89)90099-X, https://www.sciencedirect.com/science/article/pii/030442388990099X?via%3Dihub.

602   Samir C. Debnath, "Propagation Strategies and Genetic Fidelity in Strawberries," *International Journal of Fruit Science* 13, no. 1–2 (2013): 3–18, Proceedings of the 2011 North American Strawberry Symposium, https://www.tandfonline.com/doi/full/10.1080/15538362.2012.696520.

603   Brian Charlesworth, "Why Bother? The Evolutionary Genetics of Sex," *Daedalus* 136, no. 2 (2007): 37–46, https://direct.mit.edu/daed/article/136/2/37/26680/Why-bother-The-evolutionary-genetics-of-sex.

604   Richard Dawkins, *The Selfish Gene* (New York: Oxford University Press, 1976).

605   Gafni, Marc, *Your Unique Self: The Radical Path to Personal Enlightenment* (Integral Publishers, 2012).

606   I. Johnston, "Muscle Metabolism and Growth in Antarctic Fishes (Suborder Notothenioidei): Evolution in a Cold Environment," *Comparative Biochemistry and Physiology. Part B, Biochemistry & Molecular Biology* 136, no. 4 (2003): 701–713, https://doi.org/10.1016/S1096-4959(03)00258-6,

https://www.sciencedirect.com/science/article/pii/S1096495903002586. C. Cheng and H. Detrich, "Molecular Ecophysiology of Antarctic Notothenioid Fishes," *Philosophical Transactions of the Royal Society of London, Series B, Biological Sciences* 362, no. 1488 (2007): 2215–2232, https://doi.org/10.1098/RSTB.2006.1946.

607 J. Daane and H. Detrich, "Adaptations and Diversity of Antarctic Fishes: A Genomic Perspective," *Annual Review of Animal Biosciences*, https://doi.org/10.1146/annurev-animal-081221-064325.

608 UC Davis, "Antarctic Sea Ice," 17 February 2023, https://www.ucdavis.edu/climate/definitions/antarctic-sea-ice.

609 A. Orlov and A. Tokranov, "Checklist of Deep-Sea Fishes of the Russian Northwestern Pacific Ocean Found at Depths Below 1000 m," *Progress in Oceanography*, https://doi.org/10.1016/J.POCEAN.2019.102143.

610 Gabriel Amir, Boris Rubinsky, Sheick Yousif Basheer, Liana Horowitz, Leor Jonathan, Micha S. Feinberg, Aram K. Smolinsky, Jacob Lavee, "Improved Viability and Reduced Apoptosis in Sub-Zero 21-Hour Preservation of Transplanted Rat Hearts Using Anti-Freeze Proteins," *The Journal of Heart and Lung Transplantation* 24, no. 11 (2005): 1915–1929, https://doi.org/10.1016/j.healun.2004.11.003, https://www.jhltonline.org/article/S1053-2498(04)00606-0/abstract, https://www.sciencedirect.com/science/article/pii/S1053249804006060.

611 Mary Colvard, Aleeza Oshry, Esther Shyu, and Laura Bonetta, "Icefish Blood Adaptations: Antifreeze Protein," Evolution: Biointeractive.org, Howard Hughes Medical Institute, https://www.biointeractive.org/sites/default/files/IcefishAdaptationsAntifreeze-Educator-act.pdf.

612 Ibid. See also: Arthur L. DeVries and C.-H. Christina Cheng, "Antifreeze Proteins and Organismal Freezing Avoidance in Polar Fishes," *Fish Physiology* 22 (2005): 155–201, https://doi.org/10.1016/S1546-5098(04)22004-0, https://www.sciencedirect.com/science/article/pii/S1546509804220040.

613 Mario La Mesa et al., "Parental Care and Reproductive Strategies in Notothenioid Fishes," *Fish and Fisheries* 22, no. 2 (2021): 356–376, https://doi.org/10.1111/faf.12523. Jacob Daane and H. William Detrich, "Adaptations and Diversity of Antarctic Fishes: A Genomic Perspective," *Annual Review of Animal Biosciences* 10, no. 1 (February 2022), https://doi.org/10.1146/annurev-animal-081221-064325, https://www.annualreviews.org/content/journals/10.1146/annurev-animal-081221-064325. Autun Purser, Laura Hehemann, Lilian Boehringer, Sandra Tippenhauer, Mia Wege, Horst Bornemann, Santiago EA Pineda-Metz et al., "A Vast Icefish Breeding Colony Discovered in the Antarctic," *Current Biology* 32, no. 4 (2022): 842–850. Katie Hunt, "An Icefish Colony Discovered in Antarctica Is World's Largest Fish

Breeding Ground," CNN (13 January 2022), https://www.cnn.com/2022/01/13/world/icefish-colony-discovery-scn/index.html.

614   Katie Serena, "Antarctic Fish Have 'Antifreeze' Proteins in Their Blood," in "Frogs That Can Freeze Their Bodies and 6 Other Crazy Ways That Animals Survive Their Treacherous Environments," (18 January 2018), slide 3, https://www.businessinsider.in/Frogs-that-can-freeze-their-bodies-and-6-other-crazy-ways-that-animals-survive-their-treacherous-environments/Antarctic-fish-have-antifreeze-proteins-in-their-blood-/slideshow/53231167.cms.

615   J. Donnelly, J. Torres, T. Sutton, and C. Simoniello, "Fishes of the Eastern Ross Sea, Antarctica," *Polar Biology* 27 (2004): 637–650, https://doi.org/10.1007/s00300-004-0632-2. J. Daane and H. Detrich, "Adaptations and Diversity of Antarctic Fishes: A Genomic Perspective," *Annual Review of Animal Biosciences*, https://doi.org/10.1146/annurev-animal-081221-064325, https://www.annualreviews.org/content/journals/10.1146/annurev-animal-081221-064325. T. Near, J. Pesavento, and C. Cheng, "Phylogenetic Investigations of Antarctic Notothenioid Fishes (Perciformes: Notothenioidei) Using Complete Gene Sequences of the Mitochondrial Encoded 16S rRNA," *Molecular Phylogenetics and Evolution* 32, no. 3 (2004): 881–891, https://doi.org/10.1016/J.YMPEV.2004.01.002, https://www.sciencedirect.com/science/article/pii/S1055790304000387?via%3Dihub.

616   Notothenoids eat primarily Pleuragrama antarticum, Antarctic silverfish, the minnows of the region around the South Pole. But those silverfish are, guess what? Even smaller notothenioids. Those smaller fish, notothenioid silver fish, in turn, eat crystal krill. The crystal krill eat phytoplankton, protozoans, and copepods. The phytoplankton eat sunlight. They are photosynthesizers. T. Hopkins, D. Ainley, J. Torres, and T. Lancraft, "Trophic Structure in Open Waters of the Marginal Ice Zone in the Scotia-Weddell Confluence Region During Spring (1983)," *Polar Biology* 13 (1993): 389–397, https://doi.org/10.1007/BF01681980, https://link.springer.com/article/10.1007/BF01681980. D. Ainley, G. Ballard, and K. Dugger, "Competition Among Penguins and Cetaceans Reveals Trophic Cascades in the Western Ross Sea, Antarctica," *Ecology* 87, no. 8 (2006): 2080–2093, https://doi.org/10.1890/0012-9658(2006)87[2080:CAPACR]2.0.CO;2, https://esajournals.onlinelibrary.wiley.com/doi/full/10.1890/0012-9658%282006%2987%5B2080%3ACAPACR%5D2.0.CO%3B2. J. Eastman, "Pleuragramma Antarcticum (Pisces, Nototheniidae) as Food for Other Fishes in McMurdo Sound, Antarctica," *Polar Biology* 4 (1985): 155–160, https://doi.org/10.1007/BF00263878. M. Mesa and J. Eastman, "Antarctic Silverfish: Life Strategies of a Key Species in the High-Antarctic Ecosystem," *Fish and Fisheries* 13 (2012): 241–266, https://doi.org/10.1111/J.1467-2979.2011.00427.X, https://onlinelibrary.wiley.com/doi/10.1111/j.1467-2979.2011.00427.x.

617   J. Eastman, "The Nature of the Diversity of Antarctic Fishes," *Polar Biology* 28 (2004): 93–107, https://doi.org/10.1007/s00300-004-0667-4.

618   K. O'Brien and E. Crockett, "The Promise and Perils of Antarctic Fishes," *EMBO Reports* 14, European Molecular Biology Organization, https://doi.org/10.1038/embor.2012.203, https://link.springer.com/article/10.1007/s00300-004-0667-4.

619   Sandra Nagl et al., "Persistence of Neutral Polymorphisms in Lake Victoria Cichlid Fish," *Proceedings of the National Academy of Sciences* 95, no. 24 (1998): 14238–14243.

620   Lake Malawi is also known as Lake Nyasa.

621   George Barlow, *The Cichlid Fishes: Nature's Grand Experiment In Evolution* (Basic Books, 2008).

622   T.R. Funnell, R.J. Fialkowski, and P.D. Dijkstra, "Social Dominance Does Not Increase Oxidative Stress in a Female Dominance Hierarchy of an African Cichlid Fish," *Ethology* 128 (2022): 15–25, https://doi.org/10.1111/eth.13232, https://onlinelibrary.wiley.com/doi/epdf/10.1111/eth.13232.

623   L.A. Dugatkin, "Interface Between Culturally Based Preferences and Genetic Preferences: Female Mate Choice in Poecilia Reticulata," *Proceedings of the National Academy of Sciences* 93, no. 7 (2 April 1996): 2770–2773, https://doi.org/10.1073/pnas.93.7.2770, https://www.pnas.org/doi/abs/10.1073/pnas.93.7.2770.

624   Escobar-Camacho D, Carleton K.L., "Sensory Modalities in Cichlid Fish Behavior," *Current Opinion in Behavioral Sciences* 6 (December 1, 2015): 115–124, https://doi.org/10.1016/j.cobeha.2015.11.002, https://www.sciencedirect.com/science/article/pii/S2352154615001424.

625   Personal correspondence with Kelly Kissane, 1998-2002.

626   Oxford English Dictionary, s.v. "schismogenesis (n.)," https://doi.org/10.1093/OED/1144658883.

627   H. Whitehead, "Sperm Whale Clans and Human Societies," *Royal Society Open Science* 11 (2024): 231353.5, https://royalsocietypublishing.org/doi/10.1098/rsos.231353.

628   See also: R.T. Pereyra, L. Bergström, L. Kautsky, et al., "Rapid Speciation in a Newly Opened Postglacial Marine Environment, the Baltic Sea," *BMC Evolutionary Biology* 9, no. 70 (2009), https://doi.org/10.1186/1471-2148-9-70, https://bmcecolevol.biomedcentral.com/articles/10.1186/1471-2148-9-70. W. Dominey, "Effects of Sexual Selection and Life History on Speciation: Species Flock in African Cichlids and Hawaiian Drosophila," in *Evolution of Fish Species Flocks*, eds. Echelle AA and Kornfield I (Orono, ME: University of Maine at Orono Press, 1984), 231–250. O. Seehausen, J.J.M. van Alphen, and F. Witte, "Cichlid Fish Diversity Threatened by Eutrophication That Curbs Sexual Selection," *Science* 277 (1997): 1808–1811, https://doi.org/10.1126/science.277.5333.1808, https://bmcecolevol.biomedcentral.com/

articles/10.1186/1471-2148-9-70. W. Salzburger, "The Interaction of Sexually and Naturally Selected Traits in the Adaptive Radiations of Cichlid Fishes," *Molecular Ecology* 18 (2009): 169–185, https://doi.org/10.1111/j.1365-294X.2008.03981.x, https://onlinelibrary.wiley.com/doi/10.1111/j.1365-294X.2008.03981.x. Karen P. Maruska, Uyhun S. Ung, and Russell D. Fernald, "The African Cichlid Fish Astatotilapia Burtoni Uses Acoustic Communication for Reproduction: Sound Production, Hearing, and Behavioral Significance," *PLoS ONE* 7, no. 5 (2012), https://journals.plos.org/plosone/article?id=10.1371/journal.pone.0037612. Stanford University, "African Cichlid's Noisy Courtship Ritual," *ScienceDaily* (13 June 2012), www.sciencedaily.com/releases/2012/06/120613153339.htm.

629   Personal correspondence with Kelly Kissane, 1998-2002.

630   The traditional view, promoted by Ernst Mayr, is that groups need to be separated by a considerable distance to develop the genetic alterations that lead to speciation—that lead to the inability to cross-breed. However that model has proven to be incorrect, especially among fish. Ernst Mayr, *Populations, Species, and Evolution* (Cambridge, MA: Harvard University Press, 1970). Tom Tregenza and Roger K. Butlin, "Speciation Without Isolation," *Nature* (22 July 1999): 311–312. Virginia Morell, "Ecology Returns to Speciation Studies," *Science* (25 June 1999): 2106–2108. Joana I. Meier et al., "Cycles of Fusion and Fission Enabled Rapid Parallel Adaptive Radiations in African Cichlids," *Science* 381, no. 6665 (22 September 2023), https://www.science.org/doi/abs/10.1126/science.ade2833.

631   Y. Won, A. Sivasundar, Y. Wang, and J. Hey, "On the Origin of Lake Malawi Cichlid Species: A Population Genetic Analysis of Divergence," *Proceedings of the National Academy of Sciences of the United States of America* 102 (2005): 6581–6586, https://doi.org/10.1073/PNAS.0502127102, https://www.pnas.org/doi/full/10.1073/pnas.0502127102.

632   Christian Sturmbauer, Sanja Baric, Walter Salzburger, Lukas Rüber, and Erik Verheyen, "Lake Level Fluctuations Synchronize Genetic Divergences of Cichlid Fishes in African Lakes," *Molecular Biology and Evolution* 18, no. 2 (February 2001): 144–154, https://doi.org/10.1093/oxfordjournals.molbev.a003788, https://academic.oup.com/mbe/article/18/2/144/1079219.

633   I. Kornfield and P. Smith, "African Cichlid Fishes: Model Systems for Evolutionary Biology," *Annual Review of Ecology, Evolution, and Systematics* 31 (2000): 163–196, https://doi.org/10.1146/Annurev.Ecolsys.31.1.163, https://www.annualreviews.org/content/journals/10.1146/annurev.ecolsys.31.1.163. Virginia Morell, "Ecology Returns to Speciation Studies," *Science* (1999): 2106–2108. A.J. Ribbink, "Cuckoo Among Lake Malawi Cichlid Fish," *Nature* 267, no. 5608 (1977): 243–244. George Barlow, *The Cichlid Fishes: Nature's Grand Experiment In Evolution* (New York: Basic Books, 2008). M.E. Maan and K.M. Sefc, "Colour Variation in Cichlid Fish: Developmental Mechanisms, Selective Pressures and Evolutionary Consequences," *Seminars in Cell and*

Developmental Biology 24 (2013): 516–528, https://doi.org/10.1016/j.
semcdb.201305.003, https://www.ncbi.nlm.nih.gov/pmc/articles/PMC3778878.
R.T. Pereyra, L. Bergström, L. Kautsky, et al., "Rapid Speciation in a Newly
Opened Postglacial Marine Environment, the Baltic Sea," *BMC Evolutionary
Biology* 9, no. 70 (2009), https://doi.org/10.1186/1471-2148-9-70, https://
bmcecolevol.biomedcentral.com/articles/10.1186/1471-2148-9-70. W. Dominey,
"Effects of Sexual Selection and Life History on Speciation: Species Flocks in
African Cichlids and Hawaiian Drosophila," in *Evolution of Fish Species Flocks*,
eds. Echelle AA and Kornfield I (Orono, ME: University of Maine at Orono
Press, 1983), 231–250. O. Seehausen, J.J.M. van Alphen, and F. Witte, "Cichlid
Fish Diversity Threatened by Eutrophication That Curbs Sexual Selection,"
Science 277 (1997): 1808–1811, https://doi.org/10.1126/science.277.5333.1808,
https://www.science.org/doi/10.1126/science.277.5333.1808. W. Salzburger,
"The Interaction of Sexually and Naturally Selected Traits in the Adaptive
Radiations of Cichlid Fishes," *Molecular Ecology* 18 (2009): 169–185, https://
doi.org/10.1111/j.1365-294X.2008.03981.x, https://onlinelibrary.wiley.com/
doi/10.1111/j.1365-294X.2008.03981.x. Ben Crair, "The Fishy Mystery of
Lake Malawi," *Smithsonian Magazine*, March 2019. I.P. Farias, G. Ortí, and
A. Meyer, "Total Evidence: Molecules, Morphology, and the Phylogenetics
of Cichlid Fishes," *Journal of Experimental Zoology* 288 (2000): 76–92,
https://doi.org/10.1002/(SICI)1097-010X(20000415)288:1<76::AID-
JEZ8>3.0.CO;2-P, https://onlinelibrary.wiley.com/doi/10.1002/(SICI)1097-
010X(20000415)288:1%3C76::AID-JEZ8%3E3.0.CO;2-P.

634   C. Sturmbauer and A. Meyer, "Genetic Divergence, Speciation and
Morphological Stasis in a Lineage of African Cichlid Fishes," *Nature* (13 August
1992): 578. Malcolm T. Smith and Robert Layton, "Still Human After All These
Years," *The Sciences* 10 (January–February 1989). O. Seehausen, J.J.M. van
Alphen, and F. Witte, "Cichlid Fish Diversity Threatened by Eutrophication That
Curbs Sexual Selection," *Science* 277 (19 September 1997): 1808–1810.

635   "BMP4 Bone Morphogenetic Protein 4 [Homo sapiens (human)] Gene ID:
652," (5 March 2023), National Library of Medicine, https://www.ncbi.nlm.nih.
gov/gene/652.

636   R. Albertson and T. Kocher, "Genetic and Developmental Basis of Cichlid
Trophic Diversity," *Heredity* 97 (2006): 211–221, https://doi.org/10.1038/
sj.hdy.6800864, https://www.nature.com/articles/6800864.

637   Arhat Abzhanov et al., "Bmp4 and Morphological Variation of Beaks in
Darwin's Finches," *Science* 305 (2004): 1462–1465, https://doi.org/10.1126/
science.1098095, https://www.science.org/doi/10.1126/science.1098095.

638   "BMP4 Bone Morphogenetic Protein 4 [Mus Musculus (House Mouse)]
Gene ID: 12159," (2 August 2023), National Library of Medicine, National Center
for Biotechnology Information, https://www.ncbi.nlm.nih.gov/gene/12159.

639   R. Lichtneckert and H. Reichert, "Insights into the Urbilaterian Brain: Conserved Genetic Patterning Mechanisms in Insect and Vertebrate Brain Development," *Heredity* 94 (2005): 465–477, https://doi.org/10.1038/sj.hdy.6800664, https://www.nature.com/articles/6800664. "BMP4 Bone Morphogenetic Protein 4 [Homo sapiens (Human)] Gene ID: 652," updated on (1 August 2023), National Library of Medicine, National Center for Biotechnology Information, National Institutes of Health, https://www.ncbi.nlm.nih.gov/gene/652.

640   Yuer Ye, Zhiwei Jiang, Yiqi Pan, Guoli Yang, and Ying Wang, "Role and Mechanism of BMP4 in Bone, Craniofacial, and Tooth Development," *Archives of Oral Biology* 140 (2022): 105465, ISSN 0003-9969, https://doi.org/10.1016/j.archoralbio.2022.105465, https://www.sciencedirect.com/science/article/pii/S0003996922001224. Science Direct, "Bone Morphogenetic Protein," https://www.sciencedirect.com/topics/medicine-and-dentistry/bone-morphogenetic-protein. M.N. Davies, M. Volta, R. Pidsley, et al., "Functional Annotation of the Human Brain Methylome Identifies Tissue-Specific Epigenetic Variation Across Brain and Blood," *Genome Biology* 13, R43 (2012), https://doi.org/10.1186/gb-2012-13-6-r43, https://genomebiology.biomedcentral.com/articles/10.1186/gb-2012-13-6-r43.

641   Lakes Tanganyika, Malawi, and Victoria. W. Salzburger, "The Interaction of Sexually and Naturally Selected Traits in the Adaptive Radiations of Cichlid Fishes," *Molecular Ecology* 18 (2009): 169–185, https://doi.org/10.1111/j.1365-294X.2008.03981.x, https://onlinelibrary.wiley.com/doi/10.1111/j.1365-294X.2008.03981.x. G.F. Turner et al., "How Many Species of Cichlid Fishes Are There in African Lakes?" *Molecular Ecology* 10, no. 3 (2001): 793–806, https://onlinelibrary.wiley.com/doi/abs/10.1046/j.1365-294x.2001.01200.x. Axel Meyer, "The Extraordinary Evolution of Cichlid Fishes," *Scientific American* (April 2015), https://www.scientificamerican.com/article/the-extraordinary-evolution-of-cichlid-fishes.

642   Hannes Svardal et al., "Ancestral Hybridization Facilitated Species Diversification in the Lake Malawi Cichlid Fish Adaptive Radiation," *Molecular Biology and Evolution* 37, no. 4 (2020): 1100–1113, https://academic.oup.com/mbe/article/37/4/1100/5671705.

643   Manpreet Kohli et al., "Evolutionary History and Divergence Times of Odonata (Dragonflies and Damselflies) Revealed through Transcriptomics," *iScience* 24, no. 11 (19 November 2021): 103324, https://doi.org/10.1016/j.isci.2021.103324. "The Order Odonata," NatureScot, "Plants, Animals, and Fungi: Invertebrates, Freshwater Invertebrates, Dragonflies and Damselflies," Scottish Natural Heritage, https://www.nature.scot/plants-animals-and-fungi/invertebrates/freshwater-invertebrates/dragonflies-and-damselflies.

644   Galveston County Master Gardeners, "Beneficial Damselflies," Texas A&M AgriLife Extension Service, Quoted in Kathy Keatley Garvey, "Damselflies: Long,

Slender and Delicate," *Bug Squad: Happenings in the Insect World*, ANR Blogs, University of California Agriculture and Natural Resources, https://ucanr.edu/blogs/blogcore/postdetail.cfm?postnum=54931.

645   The Editors of Encyclopaedia Britannica, "Damselfly, Types, Characteristics & Behavior," *Britannica* (23 January 2024), https://www.britannica.com/animal/damselfly.

646   J. Swaegers, J. Mergeay, L. Therry, et al., "Rapid Range Expansion Increases Genetic Differentiation While Causing Limited Reduction in Genetic Diversity in a Damselfly," *Heredity* 111 (2013): 422–429, https://doi.org/10.1038/hdy.2013.64, https://www.nature.com/articles/hdy201364.

647   German Centre for Integrative Biodiversity Research (iDiv), Halle-Jena-Leipzig, "Dragonfly Species Losses and Gains in Germany," *Phys.org*, (June 2021), https://phys.org/news/2021-06-dragonfly-species-losses-gains-germany.html. Diana E. Bowler et al., "Winners and Losers Over 35 Years of Dragonfly and Damselfly Distributional Change in Germany," *Diversity and Distributions* 27, no. 8 (2021): 1353–1366. J. Ott, "Dragonflies and Climatic Change–Recent Trends in Germany and Europe," *BioRisk* 5 (2010): 253–286.

648   M. Wasscher and K. Goudsmits, "De Gaffelwaterjuffer (Coenagrion Scitulum), Terug van Weggeweest in Noordwest-Europa," *Brachytron* 13, no. 1/2 (2010): 19–25.

649   Chris Goforth, "Dragonfly Territoriality," *The Dragonfly Woman* (8 January 2010), https://thedragonflywoman.com/2010/01/08/dragonfly-territoriality.

650   Ruprecht Fadem, *Silence and Articulacy in the Poetry of Medbh McGuckian* (United States: Lexington Books, 2019). Matt Bernstein Sycamore, ed., *Nobody Passes: Rejecting the Rules of Gender and Conformity* (United States: Basic Books, 2010).

651   Andrew Clarke and Ian A. Johnston, "Evolution and Adaptive Radiation of Antarctic Fishes," *Trends in Ecology & Evolution* 11, no. 5 (1996): 212–218.

652   E.I. Svensson, F. Eroukhmanoff, and M. Friberg, "Effects of Natural and Sexual Selection on Adaptive Population Divergence and Premating Isolation in a Damselfly," *Evolution* 60 (2006): 1242–1253, https://doi.org/10.1111/j.0014-3820.2006.tb01202.x, https://academic.oup.com/evolut/article/60/6/1242/6756280.

653   Philippe Bertrand and Louis Legendre, *Earth, Our Living Planet* (Springer, 2021). Michael Schirber, "'Snowball Earth' Might Have Been Slush," *Astrobiology Magazine* (August 2015), Reprinted on the webpages of Goddard Institute for Space Studies, NASA (National Aeronautics and Space Administration), https://www.giss.nasa.gov/research/features/201508_slushball.

654   Robert Fulghum, *True Love: Stories Told To and By Robert Fulghum* (New York: HarperCollins Publishers, 1997), 98.

655   Dr. Cristina Diaz, "Sponges: Time-Lapse of Sponge Cells Recombining," *The Shape of Life*, PBS LearningMedia, (13 June 2017), https://www.shapeoflife.org/video/sponges-time-lapse-sponge-cells-recombining.

656   Patricia R. Bergquist, *Sponges* (Berkeley, CA: University of California Press, 1978), 74. For the footage of the red versus the yellow sponge cells, see Lewis Thomas and Robin Bates, "Notes of a Biology Watcher," produced and directed by Robin Bates, Nova program #818, TV script, (Boston: WGBH, 1981): 34. Eric Jantsch, *The Self Organizing Universe: Scientific and Human Implications of the Emerging Paradigm of Evolution* (Oxford: Pergamon Press, 1980), 128. Members of two myxobacterial colonies sieved together will also seek out their mates and reconstitute themselves as two separate and opposing fruiting bodies... without the telltale identifier of different cell coloring. (Daniel R. Smith and Martin Dworkin, "Territorial Interactions Between Two Myxococcus Species," *Journal of Bacteriology* 176, no. 4 (1994): 1201–1205.)

657   Gil Sharon et al., "Commensal Bacteria Play a Role in Mating Preference of *Drosophila melanogaster*," *Proceedings of the National Academy of Sciences* 107, no. 46 (2010): 20051–20056, https://www.pnas.org/doi/abs/10.1073/pnas.1009906107, https://www.pnas.org/doi/10.1073/pnas.1009906107. Ed Yong, "Gut Bacteria Change the Sexual Preferences of Fruit Flies," *National Geographic* (1 November 2010), https://www.nationalgeographic.com/science/article/gut-bacteria-change-the-sexual-preferences-of-fruit-flies. UC Museum of Paleontology, "Evidence for Speciation - Understanding Evolution," in *Understanding Evolution*, the University of California Museum of Paleontology and the National Center for Science Education, (2024), https://evolution.berkeley.edu/evolution-101/speciation/evidence-for-speciation. Diane M. B. Dodd, "Reproductive Isolation as a Consequence of Adaptive Divergence in Drosophila pseudoobscura," *Evolution* 43, no. 6 (Sep. 1989): 1308–1311, https://doi.org/10.2307/2409365.

658   We humans have between 200 and a thousand species of bacteria in our gut. Jing Yang et al., "Species-Level Analysis of Human Gut Microbiota with Metataxonomics," *Frontiers in Microbiology* 11 (25 August 2020): 2029, https://doi.org/10.3389/fmicb.2020.02029, https://www.frontiersin.org/journals/microbiology/articles/10.3389/fmicb.2020.02029/full.

659   M.J. Morowitz, E.M. Carlisle, and J.C. Alverdy, "Contributions of Intestinal Bacteria to Nutrition and Metabolism in the Critically Ill," *Surgical Clinics of North America* 91, no. 4 (2011): 771–vii, https://doi.org/10.1016/j.suc.2011.05.001, https://www.sciencedirect.com/science/article/pii/S0039610911000600.

660   Salk Institute for Biological Studies, "Fruit Fly Intestine May Hold Secret to the Fountain of Youth," News Release (31 January 2023), https://www.salk.edu/news-release/fruit-fly-intestine-may-hold-secret-to-the-fountain-of-youth.

661   Chris D. Jiggins, Igor Emelianov, and James Mallet, "Assortative Mating and Speciation as Pleiotropic Effects of Ecological Adaptation: Examples in Moths and Butterflies," *Insect Evolutionary Ecology* (2005): 451–473, https://www.cabidigitallibrary.org/doi/abs/10.1079/9780851998121.0455. Chris D. Jiggins, *The Ecology and Evolution of Heliconius Butterflies* (Oxford University Press, 2017).

662   Elio de Almeida Borghezan, Kalebe da Silva Pinto, Jansen Zuanon, and Tiago Henrique da Silva Pires, "Someone Like Me: Size-Assortative Pairing and Mating in an Amazonian Fish, Sailfin Tetra Crenuchus spilurus," *PLoS One* 14, no. 9 (2019), https://doi.org/10.1371/journal.pone.0222880.

663   F.R. McCully and P.E. Rose, "Individual Personality Predicts Social Network Assemblages in a Colonial Bird," *Scientific Reports* 13, 2258 (1 March 2023), https://doi.org/10.1038/s41598-023-29315-3, https://www.nature.com/articles/s41598-023-29315-3. Laura Baisas, "Flamingoes Have Big Personalities—and Their Friendships Prove It. The Iconic Pink Birds Can Be Very Picky About Who They Hang Out With," *Popular Science*, (1 March 2023), https://www.popsci.com/environment/flamingo-clique.

664   Harry F. Harlow, *Learning To Love* (New York: Jason Aronson, 1974), 85.

665   S.J. Suomi, H.F. Harlow, and J.K. Lewis, "Effect of Bilateral Frontal Lobectomy on Social Preferences of Rhesus Monkeys," Journal of Comparative and Physiological Psychology 70, no. 3, Pt. 1 (1970): 448–453, https://doi.org/10.1037/h0028704.

666   Harlow, *Learning to Love*, 142–143. S.J. Suomi, H.F. Harlow, J.K. Lewis, "Effect of Bilateral Frontal Lobectomy on Social Preferences of Rhesus Monkeys," *Journal of Comparative Physiology* (March 1970): 448–453.

667   F.H. Farley and C.B. Mueller, "Arousal, Personality, and Assortative Mating in Marriage: Generalizability and Cross-Cultural Factors," *Journal of Sex and Marital Therapy* (Spring 1978): 50–53.

668   T.A. Rizzo and W.A. Corsaro, "Social Support Processes in Early Childhood Friendship: A Comparative Study of Ecological Congruences in Enacted Support," *American Journal of Community Psychology* (June 1995): 389–417.

669   L. Warren, K. Pearson, S. Lutz, and J. Lee, "Assortative Mating in Man: A Cooperative Study," *Biometrika* 2, no. 4 (1 January 1903): 481–498, https://www.jstor.org/stable/2331510.

670   F.B. de Waal and L.M. Luttrell, "The Similarity Principle Underlying Social Bonding Among Female Rhesus Monkeys," *Folia Primatologica* 46, no. 4 (1986):

215–234, https://doi.org/10.1159/000156255, https://brill.com/view/journals/ijfp/46/4/article-p215_3.xml.

671   For example, "Humans often mate with those resembling themselves, a phenomenon described as positive assortative mating (PAM)." Tom MM Versluys, Alex Mas-Sandoval, Ewan O. Flintham, and Vincent Savolainen, "Why Do We Pick Similar Mates, or Do We?" *Biology Letters* 17, no. 11 (2021): 20210463, https://royalsocietypublishing.org/doi/full/10.1098/rsbl.2021.0463. H. Richard Johnston, Bronya J.B. Keats, and Stephanie L. Sherman, "Population Genetics," in *Emery and Rimoin's Principles and Practice of Medical Genetics and Genomics* (7th ed.; Academic Press, 2019): 359–373, https://dokumen.pub/emery-and-rimoins-principles-and-practice-of-medical-genetics-and-genomics-foundations-hardcovernbsped-0128125373-9780128125373.html. Encyclopedia of Evolutionary Biology, Academic Press, 2016. E.O. Wilson, *Sociobiology: The New Synthesis* (Cambridge, MA: The Belknap Press of Harvard University, 1975).

672   M. Kaplan, "Earliest Feathered Dinosaur Discovered," *Nature* (18 March 2009), https://doi.org/10.1038/news.2009.172, https://www.nature.com/articles/news.2009.172. Katie Pavid, "How Dinosaurs Evolved into Birds," National History Museum, London, https://www.nhm.ac.uk/discover/how-dinosaurs-evolved-into-birds.html.

673   K. Salomo, J.F. Smith, T.S. Feild, M.S. Samain, L. Bond, C. Davidson, J. Zimmers, C. Neinhuis, S. Wanke, "The Emergence of Earliest Angiosperms May Be Earlier Than Fossil Evidence Indicates," *Systematic Botany* 42, no. 4 (2017): 607–619, https://doi.org/10.1600/036364417X696438.

674   Robert T. Bakker, "Dinosaur Renaissance," *Scientific American* 232, no. 4 (1975), 58–79.

675   Susan Sprecher, "Does (Dis)Similarity Information About a New Acquaintance Lead to Liking or Repulsion? An Experimental Test of a Classic Social Psychology Issue," *Social Psychology Quarterly* (15 July 2019), https://doi.org/10.1177/0190272519855954.

676   W. Scott Persons and Philip J. Currie, "Feather Evolution Exemplifies Sexually Selected Bridges Across the Adaptive Landscape," *Evolution* (30 July 2019), https://doi.org/10.1111/evo.13795.

677   Zhonghe Zhou and Fucheng Zhang, "Origin of Feathers—Perspectives from Fossil Evidence," *Science Progress* 84, no. 2 (2001): 87–104, https://journals.sagepub.com/doi/pdf/10.3184/003685001783239023. Erik Stokstad, "Feathers, or Flight of Fancy?" *Science* 288, no. 5474 (2000): 2124–2125, https://www.science.org/doi/10.1126/science.288.5474.2124.

678   Pascal Godefroit et al., "A Jurassic Ornithischian Dinosaur from Siberia with Both Feathers and Scales," *Science* 345 (2014): 451–455, https://doi.org/10.1126/science.1253351. Michael Balter, "Earliest Dinosaurs May Have Sported Feathers: First Feathers May Have Been for Insulation or Sexual

Display," *Science* (24 July 2014), https://www.science.org/content/article/earliest-dinosaurs-may-have-sported-feathers.

679   L. Xu, M. Wang, R. Chen, et al., "A New Avialan Theropod from an Emerging Jurassic Terrestrial Fauna," *Nature* (2023), https://doi.org/10.1038/s41586-023-06513-7.

680   M. Kaplan, "Fossil Feathers Reveal Dinosaurs' True Colours," *Nature* (2010), https://doi.org/10.1038/news.2010.39, https://www.nature.com/articles/news.2010.39.

681   Yuto Momohara, Akihiro Kanai, and Toshiki Nagayama, "Aminergic Control of Social Status in Crayfish Agonistic Encounters," *PLoS One* 8, no. 9 (18 September 2013), https://journals.plos.org/plosone/article?id=10.1371/journal.pone.0074489.

682   Shih-Rung Yeh, Russell Fricke, and Donald Edwards, "The Effect of Social Experience on Serotonergic Modulation of the Escape Circuit of Crayfish," *Science* (19 January 1996): 366–369. Shih-Rung Yeh, Barbara E. Musolf, and Donald H. Edwards, "Neuronal Adaptations to Changes in the Social Dominance Status of Crayfish," Journal of Neuroscience (January 1997): 697–708.

683   Neil Greenberg, "A Neuroethological Study of Display Behavior in the Lizard Anolis Carolinensis (Reptilia, Lacertilia, Iguanidae)," *American Zoology* 17 (1977): 191–201, https://academic.oup.com/icb/article/17/1/191/172073. Neil Greenberg, Personal Correspondence, 1998–2002.

684   "No Author," "How Many Species Related to Dinosaurs Are Left in the World Today, and Where Do They Live?" UCSB ScienceLine (8 June 2008), University of South Carolina Beaufort, http://scienceline.ucsb.edu/getkey.php?key=1796.

685   W. Wilczynski, M.P. Black, S.J. Salem, and C. Ezeoke, "Behavioural Persistence During an Agonistic Encounter Differentiates Winners from Losers in Green Anole Lizards," *Behaviour* 152, no. 5 (2015): 563–591, https://doi.org/10.1163/1568539X-00003243.

686   Irby J. Lovette and John W. Fitzpatrick, eds., *Handbook of Bird Biology* (John Wiley & Sons, 2016).

687   Sankar Chatterjee and R.J. Templin, *Posture, Locomotion, and Paleoecology of Pterosaurs* 376, Geological Society of America (2004).

688   Y. Uno and T. Hirasawa, "Origin of the Propatagium in Non-Avian Dinosaurs," *Zoological Letters* 9, no. 4 (2023), https://doi.org/10.1186/s40851-023-00204-x. Hans Ce Larsson, T. Alexander Dececchi, and Michael B. Habib, "Navigating Functional Landscapes: A Bird's Eye View of the Evolution of Avialan Flight," *Bulletin of the American Museum of Natural History* 440 (2020): 321–332, https://digitallibrary.amnh.org/server/api/core/bitstreams/6d57a102-c87d-4fa2-ba7e-dd53eebe8d5e/content. Lewis Doty, "The Evolution of Animal

Flight Understanding a Major Transition in Ecology," *Ecology Center* (2 September 2023), https://www.ecologycenter.us/evolutionary-ecology/the-evolution-of-animal-flight-understanding-a-major-transition-in-ecology.html. CBC News, "Gliding Dinosaur Discovery Leads Researchers to Rethink Evolution of Birds, McGill Prof Says," *Canadian Broadcasting Company* (4 November 2020), https://www.cbc.ca/news/canada/montreal/dinosaur-discovery-redpath-museum-1.5788052.

689   R.O. Prum and A.H. Brush, "Dinosaur Feathers Came Before Birds and Flight," *Scientific American* 288, no. 2 (2003): 76–85, https://www.scientificamerican.com/article/dinosaur-feathers-came-before-birds-and-flight. Alan Brush, "Which Came First, the Feather or the Bird: A Long-Cherished View of How and Why Feathers Evolved Has Now Been Overturned," (2003), https://www.academia.edu/24178180/A_long_cherished_view_of_how_and_why_feathers_evolved_has_now_been_overturned.

690   E.O. Wilson, *Sociobiology: The New Synthesis* (Harvard University Press, 1975), 141.

691   J. Scholtens and N. Poll, "Behavioral Consequences of Agonistic Experiences in the Male S3 (Tryon Maze Dull) Rat," *Aggressive Behavior* 13 (1987): 213–226, https://doi.org/10.1002/1098-2337(1987)13:4<213::AID-AB2480130405>3.0.CO;2-S, https://psycnet.apa.org/record/1988-31813-001.

692   E.G. Brun, "Who Is the Top Dog in Ant Communities? Resources, Parasitoids, and Multiple Competitive Hierarchies," *Community Ecology* (30 November 2004) 142: 643–652, https://doi.org/10.1007/s00442-004-1763-4, https://link.springer.com/article/10.1007/s00442-004-1763-4. Louis Lefebvre, "Grooming in Crickets: Timing and Hierarchical Organization," *Animal Behaviour* 29, no. 4 (1981): 973–984, https://www.sciencedirect.com/science/article/pii/S0003347281800504.

693   US Geological Survey, "When Did Dinosaurs Become Extinct?" (17 May 2001), https://www.usgs.gov/faqs/when-did-dinosaurs-become-extinct.

694   Tyler Volk and Dorion Sagan, *Death & Sex* (Chelsea Green, 2009), 53.

695   Sonia Madaan, "Tropical Rainforest Biome: Climate, Precipitation, Location, Seasons, Plants and Animals," *Earth Eclipse* (July 2020), https://eartheclipse.com/environment/ecosystem/tropical-rainforest-biome.html.

696   H. Song, D.B. Kemp, L. Tian, et al., "Thresholds of Temperature Change for Mass Extinctions," *Nature Communications* 12, no. 4694 (4 August 2021), https://doi.org/10.1038/s41467-021-25019-2.

697   Brian T. Huber et al., "The Rise and Fall of the Cretaceous Hot Greenhouse Climate," *Global and Planetary Change* 167 (2018): 1–23, https://doi.org/10.1016/j.gloplacha.2018.04.004.

698   Brian T. Huber, Richard D. Norris, and Kenneth G. MacLeod, "Deep-Sea Paleotemperature Record of Extreme Warmth During the Cretaceous," *Geology* 30, no. 2 (2002): 123–126. Michon Scott and Rebecca Lindsey, "What's the Hottest Earth's Ever Been?" *Climate.gov* (22 November 2023), https://www.climate.gov/news-features/climate-qa/whats-hottest-earths-ever-been.

699   Christina Nunez, "Carbon Dioxide Levels Are at a Record High. Here's What You Need to Know," *National Geographic* (13 May 2019), https://www.nationalgeographic.com/environment/article/greenhouse-gases.

700   Oxford University Museum of Natural History, "The Breath of Life," University of Oxford, (6 November to 15 December 2014), https://oumnh.ox.ac.uk/breath-life.

701   Michon Scott and Rebecca Lindsey, "What's the Hottest Earth's Ever Been?" National Oceanic and Atmospheric Administration, Climate.gov, (22 November 2023), https://www.climate.gov/news-features/climate-qa/whats-hottest-earths-ever-been.

702   "Atmospheric pCO_2 levels reached as high as about 2,000 ppmv, average temperatures were roughly 5°C–10°C higher than today, and sea levels were 50–100 meters higher." In C. Wang, Y. Gao, D.E. Ibarra, H. Wu, and P. Wang, "An Unbroken Record of Climate During the Age of Dinosaurs—A Scientific Drilling Project in China Has Retrieved a Continuous History of Conditions from Earth's Most Recent 'Greenhouse' Period That May Offer Insights About Future Climate Scenarios," *EOS* (17 May 2021), American Geophysical Union, https://eos.org/science-updates/an-unbroken-record-of-climate-during-the-age-of-dinosaurs. Sung Kyung Hong and Yong Il Lee, "Evaluation of Atmospheric Carbon Dioxide Concentrations During the Cretaceous," *Earth and Planetary Science Letters* 327 (2012): 23–28, https://www.sciencedirect.com/science/article/pii/S0012821X12000246.

703   "Paleo [CO2]atm values did not persist above 1,500 ppmV during the past 400 million years." D.O. Breecker, Z.D. Sharp, and L.D. McFadden, "Atmospheric CO2 Concentrations During Ancient Greenhouse Climates Were Similar to Those Predicted for AD 2100," *Proceedings of the National Academy of Sciences* 107, no. 2 (2010): 576–580, https://www.pnas.org/doi/abs/10.1073/pnas.0902323106.

704   E. Nisbet and N. Sleep, "The Habitat and Nature of Early Life," *Nature* 409 (2001): 1083–1091, https://doi.org/10.1038/35059210.

705   K. Moran, J. Backman, J. Brinkhuis, et al., "The Cenozoic Palaeoenvironment of the Arctic Ocean," *Nature* 441 (2006): 601–605, https://doi.org/10.1038/nature04800.

706   Michon Scott and Rebecca Lindsey, "What's the Hottest Earth's Ever Been?" Climate.gov (22 November 2023), https://www.climate.gov/news-features/climate-qa/whats-hottest-earths-ever-been.

707 D. Vandermark, J. Tarduno, and D. Brinkman, "A Fossil Champsosaur Population from the High Arctic: Implications for Late Cretaceous Paleotemperatures," *Palaeogeography, Palaeoclimatology, Palaeoecology* 248 (2007): 49–59, https://doi.org/10.1016/J.PALAEO.2006.11.008.

708 Carolyn Gramling, "Roughly 90 Million Years Ago, a Rainforest Grew Near the South Pole," *Science News* (1 April 2020), https://www.sciencenews.org/article/rainforest-antarctica-south-pole-roughly-90-million-years-ago.

709 "New Dinosaur Discovery: Ugrunaaluk kuukpikensis or, 'Ancient Grazer,'" *Alaska Geographic* (24 June 2016), https://www.akgeo.org/new-dinosaur-discovery-ugrunaaluk-kuukpikensis-or-ancient-grazer.

710 Anthony R. Fiorillo, Ronald S. Tykoski, Philip J. Currie, Paul J. McCarthy, and Peter Flaig, "Description of Two Partial Troodon Braincases from the Prince Creek Formation (Upper Cretaceous), North Slope Alaska," Journal of Vertebrate Paleontology 29, no. 1 (2009): 178–187, https://doi.org/10.1080/02724634.2009.10010370.

B.T. Huber, K.G. MacLeod, and S.L. Wing, eds., *Warm Climates in Earth History* (Cambridge University Press, 2000). Peter Skelton, ed., *The Cretaceous World* (Cambridge University Press, 2003). Alaska Geological & Geophysical Surveys, "Fossils & Dinosaurs," Alaska Department of Natural Resources, (2021), https://dggs.alaska.gov/popular-geology/fossils-dinosaurs.html. Bob Strauss, "10 Facts About Troodon," *ThoughtCo.com* (15 August 2019), https://www.thoughtco.com/things-to-know-troodon-1093803. Michon Scott and Rebecca Lindsey, "What's the Hottest Earth's Ever Been?" Climate.gov, https://www.climate.gov/news-features/climate-qa/whats-hottest-earths-ever-been. "Global Climate in Cretaceous," *Climate Policy Watcher*, https://www.climate-policy-watcher.org/global-climate-2/cretaceous-era.html.

711 Stack Exchange, "For What Percentage of the Earth's History Has There Been Permanent Ice?" Earth Science, StackExchange, (2019), https://earthscience.stackexchange.com/questions/5376/for-what-percentage-of-the-earths-history-has-there-been-permanent-ice.

W. Buggish, M.M. Joachimski, G. Sevastopulo, and J.R. Morrow, "Mississippian δ13Ccarb and Conodont Apatite δ18O Records—Their Relation to the Late Palaeozoic Glaciation," *Palaeogeography, Palaeoclimatology, Palaeoecology* 268 (2008): 273–292. R.A. Cooper and P.M. Sadler, "The Ordovician Period," in Gradstein et al. *The Geologic Time Scale* (2012), 489–523. R.M. DeConto and D. Pollard, "Rapid Cenozoic Glaciation of Antarctica Induced by Declining Atmospheric CO2," *Nature* 421 (2003): 245–249, https://www.nature.com/articles/nature01290. G. Dromart, J.P. Garcia, S. Picard, F. Atrops, C. Lécuyer, S.M.F. Sheppard, "Ice Age at the Middle–Late Jurassic Transition?" *Earth and Planetary Science Letters* 213 (2003): 205–220. C.R. Fielding, T.D. Frank, L.P. Birgenheier, M.C. Rygel, A.T. Jones, J. Roberts, "Stratigraphic Imprint of the Late Palaeozoic Ice Age in Eastern Australia: A Record of Alternating Glacial and Nonglacial Climate Regimes," *Journal of the Geological Society* 165 (2008): 129–140. C. Korte, P.J. Jones, U. Brand, D. Mertmann, J. Veizer, "Oxygen Isotope Values from High-Latitudes: Clues for Permian Sea-Surface Temperature Gradients and Late Palaeozoic Deglaciation," *Palaeogeography, Palaeoclimatology, Palaeoecology* 269 (2008): 1–16. G.A. Shields-Zhou, A.C. Hill, B.A. Macgabhann, "The Cryogenian Period," in Gradstein et al. *The Geologic Time Scale* (2012), 393–411. M.J. Van Kranendonk, "A Chronostratigraphic Division of the Precambrian," in Gradstein et al. *The Geologic Time Scale* (2012), 299–392.

712   Utah Geological Survey, "Ice Ages – What Are They and What Causes Them?" (22 December 2011), https://geology.utah.gov/ice-ages-what-are-they-and-what-causes-them.

713   E.J. Rohling et al., "Comparison Between Holocene and Marine Isotope Stage-11 Sea-Level Histories," *Earth and Planetary Science Letters* (2010), https://doi.org/10.1016/j.epsl.2009.12.054. Anthony Watts, "Tracking the Earth's Orbit: Looking for Warming Signs," *Watts Up With That?* (7 February 2010), https://wattsupwiththat.com/2010/02/07/tracking-the-earths-orbit-looking-for-for-warming-signs.

714   M. Tigchelaar, A.S. von der Heydt, and H.A. Dijkstra, "A New Mechanism for the Two-Step δ18O Signal at the Eocene-Oligocene Boundary," *European Geosciences Union, Climate of the Past* 7, no. 1 (9 March 2011): 235–247, https://doi.org/10.5194/cp-7-235-2011, https://cp.copernicus.org/articles/7/235/2011/.

715   "Chicxulub Impact Event, Regional Effects," Lunar and Planetary Institute, (2024), https://www.lpi.usra.edu/science/kring/Chicxulub/regional-effects.

716   Rachel E. Brown, Christian Koeberl, Alessandro Montanari, and David M. Bice, "Evidence for a Change in Milankovitch Forcing Caused by Extraterrestrial Events at Massignano, Italy, Eocene-Oligocene Boundary GSSP," in *The Late Eocene Earth: Hothouse, Icehouse, and Impacts*, eds. C. Koeberl and A. Montanari, Geological Society of America, Geological Society Special Publication 452 (2009): 119–137. NASA Science Editorial Team, "Milankovitch (Orbital) Cycles and Their Role in Earth's Climate," *NASA News* (27 February 2020), https://climate.nasa.gov/news/2948/milankovitch-orbital-cycles-and-their-role-in-earths-climate.

717   Tyler Kukla, Kimberly V. Lau, Daniel Enrique Ibarra, and Jeremy KC Rugenstein, "Deterministic Icehouse and Greenhouse Climates Throughout Earth History," *EarthArXiv* (October 2022), https://eartharxiv.org/repository/view/3638. James S. Crampton, Roger A. Cooper, Peter M. Sadler, and Michael Foote, "Greenhouse–Icehouse Transition in the Late Ordovician Marks a Step Change in Extinction Regime in the Marine Plankton," *Proceedings of the National Academy of Science* 113, no. 6 (25 January 2016): 1498–1503, https://

www.pnas.org/doi/abs/10.1073/pnas.1519092113. Appy Sluijs, Jörg Pross, and Henk Brinkhuis, "From Greenhouse to Icehouse; Organic-Walled Dinoflagellate Cysts as Paleoenvironmental Indicators in the Paleogene," *Earth-Science Reviews* 68, no. 3–4 (January 2005): 281–315, https://doi.org/10.1016/j.earscirev.2004.06.001.

718   Bob Strauss, "Facts of the Pre-Historic Predator Hyaenodon," *ThoughtCo* (25 August 2020), http://thoughtco.com/hyaenodon-hyena-tooth-1093221.

719   Riley Black, "The Jaws That Bite, the Claws That Catch," *National Geographic* (14 June 2015), https://www.nationalgeographic.com/science/article/the-jaws-that-bite-the-claws-that-catch.

720   Kenneth D. Rose, *The Beginning of the Age of Mammals* (Johns Hopkins University Press, 2006). See illustration of Hyaenadon teeth on page 126.

721   T.I. Pollock, D.P. Hocking, and A.R. Evans, "The Killer's Toolkit: Remarkable Adaptations in the Canine Teeth of Mammalian Carnivores," *Zoological Journal of the Linnean Society* 196, no. 3 (November 2022): 1138–1155, https://doi.org/10.1093/zoolinnean/zlab064. M. Borths and N. Stevens, "Deciduous Dentition and Dental Eruption of Hyainailouroidea (Hyaenodonta, 'Creodonta,' Placentalia, Mammalia)," *Palaeontologia Electronica* 20 (2017): 1–34, https://doi.org/10.26879/776.

722   "Hyaenodon," Prehistoric Earth: A Natural History Wiki, https://prehistoric-earth-a-natural-history.fandom.com/wiki/Hyaenodon. Verdugo, "Bite Force: Hyaenodon gigas & Megistotherium, Carnivora," https://carnivora.net/bite-force-hyaenodon-gigas-megistotherium-t5526.html.

723   Rachel E. Brown, Christian Koeberl, Alessandro Montanari, and David M. Bice, "Evidence for a Change in Milankovitch Forcing Caused by Extraterrestrial Events at Massignano, Italy, Eocene-Oligocene Boundary GSSP," in *The Late Eocene Earth: Hothouse, Icehouse, and Impacts*, eds. C. Koeberl and A. Montanari, Geological Society of America, *Geological Society Special Publication* 452 (2009): 119–137, Special Paper 452, https://vtechworks.lib.vt.edu/items/db5b4873-f708-4899-9db0-eae5d2843219.

724   P. Sexton, R. Norris, P. Wilson, et al., "Eocene Global Warming Events Driven by Ventilation of Oceanic Dissolved Organic Carbon," *Nature* 471 (2011): 349–352, https://doi.org/10.1038/nature09826.

725   National Oceanic and Atmospheric Administration, "Daily CO2," CO2 Earth, (23 November 2023), https://www.co2.earth/daily-co2.

726   Yi Ge Zhang, Mark Pagani, Zhonghui Liu, Steven M. Bohaty, and Robert DeConto, "A 40-Million-Year History of Atmospheric CO2," *Philosophical Transactions of the Royal Society A, Mathematical, Physical and Engineering Sciences* 371, no. 1993 (28 October 2013), https://doi.org/10.1098/rsta.2013.0096.

727   Bethan Davies, "Antarctic Ice Sheet 37–34 Mya, Pre-Quaternary Antarctic Peninsula, Ice Sheet Evolution, Palaeogene (65.5 to 23.03 Ma)," Antarcticglaciers. org, https://www.antarcticglaciers.org/glacial-geology/antarctic-ice-sheet/ icesheet_evolution.

728   Werner U. Ehrmann and Andreas Mackensen, "Sedimentological Evidence for the Formation of an East Antarctic Ice Sheet in Eocene/Oligocene Time," *Palaeogeography, Palaeoclimatology, Palaeoecology* 93, no. 1–2 (1992): 85–112, https://doi.org/10.1016/0031-0182(92)90185-8. Charlotte L. O'Brien, Matthew Huber, Ellen Thomas, Mark Pagani, James R. Super, Leanne E. Elder, and Pincelli M. Hull, "The Enigma of Oligocene Climate and Global Surface Temperature Evolution," *Proceedings of the National Academy of Sciences* 117, no. 41 (28 September 2020): 25302–25309, https://www.pnas.org/doi/ abs/10.1073/pnas.2003914117.

729   David K. Hutchinson et al., "The Eocene–Oligocene Transition: A Review of Marine and Terrestrial Proxy Data, Models and Model–Data Comparisons," *European Geosciences Union, Climate of the Past* 17 (28 January 2021): 269–298, https://doi.org/10.5194/cp-17-269-2021. J.-B. Ladant, Y. Donnadieu, V. Lefebvre, and C. Dumas, "The Respective Role of Atmospheric Carbon Dioxide and Orbital Parameters on Ice Sheet Evolution at the Eocene-Oligocene Transition," *Paleoceanography* 29 (2014): 810–823, https://doi. org/10.1002/2013PA002593. Michon Scott, "Understanding Climate: Antarctic Sea Ice Extent," Climate.gov (14 March 2023), https://www.climate.gov/news-features/understanding-climate/understanding-climate-antarctic-sea-ice-extent. Nicholas B. Sullivan et al., "Millennial-Scale Variability of the Antarctic Ice Sheet During the Early Miocene," *Proceedings of the National Academy of Sciences* 120, no. 39 (2023), https://www.pnas.org/doi/abs/10.1073/pnas.2304152120.

730   J.-B. Ladant, Y. Donnadieu, V. Lefebvre, and C. Dumas, "The Respective Role of Atmospheric Carbon Dioxide and Orbital Parameters on Ice Sheet Evolution at the Eocene-Oligocene Transition," *Paleoceanography* 29 (2014): 810–823, https://doi.org/10.1002/2013PA002593.

731   Michon Scott, "Understanding Climate: Antarctic Sea Ice Extent," Climate. gov (14 March 2023), https://www.climate.gov/news-features/understanding-climate/understanding-climate-antarctic-sea-ice-extent.

732   Nicholas B. Sullivan et al., "Millennial-Scale Variability of the Antarctic Ice Sheet During the Early Miocene," *Proceedings of the National Academy of Sciences* 120, no. 39 (2023), https://www.pnas.org/doi/abs/10.1073/ pnas.2304152120. Omid Falahatkhah et al., "Recognition of Milankovitch Cycles During the Oligocene–Early Miocene in the Zagros Basin, SW Iran: Implications for Paleoclimate and Sequence Stratigraphy," *Sedimentary Geology* 421 (15 July 2021), https://www.sciencedirect.com/science/article/pii/S0037073821001093.

733   Simone Galeotti et al., "Antarctic Ice Sheet Variability Across the Eocene-Oligocene Boundary Climate Transition," *Science* 352 (2016): 76–80, https://doi.org/10.1126/science.aab0669. Werner U. Ehrmann and Andreas Mackensen, "Sedimentological Evidence for the Formation of an East Antarctic Ice Sheet in Eocene/Oligocene Time," *Palaeogeography, Palaeoclimatology, Palaeoecology* 93, no. 1–2 (1992): 85–112, https://doi.org/10.1016/0031-0182(92)90185-8, https://www.sciencedirect.com/science/article/pii/0031018292901858. Ehrmann et al., "The East Antarctic Continent Was More or Less Totally Buried Beneath the Ice During Oligocene Time."

734   Buerki S., F. Forest, T. Stadler, and N. Alvarez, "The Abrupt Climate Change at the Eocene-Oligocene Boundary and the Emergence of South-East Asia Triggered the Spread of Sapindaceous Lineages," *Annals of Botany* 112, no. 1 (July 2013): 151–160, https://doi.org/10.1093/aob/mct106.

735   Colin Schultz, "Ancient Climate Change Meant Antarctica Was Once Covered with Palm Trees," *Smithsonian Magazine* (6 April 2023), https://www.smithsonianmag.com/smart-news/ancient-climate-change-meant-antarctica-was-once-covered-with-palm-trees-12098835.

736   R. Chazdon, "Light Variation and Carbon Gain in Rain Forest Understorey Palms," *Journal of Ecology* 74 (1986): 995–1012, https://doi.org/10.2307/2260229. P. Barry Tomlinson, "The Uniqueness of Palms," *Botanical Journal of the Linnean Society* 151 (May 2006): 5–14, https://doi.org/10.1111/J.1095-8339.2006.00520.X.

737   J. Eldrett, D. Greenwood, I. Harding, et al., "Increased Seasonality Through the Eocene to Oligocene Transition in Northern High Latitudes," *Nature* 459 (2009): 969–973, https://doi.org/10.1038/nature08069, https://www.nature.com/articles/nature08069. Natural Sciences and Engineering Research Council, "Ancient Climate Change: When Palm Trees Gave Way to Spruce Trees," *ScienceDaily*, https://www.sciencedaily.com/releases/2009/06/090617131356.htm.

738   M.J. Butrim and D.L. Royer, "Leaf-Economic Strategies Across the Eocene–Oligocene Transition Correlate with Dry Season Precipitation and Paleoelevation," *American Journal of Botany* 107, no. 12 (2020): 1772–1785.

739   R. Méndez-Alonzo, H. Paz, R. Zuluaga, J. Rosell, and M. Olson, "Coordinated Evolution of Leaf and Stem Economics in Tropical Dry Forest Trees," *Ecology* 93, no. 11 (November 2012): 2397–2406, https://doi.org/10.1890/11-1213.1. R. Karban, "Deciduous Leaf Drop Reduces Insect Herbivory," *Oecologia* 153 (2007): 81–88, https://doi.org/10.1007/s00442-007-0709-z, https://link.springer.com/article/10.1007/s00442-007-0709-z. M. Tyree, H. Cochard, P. Cruiziat, B. Sinclair, and T. Améglio, "Drought-Induced Leaf Shedding in Walnut: Evidence for Vulnerability Segmentation," *Plant*

Cell and Environment 16, no. 9 (September 1993): 879–882, https://doi.org/10.1111/J.1365-3040.1993.TB00511.X.

740   Laurent Augusto et al., "Influences of Evergreen Gymnosperm and Deciduous Angiosperm Tree Species on the Functioning of Temperate and Boreal Forests," *Biological Reviews* 90, no. 2 (11 June 2014): 444–466, https://onlinelibrary.wiley.com/doi/full/10.1111/brv.12119.

741   Eivind O. Straume, Aleksi Nummelin, Carmen Gaina, and Kerim H. Nisancioglu, "Climate Transition at the Eocene–Oligocene Influenced by Bathymetric Changes to the Atlantic–Arctic Oceanic Gateways," *Proceedings of the National Academy of Sciences* 119, no. 17 (21 April 2022), https://www.pnas.org/doi/abs/10.1073/pnas.2115346119.

742   "Oligocene Epoch: Life," University of California Museum of Paleontology, https://ucmp.berkeley.edu/tertiary/oli/olilife.html.

743   "Oligocene," in *Wikipedia, The Free Encyclopedia* (2021, October 12), https://en.wikipedia.org/wiki/Oligocene.

744   C.M. Janis and P.B. Wilhelm, "Were There Mammalian Pursuit Predators in the Tertiary? Dances with Wolf Avatars," *Journal of Mammalian Evolution* 1 (1993): 103–125, https://doi.org/10.1007/BF01041590, https://link.springer.com/article/10.1007/BF01041590.

745   W. Jacquelyne Kious and R.I. Tilling, "The Himalayas: Two Continents Collide," *This Dynamic Earth: The Story of Plate Tectonics*, US Geological Survey, (1996), https://pubs.usgs.gov/gip/dynamic/himalaya.html.

746   "Carbon Cycle and the Earth's Climate," Columbia University, http://www.columbia.edu/~vjd1/carbon.htm.

747   J. Ries, "Acid Ocean Cover Up," *Nature Climate Change* 1 (21 August 2011): 294–295, https://doi.org/10.1038/nclimate1204.

748   P.A.E. Pogge von Strandmann, A. Desrochers, M.J. Murphy, A.J. Finlay, D. Selby, and T.M. Lenton, "Global Climate Stabilisation by Chemical Weathering During the Hirnantian Glaciation," *Geochemical Perspectives Letters* 3 (15 June 2017): 230–237, https://doi.org/10.7185/geochemlet.1726. Ohio State University, "Appalachian Mountains, Carbon Dioxide Caused Long-Ago Global Cooling," *ScienceDaily*, (26 October 2006), www.sciencedaily.com/releases/2006/10/061025185539.htm.

749   Eustoquio Molina, Concepción Gonzalvo, Silvia Ortiz, and Luis E. Cruz, "Foraminiferal Turnover Across the Eocene–Oligocene Transition at Fuente Caldera, Southern Spain: No Cause–Effect Relationship Between Meteorite Impacts and Extinctions," *Marine Micropaleontology* 58, no. 4 (2006): 270–286, https://doi.org/10.1016/j.marmicro.2005.11.006, https://www.sciencedirect.com/science/article/pii/S0377839805001404.

750 Bobby Azarian, *The Romance of Reality: How the Universe Organizes Itself to Create Life, Consciousness, and Cosmic Complexity* (Dallas: BenBella Books, 2022).

751 Stephon Alexander, William J. Cunningham, Jaron Lanier, Lee Smolin, Stefan Stanojevic, Michael W. Toomey, Dave Wecker, "The Autodidactic Universe," arXiv:2104.03902 (2 September 2021), https://doi.org/10.48550/arXiv.2104.03902.

752 Ibid.

753 P. Clawson and Michael Levin, "Endless Forms Most Beautiful 2.0: Teleonomy and the Bioengineering of Chimaeric and Synthetic Organisms," *Biological Journal of the Linnean Society* (2022), https://doi.org/10.1093/biolinnean/blac073.

754 "Telos," *Oxford English Dictionary*, https://www.oed.com/search/dictionary/?scope=Entries&q=telos.

755 Ludwig Buchner, *Force and Matter or Principles of the Natural Order of the Universe: With a System of Morality Based Thereon* (Whitefish, Montana: Kessinger Publishing, LLC, 2006), originally published 1855. David Darling, "Buchner, Ludwig (1824-1899)," https://www.daviddarling.info/encyclopedia/B/Buchner.html.

756 Clawson and Levin, "Endless Forms Most Beautiful 2.0."

757 Matthew 25:29.

758 Herbert Spencer, *The Principles of Biology In Two Volumes, Volume I* (New York and London: D. Appleton and Company, 1866). Darwin wrote three years later, "The expression often used by Mr. Herbert Spencer of the Survival of the Fittest is more accurate [than 'Struggle for Existence'], and is sometimes equally convenient." Charles Darwin, *On the Origin of Species* (fifth edition, 1869), ch. 3, Susan Ratcliffe, ed., *Oxford Essential Quotations* (4th ed.; Oxford University Press).

759 UCNτ Collaboration; Gonzalez, F. M., Fries, E. M., Cude-Woods, C., Bailey, T., Blatnik, M., Broussard, L. J., Callahan, N. B., Choi, J. H., Clayton, S. M., Currie, S. A., "Improved Neutron Lifetime Measurement with UCNτ," *Physical Review Letters* 127, no. 16 (13 October 2021): 162501, arXiv:2106.10375, https://doi.org/10.1103/PhysRevLett.127.162501. Particle Data Group, "Neutron Mean Life (Report)," *Review of Particle Physics*, Berkeley, CA, Lawrence Berkeley Laboratory, 2020. K. Heyde, *Beta-decay: Basic Ideas and Concepts in Nuclear Physics: An Introductory Approach*, ch. 5 (Taylor & Francis, 2004), https://doi.org/10.1201/9781420054941. Wilson, J.T., et al., "Measurement of the Free Neutron Lifetime Using the Neutron Spectrometer on NASA's Lunar Prospector Mission," *Physical Review C* 104, no. 4 (2021): 045501. Brian Koberlein, "Understanding the Early Universe Depends on Estimating the Lifespan of Neutrons," *Universe Today* (12 November 2021), https://www.universetoday.

com/153289/understanding-the-early-universe-depends-on-estimating-the-lifespan-of-neutrons.

760   B. Ripperda, M. Liska, K. Chatterjee, G. Musoke, A.A. Philippov, S.B. Markoff, A. Tchekhovskoy, Z. Younsi, "Black Hole Flares: Ejection of Accreted Magnetic Flux Through 3D Plasmoid-mediated Reconnection," *The Astrophysical Journal Letters* 924, no. 2 (2022): L32, https://doi.org/10.3847/2041-8213/ac46a1. Simons Foundation, "Origin of Supermassive Black Hole Flares Identified: Largest-ever Simulations Suggest Flickering Powered by Magnetic 'Reconnection,'" *Science Daily*, (3 February 2022), https://www.sciencedaily.com/releases/2022/02/220203161225.htm. Monisha Ravisetti and Eric Mack, "Space Flash Is Revealed as Black Hole Spewing the Light of 1,000 Trillion Suns," *Cnet* (2 December 2022), https://www.cnet.com/science/space/mysterious-space-flash-revealed-as-black-hole-spewing-the-light-of-1000-trillion-suns/.

761   Dinosaur Jungle, "Aetosaurs Facts—Information About the Extinct, Prehistoric Animal, Aetosaurs," *Dinosaur Jungle*, https://www.dinosaurjungle.com/prehistoric_animals_aetosaurs.php.

762   John R. Horner et al., "A New Pachycephalosaurid from the Hell Creek Formation, Garfield County, Montana, U.S.A.," *Journal of Vertebrate Paleontology* (2023), https://doi.org/10.1080/02724634.2023.2190369. Robert Sanders, "Newly Described Species of Dome-Headed Dinosaur May Have Sported Bristly Headgear," *PhysOrg* (23 May 2023), https://phys.org/news/2023-05-newly-species-dome-headed-dinosaur-sported.html.

763   Bernard J. Le Boeuf, *Elephant Seals: Pushing the Limits on Land and at Sea* (Published online by Cambridge University Press, 23 September 2021), https://www.cambridge.org/core/books/elephant-seals/origins-misnomers-and-bottleneck/C12D57A04AA7F30FF6CC476521281043.

764   B. Le Boeuf, "Sexual Behavior in the Northern Elephant Seal Mirounga angustirostris," *Behaviour* 41, no. 1 (1972): 1–26, https://doi.org/10.1163/156853972X00167.

765   For a map of male and female elephant seal feeding zones see: Point Reyes National Seashore, "The Northern Elephant Seal: A Life of Singular Extremes," US Department of the Interior, National Park Service, http://npshistory.com/brochures/pore/elephant-seal.pdf.

766   Point Reyes National Seashore, "The Northern Elephant Seal: A Life of Singular Extremes," Resource Newsletter, National Park Service, https://www.nps.gov/pore/learn/upload/resourcenewsletter_elephantseals.pdf.

767   Kathleen McAuliffe, "Elephant Seals, the Champion Divers of the Deep," *Smithsonian Magazine* (September 1995), https://www.smithsonianmag.com/science-nature/elephant-seals-the-champion-divers-of-the-deep-1-35587919. American Museum of Natural History, "Fast Facts: Elephant Seals," Mar 13,

2015, https://www.amnh.org/explore/news-blogs/on-exhibit-posts/fast-facts-elephant-seals. National Park Service, "Elephant Seals," Point Reyes National Seashore, last modified 2022, https://www.nps.gov/pore/learn/nature/elephant_seals.htm.

768   Lee Dugatkin, Google Scholar, https://scholar.google.com/citations?user=gbGT5rIAAAAJ.

769   Lee Alan Dugatkin, *Power in the Wild* (University of Chicago Press, 2022), 6.

770   Burney J. Le Boeuf, et al., "The Northern Elephant Seal (Mirounga angustirostris) Rookery at Año Nuevo: A Case Study in Colonization," *Aquatic Mammals* 37, no. 4 (2011): 486. Friends of the Elephant Seal, "Birthing and Breeding," https://elephantseal.org.

771   C. Deutsch, M. Haley, and B. LeBoeuf, "Reproductive Effort of Male Northern Elephant Seals: Estimates from Mass Loss," *Canadian Journal of Zoology* 68 (December 1990): 2580–2593, https://doi.org/10.1139/Z90-360, https://cdnsciencepub.com/doi/10.1139/z90-360.

772   Dugatkin, *Power in the Wild*, 5.

773   Genny Anderson, Rebecca Martin, "Elephant Seals: Reproduction," Marine Science, MarineBio.net, 2 July 2004, Clark College, Vancouver, http://marinebio.net/marinescience/05nekton/esrepro.htm.

774   T.R Spraker, T.A. Kuzmina, and R.L. DeLong, "Causes of Mortality in Northern Elephant Seal Pups on San Miguel Island, California," *Journal of Veterinary Diagnostic Investigation* 32, no. 2 (March 2020): 312–316, https://doi.org/10.1177/1040638720907100.

775   Burney J. Le Boeuf and Richard S. Peterson, "Social Status and Mating Activity in Elephant Seals," *Science* 163, no. 3862 (3 January 1969): 91–93, https://www.science.org/doi/abs/10.1126/science.163.3862.91. Genny Anderson, Rebecca Martin, "Elephant Seals: Reproduction," Marine Science, MarineBio.net, 2 July 2004, Clark College, Vancouver, http://marinebio.net/marinescience/05nekton/esrepro.htm.

776   B. J. Le Boeuf, et al., "Sex Differences in Diving and Foraging Behaviour of Northern Elephant Seals," *Symposia of the Zoological Society of London* 66 (1993), https://citeseerx.ist.psu.edu/document?repid=rep1&type=pdf&doi=97a0b9a6c785c38b3e6de791300fd7ae1172e622.

777   Burney J. Le Boeuf and Kathy J. Panken, "Elephant Seals Breeding on the Mainland in California," No. 59, California Academy of Sciences, 1977.

778   Dugatkin, *Power in the Wild*, 5–7.

779   Michael L. McKinney, "How Do Rare Species Avoid Extinction? A Paleontological View," in W. E. Kunin and K. J. Gaston, eds., *The Biology of Rarity* (1997): 110–129, http://doi.org/10.1007/978-94-011-5874-9. Wikipedia

contributors, "Extinction," in *Wikipedia, The Free Encyclopedia* (19 October 2021), https://en.wikipedia.org/wiki/Extinction.

780   Robert M. May, "Biological Diversity: Differences Between Land and Sea," *Philosophical Transactions of the Royal Society of London. Series B: Biological Sciences* 343, no. 1303 (1994): 105–111. Camila Rada, "For What Percentage of the Earth's History Has There Been Permanent Ice?," *Earth Science Stack Exchange* (13 March 2019), https://earthscience.stackexchange. com/questions/5376/for-what-percentage-of-the-earths-history-has-there-been-permanent-ice.

781   W. Buggish, M.M. Joachimski, G. Sevastopulo, and J.R. Morrow, "Mississippian δ13Ccarb and Conodont Apatite δ18O Records — Their Relation to the Late Palaeozoic Glaciation," *Palaeogeography, Palaeoclimatology, Palaeoecology* 268 (2008): 273–292. R. A. Cooper and P. M. Sadler, "The Ordovician Period," in Gradstein et al., *The Geologic Time Scale* (2012): 489–523. R.M. DeConto and D. Pollard, "Rapid Cenozoic Glaciation of Antarctica Induced by Declining Atmospheric CO2," *Nature* 421 (2003): 245–249. G. Dromart, J.P. Garcia, S. Picard, F. Atrops, C. Lécuyer, and S. M.F. Sheppard, "Ice Age at the Middle–Late Jurassic Transition?," *Earth and Planetary Science Letters* 213 (2003): 205–220. C.R. Fielding, T.D. Frank, L.P. Birgenheier, M.C. Rygel, A.T. Jones, and J. Roberts, "Stratigraphic Imprint of the Late Palaeozoic Ice Age in Eastern Australia: A Record of Alternating Glacial and Nonglacial Climate Regimes," *Journal of the Geological Society* 165 (2008): 129–140. C. Korte, P. J. Jones, U. Brand, D. Mertmann, and J. Veizer, "Oxygen Isotope Values from High-Latitudes: Clues for Permian Sea-Surface Temperature Gradients and Late Palaeozoic Deglaciation," *Palaeogeography, Palaeoclimatology, Palaeoecology* 269 (2008): 1–16. G.A. Shields-Zhou, A.C. Hill, and B.A. Macgabhann, "The Cryogenian Period," in Gradstein et al., *The Geologic Time Scale* (2012): 393–411. M.J. Van Kranendonk, "A Chronostratigraphic Division of the Precambrian," in Gradstein et al., *The Geologic Time Scale* (2012): 299–392. National Park Service, "Milkweed and Monarchs," last modified 20 March 2023, https://www.nps.gov/articles/000/milkweed-and-monarchs.htm.

782   Niklas Wahlberg, Christopher W. Wheat, and Carlos Peña, "Timing and Patterns in the Taxonomic Diversification of Lepidoptera (Butterflies and Moths)," *PLOS One* 8, no. 11 (2013), https://journals.plos.org/plosone/article?id=10.1371/journal.pone.0080875. A. Kawahara, et al., "Phylogenomics Reveals the Evolutionary Timing and Pattern of Butterflies and Moths," *Proceedings of the National Academy of Sciences of the United States of America* 116 (2019): 22657–22663, https://doi.org/10.1073/pnas.1907847116.

783   Acadia National Parks, "Milkweed and Monarchs," National Park Service, 20 March 2023, https://www.nps.gov/articles/000/milkweed-and-monarchs. htm.

Brian D. Farrell, "Evolutionary Assembly of the Milkweed Fauna: Cytochrome Oxidase I and the Age of Tetraopes Beetles," *Molecular Phylogenetics and Evolution* 18, no. 3 (2001): 467–478, https://doi.org/10.1006/mpev.2000.0888.

784   James Wallace (ed.), *Biochemical Interaction Between Plants and Insects*, Vol. 10 (Springer Science & Business Media, 2013).

785   M. Fishbein, S.C.K. Straub, J. Boutte, K. Hansen, R.C. Cronn, and A. Liston, "Evolution at the Tips: Asclepias Phylogenomics and New Perspectives on Leaf Surfaces," *American Journal of Botany* 105, no. 3 (2018): 514–524, https://bsapubs.onlinelibrary.wiley.com/doi/full/10.1002/ajb2.1062.

786   Steven M. Reppert, Haisun Zhu, and Richard H. White, "Polarized Light Helps Monarch Butterflies Navigate," *Current Biology* 14, no. 2 (2004): 155–158. Evandro G. Oliveira, Robert B. Srygley, and Robert Dudley, "Do Neotropical Migrant Butterflies Navigate Using a Solar Compass?" *Journal of Experimental Biology* 201, no. 24 (1998): 3317–3331.

787   Patrick A. Guerra and Steven M. Reppert, "Sensory Basis of Lepidopteran Migration: Focus on the Monarch Butterfly," *Current Opinion in Neurobiology* 34 (2015): 20–28. R. Muheim, "The Light-Dependent Magnetic Compass," in *Photobiology*, ed. L.O. Björn (Springer, New York, NY), https://doi.org/10.1007/978-0-387-72655-7_17.

788   S. Zhan, W. Zhang, K. Niitepõld, et al., "The Genetics of Monarch Butterfly Migration and Warning Colouration," *Nature* 514 (October 2014): 317–321, https://doi.org/10.1038/nature13812.

789   J. Pleasants and K. Oberhauser, "Milkweed Loss in Agricultural Fields Because of Herbicide Use: Effect on the Monarch Butterfly Population," *Insect Conservation and Diversity* 12 (March 2012): 6, https://doi.org/10.1111/j.1752-4598.2012.00196.x.

790   A. Agrawal, *Monarchs and Milkweed: A Migrating Butterfly, a Poisonous Plant, and Their Remarkable Story of Coevolution* (Princeton University Press, 2017).

791   Steven M. Reppert and Jacobus C. de Roode, "Demystifying Monarch Butterfly Migration," *Current Biology* 28, no. 17 (2018): R1009–R1022, https://doi.org/10.1016/j.cub.2018.02.067, https://www.sciencedirect.com/science/article/pii/S0960982218302537.

792   Ibid.

793   Stephen Jenkins, *Monarch Butterflies, Milkweed, and Migration: The Law of Unintended Consequences* (Reno: University of Nevada, 2023).

794   S. Zhan, W. Zhang, K. Niitepõld, et al., "The Genetics of Monarch Butterfly Migration and Warning Colouration," *Nature* 514 (October 2014): 317–321, https://doi.org/10.1038/nature13812.

795 James L. Gould and Carol Grant Gould, *Nature's Compass: The Mystery of Animal Navigation* (Princeton University Press, 2012).

796 Douglas J. Blackiston, Elena Silva Casey, and Martha R. Weiss, "Retention of Memory Through Metamorphosis: Can a Moth Remember What It Learned as a Caterpillar?" *PLoS One* 3, no. 3 (2008), https://journals.plos.org/plosone/article?id=10.1371/journal.pone.0001736.

797 "Insects on Plants, Chemical Ecology, and Coevolution," website of the Phytophagy Lab at Cornell University, led by Anurag Agrawal, James A. Perkins, 26 April 2017, https://agrawal.eeb.cornell.edu/2017/04/26/a-primer-on-coevolution-monarch-milkweeds.

798 Steven M. Reppert, Robert J. Gegear, and Christine Merlin, "Navigational Mechanisms of Migrating Monarch Butterflies," *Trends in Neurosciences* 33, no. 9 (2010): 399–406.

799 Steven M. Reppert, Patrick A. Guerra, Christine Merlin, "Neurobiology of Monarch Butterfly Migration," Annual Review of Entomology 61, no. 1 (2016): 25–42.

800 Steven M. Reppert and Jacobus C. de Roode, "Demystifying Monarch Butterfly Migration," *Current Biology* 28, no. 17 (September 2018): R1009–R1022, https://doi.org/10.1016/j.cub.2018.02.067, https://www.sciencedirect.com/science/article/pii/S0960982218302537.

801 Ron Sender, Shai Fuchs, and Ron Milo, "Revised Estimates for the Number of Human and Bacteria Cells in the Body," *PLoS Biology* 14, no. 8 (2016), https://journals.plos.org/plosbiology/article?id=10.1371/journal.pbio.1002533&mod=article_inline. Alison Abbott, "Scientists Bust Myth That Our Bodies Have More Bacteria Than Human Cells," *Nature*, 8 January 2016, https://www.nature.com/articles/nature.2016.19136.pdf. Mun-Keat Looi, "The Human Microbiome: Everything You Need to Know About the 39 Trillion Microbes That Call Our Bodies Home," *BBC Science Focus*, 14 July 2020, https://www.sciencefocus.com/the-human-body/human-microbiome.

802 Doris Zumpe and Richard P. Michael, *Notes on the Elements of Behavioral Science* (Springer Science & Business Media, 2001), 127.

803 Manuel de Landa, *A Thousand Years of Nonlinear History* (New York: Zone Books, 1997).

804 Genomes are spooled around master organizers, histones. See: Michael Grunstein, "Histones as Regulators of Genes," *Scientific American* 267, no. 4 (1992): 68–75, http://www.jstor.org/stable/24939255.

805 Wim Hordijk, "Exploring the Origins of Life with Autocatalytic Sets," *Research Outreach*, 25 February 2020, https://researchoutreach.org/articles/exploring-origins-life-autocatalytic-sets/. J. Craig Venter, Hamilton O. Smith, and Mark D. Adams, "The Sequence of the Human Genome," *Clinical Chemistry*

61, no. 9 (2015): 1207–1208, https://doi.org/10.1373/clinchem.2014.237016. T. Lencz and A. Darvasi, "Single Nucleotide Polymorphisms (SNPs)," *Reference Module in Life Sciences*, 2017, https://doi.org/10.1016/B978-0-12-809633-8.07157-0. Robert A. Lue, "DNA Structure & Chemistry," *Life Sciences 1A*, Harvard University, 24 September 2018, https://projects.iq.harvard.edu/files/lifesciences1abookv1/files/8_-_dna_replication_revised_9-24-2018.pdf.

806   Sarah Zielinski, "The Evolution of the Orchid and the Orchid Bee: Which Came First—the Plant or Its Pollinator?," *Smithsonian Magazine*, 23 September 2011, https://www.smithsonianmag.com/science-nature/the-evolution-of-the-orchid-and-the-orchid-bee-87336709.

807   Hayley Nolan, *Anne Boleyn: 500 Years of Lies* (Little A, 2019), 95.

808   Terry Breverton, *Owen Tudor: Founding Father of the Tudor Dynasty* (Gloucestershire, UK: Amberley Publishing, 2017). Editors of Wikipedia, "Tudors of Penmynydd," https://en.wikipedia.org/wiki/Tudors_of_Penmynydd.

809   David E. Thornton, "Rhodri Mawr (b. before 844, d. 878)," *Oxford Dictionary of National Biography*, 23 September 2004, https://doi.org/10.1093/ref:odnb/23456. Kari Maund, *Welsh Kings Warriors, Warlords and Princes* (History Press, 2011).

810   Charles Darwin, *The Descent of Man and Selection in Relation to Sex, Vol. 2* (London: John Murray, 1871), 46.

811   Ibid., 97.

812   Ibid.

813   Shagun Popli, "Chandni Chowk Simplified—10 Bazaars To Understand The Oldest Market In Delhi," *Tripoto(* 27 September 2018), https://www.tripoto.com/new-delhi/trips/chandni-chowk-simplified-10-bazaars-to-understand-the-oldest-market-in-delhi-5badf5775989d.

814   Shah Jahan may have spent $839 million on the Taj Mahal. That's close to a billion dollars. Harini Balasubramanian, "Cost of Taj Mahal: Shah Jahan May Have Spent Nearly Rs 70 Billion to Build the Taj Mahal," Housing.com, 12 January 2024, https://housing.com/news/shah-jahan-may-have-spent-nearly-rs-70-billion-to-build-the-taj-mahal.

815   Raghbendra Jha, "Islamic Invasion and Occupation of India," in *Facets of India's Economy and Her Society Volume I: Recent Economic and Social History and Political Economy* (2018): 107–123.

816   Ali Anooshahr, "Mughal Historians and the Memory of the Islamic Conquest of India," *The Indian Economic & Social History Review* 43, no. 3 (2006): 275-300.

817   "Nadir Shah made a mountain of the skulls of Hindus he killed in Delhi alone. Babur raised towers of Hindu skulls at Khanau when he defeated Rana Sanga in 1527 and later he repeated th same horrors after capturing the fort of

Chanderi. Akbar ordered a general massacre of 30,000 Rajputs after he captured Chitor in 1568. The Bahamani Sultans had an annual agenda of killing of 100,000 Hindus every year." "SikhNet, 'Islamic India: The Biggest Holocaust in World History Whitewashed from History Books,'" SikhNet, 27 January 2015, https://www.sikhnet.com/news/islamic-india-biggest-holocaust-world-history. M. Axworthy, *Sword of Persia: Nader Shah, from Tribal Warrior to Conquering Tyrant* (Bloomsbury Publishing, 2010).

818   SikhNet. "Islamic India."

819   Michael Turtle, "Love and Power at the Taj Mahal," *Time Travel Turtle*, 14 February 2024, https://www.timetravelturtle.com/visiting-taj-mahal-india.

820   Lucy L. Brown, Ph.D. Clinical Professor, The Saul R. Korey Department of Neurology, Dominick P. Purpura Department of Neuroscience. Brown's area of research includes: basal ganglia neuroanatomy and function; reward systems; neuroscience of romantic love and attachment, https://www.einsteinmed.edu/faculty/312/lucy-brown.

821   Lucy Scott Brown and John Wright, "The Relationship Between Attachment Strategies and Psychopathology in Adolescence," *Psychology and Psychotherapy: Theory, Research and Practice* 76, no. 4 (2003): 351–367. Arthur Aron, Helen Fisher, Debra J. Mashek, Greg Strong, Haifang Li, and Lucy L. Brown, "Reward, Motivation, and Emotion Systems Associated with Early-Stage Intense Romantic Love," *Journal of Neurophysiology* 94, no. 1 (2005): 327–337. Helen E. Fisher, Lucy L. Brown, Arthur Aron, Greg Strong, and Debra Mashek, "Reward, Addiction, and Emotion Regulation Systems Associated with Rejection in Love," *Journal of Neurophysiology* 104, no. 1 (2010): 51–60.

822   William Roscoe, *The Life of Lorenzo de Medici, Vol. 1* (Basil: J.J. Tourneisen, 1799), 23. Captain James Edward Alexander, *Travels to the Seat of War in the East, Through Russia and the Crimea, in 1829: With Sketches of the Imperial Fleet and Army, Personal Adventures, and Characteristic Anecdotes, Vol. 2* (London: Henry Colburn and Richard Bentley, 1830). Nicholas D. Proksch, Luther's Eschatology and the Turks (Bethany Lutheran Theological Seminary, 2010), *International Congress on Medieval Studies*, Kalamazoo, MI, 1 February 2016, https://web.augsburg.edu/~mcguire/Proksch_Luther_Turks.pdf. Sarah Henrich and James L. Boyce, "Martin Luther—Translations of Two Prefaces on Islam: Preface to the Libellus de Ritu et Moribus Turcorum (1530), and Preface to Bibliander Edition of the Qur'an (1543)," *Word & World* 16, no. 2 (Spring 1996): 250–251. Peter O'Brien, *European Perceptions of Islam and America from Saladin to George W. Bush* (Palgrave Macmillan, 2008), 75–76.

823   Franklin L. Baumer, "England, the Turk, and the Common Corps of Christendom," *The American Historical Review* 50, no. 1 (October 1944): 26–48. John W. Bohnstedt, "The Infidel Scourge of God: The Turkish Menace as Seen by German Pamphleteers of the Reformation Era," *Transactions of the*

American Philosophical Society 58, no. 9 (1968): 1–58. Robert O. Smith, "Luther, the Turks, and Islam," *Currents in Theology and Mission* 34, no. 5 (October 2007): 351, Gale Academic OneFile, link.gale.com/apps/doc/A169989145/AONE?u=nysl_oweb&sid=googleScholar&xid=da0835a2.

824   Nolan, *Anne Boleyn*. Darwin, *The Descent of Man*.

825   Amit Sengupta, Head of Press & Communications—British High Commission in India, "Tropical Cyclone, Hurricane, Storm Formation Explained | Cyclone Biparjay in Arabian Sea, Gujarat," https://www.youtube.com/watch?v=W2UDbDXXYGE.

826   Rhett Herman, "How Fast Is the Earth Moving?" Scientific American, 26 October 1998, https://www.scientificamerican.com/article/how-fast-is-the-earth-mov.

827   NASA, "How Do Hurricanes Form?" NASA Science, Space Place, 28 December 2023, https://www.nasa.gov/audience/forstudents/k-4/stories/nasa-knows/what-are-hurricanes-k4.html.

828   NASA/Goddard Space Flight Center, "NASA Looks at a Hurricane's Temperature in the Eye," EurekAlert!, 30 April 2002, https://www.eurekalert.org/news-releases/739107.

829   National Ocean Service, "How Do Hurricanes Form?" National Oceanic and Atmospheric Administration's National Ocean Service, 11 July 2019, https://oceanservice.noaa.gov/facts/how-hurricanes-form.html.

830   Typhoon Tip measured 1,380 miles across. Meghan Evans & Accuweather, "Earth's Strongest, Most Massive Storm Ever," *Scientific American*, 12 October 2012, https://www.scientificamerican.com/article/earths-strongest-most-massive-storm-ever. Editors of Wikipedia, "Typhoon Tip," Wikipedia, https://en.wikipedia.org/wiki/Typhoon_Tip. National Weather Service, "Hurricane Facts," https://www.weather.gov/source/zhu/ZHU_Training_Page/tropical_stuff/hurricane_anatomy/hurricane_anatomy.html.

831   "During Just One Hurricane, Raging Winds Can Churn Out About Half as Much Energy as the Electrical Generating Capacity of the Entire World, While Cloud and Rain Formation from the Same Storm Might Release a Staggering 400 Times That Amount." So Says the National Oceanic and Atmospheric Administration in "How Hurricanes Form," https://oceanservice.noaa.gov/facts/how-hurricanes-form.html.

832   "Great Red Spot," *Encyclopedia Britannica*, 14 Sep. 2023, https://www.britannica.com/place/Great-Red-Spot.

833   Steven Lawson, "The Reformation and the Men Behind It," *Ligonier Ministries*, https://www.ligonier.org/learn/articles/reformation-and-men-behind-it.

834   Anne Boleyn was born in 1501. "Anne Boleyn," Wikipedia, https://en.wikipedia.org/wiki/Anne_Boleyn.

835   Catherine Nixey, *The Darkening Age: The Christian Destruction of the Classical World* (Houghton Mifflin Harcourt, 2018), 99. "Timeline of the Catholic Church," Wikipedia, https://en.wikipedia.org/wiki/Timeline_of_the_Catholic_Church#313%E2%80%93476. "Early Christians," *The Roman Empire in the First Century*, PBS, https://www.pbs.org/empires/romans/empire/christians.html.

836   David Potter, *Constantine the Emperor* (Oxford University Press, 2015), 155.

837   Christopher Lascelles, *Pontifex Maximus: A Short History of the Popes* (Crux Publishing Ltd, 2017).

838   Luther's visit to Rome was in 1510 or 1511.

839   Hans J. Hillerbrand, "Martin Luther," *Encyclopedia Britannica*, 14 February, 2024, https://www.britannica.com/biography/Martin-Luther.

840   Peter O'Brien, *European Perceptions of Islam and America from Saladin to George W. Bush* (Palgrave Macmillan, 2008), 75-76. Norman Housley, *Crusading and the Ottoman Threat, 1453–1505* (Oxford University Press, 2012), 3. "Luther and the Turks, Luther: Widerrufsverweigerung Worms 1521," https://www.worms.de/en/web/luther/Lutherkritik/Luther_Tuerken.php Luther, Martin. "On War Against the Turk (Vom Kriege wider die Türken)," 1528, https://bible-quran.com/martin-luther-on-war-against-the-turk-vom-kriege-wider-die-turken-1528-2.

841   Janelle Zara, "How Michelangelo Spent His Final Years Designing St. Peter's Basilica in Rome," *Architectural Digest*, June 25, 2019, https://www.architecturaldigest.com/story/how-michelangelo-spent-final-years-designing-st-peters-basilica-rome.

842   Sandra Feder, "Stanford Professor Sees Hagia Sophia as a 'Time Tunnel' Linking Ottomans to the Roman Empire," *Stanford News*, 7 August 2020, https://news.stanford.edu/2020/08/07/hagia-sophias-continuing-legacy/.

843   Blake Ehrlich, "Istanbul," *Encyclopaedia Britannica*, February 27, 2024, https://www.britannica.com/place/Istanbul.

844   "Templo Mayor," *Wikipedia*, https://en.wikipedia.org/wiki/Templo_Mayor.

845   N. Karnam, "8 Beautiful Temples In China You Cannot Afford To Miss," Travel.Earth (8 September 2020), https://travel.earth/beautiful-temples-in-china.

846   "St. Peter's Basilica," *Encyclopaedia Britannica*, February 18, 2024, https://www.britannica.com/topic/Saint-Peters-Basilica.

847   Gerald Posner, *God's Bankers: A History of Money and Power at the Vatican* (Simon and Schuster, 2015), 9.

848   Dante Alighieri, *Dante's Purgatory* (Indiana University Press, 1981), 74. Originally published in the early 1300s.

849   Dan Graves, MSL, "Infamous Indulgence Led to Reformation," *Christianity Plus*, 12 December 2022, https://www.christianity.com/church/church-history/timeline/1501-1600/infamous-indulgence-led-to-reformation-11629920.html.

850   "the Pope has power…to apply the benefits of an Indulgence to the souls in Purgatory. Moreover, to say the Pope cannot absolve the least venial sin is erroneous." Thus wrote Von Valentin Gröne in his 1867 article: "John Tetzel," *Dublin Review* (July–October 1867): 38–42.

851   Thomas James Dandelet and John A. Marino, *Spain in Italy: Politics, Society, and Religion 1500-1700, American Academy in Rome* (Brill, 2007), 183.

852   Ibid.

853   J Merle D'Aubigne, *The Life and Times of Martin Luther* (Moody Publishers, 1978). Editors of Wikipedia, "Johann Tetzel," https://en.wikipedia.org/wiki/Johann_Tetzel.

854   Martin Luther, The 95 Theses, www.luther.de, https://www.luther.de/en/95thesen.html, see thesis 51.

855   Nolan, *Anne Boleyn: 500 Years of Lies*, 30. See also "the pope could even forgive one who had had carnal intercourse with the Holy Virgin." In John Fletcher Hurst, *History of the Christian Church, Volume 2* (Eaton & Mains, 1900), 157.

856   Carol Zaleski, *Otherworld Journeys: Accounts of Near-Death Experience in Medieval and Modern Times* (Oxford University Press, 1989).

857   Martin Luther, "Luther's Correspondence and Other Contemporary Letters: 1521-1530" (Lutheran Publication Society, 1918), 73, https://rpmministries.org/2011/10/reformation-sunday-martin-luthers-story-part-4-clothed-by-christ.

858   Martin Luther, The 95 Theses, https://www.luther.de/en/95thesen.html

859   Nolan, *Anne Boleyn*, 30.

860   "Columbus Reports on His First Voyage, 1493: A Spotlight on a Primary Source by Christopher Columbus," The Gilder Lehrman Institute of American History, https://www.gilderlehrman.org/history-resources/spotlight-primary-source/columbus-reports-his-first-voyage-1493. Daniel J. Boorstin, *The Discoverers: A History of Man's Search To Know His World and Himself* (New York: Vintage Books, 1985).

861   Donald K. McKim, *The Cambridge Companion to Martin Luther* (Cambridge University Press, 2003), 182.

862   Uta-Renate Blumenthal, *The Investiture Controversy: Church and Monarchy from the Ninth to the Twelfth Century* (University of Pennsylvania Press, 2010), 125.

863   Ibid., 5.

864   Vatican Information Service, "The College of Cardinals," Vatican City, 23 January 2001, https://www.ewtn.com/catholicism/library/college-of-cardinals-1582.

865   Thomas Henry Dyer, *The History of Modern Europe* (London: John Murray, 1861), 512–514. See the League of Torgau 1526 and the Schmalkaldic League 1531.

866   Mark Greengrass, *Christendom Destroyed: Europe 1517–1648* (Penguin, 2014). In the opinion of Joseph Ruane, Europe's religious wars continued from the days of Martin Luther in the 1500s to the 20th century. See: Joseph Ruane (2021), "Long Conflict and How It Ends: Protestants and Catholics in Europe and Ireland," *Irish Political Studies* 36, no. 1 (2021): 109–131, https://doi.org/10.108 0/07907184.2021.1877900.

867   The religious wars began in 1522 with the Knights' Revolt in Germany. See "Knights' Revolt," Wikipedia, https://en.wikipedia.org/wiki/Knights%27_Revolt.

868   "European Wars of Religion," https://en.wikipedia.org/wiki/European_wars_of_religion.

869   Matthew Cappucci, "Explainer: The Furious Eye(wall) of a Hurricane or Typhoon," *ScienceNewsExplores*, 12 October 2018, https://www.snexplores.org/article/explainer-what-is-eyewall-of-hurricane-or-typhoon.

870   University of Rhode Island's Graduate School of Oceanography and the National Science Foundation, "Hurricanes: Science and Society, Hurricane Structure," 2012, https://hurricanescience.org/science/science/hurricanestructure/index.html. Meghan Evans & Accuweather, "Earth's Strongest, Most Massive Storm Ever," *Scientific American*, 12 October 2012, https://www.scientificamerican.com/article/earths-strongest-most-massive-storm-ever/.

871   Anne Boleyn was sent to the Court of Margaret in 1513. Sarah Gristwood, *Game of Queens: The Women Who Made Sixteenth-Century Europe* (Basic Books, 2016).

872   Hever Castle & Gardens, "Anne Boleyn Timeline: A Journey Through the Life of Anne Boleyn," https://www.hevercastle.co.uk/visit/hever-castle/timelines/anne-boleyn-timeline.

873   Nolan, *Anne Boleyn*, 12.

874   S. Bryson, *La Reine Blanche: Mary Tudor, A Life in Letters* (United Kingdom: Amberley Publishing 2018).

875   The French court's possible rival, the court of Holy Roman Emperor Charles V in Toledo, Spain, would not begin until 1519.

876   Nolan, *Anne Boleyn*, 30. Reginald Drew, *Anne Boleyn* (Boston: Sherman, French & Company, 1912), 325–326.

877 "Mary Tudor, Queen of France, First Marriage," Wikipedia, https://en.wikipedia.org/wiki/Mary_Tudor,_Queen_of_France#First_marriage:_Queen_of_France.

878 Ebsy18, "January 1st, 1515: The Death of Louis XII and a Lucky Escape for Mary," The Tudorials: History Like Your Teacher Never Taught You, https://thetudorials.com/2016/01/01/january-1st-1515-the-death-of-louis-xii-and-a-lucky-escape-for-mary/.

879 Nolan, *Anne Boleyn*, 30.

880 "Anne Boleyn: The Netherlands and France," Wikipedia, https://en.wikipedia.org/wiki/Anne_Boleyn#The_Netherlands_and_France

881 Nolan, *Anne Boleyn*, 30.

882 "Anne Boleyn: She Failed To Give Henry Viii A Son And Paid With Her Life," Historical Royal Palaces, https://www.hrp.org.uk/tower-of-london/history-and-stories/anne-boleyn/#gs.6a1hth.

883 Brian A. Pavlac, "Anne Boleyn," Women's History, King's College, Wilkes-Barre, Pennsylvania, 29 March 2007, https://departments.kings.edu/womens_history/anneboleyn.html.

884 Rebecca Larson, "Henry Percy: The Man Who Loved Anne Boleyn," Tudors Dynasty, 19 November 2015, https://tudorsdynasty.com/henry-percy-loved-anne-boleyn.

885 Claire Ridgway, "2 March 1522—A Shrovetide Joust and Unrequited Love," The Anne Boleyn Files, 2 March 2019, https://www.theanneboleynfiles.com/2-march-1522-a-shrovetide-joust-and-unrequited-love/

886 Claire Ridgway, "4 March 1522—Anne Boleyn Plays Perseverance," The Anne Boleyn Files, 1 March 2014, https://www.theanneboleynfiles.com/1st-march-1522-anne-boleyn-plays-perseverance.

887 Josephine Wilkinson, *Mary Boleyn: The True Story of Henry VIII's Favorite Mistress* (Chalford Stroud, UK, Amberley Publishing, 2009), 56.

888 For a detailed description of the pageant, see: Alison Weir, *Henry VIII: The King and His Court* (New York: Ballantine Books, 2001), 235–236. For a contemporary account of the pageant by chronicler Edward Hall from which Weir draws her details, see: Claire Ridgway, "4 March 1522—Anne Boleyn and the Chateau Vert Pageant," The Anne Boleyn Files, 4 March 2016, https://www.theanneboleynfiles.com/4-march-1522-anne-boleyn-chateau-vert-pageant.

889 Claire Ridgway, "4th March 1522—Anne Boleyn Plays Perseverance," https://www.theanneboleynfiles.com/1st-march-1522-anne-boleyn-plays-perseverance.

890 L. Smith, "Sexual Allure and the Tudors," *Journal of Family Planning and Reproductive Health Care* 32 (2006): 129–130, https://doi.org/10.1783/147118906776276413. Elizabeth Chadwick, "Standing on the

Shoulders of Giants: How Tall Were the People of Medieval England?," *The History Girls*, 24 March 2019, https://the-history-girls.blogspot.com/2019/03/standing-on-shoulders-of-giants-how.html.

891   Alison Weir, *Henry VIII: The King and His Court* (Random House, 2002). David Starkey, ed., *Rivals in Power: Lives and Letters of the Great Tudor Dynasties* (London: Macmillan, 1990).

892   T. Borman, *Thomas Cromwell: The Untold Story of Henry VIII's Most Faithful Servant* (Grove Press, 2015).

893   Ibid.

894   Claire Ridgway, "A Shrovetide Joust and Unrequited Love," https://www.theanneboleynfiles.com/2-march-1522-a-shrovetide-joust-and-unrequited-love/ Claire Ridgway, "Elle mon Coeur a Navera," *The Anne Boleyn Files*, 2 March 2019, https://www.theanneboleynfiles.com/tag/elle-mon-coeur-a-navera.

895   Only one of Anne's letters to Henry has survived. See Henry VIII and Anne Boleyn, *The Love Letters of Henry VIII to Anne Boleyn with Notes* (Boston and London: John W. Luce & Company, 1906).

896   Michele Morrical, "Love Letters from Henry VIII to Anne Boleyn," *Tudor History by Michele Morrical*, https://michelemorrical.com/love-letters-from-henry-viii-to-anne-boleyn.

897   Henry VIII had been married to Catherine of Aragon since June 11, 1509.

898   "In 1520, Catherine's nephew, the Holy Roman Emperor Charles V, paid a state visit to England, and she [Catherine] urged Henry to enter an alliance with Charles rather than with France." "Catherine of Aragon," Wikipedia, https://en.wikipedia.org/wiki/Catherine_of_Aragon.

899   British History and International Relations expert William Anthony Hay calls "Holy Roman Emperor Charles V, ruler of the world's first transatlantic empire." William Anthony Hay, "'Emperor' Review: A Sovereign on the Move," Wall Street Journal, 21 June 2019, https://www.wsj.com/articles/emperor-review-a-sovereign-on-the-move-11561153765. The AP World Wiki counts the Roman Empire as the first transoceanic empire. The Wiki is wrong. Rome did not cross an ocean to get to its farthest colony, London. It did not cross the Atlantic, the Pacific, or the Indian Ocean. https://apwhwiki.wordpress.com/transoceanic-empires. Though Romans referred to the waters they crossed as ocean, this was not true. Rome crossed a tiny slice of the sea. See: M. Fulford, "The South-West of England in Roman Times - (S.A.) Thomas on the Edge of Empire. Society in the South-West of England During the First Century BC to Fifth Century AD," The Classical Review 73 (2022): 298–300, https://doi.org/10.1017/S0009840X2200244X. Review of Siân Alyce Thomas, *On the Edge of Empire: Society in the South-West of England During the First Century BC to Fifth Century AD* (Oxford: BAR Publishing, 2021).

900   "Henry Duke of Cornwall," Wikipedia, https://en.wikipedia.org/wiki/Henry,_Duke_of_Cornwall.

901   Catherine Hanley, *Matilda: Empress, Queen, Warrior* (Yale University Press, 2019). Matthew Lewis, *Stephen and Matilda's Civil War: Cousins of Anarchy* (Pen and Sword History, 2020). Jim Bradbury, *Stephen and Matilda: The Civil War of 1139–53* (The History Press, 2011). Helen Castor, "Empress Matilda, Daughter of Henry I: A Queen in a King's World," *History Extra*, https://www.historyextra.com/period/medieval/matilda-daughter-of-henry-i-a-queen-in-a-kings-world. James Brigden, "9 Facts About 'The Anarchy': England's Dark Period of Lawlessness and War," *Sky History*, https://www.history.co.uk/articles/the-anarchy-england-s-dark-period-of-lawlessness-and-war.

902   Claire Ridgway, "Henry VIII Falls in Love with Anne Boleyn," *The Anne Boleyn Files*, 23 November 2010, https://www.theanneboleynfiles.com/henry-viii-falls-in-love-with-anne-boleyn.

903   According to William Cavendish, in his Life of Wolsey. Quoted in Elizabeth Norton, *Anne Boleyn: Henry VIII's Obsession* (Amberley Publishing Limited, 2008). "The Relationship between Henry Percy & Anne Boleyn 1523," English History, https://englishhistory.net/tudor/henry-percy-anne-boleyn-relationship.

904   Claire Ridgway, "The Negotiations for Anne Boleyn to Marry James Butler," The Anne Boleyn Files, 2 September 2014, https://www.theanneboleynfiles.com/negotiations-anne-boleyn-marry-james-butler.

905   Claire Ridgway, "Henry VIII Falls in Love with Anne Boleyn," https://www.theanneboleynfiles.com/henry-viii-falls-in-love-with-anne-boleyn.

906   Per George Cavendish. Quoted in https://www.theanneboleynfiles.com/henry-viii-falls-in-love-with-anne-boleyn.

907   George Cavendish, Thomas Wolsey, *Late Cardinal: His Life and Death Written by His Gentleman-usher, George Cavendish* (Folio Society, 1962), originally published 1641. George Cavendish, *The Life of Cardinal Wolsey*, Project Gutenberg, https://www.gutenberg.org/files/54043/54043-h/54043-h.htm.

908   Alison Weir, *Henry VIII: The King and His Court* (Ballantine, 2002), 158.

909   George Cavendish, *The Life of Cardinal Wolsey*, https://www.gutenberg.org/files/54043/54043-h/54043-h.htm.

910   Elizabeth Norton, *The Boleyn Women: The Tudor Femmes Fatales Who Changed English History* (Amberley Publishing Limited, 2013).

911   Claire Ridgway, "Henry VIII Falls in Love with Anne Boleyn," https://www.theanneboleynfiles.com/henry-viii-falls-in-love-with-anne-boleyn.

912   Hever Castle & Gardens, "Anne Boleyn Timeline," https://www.hevercastle.co.uk/visit/hever-castle/timelines/anne-boleyn-timeline.

913   Hever Castle & Gardens, "Owners of Hever Castle," https://www.hevercastle.co.uk/visit/hever-castle/owners/

914   Henry VIII, "Henry VIII to Anne Boleyn: Love Letter #4, after May 1527," in Rebecca Larson, "Love Letters from Henry VIII to Anne Boleyn," *Tudors Dynasty*, 16 March 2016, https://tudorsdynasty.com/love-letter-henry-anne/

915   Claire Ridgway, "Henry VIII Falls in Love with Anne Boleyn," https://www.theanneboleynfiles.com/henry-viii-falls-in-love-with-anne-boleyn.

916   William Baptiste Scoones, ed., *Four Centuries of English Letters: Selections from the Correspondence of One Hundred and Fifty Writers from the Period of the Paston Letters to the Present Day* (Kegan Paul, Trench & Company, 1883), 16.

917   Weir, *Henry VIII*, 257.

918   It is said that Thomas Wolsey had more money at hand than Henry VIII. Thus the king was only one of the top two richest men in England. S. Cargas, "Christianity and Genocide in Rwanda (review)," *Human Rights Quarterly* 32 (2010): 1063–1068, https://doi.org/10.1353/hrq.2010.0014.

919   Henry VIIIth, love letter number four.

920   Henry VIIIth, "Henry VIII to Anne Boleyn: Love Letter #5 (July 1527)," Tudorsdynasty, https://tudorsdynasty.com/love-letter-henry-anne.

921   Henry VIIIth, "Henry VIII to Anne Boleyn: Love Letter #5 (July 1527)," Tudorsdynasty, https://tudorsdynasty.com/love-letter-henry-anne. Walter Littlefield, *Love Letters of Famous Royalties and Commanders* (New York, The John McBride Co, 1909), 125.

922   Twycross, Meg, "Translations of the Bible," Lancaster University, 1998, https://www.lancaster.ac.uk/users/yorkdoom/palweb/week05/douai.htm.

923   Nolan, *Anne Boleyn*, 103.

924   William Tyndale, *The New Testament of our Lord and Saviour Jesus Christ*, 1526/1534, https://www.biblestudytools.com/tyn.

925   Reid Hensarling, *The Biblical Gospel: Its Significance and Impact in Spiritual Renewal* (WestBow Press, 2012), 36.

926   William Tyndale, *The Obedience of a Christian Man and How Christian Rulers Ought to Govern* (Merten de Keyser, 1528).

927   Nolan, *Anne Boleyn*, 112.

928   Ibid., 95.

929   David G Newcombe, *Henry VIII and the English Reformation* (Routledge, 2002).

930   A.L. Rowse, *The Expansion of Elizabethan England* (London: MacMillan, 1955).

931   The others who attempted to head their own churches included King Gustav Vasa of Sweden and King Christian III of Denmark-Norway. King Christian's attempt eventually failed. Henry VIII's did not. See: Paul Lockhart, *Frederik II and the Protestant Cause: Denmark's Role in the Wars of Religion, 1559–1596* (Brill, 2004), 14. See also: "Historical Development," *Eurydice*, An official website of the European Union, 27 November 2023, https://eurydice. eacea.ec.europa.eu/national-education-systems/sweden/historical-development.

932   Nolan, *Anne Boleyn*, 103.

933   Norbert Elias and Eric Dunning, *Sport and Leisure in the Civilizing Process, Volume 10* (Basil Blackwell, 1986).

934   Muzafer Sherif, O. J. Harvey, William R. Hood, Carolyn W. Sherif, Jack White, *The Robbers Cave Experiment: Intergroup Conflict and Cooperation* [Originally published as Intergroup Conflict and Group Relations] (Wesleyan University Press, 2010) [originally published 1954].

935   James Anthony Froude, *The Divorce of Catharine of Aragon: The Story As Told By The Imperial Ambassadors Resident At The Court Of Henry VIII* (New York: Charles Scribner's Sons, 1891).

936   Diarmaid MacCulloch, *The Reformation: A History* (New York: Viking, 2004).

937   Eric Ives, *The Life and Death of Anne Boleyn* (Wiley-Blackwell, 2005).

938   Teysko, Heather, "Tudor Minute June 1, 1533: Anne Boleyn's Coronation," Renaissance English History Podcast, 1 June 2022, https://www.englandcast. com/2022/06/anne-boleyn-crowned-queen. For more on Edward Hall and his contemporary accounts of Anne Boleyn, see: Elizabeth Norton, *Anne Boleyn In Her Own Words & the Words of Those Who Knew Her* (Chalford Stroud, UK, Amberley Publishing, 2011).

939   Susan Doran, *England and Europe 1485-1603, Second Edition* (Taylor and Francis, 1996).

940   "1531: Pope Clement VII Forbids King Henry VIII from Remarrying," History.com, 5 January 2021, https://www.history.com/this-day-in-history/ pope-clement-vii-forbids-king-henry-viii-from-remarrying.

941   In fact, "For most of his life Charles V was by far the most powerful man in the civilised world." According to John Julius Norwich, in Four Princes: Henry VIII, Francis I, Charles V, Suleiman the Magnificent and the Obsessions that Forged Modern Europe, New York: Grove Atlantic, 2017.

942   C. Lipp, *France and the Holy Roman Empire* (Routledge, 2022), https:// doi.org/10.4324/9780367347093-RERW23-1. Robert Knecht, "Francis I, King of France," in *Renaissance and Reformation*, ed. Margaret King, Oxford Bibliographies, last modified 28 March 2018, https://www.oxfordbibliographies. com/view/document/obo-9780195399301/obo-9780195399301-0081.xml.

943   Alison Weir, "Anne Boleyn in France: A Mysterious Episode," Guest Articles, *Tudor Times*, 15 May 2017, https://tudortimes.co.uk/guest-articles/anne-boleyn-in-france.

944   Longueville, Olivia, "Anne Boleyn and King François I of France," OliviaLongueville.com, 14 January 2016, https://olivialongueville.com/2016/01/14/anne-boleyn-and-king-francois-i-of-france.

945   Claire Ridgway, "Anne Boleyn and the French Court 1514-1521," *The Anne Boleyn Files*, 14 August 2014, https://www.theanneboleynfiles.com/anne-boleyn-french-court-1514-1521.

946   W. Wilkie, *The Cardinal Protectors of England: Rome and the Tudors before the Reformation* (Cambridge University Press, 1975). K. Lehnhof, "Incest and Empire in The Faerie Queene," *ELH (English Literary History)* 73 (2006): 215–243, https://doi.org/10.1353/elh.2006.0007.

947   Susan Bordo, *The Creation of Anne Boleyn: A New Look at England's Most Notorious Queen* (Houghton Mifflin Harcourt, 2013), 71–72.

948   Retha M. Warnicke, *The Rise and Fall of Anne Boleyn: Family Politics at the Court of Henry VIII* (Cambridge University Press, 1991), 117.

949   Sarah Clement, "Did Henry VIII Sleep With Anne Boleyn?" *The Historical Novel*, 2017, https://thehistoricalnovel.com/2018/01/03/did-henry-viii-sleep-with-anne-boleyn.

950   M. Lewis, *Stephen and Matilda's Civil War: Cousins of Anarchy* (United Kingdom: Pen & Sword Books, 2020).

951   Heather Shanette, "Elizabethan Church—Background: Mary I," www.elizabethi.org, https://www.elizabethi.org/contents/elizabethanchurch/marian.html.

952   John Foxe, *A Select History of the Lives and Sufferings of the Principal English Protestant Martyrs—Chiefly of Those Executed in the Bloody Reign of Queen Mary* (John Day, London, 1563), 42.

953   From Abraham Lincoln's notes on "A Stand Against Slavery" in the Morgan Library, see: Herbert Mitgang, "Morgan Library Trove Sheds Light on Lincoln," *New York Times*, 12 February 1987, https://www.nytimes.com/1987/02/12/arts/morgan-library-trove-sheds-light-on-lincoln.html.

954   Meilan Solly, "The Myth of 'Bloody Mary'" *Smithsonian Magazine*, 12 March 2020, https://www.smithsonianmag.com/history/myth-bloody-mary-180974221.

955   R. Mason, "Scotland, Elizabethan England and the Idea of Britain," *Transactions of the Royal Historical Society* 14 (2004): 279–293, https://doi.org/10.1017/S0080440104000106. Christopher Ivic, "Literature and Nationalism," in *The Wiley Blackwell Encyclopedia of Race, Ethnicity, and*

Nationalism, eds. A.D. Smith, X. Hou, J. Stone, R. Dennis, and P. Rizova, https://doi.org/10.1002/9781118663202.wberen035.

956   Una McIlvenna, "What Inspired Queen 'Bloody' Mary's Gruesome Nickname?," HISTORY, 8 June 2023, https://www.history.com/news/queen-mary-i-bloody-mary-reformation.

957   Heather Shanette, "Queen Elizabeth I and the Church," Elizabeth1.org, February 2018, https://www.elizabethi.org/contents/elizabethanchurch/queenandchurch.html.

958   Royal Museums Greenwich, "Queen Elizabeth I Facts and Myths," February 2019, https://www.rmg.co.uk/stories/topics/queen-elizabeth-i-facts-myths. Michael W. Simmons, *Elizabeth I: Legendary Queen of England* (independently published, 2016).

959   Ben Johnson, "Queen Elizabeth I of England," Historic UK, 2000, https://www.historic-uk.com/HistoryUK/HistoryofEngland/Queen-Elizabeth-I.

960   Picture caption [pict1952: an X-ray diffraction image of DNA was taken by Raymond Gosling in May 1952, a student supervised by Rosalind Franklin.] "History of Genetics—Early Timeline," Wikipedia, https://en.wikipedia.org/wiki/History_of_genetics#Early_timeline.

961   C. Robinson, A. Sali, and W. Baumeister, "The Molecular Sociology of the Cell," *Nature* 450 (12 December 2007): 973–982, https://doi.org/10.1038/nature06523. Carrie Arnold, "How Supergenes Beat the Odds—and Fuel Evolution," *Wired*, 11 January 2023, https://www.wired.com/story/how-supergenes-beat-the-odds-and-fuel-evolution.

962   Susan Doran, *Queen Elizabeth I* (New York: New York University Press, 2003). Susan Doran, *Elizabeth I and Religion 1558–1603* (United Kingdom: Taylor & Francis, 2002). Catherine Larson, "Her Majesty's Dignity: Secularization in the Age of Reformation," Royal Museums Greenwich, 2000, "Elizabeth I's Religious Settlement," https://www.rmg.co.uk/stories/topics/elizabeth-religious-settlement.

963   D.M. Palliser, *The Age of Elizabeth: England Under the Later Tudors* (Taylor and Francis, 2014), 424.

964   In December 1574 the Common Council of London, under the influences of puritanical factions, issued a statement describing: "great disorder rampant in the city by the inordinate haunting of great multitudes of people, especially youth, to plays, interludes, namely occasion of frays and quarrels, evil practices of incontinency in great inns having chambers and secret places adjoining to their open stages and galleries, inveigling and alluring of maids, especially of orphans and good citizens' children under age, to privy and unmeet contracts, the publishing of unchaste, uncomely, and unshamefast speeches and doings . . . uttering of popular, busy, and seditious matters, and many other corruptions of youth and other enormities . . . [Thus] from henceforth no play, comedy,

tragedy, interlude, not public show shall be openly played or showed within the liberties of the City . . . and that no innkeeper, tavernkeeper, nor other person whatsoever within the liberties of this City shall openly show or play . . . any interlude, comedy, tragedy, matter, or show which shall not be first perused and allowed." William Shakespeare Info (2000), "The Globe Theatre," https://www.william-shakespeare.info/william-shakespeare-globe-theatre.htm Also in John Pendergast, *Shakespeare's World: The Comedies: A Historical Exploration of Literature* (United States: Bloomsbury Publishing, 2019).

965   Leonard Tennenhouse, "Strategies of State and Political Plays: A Midsummer Night's Dream, Henry IV, Henry V, Henry VIII," in *Political Shakespeare: Essays in Cultural Materialism*, eds. Jonathan Dollimore and Alan Sifield (Manchester, UK: Manchester University Press), 115–116. Alison Weir, *The Life of Elizabeth I* (New York: Ballantine Books, 1998), 250.

966   William Shakespeare, ed. Thomas Parry, *Shakespeare's King John* (London: Longmans, Green, and Co., 1884), 5.

967   Irvin Leigh Matus, *Shakespeare In Fact* (Mineola, NY: Dover Publications, 1994), 333.

968   Harold Bloom, *Shakespeare: The Invention of the Human* (Riverhead Books, 1998), xviii–xix.

969   Jeffrey L. Singman, "Elizabethan Life for a Middle Class Townsperson," Chicostume.org, The Internet Home of the Chicagoland Costumers' Guild, http://chicostume.org/handouts/ElizabethanLifeforaMiddleClassTownsperson.pdf.

970   "The first generation of Anglican churchmen were unabashedly nationalistic, even jingoistic, about their religion and their national identity. John Aylmer, the bishop of London at the end of the sixteenth century, stated unequivocally that 'God is English.'" Miles Smith IV, "Anglicanism: A Better Christian Nationalism," *The North American Anglican*, 7 July 2021, https://northamanglican.com/anglicanism-a-better-christian-nationalism.

971   Amy Cook, "Past/Future, Microscope/Telescope, Performance/Science," *Shakespearean Neuroplay: Reinvigorating the Study of Dramatic Texts and Performance through Cognitive Science* (New York: Palgrave Macmillan US, 2010), 123–148.

972   Andre S. Chanderbali, et al., "Evolving Ideas on the Origin and Evolution of Flowers: New Perspectives in the Genomic Era," *Genetics* 202.4 (2016): 1255–1265. University of Bristol, "The Impact of Flowering Plants on the Evolution of Life on Earth," Phys.org, 17 November 2021, https://phys.org/news/2021-11-impact-evolution-life-earth.html.

973   Letter from Charles Darwin to Joseph Dalton Hooker, 22 July 1879, page 3. The letter appears in William E. Friedman, "The Meaning of Darwin's 'Abominable Mystery,'" *American Journal of Botany* 96, no. 1 (January 2009): 5–21, https://bsapubs.onlinelibrary.wiley.com/doi/full/10.3732/ajb.0800150.

Helen Briggs, "New Light Shed on Charles Darwin's 'Abominable Mystery,'" BBC, 23 January 2021, https://www.bbc.com/news/science-environment-55769269.

974   M.J. Benton, P. Wilf, and H. Sauquet, "The Angiosperm Terrestrial Revolution and the Origins of Modern Biodiversity," *New Phytologist* 233 (2022): 2017–2035, https://doi.org/10.1111/nph.17822.

975   Howard Bloom, "Instant Evolution: The Influence of the City on Human Genes: A Speculative Case," *New Ideas in Psychology* 19, no. 3 (2001): 203–220.

976   Aristotle, *The Logic of Science: A Translation of the Posterior Analytics of Aristotle, with Notes and an Introduction by Edward Poste, M.A. Fellow of Oriel College* (Oxford: Francis Macpherson, 1850).

977   Elizabeth Gibney and Davide Castelvecchi, "CERN's Supercollider Plan: $17-Billion 'Higgs Factory' Would Dwarf LHC," *Nature*, 6 February 2024, https://www.nature.com/articles/d41586-024-00353-9.

978   1896 is the date of Boltzmann's Lectures on Gas Theory.

979   Andrea Holstein, "Faustus' England: Marlowe's Representation of Individualism and Spiritual Authority in Elizabethan England in The Tragical History of Doctor Faustus," *Western Libraries*, 2016. Peggy Thompson, *Radical Individualism in Seventeenth-Century Drama* (Renaissance, Restoration, England) (Indiana University, 1985).

980   C. Ivic, *Literature and Nationalism*, 2015, https://doi.org/10.1002/9781118663202.WBEREN035.

981   Pamela O. Long, "Review of Power, Knowledge, and Expertise in Elizabethan England," *Journal of Interdisciplinary History* 38, no. 2 (Autumn 2007): 270–271, muse.jhu.edu/article/219511. Rudolph P. Almasy, review of Eric H. Ash, *Power, Knowledge, and Expertise in Elizabethan England*, *The Sixteenth Century Journal* 37, no. 2 (2006): 536–537, https://doi.org/10.2307/20477912.

982   Ibid.

983   Stephen Johnston, 'Making Mathematical Practice: Gentlemen, Practitioners and Artisans in Elizabethan England' (Ph.D. dissertation, Cambridge, 1994), 1–49. B. Jardine, "Instruments of Statecraft: Humphrey Cole, Elizabethan Economic Policy and the Rise of Practical Mathematics," *Annals of Science* 75 (17 Oct 2018): 304–329, https://doi.org/10.1080/00033790.2018.1528510.

984   W. Scaife, *From Galaxies to Turbines: Science, Technology and the Parsons Family* (Boca Raton, FL: CRC Press, 1998), https://doi.org/10.1201/9781420046922.

985   Eric H. Ash, *Power, Knowledge, and Expertise in Elizabethan England* (Johns Hopkins University Press, 2005).

986   Yale University's summary of Annabel Patterson's book *Reading Holinshed's Chronicles* (University of Chicago Press, 1994). See https://english. yale.edu/publications/reading-holinsheds-chronicles.

987   T. Savery, The Miner's Friend: Or, an Engine to Raise Water (London: S. Crouch, 1827). David Voss, Leah Poffenberger, Alaina G. Levine, "July 2, 1698: Thomas Savery Patents an Early Steam Engine," *This Month in Physics History, American Physical Society*, APS News, https://www.aps.org/publications/ apsnews/201807/history.cfm. Rosie Lesso, "5. The Steam Engine: One of the Most Important Inventions of the Renaissance," in What Were the Best Inventions of the Renaissance? (Top 5), *The Collector*, 10 February 2022, https:// www.thecollector.com/best-renaissance-inventions.

988   Ingo Muller, *A History of Thermodynamics: The Doctrine of Energy and Entropy* (Berlin: Springer, 2007), 48. W.A. Young, "Thomas Newcomen, Ironmonger: The Contemporary Background," *Transactions of the Newcomen Society* 20.1 (1939): 1–15, http://doi.org/10.1179/tns.1939.001. https://www. tandfonline.com/doi/pdf/10.1179/tns.1939.001. "Thomas Newcomen," in *Famous Scientists: The Art of Genius*, https://www.famousscientists.org/ thomas-newcomen.

989   "Newcomen atmospheric engine, Newcomen engine fact file," National Museums of Scotland, 2000, https://www.nms.ac.uk/explore-our-collections/ stories/science-and-technology/newcomen-engine.

990   Robert Greenhalgh Albion, *The Timber Problem Of The Royal Navy, 1652-1862* (Cambridge, MA: Harvard University Press, 1926).

991   Dr Paul Hunneyball, "Parliament and the Elizabethan Energy Crisis," The History of Parliament, 26 January 2023, https://thehistoryofparliament. wordpress.com/2023/01/26/parliament-and-the-elizabethan-energy-crisis.

992   Richard Rhodes, "Why Nuclear Power Must Be Part of the Energy Solution," *Yale Environment* 360 (2018): 19.

993   Actually, coal use began 3,600 years ago in China. Menghan Qiu et al., "Earliest Systematic Coal Exploitation for Fuel Extended to ~3600 B.P.," *Science Advances* 9 (26 July 2023), http://doi.org/10.1126/sciadv.adh0549. Contributors to Wikipedia, "Coal Mining in the United Kingdom," Wikipedia, https:// en.wikipedia.org/wiki/Coal_mining_in_the_United_Kingdom.

994   P. W. King, "Dud Dudley," *The Oxford Dictionary of National Biography*, (Oxford University Press, 2004), http://doi.org/10.1093/ref:odnb/8146 "Dud Dudley," Wikipedia, https://en.wikipedia.org/wiki/Dud_Dudley

995   Historic England, "8 Things to Know About the Black Country – the Historic England Blog," Heritagecalling.com, 14 July 2020, https:// heritagecalling.com/2020/07/14/8-things-to-know-about-the-black-country.

996   Dud Dudley, *Dud Dudley's Metallum Martis: Or, Iron Made with Pit-coale, Sea-coale &c*, Google Books, originally published 1665, https://www.google.com/books/edition/Dud_Dudley_s_Metallum_Martis/qJkwo0iujccC.

997   James Burke, *Connections* (Boston: Little, Brown and Company, 1978). James Burke, *The Day the Universe Changed* (Boston: Little, Brown and company, 1985).

998   Trevor Gledhill, "Glass Making Raw Materials: 17th to 19th Century," *Glass Technology—European Journal of Glass Science and Technology Part A* 63, no. 6 (2022): 183–188, https://www.ingentaconnect.com/content/sgt/gta/2022/00000063/00000006/art00011.

999   J. Farey, *A Treatise on the Steam Engine: Historical, Practical, and Descriptive Illustrated by Numerous Engravings and Diagrams* (United Kingdom: Longman, Rees, Orme, Brown, and Green, 1827), 272.

1000 "Mineral Industry of Europe," Wikipedia, https://en.wikipedia.org/wiki/Mineral_industry_of_Europe.

1001 "Water Pumping," https://en.wikipedia.org/wiki/Water_pumping.

1002 See Paul Sen, *Einstein's Fridge: How the Difference Between Hot and Cold Explains the Universe* (Scribner, 2021).

1003 See illustration at "Newcomen Engine, Historical Landmark—ASME 1712," The American Society Of Mechanical Engineers, https://www.asme.org/about-asme/engineering-history/landmarks/70-newcomen-engine.

1004 "List of Revolutionary War Battles for 1782, Raids & Skirmishes • American Revolutionary War," RevolutionaryWar.us, https://revolutionarywar.us/year-1782.

1005 "History of Birmingham," https://en.wikipedia.org/wiki/History_of_Birmingham.

1006 Jennifer S. Uglow, *The Lunar Men: Five Friends Whose Curiosity Changed the World* (Farrar, Strauss, and Giroux, 2002).

1007 State University of New York HCC, "Boulton and Watt," *World History 2*, https://courses.lumenlearning.com/suny-hccc-worldhistory2/chapter/boulton-and-watt.

1008 Eric Robinson, "Matthew Boulton and the Art of Parliamentary Lobbying," *The Historical Journal* 7, no. 2 (1964): 209–29, http://www.jstor.org/stable/3020351.

1009 "Matthew Boulton," Wikipedia, https://en.wikipedia.org/wiki/Matthew_Boulton.

1010 Jayanta K. Nanda, *Management Thought* (New Delhi: Sarup & Sons, 2006), 40–41. "Soho Foundry," Wikipedia, https://en.wikipedia.org/wiki/Soho_Foundry.

1011 Ibid.

1012 Mark Cartwright, "Steam Hammer," *World History Encyclopedia*, 14 February 2023, https://www.worldhistory.org/Steam_Hammer.

1013 Mark Cartwright, "The Steam Engine in the British Industrial Revolution," *World History Encyclopedia*, 8 Febuary 2023, https://www.worldhistory.org/article/2166/the-steam-engine-in-the-british-industrial-revolut.

1014 In the world of Seventh century Islam, the clothes, armor, and weapons stripped from just one dead enemy warrior could be sold back in Medina for enough money to buy a small date-palm grove. A. Guillaume, *The Life of Muhammad: A Translation of Ibn Ishaq's Sirat Rasul Allah* (New York: Oxford University Press, 2004), 571. Eve Fisher, "The $3500 Shirt - A History Lesson in Economics," *SleuthSayers*, 6 June 2013, https://www.sleuthsayers.org/2013/06/the-3500-shirt-history-lesson-in.html. Howard Bloom, *The Muhammad Code: How a Desert Prophet Brought You ISIS, Al Qaeda, and Boko Haram* (Feral House, 2016).

1015 Jeffrey L. Singman, "Elizabethan Life for a Middle Class Townsperson," Chicostume.org, The Internet Home of the Chicagoland Costumers' Guild, http://chicostume.org/handouts/ElizabethanLifeforaMiddleClassTownsperson.pdf.

1016 Joan Thirsk and F.J. Fisher, *Industries in the Countryside* (Wiley, 1994). J. Chartres, *The Industrial Revolutions, Volume 1: Pre-Industrial Britain* (Wiley, 1994).

1017 Gregory Clark, *The British Industrial Revolution, 1760-1860, World Economic History* (University of California at Davis, 2005), 7, https://faculty.econ.ucdavis.edu/faculty/gclark/ecn110b/readings/chapter2-2002.pdf.

1018 S.G. Stephens, "Cotton Growing in the West Indies During the 18th and 19th Centuries," Trinidad and Tobago: University of the West Indies, https://journals.sta.uwi.edu/ojs/index.php/ta/article/view/5113.

1019 S. Beckert, *Empire of Cotton: A Global History* (Knopf Doubleday Publishing Group, 2015).

1020 "James Hargreaves," *Encyclopedia Britannica*, 18 Apr. 2023, https://www.britannica.com/biography/James-Hargreaves.

1021 "Water Frame, Textile Technology," *Encyclopedia Britannica*, 19 Sep. 2019, https://www.britannica.com/technology/water-frame.

1022 "Edmund Cartwright," *Encyclopedia Britannica*, 20 Apr. 2023, https://www.britannica.com/biography/Edmund-Cartwright.

1023 The British National Archives, "Why Did the Luddites Protest? Political reform in 19th century Britain," the National Archives, Kew, Richmond, https://www.nationalarchives.gov.uk/education/resources/why-did-the-luddites-protest.

1024 Ibid.

1025 "Steam Engine, Machine," *Encyclopedia Britannica*, 8 June 2023, https://www.britannica.com/technology/steam-engine.

1026 Paul Kennedy, *The Rise and Fall of Great Powers: Economic Change and Military Conflict from 1500 to 2000* (Knopf Doubleday, 2010). Bank of England, "How Has Growth Changed Over Time?," Bank of England, 10 January 2019, https://www.bankofengland.co.uk/explainers/how-has-growth-changed-over-time.

1027 Paul Kennedy, *The Rise and Fall of Great Powers.*

1028 Clark Nardinelli, "Industrial Revolution and the Standard of Living," Encyclopedia, The Library of Economics and Liberty, Liberty Fund Network, https://www.econlib.org/library/Enc/IndustrialRevolutionandtheStandardofLiving.html.

1029 Bank of England, "How Has Growth Changed Over Time," Bank of England, https://www.bankofengland.co.uk/explainers/how-has-growth-changed-over-time.

1030 Ibid.

1031 "Prices and Wages by Decade: 1800-1809," in Edward T. Williams, *Niagara County, New York... a concise record of her progress and people, 1821-1921, published during its centennial year. Volume 1*, 148, in Libraries of the University of Missouri, https://libraryguides.missouri.edu/pricesandwages/1800-1809.

1032 David I. Jeremy, "Damming the Flood: British Government Efforts to Check the Outflow of Technicians and Machinery, 1780-1843," *The Business History Review* 51, no. 1, Spring (1977): 1-34, https://www.jstor.org/stable/3112919.

1033 Ingo Muller, *A History of Thermodynamics: The Doctrine of Energy and Entropy* (Berlin: Springer, 2007).

1034 Hillman, Larry H, "Vaugelas and the 'Cult of Reason,'" *Philological Quarterly* 55, no. 2 (1976): 211.

1035 Napoleon was a mere 5'2" according to Thierry Lentz, "Bullet Point #15- Was Napoleon Small?," Napoleon.org, https://www.napoleon.org/en/history-of-the-two-empires/articles/was-napoleon-small.

1036 D.G. Chandler, a scholar "regarded as the greatest interpreter in modern times of the Napoleonic Era," refers to Napoleon's "majestic intellect." D.G. Chandler, *The Campaigns of Napoleon* (United Kingdom: Scribner, 2009).

1037 L. Carnot, *Reply of L N M Carnot, Citizen of France, One of the Founders of the Republic, and Constitutional Member of the Executive Directory: To the Report Made on the Conspiracy of the 18th Fructidor 5th Year*, edited by Jaques Charles Bailleul (United States: Creative Media Partners, LLC, 2018; originally published 1799).

1038 Jan Voerman, "The Reign of Terror," *Andrews University Seminary Studies (AUSS)* 47, no. 1 (2009): 7, https://digitalcommons.andrews.edu/cgi/viewcontent.cgi?article=3055&context=auss "The Reign of Terror," The Core For Building Knowledge, https://www.coreknowledge.org/wp-content/uploads/2018/04/CKHG_G6_U4_French-Revolution-and-Romanticism_WTNK_C10_ReignOfTerror.pdf. "1794: Robespierre Overthrown in France," History Channel, https://www.history.com/this-day-in-history/robespierre-overthrown-in-france.

1039 Nathan D. Jensen, "Napoleonic Biographies 1789–1815," FrenchEmpire.net, https://www.frenchempire.net/biographies.

1040 "Lazare Carnot," *Wikipedia*, https://en.wikipedia.org/wiki/Lazare_Carnot. "Lazare Carnot: French military engineer," *Encyclopedia Britannica*, https://www.britannica.com/biography/Lazare-Carnot.

1041 Frank J. Swetz, "Mathematical Treasure: Lazare Carnot's Geometry," *Convergence*, June 2014, MAA Publications, Mathematical Association of America, https://maa.org/press/periodicals/convergence/mathematical-treasure-lazare-carnot-s-geometry.

1042 Lazare Carnot, "Essay on Machines in General" (1786), in *Text, Translations and Commentaries, Lazare Carnot's Mechanics—Volume 1*, edited by Raffaele Pisano, Jennifer Coopersmith, and Murray Peake (Springer 2021).

1043 Charles Coulston Gillispie, *Raffaele Pisano, Lazare and Sadi Carnot: A Scientific and Filial Relationship* (Springer, 2014), 58.

1044 Thomas G. Chondros, "Archimedes Life Works and Machines," *Mechanism and Machine Theory* 45, no. 11 (November 2010): 1766-1775, https://doi.org/10.1016/j.mechmachtheory.2010.05.009, https://www.sciencedirect.com/science/article/pii/S0094114X10000959.

1045 Archimedes, *The Works of Archimedes*, edited by T.L. Heath (Cambridge University Press, 1897).

1046 Richard S. Westfall and Samuel Devons, *Never At Rest: A Biography of Isaac Newton* (Cambridge University Press, 1981), 988–991.

1047 Nicholas Léonard, *Sadi Carnot, Reflections On The Motive Power Of Fire And On Machines Fitted To Develop That Power*, originally published in 1824, online version, The American Society of Mechanical Engineers, https://www.asme.org/getmedia/e29ff2e8-c1b0-419b-8a15-7648a258aa65/carnot-brochure-final-2021.pdf.

1048 Peter Mander, "Carnot's Dilemma," CarnotCycle, the classical blog on thermodynamics, 9 August 2012, https://carnotcycle.wordpress.com/2012/08/09/carnots-dilemma.

1049 N.L.S. Carnot, *On the Motive Power of Heat*, edited by Robert Henry Thurston (London: Chapman & Hall, Limited, 1897).

1050 Sadi Carnot, edited by Robert Fox, *Reflexions on the Motive Power of Fire: A Critical Edition with the Surviving Scientific Manuscripts* (Manchester University Press, 1986).

1051 Rudolf Clausius, "The Mechanical Theory Of Heat, With Its Applications To The Steam-Engine And To The Physical Properties Of Bodies" (London: John Van Voorst, 1867).

1052 Sadi Carnot, Hippolyte Carnot, Baron Kelvin, and William Thomson, *Reflections on the Motive Power of Heat*, edited by Robert Henry Thurston (United Kingdom: John Wiley, 1897).

1053 Alan Chodos, "This Month in Physics History: June 12, 1824: Sadi Carnot publishes treatise on heat engines," *American Physical Society News* 18, no. 6 (June 2009).

1054 Stephen G. Brush, Nancy S. Hall (eds.), *The Kinetic Theory Of Gases: An Anthology Of Classic Papers With Historical Commentary* (World Scientific Publishing Company, 2003), 561. This book reports that Sadi Carnot's brother claimed Sadi had died of cholera to cover up the fact that he'd died in an insane asylum. Carnot did not die of cholera. But the cholera story persists. See: Logan Chipkin, "From Death and Disgrace to Revolution and Reverence," Substack, 18 February 2020, https://chipkin-logan.medium.com/from-death-and-disgrace-to-revolution-and-reverence-428851864c5d, originally published in History Magazine. And see Paul Sen, *Einstein's Fridge: How the Difference Between Hot and Cold Explains the Universe* (Scribner, Kindle Edition), 21.

1055 Crosbie W. Smith, "William Thomson and the Creation of Thermodynamics: 1840–1855," *Archive for History of Exact Sciences* 16, no. 3 (1977): 231–88, http://www.jstor.org/stable/41133471.

1056 "William Thomson (Lord Kelvin)," MacTutor, School of Mathematics and Statistics, University of St Andrews, Scotland, https://mathshistory.st-andrews.ac.uk/Biographies/Thomson.

1057 Thomson died 33 miles outside of Glasgow at his main residence at Largs. MacTutor, https://mathshistory.st-andrews.ac.uk/Biographies/Thomson/

1058 H. Addington Bruce, "New Ideas in Child Training," *American Illustrated Magazine* 72 (1911): 294, https://www.google.com/books/edition/American_Illustrated_Magazine/LGhEAQAAMAAJ?hl=en&gbpv=1&dq=how+old+was+william+thomson+when+his+father+began+to+teach+him+math&pg=PA294&printsec=frontcover.

1059 MagLab, "William Thomson, Lord Kelvin," The National High Magnetic Field Laboratory, MagLab, National Science Foundation and the State of Florida, https://nationalmaglab.org/magnet-academy/history-of-electricity-magnetism/pioneers/william-thomson-lord-kelvin.

1060 Robert H. Silliman, "Fresnel and the Emergence of Physics as a Discipline," *Historical Studies in the Physical Sciences* , 1974, Vol. 4 (1974), pages 137-162, https://www.jstor.org/stable/27757329

1061 Charles Gibson, "Lord Kelvin and His Brother James Kelvin 1824—1907," in Stories of Great Scientists, Heritage History Electronic Library, https://www.heritage-history.com/index.php?c=read&author=gibson&book=scientists&story=kelvin.

1062 Chantae Reden, "How Transatlantic History Shaped The World As We Know It," Royal Caribbean International, 25 April 2022, https://www.royalcaribbean.com/guides/transatlantic-history-crossing-cruise.

1063 John Wilson, "The Voyage Out," *Te Ara: the Encyclopedia of New Zealand*, https://teara.govt.nz/en/the-voyage-out.

1064 Paul Sen, *Einstein's Fridge: How the Difference Between Hot and Cold Explains the Universe* (Scribner), 36.

1065 "Nearly one packet [ship] in six was totally lost in service. This means that out of 6,000 crossings, about 22 ended in such wrecks." Robert Greenhalgh Albion, *Square-riggers on Schedule: The New York Sailing Packets to England, France, and the Cotton Ports* (Princeton: Princeton University Press, 1938), 202. Also see Shannon Selin, "The Wreck of the Packet Ship Albion," Imagining The Bounds of History, Shannon Selin.com, https://shannonselin.com/2017/04/wreck-packet-ship-albion.

1066 Sen, *Einstein's Fridge,* 33.

1067 S.P. Thomson, *The Life of Lord Kelvin* (London, 1976), cited in https://mathshistory.st-andrews.ac.uk/Biographies/Thomson.

1068 Thomson's defense of Fourier was his paper "On the Uniform Motion of Heat and Its Connection with the Mathematical Theory of Electricity," *Cambridge Mathematical Journal* 3 (1842): 25—27. Sen, *Einstein's Fridge*, 33-34.

1069 William Thomson, "Mechanical Integration of the Linear Differential Equations of the Second Order with Variable Coefficients," *Proceedings of the Royal Society of London* 24, 1875-1876, pages 269-271, https://www.jstor.org/stable/113222 Peter Weinberger "Laplace and the Era of Differential Equations," *Philosophical Magazine* 92, no. 32 (2012): 3882-3890, http://doi.org/10.1080/14786435.2012.699690.

1070 Adrian Rice and Eugene Seneta, "De Morgan in the Prehistory of Statistical Hypothesis Testing," *Journal of the Royal Statistical Society Series A: Statistics in Society* 168, no. 3 (2005): 615—627.

1071 S.S. Demidov, "On the History of the Theory of Linear Differential Equations," Archive for History of Exact Sciences (1983): 369—387.

1072 It was late 19th century British prime minister Benjamin Disraeli who first called India "the jewel in the imperial crown," the jewel in the crown. Gilmour,

David, *The Ruling Caste: Imperial Lives in the Victorian Raj* (Farrar, Straus and Giroux, 2007).

1073 Adrian Rice, Raymond Flood, Robin Wilson, eds., *Mathematics in Victorian Britain* (Oxford University Press, 2011), 191.

1074 Crosbie W. Smith, William Thomson and the Creation of Thermodynamics: 1840–1855, University of Notre Dame, 234, https://www3.nd.edu/~powers/ ame.20231/cwsmith1977.pdf.

1075 MacTutor, "William Thomson," https://mathshistory.st-andrews.ac.uk/ Biographies/Thomson.

1076 K. Kenyon, "Science and Celebrity: Humphry Davy's Rising Star," *Distillations Magazine*, 23 December 2008, https://sciencehistory.org/stories/ magazine/science-and-celebrity-humphry-davys-rising-star.John B. West, "Humphry Davy, Nitrous Oxide, the Pneumatic Institution, and the Royal Institution," *American Journal of Physiology-Lung Cellular and Molecular Physiology* 307, no. 9 (2014), https://journals.physiology.org/doi/full/10.1152/ ajplung.00206.2014. "Showing Off: Scientific Lecturing in the 19th Century," Digital Museum, The Dickinsonia History Project, Digital Humanities, Dickinson College, https://itech.dickinson.edu/dh-archive/digitalmuseum/exhibit-artifact/ making-the-invisible-visible/showing-scientific-lecturing-19th-century.html. Lauren Young, "The Real Electric Frankenstein Experiments of the 1800s," *Atlas Obscura*, 31 October 2016, https://www.atlasobscura.com/articles/the-real-electric-frankenstein-experiments-of-the-1800s.

1077 Sen, *Einstein's Fridge*, 34.

1078 Michael V. Volkenstein, "Reflections on the Motive Power of Fire...," in *Entropy and Information—Progress in Mathematical Physics, vol. 57* (Birkhäuser Basel, 2009), https://doi.org/10.1007/978-3-0346-0078-1_1, https://link. springer.com/chapter/10.1007/978-3-0346-0078-1_1.

1079 MacTutor, "Benoit Paul Émile Clapeyron," https://mathshistory.st-andrews.ac.uk/Biographies/Clapeyron. William Nuttle, "Birth of Thermodynamics — Benôit Paul Émile Clapeyron," in *The Geometry of Thermodynamics*, https:// medium.com/eiffels-paris-an-engineers-guide/birth-of-thermodynamics-clapeyron-481307764481.

1080 I've taken the liberty of translating some of the booksellers' French in this quote into English. William Thomson, *Popular Lectures and Addresses, 3 vols.*, 2:458n (London: Macmillan, 1892), also quoted in Libb Thims and Georgi Gladyshev, *Encyclopedia of Human Thermodynamics*, https://www.eoht.info/ page/Thomson%E2%80%99s%20search%20for%20Carnot%E2%80%99s%20 Reflections.

1081 Lord Kelvin (William Thomson), "On an Absolute Thermometric Scale founded on Carnot's Theory of the Motive Power of Heat and calculated from Regnault's Observations," *Philosophical Magazine*, October 1848. In Sir William

Thomson, *Mathematical and Physical Papers, vol. 1* (Cambridge University Press, 1882), 100–106, https://zapatopi.net/kelvin/papers/on_an_absolute_thermometric_scale.html.

1082 "William Thomson, Lord Kelvin (1824-1907) Physicist and Engineer," University of Aberdeen, University Collections, https://exhibitions.abdn.ac.uk/university-collections/exhibits/show/connecting-collections/william-thomson-lord-kelvin.

1083 Brian C. Shipley, "'Had Lord Kelvin a Right?'," in John Perry, *Natural Selection, and the Age of the Earth* (1895): 91–105, https://www.lyellcollection.org/doi/abs/10.1144/GSL.SP.2001.190.01.08. "Lord Kelvin," *Nature* 2855, no. 114 (19 July 1924): 77–78, https://doi.org/10.1038/114077a0. https://www.nature.com/articles/114077a0. This is an unsigned article on the hundredth anniversary of the birth of William Thomson, Baron Kelvin of Largs.

1084 Libb Thims, Georgi Gladyshev, *Encyclopedia of Human Thermodynamics* , https://www.eoht.info/page/Thomson%E2%80%99s%20search%20for%20Carnot%E2%80%99s%20Reflections.

1085 William Thomson, "An Account of Carnot's Theory of the Motive Power of Heat, with Numerical Results Deduced from Regnault's Experiments on Steam," *Transactions of the Royal Society of Edinburgh* 16 (1849): 541. Wayne M. Saslow, "A History of Thermodynamics: The Missing Manual," *Entropy* 22, no. 1 (2020): 77, https://www.ncbi.nlm.nih.gov/pmc/articles/PMC7516509/, https://www.mdpi.com/1099-4300/22/1/77.

1086 Laura J. Snyder, "William Whewell," The Stanford Encyclopedia of Philosophy (Summer 2022 Edition), ed. Edward N. Zalta, https://plato.stanford.edu/archives/sum2022/entries/whewell.

1087 Thomson says that he first communicated his "theory of the dissipation of energy... to the Royal Society of Edinburgh in 1852, in a paper entitled 'On a Universal Tendency in Nature to the Dissipation of Mechanical Energy.'" See William Thomson, "Kinetic Theory of the Dissipation of Energy," *Nature* 9 (1874): 441–444, https://doi.org/10.1038/009441c0.

1088 A.C. Crombie, "Helmholtz," *Scientific American* 198, no. 3 (March 1958): 94–103, https://www.jstor.org/stable/10.2307/24940945.

1089 Philip E. B. Jourdain, "The Nature and Validity of the Principle of Least Action," *The Monist* 23 (1913). William Thomson (Lord Kelvin), "On a Universal Tendency in Nature to the Dissipation of Mechanical Energy," *Proceedings of the Royal Society of Edinburgh* (1857), published online by Cambridge University Press, 16 March 2015, https://www.cambridge.org/core/journals/proceedings-of-the-royal-society-of-edinburgh/article/2-on-a-universal-tendency-in-nature-to-the-dissipation-of-mechanical-energy/862309E0AF0924FA-7C0AA7FA24B74F6F. Hermann von Helmholtz, *Über die Erhaltung der Kraft (On the Conservation of Force)*, 1847. Hermann Helmholtz, "On the Conservation of Force; A Physical Memoir," in *Scientific Memoirs, Selected from the Transactions of Foreign Academies of Science, and from Foreign Journals, Natural Philosophy*, edited by John Tyndall and William Francis (London: Taylor & Francis, 1853), 114–162. Crosbie Smith and M. Norton Wise, *Energy and Empire: A Biographical Study of Lord Kelvin* (Cambridge University Press, 1989), 500. "Heat Death of the Universe," ChemEurope.com, https://www.chemeurope.com/en/encyclopedia/Heat_death_of_the_universe.html#Origins_of_the_idea. ChemEurope.com bills itself as "The Leading International Platform for the Chemical Industry from Laboratory to Process." David Cahan, "Helmholtz and the British Scientific Elite: From Force Conservation to Energy Conservation," *Notes and Records: The Royal Society Journal of the History of Science*, 16 November 2011, https://doi.org/10.1098/rsnr.2011.0044. A.C. Crombie, Helmholtz, *Scientific American* 198, no. 3 (March 1958): 94–103, https://www.jstor.org/stable/10.2307/24940945. S.S. Thipse, *Advanced Thermodynamics* (Alpha Science International Limited, 2013). Michael W. Collins and Richard C. Dougal, *Kelvin, Thermodynamics, and the Natural World*, eds. Carola S. König and Ivan S. Ruddock (UK: WIT Press, 2016), 266.

1090 For a typical example of the use of the sugar cube, see "Thermodynamics," *Science Clarified, Encyclopedia*, http://www.scienceclarified.com/Sp-Th/Thermodynamics.html.

1091 William Thomson, "On a Universal Tendency in Nature to the Dissipation of Mechanical Energy," *Proceedings of the Royal Society of Edinburgh*, 19 April 1852, also, October 1852. Additionally in William Thomson, *Mathematical and Physical Papers, Volume 1* (Cambridge University Press, 2015). Available online in full at: https://zapatopi.net/kelvin/papers/on_a_universal_tendency.html.

1092 By 1862, William Thomson was already using the term "the second great law of thermodynamics." Not to mention preaching his vision of heat death to the masses—without using the term "heat death." See Sir William Thomson (Lord Kelvin), "On the Age of the Sun's Heat," *Macmillan's Magazine* 5 (5 March 1862), 388–393. Available in William Thomson, *Popular Lectures and Addresses, vol. 1, 2nd edition* (Macmillan and Company, 1910), 356-375, https://zapatopi.net/kelvin/papers/on_the_age_of_the_suns_heat.html. For the online version of Thomson's *Popular Lectures and Addresses*, see https://www.google.com/books/edition/Popular_Lectures_and_Addresses/MInJzBogXygC.

1093 Rudolf Clausius, *The Mechanical Theory of Heat, With Its Applications to the Steam-Engine and to the Physical Properties of Bodies* (London: John Van Voorst, 1867), 865.

1094 Ludwig Boltzmann, "On the Relationship between the Second Fundamental Theorem of the Mechanical Theory of Heat and Probability Calculations Regarding the Conditions for Thermal Equilibrium," Sitzungberichte der Kaiserlichen Akademie der Wissenschaften, Mathematisch-Naturwissen Classe, Abt. II LXXVI (1877): 373–435. Reprinted in Wissenschaftliche Abhandlungen, vol.

II, reprint 42, 164–223, Barth, Leipzig, 1909, https://www.mdpi.com/1099-4300/17/4/1971.

1095 Maurik Holtrop, "Boltzmann's Entropy Equation," Lecture, Spring 2003, University of New Hampshire, unh-npg Nuclear and Particle Physics Group, http://nuclear.unh.edu/~maurik/Phys408_Spring2003_Holtrop/Lectures/Lecture14/Lecture14.pdf.

1096 David and Julia Bart, "Sir William Thomson, on the 150th Anniversary of the Atlantic Cable," *The Antique Wireless Association Review* 21 (2008), https://atlantic-cable.com/CablePioneers/Kelvin.

1097 Lord Kelvin (William Thomson), "On the Secular Cooling of the Earth," *Transactions of the Royal Society of Edinburgh* 23 (1864): 167–169. Read 28 April 1862. From William Thomson, *Mathematical and Physical Papers, vol. III*, 1890, 295. See also: Evelyn Lamb, "Lord Kelvin and the Age of the Earth," *Scientific American*, 26 June 2013, https://blogs.scientificamerican.com/roots-of-unity/lord-kelvin-age-of-the-earth.

1098 P.C. England, P. Molnar, and F.M. Richter, "Kelvin, Perry and the Age of the Earth: Had Scientists Better Appreciated One of Kelvin's Contemporary Critics, the Theory of Continental Drift Might Have Been Accepted Decades Earlier," *American Scientist* 95, no. 4 (2007): 342–349, https://www.americanscientist.org/article/kelvin-perry-and-the-age-of-the-earth. Jean Paul Poirier, "About the Age of the Earth," *Comptes Rendus Geoscience* 349, no. 5 (September 2017): 223–225, https://www.sciencedirect.com/science/article/pii/S1631071317300731. Iain Stewart, "Men of Rock, Deep Time, Hot Rocks," BBC Two, https://www.bbc.co.uk/programmes/p00ccq58. Crosbie Smith and M. Norton Wise, *Energy and Empire: A Biographical Study of Lord Kelvin*, 606. Mario Livio, *Brilliant Blunders: From Darwin to Einstein—Colossal Mistakes by Great Scientists That Changed Our Understanding of Life and the Universe* (Simon and Schuster, 2014), 73.

1099 Jean-Paul Poirier, "About the Age of the Earth," *Comptes Rendus Geoscience*, June 2017, https://comptes-rendus.academie-sciences.fr/geoscience/articles/en/10.1016/j.crte.2017.08.002. P.C. England, P. Molnar, and F.M. Richter, "Kelvin, Perry and the Age of the Earth," 342, https://www.americanscientist.org/article/kelvin-perry-and-the-age-of-the-earth. Peter Lynch, "How Joseph Fourier Discovered the Greenhouse Effect," *Irish Times*, 21 March 2019, https://www.irishtimes.com/news/science/how-joseph-fourier-discovered-the-greenhouse-effect-1.3824189. For Thomson's equations on the age of the earth, see: Gallica—The BnF Digital Library, Bibliothèque Nationale de France, http://visualiseur.bnf.fr/CadresPage?O=NUMM-95120&I=313&M=tdm&T=&Y=Image.

1100 Huw Price, "Boltzmann's Time Bomb," *The British Journal for the Philosophy of Science* 53, no. 1 (2002): 83–119.

1101 Sir Arthur Eddington, *The Nature of the Physical World: The Gifford Lectures 1927*, edited by Klaus-Dieter Sedlacek (Books on Demand, 2021), 60. Originally published 1929.

1102 Lewis Carroll, *Through the Looking-Glass, and What Alice Found There* (Macmillan & Co, Christmas 1871). *Through the Looking-Glass*, The Project Gutenberg eBook, https://www.gutenberg.org/files/12/12-h/12-h.htm.

1103 H.W. Koch, *A History of Prussia* (New York: Dorset Press, 1978). Geoffrey Barraclough, *The Origins of Modern Germany* (New York: W.W. Norton, 1984). Jennifer Llewellyn and Steve Thompson, "Militarism as a Cause of World War I," *Alpha History*, 21 September 2020, https://alphahistory.com/worldwar1/militarism.

1104 Sen. *Einstein's Fridge*, 43.

1105 R. Steven Turner, "The Growth of Professorial Research in Prussia, 1818 to 1848—Causes and Context," Historical Studies in the Physical Sciences 3 (1971): 137–82, https://doi.org/10.2307/27757317.

1106 J.J, O'Connor and E.F. Robertson, "Rudolf Julius Emmanuel Clausius," *MacTutor*, https://mathshistory.st-andrews.ac.uk/Biographies/Clausius.

1107 "the unitary school system of Prussia, designed in 1812 by Wilhelm von Humboldt (1767–1835) and Johann Wilhelm Süvern (1775–1829), culminated in the gymnasium...." In "Gymnasium Schooling," Cengage, *Encyclopedia.com*, https://www.encyclopedia.com/children/encyclopedias-almanacs-transcripts-and-maps/gymnasium-schooling.

1108 M. Crosland, "A Science Empire in Napoleonic France," *History of Science* 44 (2006): 29–48, https://doi.org/10.1177/007327530604400102. David A. Bell, "'The Emperor, 1804–1812,'" in *Napoleon: A Very Short Introduction* (New York, 2018), https://doi.org/10.1093/actrade/9780199321667.003.0005. Robert Wilde, "When and How the French Revolution Ended," *ThoughtCo.com*, 1 March 2018, https://www.thoughtco.com/when-did-the-french-revolution-end-1221875.

1109 Jincan Chen, "The Maximum Power Output and Maximum Efficiency of an Irreversible Carnot Heat Engine," *Journal of Physics D: Applied Physics* 27, no. 6 (1994): 1144. Tan Wang et al., "Performance Analysis and Optimization of an Irreversible Carnot Heat Engine Cycle for Space Power Plant," *Energy Reports* 8 (2022): 6593–6601.

1110 Michael Fowler, "Early Attempts to Understand Heat: Is It a Fluid, or What?," *Physics 152: Heat and Thermodynamics*, University of Virginia, https://galileo.phys.virginia.edu/classes/152.mf1i.spring02/What%20is%20Heat.htm. Robert J. Morris, "Lavoisier and the Caloric Theory," *The British Journal for the History of Science* 6, no. 1 (June 1972): 1–38, https://www.cambridge.org/core/journals/british-journal-for-the-history-of-science/article/abs/lavoisier-and-the-caloric-theory/F37091B910E43FF7D61373A2EE2EAFE3.

1111   Antoine Lavoisier, Elements of Chemistry: In *A New Systematic Order, Containing All the Modern Discoveries* (United States: Dover Publications, 1965; originally published 1789).

1112   Sen, *Einstein's Fridge*, 9.

1113   "Law of Conservation of Mass," https://www.chem.fsu.edu/chemlab/chm-1045lmanual/conserve/introduction.html.

1114   Antoine Lavoisier, *Elements of Chemistry, vol. I*, 140–141 (1789). Antoine Lavoisier, *Elements of Chemistry*, 186–187 (1796). Quoted in Roberto de Andrade Martins, "A Priori Components of Science: Lavoisier and the Law of Conservation of Mass in Chemical Reactions," https://www.researchgate.net/profile/Roberto-Martins-2/publication/358043470_A_priori_components_of_science_Lavoisier_and_the_law_of_conservation_of_mass_in_chemical_reactions/links/61eda4958d338833e38d302d/A-priori-components-of-science-Lavoisier-and-the-law-of-conservation-of-mass-in-chemical-reactions.pdf. Mikuláš Teich, "Circulation, Transformation, Conservation of Matter and the Balancing of the Biological World in the Eighteenth Century," *Ambix* 29, no. 1 (1982): 17–28, http://doi.org/10.1179/amb.1982.29.1.17. Richard S. Treptow, "Conservation of Mass: Fact or Fiction?" *Journal of Chemical Education* 63.2 (1 February 1986): 103, American Chemical Society, https://doi.org/10.1021/ed063p103.

1115   Seth Rogoff, "The Revolutions of 1848," Usm Open Source History Text: The World At War: World History 1914-1945, The University of Southern Mississippi.

1116   Rosemary Ashton, *142 Strand: A Radical Address in Victorian London* (London: Random House UK, 2008.

1117   Rudolf Clausius, "On the Moving Force of Heat, and the Laws Regarding the Nature of Heat Itself Which Are Deducible Therefrom," *Philosophical Magazine And Journal Of Science* (London, July 1851), 2.

1118   Rudolf Clausius, "The Mechanical Theory Of Heat, With Its Applications To The Steam-Engine And To The Physical Properties Of Bodies" (London: John Van Voorst, 1867), 17.

1119   MacTutor, "Rudolf Clausius," https://mathshistory.st-andrews.ac.uk/Biographies/Clausius.

1120   Rudolf Clausius, "The Mechanical Theory Of Heat, With Its Applications To The Steam-Engine And To The Physical Properties Of Bodies," edited by T. Archer Hirst (London, John van Voorst, 1867).

1121   Antoine Laurent Lavoisier, *Traité élémentaire de chimie* (Paris: Chez Cuchet, 1789). George B. Kauffman, "The Making of Modern Chemistry," *Nature* 338 (27 April 1989), https://doi.org/10.1038/338699a0. "Conservation of Mass—There Is No New Matter," The LibreTexts Project, University of California, Davis, https://chem.libretexts.org/Bookshelves/Introductory_Chemistry/Introductory_Chemistry/03%3A_Matter_and_Energy/3.07%3A_Conservation_of_Mass_-_There_is_No_New_Matter.

1122 David Sedley, "Lucretius," *The Stanford Encyclopedia of Philosophy Archive*, Edward N. Zalta (editor), https://plato.stanford.edu/archives/win2018/entries/lucretius.

1123 "Marcus Tullius Cicero, To His Brother Quintus, In The Country, Rome (February)," *The Letters Of Cicero (c. 45 CE)*, trans. E. S. Shuckburgh, Friedrich Von Steuben Metropolitan Science Center. See also: Marcus Tullius Cicero, *Letters to his brother Quintus 2.9* (CXXXI), Wikisource, https://en.wikisource.org/wiki/Letters_to_his_brother_Quintus/2.9.

1124 H.H. Kubbingo, "The First 'Molecular' Theory (1620): Isaac Beeckman (1588–1637)," *Journal of Molecular Structure: Theochem* 181, no. 3–4 (December 1988): 205–218, https://doi.org/10.1016/0166-1280(88)80487-1. https://www.sciencedirect.com/science/article/pii/0166128088804871.

1125 Ibid.

1126 Line Cottegnies, "Michel de Marolles's 1650 French Translation of Lucretius and Its Reception in England," in *Lucretius and the Early Modern*, ed. David Norbrook, Stephen Harrison, and Philip Hardie, Classical Presences (Oxford: Oxford University Press, 2015), https://doi.org/10.1093/acprof:oso/9780198713845.003.0008.

1127 Saul Fisher, "Pierre Gassendi," *The Stanford Encyclopedia of Philosophy Archive*, Edward N. Zalta (editor), https://plato.stanford.edu/archives/spr2014/entries/gassendi.

1128 The math Newton used in his Principia was geometry. Two years later it was Leibnitz who first translated Newton's geometry into equations. Gottfried Wilhelm Leibniz, Marginalia in *Newtoni Principia Mathematica*, ed. E.A. Fellmann (Paris: Vrin, 1973). Michael Nauenberg, "The Reception of Newton's Principia," http://physics.ucsc.edu/~michael/newtonreception6.pdf. "Why Is Calculus Missing from Newton's Principia?," *History of Science and Mathematics Stack Exchange*, https://hsm.stackexchange.com/questions/2362/why-is-calculus-missing-from-newtons-principia.

1129 Mark Goldie, "Isaac Newton and John Locke: in Public and in Private," *Newton & the Mint*, https://newtonandthemint.history.ox.ac.uk/economic-theories/newton-and-locke. Norriss S. Hetherington, "Isaac Newton's Influence on Adam Smith's Natural Laws in Economics," *Journal of the History of Ideas* 44, no. 3 (1983): 497–505, https://www.jstor.org/stable/2709178. Bruce Dickerson, "The Age of Enlightenment," Indian Hills Community College, https://webcontent.indianhills.edu/_myhills/courses/HIS111/documents/lu01_age_enlightenment.pdf.

1130 Donald F. Lach, "Leibniz and China," *Journal of the History of Ideas* (1945): 436–455.

1131 Isaac Newton, *The Principia: The Authoritative Translation and Guide: Mathematical Principles of Natural Philosophy*, trans. I. Bernard Cohen and Anne Whitman (University of California Press, 2016), https://www.google.com/books/edition/The_Principia_The_Authoritative_Translat/HN4kDQAAQBAJ.

1132 Marij van Strien, "On the Origins and Foundations of Laplacian Determinism," *Studies in History and Philosophy of Science* 45, no. 1 (March 2014): 24–31, https://hal.science/hal-01610331/document.

1133 Pierre Simon Marquis de Laplace, *A Philosophical Essay on Probability*, trans. Frederick Wilson Truscott and Frederick Lincoln Emory (New York: John Wiley & Sons, 1902; original publication 1825).

1134 Isaac Newton, T*he Three First Sections and Part of the Seventh Section of Newton's Principia, with a Preface Recommending a Geometrical Course of Mathematical Reading, and an Introduction on the Atomic Constitution of Matter, and the Laws of Motion*, ed. George Leigh Cooke (Oxford: John Henry Parker, 1850), 24–25, 98. Greg Poole, "The Long Arc of Light and Gravity," *Journal of Physics & Astronomy Review* 7, no. 3 (15 August 2019).

1135 Keith Devlin, "The Pascal Fermat Correspondence: How Mathematics is Really Done," *Mathematics Teacher* 103, no. 8, April 2010, https://www.maa.org/sites/default/files/images/upload_library/46/NCTM/The-Pascal-Fermat-Correspondence.pdf.

1136 Robin L. Plackett, "Herschel on Estimation," *Journal of the Royal Statistical Society: Series A* 155 (January 1991): 29–35, https://www.jstor.org/stable/2982667.

1137 "A Timeline of the Life of Charles Darwin," Christ's College, Cambridge, https://www.christs.cam.ac.uk/timeline-life-charles-darwin.

1138 "Young Naturalist," Part of the Darwin exhibition, American Museum of Natural History, https://www.amnh.org/exhibitions/darwin/young-naturalist.

1139 Kerry Lotzof, "Charles Darwin: History's Most Famous Biologist," Natural History Museum, London, https://www.nhm.ac.uk/discover/charles-darwin-most-famous-biologist.html.

1140 José María Rodríguez García, "Exiles and Arrivals in Christopher Columbus and William Bradford," in *Explorations in Renaissance Culture* 28, ed. Andrew J. Fleck (2002): 75–98, https://doi.org/10.1163/23526963-90000244.

1141 Gerd Kohlhepp, "Scientific Findings of Alexander von Humboldt's Expedition into the Spanish-American Tropics (1799-1804) from a Geographical Point of View," Earth Sciences, *Anais da Academia Brasileira de Ciencias* 77, no. 2 (June 2005), https://doi.org/10.1590/S0001-37652005000200010.

1142 Smithsonian American Art Museum, "Alexander Von Humboldt's Influence on American Art, Nature and the American Identity," *Impact* 6, no. 2 (April 2020), Smithsonian Institute, https://www.si.edu/support/impact/humboldt.

1143 The set of instruments von Humboldt took on his expedition was so extensive that when he sat down to list them, his inventory came to six pages. Timothy Rooks, "How Humboldt Put South America on the Map," *Deutsche Welle*, DW.com, 12 July 2019, https://www.dw.com/en/how-scientist-alexander-von-humboldt-put-spanish-south-america-on-the-global-map/a-46693502.

1144 Stanley Finger, Marco Piccolino, and Frank W. Stahnisch, "Alexander von Humboldt: Galvanism, Animal Electricity, and Self-Experimentation Part 2: The Electric Eel, Animal Electricity, and Later Years," *Journal of the History of the Neurosciences* 22, no. 4 (2013): 327–352, https://doi.org/10.1080/096470 4X.2012.732728.

1145 John Delaney, "Alexander von Humboldt, 1769–1859: German Naturalist Alexander von Humboldt Lived to See the New Direction Taken by Geographic Studies, Which His Own Work Had Initiated," *First X, Then Y, Now Z: Landmark Thematic Maps*, Historic Maps Collection, Princeton University Library, https://library.princeton.edu/visual_materials/maps/websites/thematic-maps/humboldt/humboldt.html.

1146 Gerard Helferich, *Humboldt's Cosmos: Alexander von Humboldt and the Latin American Journey That Changed the Way We See the World* (Tantor Media, Incorporated, 2011).

1147 Daniel Kehlmann, *Measuring The World: A Novel*, trans. Carol Brown Janeway (Knopf Doubleday Publishing Group, 2009), https://www.krabarchive.com/ralphmag/EQ/humboldt.html.

1148 The first scientific expedition was apparently carried out over one hundred years earlier by French astronomer Jean Richer. See John W. Olmsted, "The Scientific Expedition of Jean Richer to Cayenne (1672–1673)," *Isis* 34, no. 2 (1942): 117–128, https://www.journals.uchicago.edu/doi/pdf/10.1086/347762.

1149 David Kidd, "The Humboldt Map: Natural Features and Human Artefacts that Commemorate the Great Explorer and Scientist Alexander von Humboldt (1768–1859)," ArcGIS StoryMaps, 10 February 2020, https://storymaps.arcgis.com/stories/79eeffa9f54d429687c17fa8267d3ba2.

1150 Alexander von Humboldt, *Kosmos: Entwurf einer physischen Weltbeschreibung, ed. Ottmar Ette and Oliver Lubrich* (Berlin: Die Andere Bibliothek, 2014), ISBN 978-3-8477-0014-2.

1151 Andrea Wulf, *The Iinvention of Nature: Alexander Von Humboldt's New World* (Knopf, 2015).

1152 "A Five-Year Journey," part of the Darwin exhibition, American Museum of Natural History, https://www.amnh.org/exhibitions/darwin/a-trip-around-the-world/a-five-year-journey.

1153 National Geographic Society, "HMS Beagle: Darwin's Trip around the World," 2 August 2023, https://education.nationalgeographic.org/resource/hms-beagle-darwins-trip-around-world.

1154 Discovering Galapagos, "The Five Year Voyage," Galapagos Conservation Trust, http://evolution.discoveringgalapagos.org.uk/evolution-zone/discovering-darwin/voyage-of-the-beagle/the-five-year-voyage.

1155 Frank N. Egerton, "A History of the Ecological Sciences, Part 32: Humboldt, Nature's Geographer," *Bulletin, Ecological Society of America* (1 July 2009), https://doi.org/10.1890/0012-9623-90.3.253.

1156 David Bressan, "Darwin the Geologist," *Scientific American*, 12 February 2012, https://blogs.scientificamerican.com/history-of-geology/darwin-the-geologist.

1157 Museum of Natural History, University of Oxford, "Charles Lyell's Friends And Family," *More Than a Dodo, Oxford University Museum of Natural History Blog*, 21 December 2016, https://morethanadodo.com/2016/12/21/lyells-friends-and-family. Charles Lyell, *Principles of Geology* (London: Penguin Books, 1997) (originally published in three volumes from 1830-1833 by John Murray, London).

1158 Wrote Darwin, "I deeply regretted that I did not proceed far enough at least to understand something of the great leading principles of mathematics, for men thus endowed seem to have an extra sense." Santa Fe Institute, "Darwin's Extra Sense: How Mathematics Is Revolutionizing Biology," Santa Fe Institute, 25 January 2013, https://www.santafe.edu/news-center/news/darwins-extra-sense-announce.

1159 E.G. Hernández-Avilez and R. Ruiz-Gutiérrez, "From One Darwin to Another: Charles Darwin's Annotations to Erasmus Darwin's 'The Temple of Nature," *Humanities and Social Sciences Communications* 10, no. 143 (2023), https://doi.org/10.1057/s41599-023-01616-y.

1160 Andrea Wulf, *Magnificent Rebels: The First Romantics and the Invention of the Self* (Hachette UK, 2022).

1161 Noel Jackson, "Rhyme and Reason: Erasmus Darwin's Romanticism," *Modern Language Quarterly* 70, no. 2 (1 June 2009): pages 171–194, https://doi.org/10.1215/00267929-2008-036 "Erasmus Darwin, 1731–1802," Poetry Foundation, https://www.poetryfoundation.org/poets/erasmus-darwin.

1162 Erasmus Darwin, *The Temple of Nature* (Outlook Verlag, 2020), 9. Originally published 1803.

1163 Michael Freeman, "Tracks to a New World: Railway Excavation and the Extension of Geological Knowledge in Mid-Nineteenth-Century Britain," *The British Journal for the History of Science* 34, no. 1 (2001): 51–65. K. Kris Hirst, "Stratigraphy: Earth's Geological, Archaeological Layers," ThoughtCo, 25 February

2019, https://www.thoughtco.com/stratigraphy-geological-archaeological-layers-172831.

1164 University of California at Berkeley, Museum of Paleontology, "The History of Evolutionary Thought: 1800s, Uniformitarianism: Charles Lyell," https://evolution.berkeley.edu/the-history-of-evolutionary-thought/1800s/uniformitarianism-charles-lyell.

1165 Dr. Terry Mortenson, "The 19th-century Scriptural Geologists: Historical Background," Creation Ministries International, https://creation.com/the-19th-century-scriptural-geologists-historical-background.

1166 Lyell says we have the record of earth's existence "millions of ages before our times." An age is a hundred years. So Lyell believed the earth was hundreds of millions of years old. See Charles Lyell, "Concluding Remarks," in *Principles of Geology: Or The Modern Changes of the Earth and Its Inhabitants Considered as Illustrative of Geology*, Project Gutenberg, https://www.gutenberg.org/cache/epub/33224/pg33224-images.html. Originally published in 1830.

1167 The University of Edinburgh, Information Services, "About Sir Charles Lyell," 5 August 2021, https://www.ed.ac.uk/information-services/library-museum-gallery/cultural-heritage-collections/crc/sir-charles-lyell-collection/about-sir-charles-lyell.

1168 Charles Darwin, "08-Sep-1832: Darwin Encounters Bolas," from *Darwin's Beagle Diary*, 8 September 1832, The Friends of Charles Darwin, http://friendsofdarwin.com/articles/darwin-encounters-bolas.

1169 Charles Darwin, *The Voyage of the Beagle* (P.F. Collier, 1909), 53, 145. Originally published 1839.

1170 "A Five-Year Journey," Part of the Darwin exhibition, American Museum of Natural History, https://www.amnh.org/exhibitions/darwin/a-trip-around-the-world/a-five-year-journey.

1171 Charles Darwin, *Darwin's Beagle Field Notebooks (1831-1836)*, Darwin Online, http://darwin-online.org.uk/EditorialIntroductions/Chancellor_fieldNotebooks.html.

1172 Charles Darwin, The Project Gutenberg eBook of On the Origin of Species, 1859 Edition, https://www.gutenberg.org/files/1228/1228-h/1228-h.htm.

1173 Carl Zimmer, "Pigeons Get a New Look," *New York Times*, 4 February 2013, https://archive.nytimes.com/www.nytimes.com/2013/02/05/science/pigeons-a-darwin-favorite-carry-new-clues-to-evolution.html.

1174 Charles Darwin, The Project Gutenberg eBook of *On the Origin of Species*, 1859 Edition, https://www.gutenberg.org/files/1228/1228-h/1228-h.htm.

1175 "Charles Darwin's 'On the Origin of Species,'" Brandeis University Library, https://www.brandeis.edu/library/archives/essays/special-collections/darwin.html.

1176 "Darwin's Book Publications," American Museum of Natural History, https://www.amnh.org/research/darwin-manuscripts/published-books.

1177 Charles Darwin, The Project Gutenberg eBook of *On the Origin of Species*, 1859 Edition, https://www.gutenberg.org/files/1228/1228-h/1228-h.htm.

1178 Herbert Spencer, *First Principles* (Cambridge University Press, 1862), http://doi.org/10.1017/CBO9780511693939.

1179 Aristotle, *Metaphysics*, in *Library of the Future* Ver. 5.0 (Irvine, CA: World Library, Inc., 1996). CD-ROM.

1180 "How Old Is the Earth?," Darwin Correspondence Project, University of Cambridge, https://www.darwinproject.ac.uk/letters/darwins-works-letters/re-writing-origin-later-editions/how-old-earth.

1181 For William Thomson's equations on the age of the earth in "On the Secular Cooling of the Earth," http://visualiseur.bnf.fr/Cadre-sPage?O=NUMM-95120&I=313&M=tdm&T=&Y=Image.

1182 P.C. England, P. Molnar, and F.M. Richter, "Kelvin, Perry and the Age of the Earth: Had Scientists Better Appreciated One of Kelvin's Contemporary Critics, the Theory of Continental Drift Might Have Been Accepted Decades Earlier," *American Scientist* (2007), https://www.americanscientist.org/article/kelvin-perry-and-the-age-of-the-earth. BBC-TV, "Lord Kelvin, the Eminent 19th and Early 20th Century Scientist, Was Determined to Work Out the Age of the Earth. A Simple Experiment with Molten Rock Gave Him Figures for His Calculations," in Two Men of Rock, https://www.bbc.co.uk/programmes/p00ccq58. Jean Paul Poirier, "About the Age of the Earth," *Comptes Rendus Geoscience* 349, no. 5 (September 2017): 223–225, https://www.sciencedirect.com/science/article/pii/S1631071317300731. Crosbie Smith and M. Norton Wise, *Energy and Empire: A Biographical Study of Lord Kelvin*, 606. Mario Livio, *Brilliant Blunders: From Darwin to Einstein—Colossal Mistakes by Great Scientists That Changed Our Understanding of Life and the Universe* (Simon and Schuster, 2014), 73.

1183 Lord Kelvin (William Thomson), "On the Secular Cooling of the Earth," *Transactions of the Royal Society of Edinburgh* 23 (1864): 167–169. Reprint in *Mathematical and Physical Papers*, vol. III (1890): 295. Thomson read this paper to the Royal Society on 28 April 1862.

1184 Graham Dolan, The Royal Observatory Greenwich, http://www.royalobser-vatorygreenwich.org/articles.php.

1185 Lord Kelvin (William Thomson), "On the Secular Cooling of the Earth." Sen, *Einstein's Fridge*, 72.

1186 James Hutton, "Theory of the Earth with Proofs and Illustrations," printed for Messers Cadell, Junior, and Davis, London; and William Creech, Edinburgh, 1795).

1187 Patrick N. Wyse Jackson, "William Thomson's Determinations of the Age of the Earth," in Raymond Flood, Mark McCartney, and Andrew Whitaker, eds., *Kelvin: Life, Labours and Legacy* (Oxford: Oxford University Press, 2008; online edition, Oxford Academic, 1 May 2008), https://doi.org/10.1093/acprof:oso/9780199231256.003.0010.

1188 "Gradualism: Phyletic gradualism and punctuated equilibrium represent the opposite extremes of a continuum," in *Encyclopedia of Biodiversity*, 2001, ScienceDirect, Elsevier, https://www.sciencedirect.com/topics/earth-and-planetary-sciences/gradualism.

1189 Michael W. Taylor, *Men Versus the State: Herbert Spencer and Late Victorian Individualism* (Oxford University Press, 1992).

1190 Mark Francis, *Herbert Spencer and the Invention of Modern Life* (London: Routledge, 2007).

1191 Inder S. Marwah, "Rethinking Resistance: Spencer, Krishnavarma, and The Indian Sociologist," in Burke A. Hendrix and Deborah Baumgold, eds., *Colonial Exchanges: Political Theory and the Agency of the Colonized* (Manchester: Manchester Scholarship Online, 18 Jan. 2018), https://doi.org/10.7228/manchester/9781526105646.003.0003.

1192 Jin, Xiaoxing, "The Evolution of Social Darwinism in China, 1895–1930," *Comparative Studies in Society and History* 64, no .3 (2022): 690–721.

1193 Michio Nagai, "Herbert Spencer in Early Meiji Japan," *The Far Eastern Quarterly* vol. 14, no. 1 (1954): 55–64, https://doi.org/10.2307/2942228

1194 Peter J. Bowler, *The Eclipse of Darwinism: Anti-Darwinian Evolution Theories in the Decades around 1900* (Baltimore: Johns Hopkins University Press), 3, 23–24.

1195 Leonard Wilson, "Religious Assumptions in Lord Kelvin's Estimates of the Earth's Age," *Earth Sciences History* 29, no. 2 (2010): 187–212, https://doi.org/10.17704/eshi.29.2.46678x0701k62j0j.

1196 Paul James-Griffiths, "Sir William Thomson, Lord Kelvin (1824–1907)," *Christian Heritage Edinburgh*, 23 August 2016, https://www.christianheritageedinburgh.org.uk/2016/08/23/sir-william-thomson-lord-kelvin-1824-1907.

1197 "About Christian Evidence," https://christianevidence.org/about.

1198 Charles Darwin, Letter to J. D. Hooker, 28 February 1866, Darwin Correspondence Project, University of Cambridge, https://www.darwinproject.ac.uk/letter/DCP-LETT-5020.xml.

1199 These quotes appear in Sen, *Einstein's Fridge*, 72.

1200 Mano Singham, "When Lord Kelvin Nearly Killed Darwin's Theory," *Scientific American*, 5 September 2021, https://www.scientificamerican.com/article/when-lord-kelvin-nearly-killed-darwins-theory1.

1201 Online Variorum of Darwin's Origin of Species, fifth British edition (1869), page 552, http://darwin-online.org.uk/Variorum/1869/1869-552-c-1872.html.

1202 A. Plutynski, "The Modern Sythesis," PhilSci-Archive, 2009, https://philpapers.org/rec/PLUTMS.

1203 E.O. Wilson, *The Social Conquest of Earth* (New York: W.W. Norton & Co., 2012).

1204 Richard Dawkins, *The Extended Phenotype* (Oxford: Oxford University Press, 2016), 8. Originally published 1982.

1205 Dawkins, Richard, The Selfish Gene, Oxford: Oxford University Press, 1989. Originally published 1976.

1206 John Scott, "Rational Choice Theory," *Understanding Contemporary Society: Theories of the Present* 129 (2000): 126–138. Steven L. Green, "Rational Choice Theory: An Overview," Baylor University Faculty Development Seminar on Rational Choice Theory, 2002.

1207 David Loye, "Darwin's Lost Theory and Its Implications for the 21st Century," *World Futures: Journal of General Evolution* 55, no. 3 (2000): 201–226.

1208 Chris Bateman, "The Big Fight: Reductionism versus Holism," 2 November 2005, https://onlyagame.typepad.com/only_a_game/2005/11/the_big_fight_r.html.

1209 Robin Fox, *Participant Observer: A Memoir of a Transatlantic Life* (UK: Routledge), 2018. The term "participant observer" science goes back to 1924, with the publication of E.C. Lindeman's *Social Discovery: An Approach to the Study of Functional Groups* (New York: The Press of the New Era Printing Company, 1924).

1210 Rudolf Clausius, *The Mechanical Theory of Heat, With Its Applications To The Steam-Engine and To The Physical Properties Of Bodies* (London: John Van Voorst, 1867), 865.

1211 Samuel J. Ling, "Entropy and Disorder," *Physics Bootcamp*, http://www.physicsbootcamp.org/Entropy-and-Disorder.html.

1212 C.F. Demoulin, Y.J. Lara, L. Cornet, et al., "Cyanobacteria Evolution: Insight from the Fossil Record," *Free Radical Biology and Medicine* 140 (2019): 206–223, https://doi.org/10.1016/j.freeradbiomed.2019.05.007, https://www.ncbi.nlm.nih.gov/pmc/articles/PMC6880289.

1213 Minnesota Pollution Control Agency, "Blue-Green Algae and Harmful Algal Blooms," https://www.pca.state.mn.us/water/blue-green-algae-and-harmful-algal-blooms.

1214 Eugene Rabinowitch, "Photosynthesis," US Atomic Energy Commission, 1949. National Drought Mitigation Center, "Soil Water Is the Limiting Factor for Photosynthesis," *Plant Growth and Development*, https://drought.unl.edu/

ranchplan/DroughtBasics/GrassesandDrought/PlantGrowthandDevelopment. aspx.

1215 Cara M. Santelli et al., "Abundance and Diversity of Microbial Life in Ocean Crust," *Nature* 453, no. 7195 (2008): 653–656. National Science Foundation, press release, "Bacteria 'Feed' on Earth's Ocean-Bottom Crust," 28 May 2008, https://www.nsf.gov/news/news_summ.jsp?cntn_id=111587.

1216 Steven Sweetman, Grant Smith, and David Martill, "Highly Derived Eutherian Mammals from the Earliest Cretaceous of Southern Britain," *Acta Palaeontologica Polonica* 62 (2017), https://researchportal.port.ac.uk/en/publications/highly-derived-eutherian-mammals-from-the-earliest-cretaceous-of-. Charles Q. Choi, "These Rodent-Like Creatures Are the Earliest Known Ancestor of Humans, Whales and Shrews," *Live Science*, 8 November 2017, https://www.livescience.com/60888-rat-creatures-were-earliest-eutherian-mammal-ancestors.html.

1217 Christopher Shields, "Aristotle," *The Stanford Encyclopedia of Philosophy* (Winter 2023 Edition), Edward N. Zalta & Uri Nodelman, eds., https://plato.stanford.edu/archives/win2023/entries/aristotle, https://plato.stanford.edu/entries/aristotle/

1218 Jack P. Cunningham and Mark Hocknull, eds., *Robert Grosseteste and the Pursuit of Religious and Scientific Learning in the Middle Ages, Vol. 18* (Springer International Publishing, 2016), 7, https://link.springer.com/content/pdf/10.1007/978-3-319-33468-4.pdf. Phil Gibbs and Sugihara Hiroshi, "What Is Occam's Razor," ResearchGate, https://www.researchgate.net/profile/Philip-Gibbs-2/publication/330171618_What_is_Occam's_Razor/links/5c313917458515a4c7109db5/What-is-Occams-Razor.pdf.

1219 Karl Jaspers, *The Origin and Goal of History*, ed. C. Thornhill (Routledge, 2021), originally published 1949.

1220 Gregory Bateson, *Naven: A Survey of the Problems suggested by a Composite Picture of the Culture of a New Guinea Tribe drawn from Three Points of View* (The University Press, 1936).

1221 Harold Bloom, *Shakespeare: The Invention of the Human* (New York: Riverhead Books, 1998).

1222 K. Weiss, "Thomas Henry Huxley (1825–1895) Puts Us in Our Place," *Journal of Experimental Zoology. Part B, Molecular and Developmental Evolution* 302, no. 3 (15 June 2004): 196–206, https://doi.org/10.1002/JEZ.B.21000. Samantha Hauserman, "Thomas Henry Huxley (1825–1895)," Embryo Project Encyclopedia (2013), https://embryo.asu.edu/pages/thomas-henry-huxley-1825-1895.

1223 E.G. Nisbet and C.M.R. Fowler, "Archaean Metabolic Evolution of Microbial Mats," *Proceedings of the Royal Society* 266 (1999): 2375–2382, https://doi.org/10.1098/rspb.1999.0934. N. Noffke, R. Hazen, and N. Nhleko, "Earth's Earliest Microbial Mats in a Siliciclastic Marine Environment (2.9 Ga Mozaan Group, South Africa)," *Geology* 31 (2003): 673–676, https://doi.org/10.1130/G19704.1.

1224 Diane McKnight et al., "Microbial Composition of Chlorophyte-Dominated Mats in Glacial Meltwater Streams in the McMurdo Dry Valleys, Antarctica," *AGU Fall Meeting Abstracts*, December 2022, B16E-08, https://ui.adsabs. harvard.edu/abs/2022AGUFM.B16E..08M. American Geophysical Union, "These Freeze-Drying Algae Can Awaken From Cryostasis, Could Help Spaceflights Go Farther," *AGU: Advancing Earth and Space Sciences*, 8 December 2022, https:// news.agu.org/press-release/these-freeze-drying-algae-can-awaken-from-cryostasis/.

1225 Roberto Guidetti and Roberto Bertolani, "Paleontology and Molecular Dating," in *Water Bears: The Biology of Tardigrades* (Springer, 2019), 131–143, https://link.springer.com/chapter/10.1007/978-3-319-95702-9_5. Caryn Babaian and Sudhir Kumar, "Adventures in Evolution: The Narrative of Tardigrada, Trundlers in Time," *The American Biology Teacher* 81, no. 8 (2019): 543–552.

1226 Kenta Sugiura and Midori Matsumoto, "Sexual Reproductive Behaviours of Tardigrades: A Review," *Invertebrate Reproduction & Development* 65, no. 4 (2021): 279–287, https://doi.org/10.1080/07924259.2021.1990142. Diane R. Nelson, "Current Status of the Tardigrada: Evolution and Ecology," *Integrative and Comparative Biology* 42, no. 3 (2002): 652–659, https://doi.org/10.1093/ icb/42.3.652.

1227 Ashleigh Papp, "Tardigrades, an Unlikely Sleeping Beauty," 60-Second Science, *Scientific American*, November 30, 2022, https://www.scientificamerican. com/podcast/episode/tardigrades-an-unlikely-sleeping-beauty.

1228 Sugiura Matsumoto "Sexual Reproductive Behaviours of Tardigrades." Kay Boatner, "Tardigrade," *National Geographic Kids*, https://kids.nationalgeographic.com/animals/invertebrates/facts/tardigrade.

1229 "Tardigrade," *Encyclopedia Britannica*, 5 March 2024, https://www.britannica.com/animal/tardigrade.

1230 Sarah Bordenstein, Marine Biological Laboratory, "Tardigrades (Water Bears)," Microbial Life, Educational Sources, Science Education Resource Center at Carleton College, https://serc.carleton.edu/microbelife/topics/tardigrade/index.html.

1231 "How Deep Is the Ocean?," National Ocean Service, National Oceanic and Atmospheric Administration, https://oceanservice.noaa.gov/facts/oceandepth. html.

1232 D. Sloan, R. Alves Batista, and A. Loeb, "The Resilience of Life to Astrophysical Events," *Scientific Reports* 7 (2017): 5419, https://doi.org/10.1038/ s41598-017-05796-x.

1233 Kristine Phillips, "A Single Jawbone Has Revealed Just How Much Radiation Hiroshima Bomb Victims Absorbed," *Washington Post*, 2 May 2018.

1234 "Gray (Unit)," *Wikipedia*, https://en.wikipedia.org/wiki/Gray_(unit).

1235 CfA Communications, "Last Survivors on Earth, As Long as the Sun Shines, Hardy Tardigrade Will Carry On," *The Harvard Gazette*, 20 July 2017, https://news.harvard.edu/gazette/story/2017/07/tardigrade-or-water-bear-will-survive-until-the-sun-dies. Diane R. Nelson, "Current Status of the Tardigrada: Evolution and Ecology," *Integrative and Comparative Biology* 42, no. 3 (July 2002): 652–659, https://doi.org/10.1093/icb/42.3.652.

1236 Ibid.

1237 Joseph Smythers, H. M. O'Dell, T. A. Clark, J. R. Crislip, B. B. Flinn, et al., "Chemobiosis Reveals Tardigrade Tun Formation Is Dependent on Reversible Cysteine Oxidation," *Plos One* 19, no. 1 (2024), https://journals.plos.org/plosone/article?id=10.1371/journal.pone.0295062.

1238 Sugiura and Matsumoto, "Sexual Reproductive Behaviours of Tardigrades."

1239 William Randolph Miller, "Tardigrades," *American Scientist*, 2010, https://www.americanscientist.org/article/tardigrades.

1240 "Tardigrade," *Wikipedia*, https://en.wikipedia.org/wiki/Tardigrade

1241 Crawford H. Greenewalt, "The Wings of Insects and Birds as Mechanical Oscillators," *Proceedings of the American Philosophical Society* 104, no. 6, (1960): pages 605–11, http://www.jstor.org/stable/985536.

1242 American Museum of Natural History, "Parts of a Flower," Plant Morphology, part of the Biodiversity Counts Collection, American Museum of Natural History, https://www.amnh.org/learn-teach/curriculum-collections/biodiversity-counts/plant-identification/plant-morphology/parts-of-a-flower.

1243 Peter Bernhardt, *Wily Violets and Underground Orchids: Revelations of a Botanist* (University of Chicago Press, 2003), 64.

1244 Kathryn A. Orr and Murray E. Fowler, "Order Trochiliiformes (Hummingbirds)," in *Biology, Medicine, and Surgery of South American Wild Animals*, ed. Murray E. Fowler and Zalmir Silvino Cubas (Iowa State University Press, 2008), 174–179.

1245 Ibid., 174. Elizabeth Donaldson, "Can Hummingbirds Walk," BackYardVisitors.com, https://hummingbirdbliss.com/can-hummingbirds-walk.

1246 Clementina González and Juan Francisco Ornelas, "Male Relatedness, Lekking Behavior Patterns, and the Potential for Kin Selection in a Neotropical Hummingbird," *The Auk* 136, no. 3 (1 July 2019): https://doi.org/10.1093/auk/ukz038.

1247 Jillian Mock, "Hummingbirds Shake Their Tail Feathers to Generate High-Pitched Sounds," *Audubon*, 15 February 2019, https://www.audubon.org/news/hummingbirds-shake-their-tail-feathers-generate-high-pitched-sounds. Audubon for Kids, "The Hummingbird Wing Beat Challenge," *National Audubon Society*, https://www.audubon.org/news/the-hummingbird-wing-beat-challenge.

1248 Elizabeth Donaldson, "Hummingbird Parents: (Mating to Nesting)," Back-YardVisitors.com, https://hummingbirdbliss.com/hummingbird-parents-mating-to-nesting. Wisconsin Pollinators, "Hummingbird Sex—XXX Rated."

1249 Eric Mohrman, "How Do Hummingbirds Mate?" Sciencing, 22 November 2019, https://sciencing.com/hummingbirds-mate-4566850.html.

1250 David Rand, "Evolution of Sex," Evolutionary Biology, Brown University Department of Biology, https://biomed.brown.edu/Courses/BIO48/19.Evol.of.Sex.HTML.

1251 Tamás L. Czárán, Rolf F. Hoekstra, and Ludo Pagie, "Chemical Warfare Between Microbes Promotes Biodiversity," *Proceedings of the National Academy of Sciences* 99, no. 2 (15 January 2002): 786–790, https://doi.org/10.1073/pnas.012399899. Daniel M. Cornforth and Kevin R. Foster, "Antibiotics and the Art of Bacterial War," *PNAS* 112, no. 35 (24 August 2015): 10827–10828, https://doi.org/10.1073/pnas.1513608112. Markus F. Weber, Gabriele Poxleitner, Elke Hebisch, Erwin Frey, and Madeleine Opitz, "Chemical Warfare and Survival Strategies in Bacterial Range Expansions," *Journal of the Royal Society Interface* 11, no. 94 (6 July 2014), https://doi.org/10.1098/rsif.2014.0172, https://royalsocietypublishing.org/doi/10.1098/rsif.2014.0172.

1252 Sarwat Saulat, *The Life of The Prophet* (Lahore: Islamic Publications Ltd.: Pakistan, 1983).

1253 Howard K. Bloom, *The Muhammad Code: How a Desert Prophet Brought You ISIS, al Qaeda, and Boko Haram* (Port Townsend, WA: Feral House, 2016).

1254 Richard Cavendish, "Latimer and Ridley Burned at the Stake: The Oxford Martyrs were killed on 16 October 1555," *History Today Volume* 55, no. 10 (October 2005), https://www.historytoday.com/archive/months-past/latimer-and-ridley-burned-stake.

1255 Meilan Solly, "The Myth of 'Bloody Mary'," 12 March 2020, https://www.smithsonianmag.com/history/myth-bloody-mary-180974221.

1256 For the way in which blood bonds people together, see: Richard A. Koenigsberg, *Nations Have the Right to Kill: Hitler, the Holocaust, and War* (Library of Social Science, 2009).

1257 John Foxe, *Foxe's Book Of Martyrs, White Fish* (Montana: Kessinger Publishing, 2010). Originally published 1563.

1258 Sandy Fitzgerald, "Trump: 'Thugs, Tyrants' Awakened a 'Sleeping Giant'," *Newsmax*, 17 September 2022, https://www.newsmax.com/us/donald-trump-rally-maga/2022/09/17/id/1087965.

1259 Ali Swenson, "Retread Scare: Trump and other Republicans evoke another era by calling Democrats 'communists'," Associated Press, June 19, 2023, https://www.pbs.org/newshour/politics/retread-scare-trump-and-other-republicans-evoke-another-era-by-calling-democrats-communists.

1260 Andrew Kaczynski, "Marjorie Taylor Greene Indicated Support for Executing Prominent Democrats in 2018 and 2019 Before Running for Congress," CNN, 26 January 2021, https://www.cnn.com/2021/01/26/politics/marjorie-taylor-greene-democrats-violence/index.html.

1261 Wwg1wga, *QAnon: An Invitation to the Great Awakening* (Dallas, TX: Relentlessly Creative Books, 2019). Mike Rothschild, *The Storm Is Upon Us: How QAnon Became a Movement, Cult, and Conspiracy Theory of Everything* (Melville House, 2021), 22. David Klepper and Ali Swenson, Associated Press, "'The Storm Is Coming': Trump Openly Embracing QAnon Conspiracy Theories," *Times of Israel*, 16 September 2022, https://www.timesofisrael.com/the-storm-is-coming-trump-openly-embracing-qanon-conspiracy-theories/. Camila Domonoske, "The QAnon 'Storm' Never Struck. Some Supporters Are Wavering, Others Steadfast," NPR, 20 January 2021, https://www.npr.org/sections/inauguration-day-live-updates/2021/01/20/958907699/the-qanon-storm-never-struck-some-supporters-are-wavering-others-steadfast.

1262 S. Smith, *Qanon For Beginners: Discover The Hidden Secrets and The Main Conspiracy Theories. Destroy The New World Order and Take The Millennial Kingdom By Force* (Youcanprint, 2021). David Klepper and Ali Swenson, Associated Press, "'The Storm Is Coming': Trump Openly Embracing QAnon Conspiracy Theories," *Times of Israel*, 16 September 2022, https://www.timesofisrael.com/the-storm-is-coming-trump-openly-embracing-qanon-conspiracy-theories.

1263 Rachel E. Greenspan, "The History of QAnon: How the Conspiracy Theory Snowballed from the Fringes of the Internet into the Mainstream," *Business Insider*, 11 February 2021, https://www.insider.com/qanon-history-who-is-q-conspiracy-theory-what-does-believe-2021-2.

1264 Coast to Coast AM with George Noory, Premiere Networks, iHeartMedia, https://www.coasttocoastam.com.

1265 Donald Trump, "Announcement of Candidacy, Trump Tower, New York, NY, June 16, 2015," P2016, Race for the White House, https://www.p2016.org/trump/trump061615sp.html.

1266 I. Eibl-Eibesfeldt, *Love and Hate: The Natural History of Behavior Patterns* (United Kingdom: Taylor & Francis, 2017), 228. J.K. Bosson, A.B. Johnson, K. Niederhoffer, and W.B. Swann, Jr., "Interpersonal Chemistry through Negativity: Bonding by Sharing Negative Attitudes about Others," *Personal Relationships* 13 (2006): 135–150, https://onlinelibrary.wiley.com/doi/abs/10.1111/j.1475-6811.2006.00109.x. Markham Heid, "How Shared Hatred Helps You Make Friends," https://forge.medium.com/how-shared-hatred-helps-you-make-friends-40f5c988c76a.

1267 R.E. Ulrich and N.H. Azrin, "Reflexive Fighting in Response to Aversive Stimulation," *Journal of the Experimental Analysis of Behavior* 5, no. 4 (1962): 511–520.

1268 Ibid.

1269 Adam Gabbatt, "Golden Escalator Ride: The Surreal Day Trump Kicked Off His Bid for President," *The Guardian*, 14 June 2019, https://www.theguardian.com/us-news/2019/jun/13/donald-trump-presidential-campaign-speech-eyewitness-memories.

1270 Wayne Kernodle, "Some Implications of the Homogamy-Complementary Needs Theories of Mate Selection for Sociological Research," *Social Forces* 38, no. 2 (1959): 145–152, https://doi.org/10.2307/2573935. R. Gaunt, "Couple Similarity and Marital Satisfaction: Are Similar Spouses Happier?," *Journal of Personality* 74, no. 5 (2006): 1401–20, https://doi.org/10.1111/j.1467-6494.2006.00414.x. D. M. Buss and M. Barnes, "Preferences in Human Mate Selection," *Journal of Personality and Social Psychology* 50, no. 3 (1986): 559–570, https://doi.org/10.1037/0022-3514.50.3.559. M. D. Botwin, D. M. Buss, and T. K. Shackelford, "Personality and Mate Preferences: Five Factors in Mate Selection and Marital Satisfaction," *Journal of Personality* 65, no. 1 (1997): 107–136, https://doi.org/10.1111/j.1467-6494.1997.tb00531.x. Alan C. Kerckhoff and Keith E. Davis, "Value Consensus and Need Complementarity in Mate Selection," *American Sociological Review* 27, no. 3 (1962): 295–303, https://doi.org/10.2307/2089791. Y. Sin, G. Annavi, C. Newman, C. Buesching, T. Burke, D. Macdonald, and H. Dugdale, "MHC Class II-Assortative Mate Choice in European Badgers (Meles meles)," *Molecular Ecology* 24 (2015), https://doi.org/10.1111/mec.13217.

1271 Gillian E. Brennan, "Papists and Patriotism in Elizabethan England," Cambridge University Press, 2015, https://www.cambridge.org/core/journals/british-catholic-history/article/abs/papists-and-patriotism-in-elizabethan-england/917EE1495A0F8D4FB2C808EF2F6D26E8.

1272 E. Fehr and S. Gächter, "Altruistic Punishment in Humans," *Nature* 415 (2002): 137–140, https://doi.org/10.1038/415137a. Sören Enge, Hendrik Mothes, Monika Fleischhauer, Andreas Reif, and Alexander Strobel, "Genetic Variation of Dopamine and Serotonin Function Modulates the Feedback-Related Negativity During Altruistic Punishment," *Scientific Reports* (2017), https://link.springer.com/content/pdf/10.1038/s41598-017-02594-3.pdf.

1273 Martin Reuter, Bernd Weber, Christian J. Fiebach, Christian Elger, and Christian Montag, "The Biological Basis of Anger: Associations with the Gene Coding for DARPP-32 (PPP1R1B) and with Amygdala Volume," *Behavioural Brain Research* 202, no. 2 (2009): 179–183, https://doi.org/10.1016/j.bbr.2009.03.032. Nancy K. Morrison and Sally K. Severino, "Moral Values: Development and Gender Influences," *Psychodynamic* 13 July 2017, https://

doi.org/10.1521/jaap.1.1997.25.2.255. Morgan Bimm, "This Is Why We Can't Have Nice Things: Tumblr Publics, John Green, and Sanctionable Girlhood," *Youth Mediations and Affective Relations* (2018): 213–231, https://doi.org/10.1007/978-3-319-98971-6_13. Jean Kim M.D., "Anger's Allure: Are You Addicted to Anger?," *Psychology Today Blog*, 25 August 2015. Daniel Taylor, "Righteous Anger and the Pleasure Centers of the Brain," Wordtaylor.com, 6 December 2017, https://www.wordtaylor.com/neither-nor-blog/righteous-anger-and-the-pleasure-centers-of-the-brain.

1274 Gianluigi Tanda et al., "Cannabinoid and Heroin Activation of Mesolimbic Dopamine Transmission by a Common μ1 Opioid Receptor Mechanism," *Science* 276 (1997): 2048–2050, http://doi.org/10.1126/science.276.5321.2048.

1275 C.K. De Dreu, L.L. Greer, M.J.J. Handgraaf, S. Shalvi, G.A. Van Kleef, M. Baas, F.S.T. Velden, E. Van Dijk, and S.W.W. Feith, "The Neuropeptide Oxytocin Regulates Parochial Altruism in Intergroup Conflict Among Humans," *Science* 328 (2010): 1408–1411, http://doi.org/10.1126/science.1189047. Julia H. Egito, Michael Nevat, Simone G. Shamay-Tsoory, Ana Alexandra C. Osório, "Oxytocin Increases the Social Salience of the Outgroup in Potential Threat Contexts," *Hormones and Behavior* 122 (2020), https://doi.org/10.1016/j.yhbeh.2020.104733. Wiet van Helmond, "Oxytocin," *Homœopathic Links* 29, no. 4 (2016): 256–258.

1276 Carsten KW De Dreu, "Oxytocin Modulates Cooperation Within and Competition Between Groups: An Integrative Review and Research Agenda," *Hormones and Behavior* 61, no. 3 (2012): 419–428. Hejing Zhang et al., "Oxytocin Promotes Coordinated Out-Group Attack During Intergroup Conflict in Humans," *eLife* 8 (2019), https://elifesciences.org/articles/40698. Egito JH, Nevat M, Shamay-Tsoory SG, Osório AAC, "Oxytocin Increases the Social Salience of the Outgroup in Potential Threat Contexts," *Hormones and Behavior* (June 2020), doi: http://doi.org/10.1016/j.yhbeh.2020.104733.

1277 Howard K. Bloom, *The Muhammad Code: How a Desert Prophet Brought You ISIS, al Qaeda, and Boko Haram* (Feral House, 2016).

1278 Pew Research Center, "Muslims," April 2, 2015, https://www.pewresearch.org/religion/2015/04/02/muslims.

1279 Pew Research Center, "Global Christianity—A Report on the Size and Distribution of the World's Christian Population," December 19, 2011, https://www.pewresearch.org/religion/2011/12/19/global-christianity-exec.

1280 The British Empire at its peak covered 13.71 million square miles and about one-quarter of the world's population. Florian Zandt, "The Biggest Empires In Human History," *Statista*, 2 November 2023, https://www.statista.com/chart/20342/peak-land-area-of-the-largest-empires. "Members of the OIC: Organization of Islamic Cooperation," World Data, https://www.worlddata.info/alliances/oic-islamic-cooperation.php. Khalid Yahya Blankinship, The End of the Jihad State, the Reign of Hisham Ibn 'Abd-al Malik and the Collapse of the Umayyads (State University of New York Press, 1994), 37. "Early Muslim Conquests," *Wikipedia*, 1 August 2023, https://en.wikipedia.org/wiki/Early_Muslim_conquests.

1281 Florian Zandt, "The Biggest Empires In Human History," *Statista*, May 25, 2020, https://www.statista.com/chart/20342/peak-land-area-of-the-largest-empires/ National Geographic Kids, "British Empire Facts!," https://www.natgeokids.com/nz/discover/history/general-history/british-empire-facts.

1282 Dylan Lyons, "How Many People Speak English, And Where Is It Spoken?" *Babbel Magazine*, 10 March 2021, https://www.babbel.com/en/magazine/how-many-people-speak-english-and-where-is-it-spoken.

1283 Harvard Divinity School,"Islam in Nigeria," Religion and Public Life, Harvard Divinity School, https://rpl.hds.harvard.edu/faq/islam-nigeria.

1284 "2022 Report on International Religious Freedom: Kenya," Office of International Religious Freedom, U.S. Department of State, https://www.state.gov/reports/2022-report-on-international-religious-freedom/kenya.

1285 MEMRI, "ISIS, Islamic State (ISIS) Weekly Editorial Exploits Unrest In Ethiopia Over Demolishing Mosques, Calls On Ethiopian Muslims To Wage Jihad Or Join ISIS In East Africa," *MEMRI—Jihad & Terror Threat Monitor*, 9 June 2023, https://www.memri.org/jttm/islamic-state-isis-weekly-editorial-exploits-unrest-ethiopia-over-demolishing-mosques-calls. Desta Heliso, "Ethiopia's Increasing Vulnerability To Islamic Extremism And What That Means For The Horn Of Africa," *Religion Unplugged*, November 17, 2020, https://religionunplugged.com/news/2020/11/17/ethiopias-increasing-vulnerability-to-islamic-extremism-and-what-that-means-for-the-horn-of-africa.

1286 Daurius Figueira, *Jihad in Trinidad and Tobago* (iUniverse, 2002). Michael Adams, Michał Pawiński, "'Caribbean Jihad': Radical Social Networks and ISIS Foreign Fighters from Trinidad and Tobago," *Small States & Territories Journal*, November 2022, https://www.um.edu.mt/library/oar/handle/123456789/44404. For the full article: https://www.researchgate.net/publication/365306630_'Caribbean_Jihad'_radical_social_networks_and_ISIS_foreign_fighters_from_Trinidad_and_Tobago.

1287 For an example of the British influence in Prussia, Immanuel Kant wrote about the English example and the importance of republican constitutions in 1795, Immanuel Kant, *Perpetual Peace: A Philosophical Sketch*, http://fs2.american.edu/dfagel/www/Class%20Readings/Kant/Immanuel%20Kant,%20_Perpetual%20Peace_.pdf.

1288 To be precise, as I wrote this in March 2024, World Data's latest tally of the Muslim portion of the global population was 24.91%. See "Members of the OIC: Organization of Islamic Cooperation," https://www.worlddata.info/alliances/oic-islamic-cooperation.php.

1289 Bill Hayton, *The Invention of China* (Yale University Press, 2020), https://doi.org/10.2307/j.ctv17z8490. James Hannam, *The Globe: How the Earth Became Round* (Reaktion Books Ltd., 2023).

1290 Tingyang Zhao and Tingyang Zhao, *The Concept of Tianxia and Its Story: Redefining A Philosophy for World Governance* (Springer, 2019), 1–19.

1291 I.C. Malhotra, *Red Fear: The China Threat* (Bloomsbury India, 2020). James A. Millward, *Beyond the Pass: Economy, Ethnicity, and Empire in Qing Central Asia, 1759–1864* (Stanford University Press, 1998), 298.

1292 T. Kim, "Actualized Stigma: The Historical Formation of Anti-Americanism in North Korea," *Modern Asian Studies* 51 (2017): 543–576, https://doi.org/10.1017/S0026749X15000396.

1293 Korean Central News Agency, "American-Style Gangster-Like Doctrine Does Not Work on Korean Peninsula," KCNA Commentary, Rodong Sinmun, 26 November 2023, http://www.rodong.rep.kp/en/index.php?MTJAMjAyMyoxM-SoyNi1IMDAyQDExQDBAYW1lcmljYUAwQDE===.

1294 Since the start of the city of Babylon in roughly 1,870. R. Boer, "Beginnings of Old Babylonian Babylon: Sumu-abum and Sumu-la-El," *Journal of Cuneiform Studies* 70 (2018): 53–86, https://doi.org/10.5615/jcunestud.70.2018.0053.

1295 Luke Kempa, Chi Xuc, Joanna Depledge, Kristie L. Ebie, Goodwin Gibbins, Timothy A. Kohlerg, Johan Rockström, Marten Scheffer, Hans Joachim Schellnhuber, Will Steffen, and Timothy M. Lenton, "Climate Endgame: Exploring Catastrophic Climate Change Scenarios," *PNAS* 119, no. 36 (2022), https://www.pnas.org/doi/abs/10.1073/pnas.2108146119.

1296 William E. Rees, "The Human Ecology of Overshoot: Why a Major 'Population Correction Is Inevitable," *World* 4, no. 3 (2023): 509–527.

1297 Luke Kemp et al., "Climate Endgame: Exploring Catastrophic Climate Change Scenarios," *Proceedings of the National Academy of Sciences* 119, no. 34 (2022), https://www.pnas.org/doi/abs/10.1073/pnas.2108146119. Damian Carring, "Climate Endgame: Risk of Human Extinction 'Dangerously Underexplored,'" *The Guardian*, 1 August 2022, https://www.theguardian.com/environment/2022/aug/01/climate-endgame-risk-human-extinction-scientists-global-heating-catastrophe.

1298 J.M. Creamean et al., "Ice Nucleating Particles Carried from Below a Phytoplankton Bloom to the Arctic Atmosphere," *Geophysical Research Letters* 46, no. 14 (2019): 8572–8581, https://doi.org/10.1029/2019GL083039.

1299 Ulrich Pöschl et al., "Rainforest Aerosols as Biogenic Nuclei of Clouds and Precipitation in the Amazon," *Science* 329, no. 5998 (2010): 1513–1516, https://www.science.org/doi/full/10.1126/science.1191056. Christina Nunez, "Rainforests Explained," *National Geographic*, 19 May 2022, https://www.nationalgeographic.org/article/rainforests-explained.

1300 Ellard Hunting and Liam J. O'Reilly, "Observed Electric Charge of Insect Swarms and Their Contribution to Atmospheric Electricity," *iScience* 25, no. 10 (2022): 105241, https://doi.org/10.1016/j.isci.2022.105241. Phys.org, "Insects Contribute to Atmospheric Electricity," *Cell Press*, 24 October 2022, https://phys.org/news/2022-10-insects-contribute-atmospheric-electricity.html.

1301 Shayla Love, "A Honeybee Swarm Has as Much Electric Charge as a Thundercloud," *Scientific American*, November 15, 2022, https://www.scientificamerican.com/podcast/episode/a-honeybee-swarm-has-as-much-electric-charge-as-a-thundercloud.

1302 S.J. Konturek, J. Bilski, J. Tasler, M. Cieszkowski, "Role of Cholecystokinin in the Inhibition of Gastric Acid Secretion in Dogs," *The Journal of Physiology* 451 (1992): 477–489, https://doi.org/10.1113/jphysiol.1992.sp019174.

1303 K.A. Keay, M.A. Argueta, D.N. Zafir, P.M. Wyllie, G.J. Michael, D.C. Boorman, "Evidence that Increased Cholecystokinin (CCK) in the Periaqueductal Gray (PAG) Facilitates Changes in Resident-Intruder Social Interactions Triggered by Peripheral Nerve Injury," *Journal of Neurochemistry* 158, no. 5 (September 2021): 1151–1171, https://doi.org/10.1111/jnc.15476, https://onlinelibrary.wiley.com/doi/pdf/10.1111/jnc.15476, https://onlinelibrary.wiley.com/doi/pdf/10.1111/jnc.15476.

1304 K. Uvnäs-Moberg, M. Petersson, "Oxytocin—Biochemical Link for Human Relations. Mediator of Antistress, Well-Being, Social Interaction, Growth, Healing...," *Lakartidningen* 101, no. 35 (2004): 2634–2639. A. Weller, R. Feldman, "Emotion Regulation and Touch in Infants: The Role of Cholecystokinin and Opioids," *Peptides* 24, no. 5 (May 2003): 779–788, https://doi.org/10.1016/s0196-9781(03)00118-9. Natalie Angier, *Woman: An Intimate Geography* (New York: Houghton Mifflin, 1999), 316.

1305 S. Harmand, J. Lewis, C. Feibel, et al., "3.3-Million-Year-Old Stone Tools from Lomekwi 3, West Turkana, Kenya," *Nature* 521 (2015): 310–315, https://doi.org/10.1038/nature14464. Karenleigh A. Overmann, Frederick L. Coolidge, eds., *Squeezing Minds from Stones: Cognitive Archaeology and the Evolution of the Human Mind* (United States: Oxford University Press, 2019). Smithsonian, "Stone Tools," Smithsonian Human Origins, https://humanorigins.si.edu/evidence/behavior/stone-tools. Rebecca Morelle, "Oldest Stone Tools Pre-Date Earliest Humans," BBC, 20 May 2015, https://www.bbc.com/news/science-environment-32804177.

1306 S. Putt, S. Wijeakumar, J. Spencer, "Prefrontal Cortex Activation Supports the Emergence of Early Stone Age Toolmaking Skill," *NeuroImage* 199 (2019): 57–69, https://doi.org/10.1016/j.neuroimage.2019.05.056. N. Toth, K. Schick, "Why Did the Acheulean Happen? Experimental Studies into the Manufacture and Function of Acheulean Artifacts," *L'Anthropologie* 123, no. 4–5 (November–December 2019): 724–768, https://doi.org/10.1016/j.anthro.2017.10.008.

1307 G. Suwa, T. Sasaki, S. Semaw, M.J. Rogers, S.W. Simpson, Y. Kunimatsu, M. Nakatsukasa, et al., "Canine Sexual Dimorphism in Ardipithecus ramidus Was Nearly Human-Like," *Proceedings of the National Academy of Sciences* 118, no. 49 (2021), https://www.pnas.org/doi/abs/10.1073/pnas.2116630118. Clare Wilson, "Canine Teeth Shrank in Human Ancestors at Least 4.5 Million Years Ago," *New Scientist*, 29 November 2021, https://www.newscientist.com/article/2299286-canine-teeth-shrank-in-human-ancestors-at-least-4-5-million-years-ago.

1308 Ralph D. Hermansen, *Down from the Trees: Man's Amazing Transition from Tree-Dwelling Ape Ancestors* (Apple Academic Press, 2018), 278–281.

1309 A. Barash, M. Bastir, E. Been, "3D Morphometric Study of the Mandibular Fossa and Its Implication for Species Recognition in Homo erectus," *Advances in Anthropology* 5 (2015): 152–163, https://doi.org/10.4236/AA.2015.53014.

1310 A. Timmermann, T. Friedrich, "Late Pleistocene Climate Drivers of Early Human Migration," *Nature* 538 (2016): 92–95, https://doi.org/10.1038/nature19365. Amanda Mascarelli, "Climate Swings Drove Early Humans Out of Africa (and Back Again)," Sapiens.org, 21 September 2016, https://www.sapiens.org/biology/early-human-migration.

1311 M. Medler, "Speculations about the Effects of Fire and Lava Flows on Human Evolution," *Fire Ecology* 7 (2011): 13–23, https://doi.org/10.4996/FIREECOLOGY.0701013. Medler derives his views from two of Richard Wrangham's works, *Catching Fire: How Cooking Made Us Human* (Basic Books, 2009). And Richard N. Carmody, R.W. Wrangham, "The Energetic Significance of Cooking," *Journal of Human Evolution* 57 (2009): 379–391, https://doi.org/10.1016/j.jhevol.2009.02.011, https://www.sciencedirect.com/science/article/abs/pii/S0047248409001262. Others believe humans did not tame fire until 400,000 years ago. See K. MacDonald, F. Scherjon, E. Veen, K. Vaesen, W. Roebroeks, "Middle Pleistocene Fire Use: The First Signal of Widespread Cultural Diffusion in Human Evolution," *Proceedings of the National Academy of Sciences of the United States of America* 118 (2021), https://doi.org/10.1073/pnas.2101108118.

1312 Richard Wrangham, *Catching Fire: How Cooking Made Us Human* (Profile Books, 2010), 85–86. S.J. Pyne, *The Pyrocene: How We Created an Age of Fire, and What Happens Next* (United States: University of California Press, 2021), 4, 21. One and a half million years ago there is evidence of the human use of fire complete with what the specialists call "pot lids." See S. Hlubik, F. Berna, C. Feibel, D. Braun, J.W.K. Harris, "Researching the Nature of Fire at 1.5 Mya on the Site of Fxjj20 AB, Koohi Fora, Kenya, Using High-Resolution Spatial Analysis and FTIR Spectrometry," *Current Anthropology* 58, Supplement 16 (2017), https://www.journals.uchicago.edu/doi/full/10.1086/692530. C.K. Brain, *The Hunters or the Hunted? An Introduction to African Cave Taphonomy* (University of Chicago Press, 1983). John A.J. Gowlett, "The Discovery of Fire by Humans:

A Long and Convoluted Process," *Philosophical Transactions of the Royal Society B: Biological Sciences* 371, no. 1696 (June 2016), https://doi.org/10.1098/rstb.2015.0164, https://doi.org/10.1098/rstb.2015.0164

1313 H. Wang, S. Ambrose, C. Liu, L. Follmer, "Paleosol Stable Isotope Evidence for Early Hominid Occupation of East Asian Temperate Environments," *Quaternary Research* 48 (1997): 228–238, https://doi.org/10.1006/qres.1997.1921.

1314 We lost our fur 1.2 million years ago. See: Rosalind Eswaran, Henry Harpending, John McCullough, "Genetic Variation at the MC1R Locus and the Time Since Loss of Human Body Hair," (2004). Nicholas Wade, "Why Humans and Their Fur Parted Ways," *New York Times*, 19 August 2003, https://www.nytimes.com/2003/08/19/science/why-humans-and-their-fur-parted-ways.html. Nina G. Jablonski, "The Naked Truth: Why Humans Have No Fur," *Scientific American*, 1 February 2010, https://www.scientificamerican.com/article/the-naked-truth-why-humans-have-no-fur. F. Ebling, "The Biology of Hair," *Dermatologic Clinics* 5.3 (1987): 467–481, https://doi.org/10.1016/S0733-8635(18)30728-9.

1315 R. Shimelmitz, S. Kuhn, A. Jelinek, A. Ronen, A. Clark, M. Weinstein-Evron, "'Fire at Will': The Emergence of Habitual Fire Use 350,000 Years Ago," *Journal of Human Evolution* 77 (2014): 196–203, https://doi.org/10.1016/j.jhevol.2014.07.005.

1316 Jürgen Ehlers, Philip Leonard Gibbard, Philip D. Hughes, "Quaternary Glaciations and Chronology," in *Past Glacial Environments*, eds. John Menzies and Jaap van der Meer (Elsevier, 2018), 77–101.

1317 M. Stahlschmidt, C. Mallol, C. Miller, "Fire as an Artifact—Advances in Paleolithic Combustion Structure Studies: Introduction to the Special Issue," *Journal of Paleolithic Archaeology* 3 (2020): 503–508, https://doi.org/10.1007/s41982-020-00074-1. A. Danchin, "Bacteria in the Ageing Gut: Did the Taming of Fire Promote a Long Human Lifespan?," *Environmental Microbiology* 20 (2018): 1966–1987, https://doi.org/10.1111/1462-2920.14255.

1318 H. Dillon, R. Carmen, G. Geher, "The Creatures of Flame: Richard Wrangham's Catching Fire," *Evolution: Education and Outreach* 4 (2011): 173–174, https://doi.org/10.1007/s12052-010-0303-4.

1319 "Fire became coded into hominin DNA; thanks to favorable conditions at the end of the last ice age, second-fire steadily spread everywhere humans did," S.J. Pyne, *The Pyrocene: How We Created an Age of Fire, and What Happens Next* (University of California Press, 2021), 4, 21.

1320 Richard Wrangham, *Catching Fire: How Cooking Made Us Human* (Profile Books, 2010), 61

1321 Ibid., 86.

1322 Ibid., 134.

1323 N. Alperson-Afil, G. Sharon, M. Kislev, Y. Melamed, I. Zohar, S. Ashkenazi, R. Rabinovich, R. Biton, R. Werker, G. Hartman, C. Feibel, N. Goren-Inbar, "Spatial Organization of Hominin Activities at Gesher Benot Ya'aqov, Israel," *Science* 326 (18 Dec 2009): 1677–1680, https://doi.org/10.1126/science.1180695.

1324 R. Shahack-Gross, F. Berna, P. Karkanas, C. Lemorini, A. Gopher, R. Barkai, "Evidence for the Repeated Use of a Central Hearth at Middle Pleistocene (300 ky ago) Qesem Cave, Israel," *Journal of Archaeological Science* 44 (April 2014): 12–21, https://doi.org/10.1016/J.JAS.2013.11.015.

1325 T.H. Van Andel, W. Davies, B. Weninger, "The Human Presence in Europe during the Last Glacial Period I: Human Migrations and the Changing Climate," Neanderthals and Modern Humans in the European Landscape during the Last Glaciation: Archaeological Results of the Stage 3 (MacDonald Institute for Archaeological Research, University of Cambridge, 2003), https://www.researchgate.net/publication/236109404_The_Human_Presence_in_Europe_during_the_Last_Glacial_Period_I_Human_Migrations_and_the_Changing_Climate.

1326 Trenton W. Holliday, "Body Size, Body Shape, and the Circumscription of the Genus Homo," *Current Anthropology* 53, S6 2012: S330–S345.

1327 M. Courty, E. Carbonell, J. Poch, R. Banerjee, "Microstratigraphic and Multi-Analytical Evidence for Advanced Neanderthal Pyrotechnology at Abric Romani (Capellades, Spain)," *Quaternary International* 247 (2012): 294–312, https://doi.org/10.1016/J.QUAINT.2010.10.031. D. Barsky, E. Carbonell, R. Sala-Ramos, J. Castro, F. García-Vadillo, "Late Acheulian Multiplicity in Manufactured Stone Culture at the End of the Middle Pleistocene in Western Europe," *Quaternary International*, https://doi.org/10.1016/J.QUAINT.2021.04.017. A.C. Sorensen, E. Claud, M. Soressi, "Neandertal Fire-Making Technology Inferred from Microwear Analysis," *Scientific Reports* 8, no. 1 (2018): 10065. "Homo heidelbergensis, What Does It Mean to Be Human," Human Origins, Smithsonian National Museum of Natural History, https://humanorigins.si.edu/evidence/human-fossils/species/homo-heidelbergensis. Dennis O'Neil, "Homo heidelbergensis," Palomar College, https://www.palomar.edu/anthro/homo2/mod_homo_1.htm. A. Burkard, "Homo heidelbergensis: The Tool to Our Success," *Auctus: The Journal of Undergraduate Research and Creative Scholarship* 47 (2016), https://scholarscompass.vcu.edu/auctus/47. S. Hlubik, et al., "Hominin Fire Use in the Okote Member at Koobi Fora, Kenya: New Evidence for the Old Debate," *Journal of Human Evolution* 133 (2019): 214–229. D. Henry, "The Palimpsest Problem, Hearth Pattern Analysis, and Middle Paleolithic Site Structure," *Quaternary International* 247 (2012): 246–266, https://doi.org/10.1016/J.QUAINT.2010.10.013. R. Shahack-Gross, F. Berna, P. Karkanas, C. Lemorini, A. Gopher, R. Barkai, "Evidence for the Repeated Use of a Central Hearth at Middle Pleistocene (300 ky ago) Qesem Cave, Israel," *Journal of Archaeological Science* 44: 12–21, https://doi.org/10.1016/J.JAS.2013.11.015.

1328 The oldest spears appear to be 375,000-400,000 years old. H. Thieme, "Lower Palaeolithic Hunting Spears from Germany," *Nature* 385 (1997): 807–810, https://doi.org/10.1038/385807A0. H. Thieme, "Die ältesten Speere der Welt – Fundplätze der frühen Altsteinzeit im Tagebau Schöningen," *Archäologisches Nachrichtenblatt* 10 (2005): 409–417. M. Baales, O. Jöris, "Zur Altersstellung der Schöninger Speere," in *Erkenntnisjäger, Kultur und Umwelt des frühen Menschen*, ed. J. Burdukiewicz et al. (Veröffentlichungen des Landesamtes für Archäologie Sachsen-Anhalt 57, 2003): 281–288. O. Jöris, "Aus einer anderen Welt – Europa zur Zeit des Neandertalers," in *Vom Neandertaler zum modernen Menschen*, ed. N. J. Conard et al. (Blaubeuren: Ausstellungskatalog, 2005), 47–70. H. Thieme, "Lower Paleolithic Hunting Spears from Germany," *Nature* 385, no. 27 (1997): 807–810, https://www.science.org/content/article/when-did-humans-begin-hurling-spears.

1329 I. Verheijen, B.M. Starkovich, J. Serangeli, T. van Kolfschoten, N.J. Conard, "Early Evidence for Bear Exploitation during MIS 9 from the Site of Schöningen 12 (Germany)," *Journal of Human Evolution* 177 (2023), https://doi.org/10.1016/j.jhevol.2022.103294, https://www.sciencedirect.com/science/article/pii/S0047248422001543.

1330 M. Hyodo, H. Nakaya, A. Urabe, H. Saegusa, X. Shunrong, Y. Jiyun, J. Xuepin, "Paleomagnetic Dates of Hominid Remains from Yuanmou, China, and Other Asian Sites," *Journal of Human Evolution* 43, no. 1 (2002): 27–41, https://doi.org/10.1006/JHEV.2002.0555. Z. Zhu, R. Dennell, W. Huang, et al., "Hominin Occupation of the Chinese Loess Plateau Since About 2.1 Million Years Ago," *Nature* 559 (2018): 608–612, https://doi.org/10.1038/s41586-018-0299-4.

1331 H. Wang, S> Ambrose, C. Liu, & L. Follmer, "Paleosol Stable Isotope Evidence for Early Hominid Occupation of East Asian Temperate Environments," *Quaternary Research* 48 (1997): 228–238, https://doi.org/10.1006/qres.1997.1921.

1332 C. Falguères, J. Bahain, Y. Yokoyama, J. Arsuaga, J. Castro, E. Carbonell, J. Bischoff, M. Dolo, "Earliest Humans in Europe: The Age of TD6 Gran Dolina, Atapuerca, Spain," *Journal of Human Evolution* 37, no. 3–4 (1999): 343–352, https://doi.org/10.1006/JHEV.1999.0326.

1333 Emma Groeneveld, "Early Human Migration," *World History Encyclopedia* 15 (2017), https://www.worldhistory.org/article/1070/early-human-migration.

1334 The edges of ice sheets are called "the periglacial environment." H. M. French, *The Periglacial Environment* (Wiley, 2017).

1335 Brian Handwerk, "Evidence of Fur and Leather Clothing, Among World's Oldest, Found in Moroccan Cave," *Smithsonian Magazine*, 16 September 2021, https://www.smithsonianmag.com/science-nature/evidence-of-fur-and-leather-clothing-among-worlds-oldest-found-in-moroccan-cave-180978689. National Museum of Denmark, "Fur in Prehistory," https://en.natmus.dk/historical-knowledge/historical-themes/the-fur-trail/fur-in-prehistory.

1336 Francois d'Errico, Laurent Doyon, Shanshan Zhang, Michael Baumann, Michaela Lázničková-Galetová, Xiaoping Gao, Fei Chen, Yujie Zhang, "The Origin and Evolution of Sewing Technologies in Eurasia and North America," *Journal of Human Evolution* 125 (2018): 71–86, https://doi.org/10.1016/j.jhevol.2018.10.004.

1337 Vadim N. Stepanchuk, "The Earliest Evidence for Dwelling Construction in the Upper Paleolithic of Eastern Europe: A 30,000-Year-Old Surface Structure from Mira Layer I," *Vita Antiqua* 13 (2021): 15–26.

1338 Jarosław Wilczyński, Piotr Wojtal, Martin Oliva, Krzysztof Sobczyk, Gary Haynes, Janis Klimowicz, György Lengyel, "Mammoth Hunting Strategies during the Late Gravettian in Central Europe as Determined from Case Studies of Milovice I (Czech Republic) and Kraków Spadzista (Poland)," *Quaternary Science Reviews* 223 (2019), https://doi.org/10.1016/j.quascirev.2019.105919, https://www.sciencedirect.com/science/article/pii/S0277379119302598

1339 Wendy Rendu, Stéphanie Costamagno, Luc Meignen, Marie Soulier, "Monospecific Faunal Spectra in Mousterian Contexts: Implications for Social Behavior," *Quaternary International* 247 (2012): 50–58, https://doi.org/10.1016/J.QUAINT.2011.01.022.

1340 François Djindjian, "Identifying the Hunter-Gatherer Systems Behind Associated Mammoth Bone Beds and Mammoth Bone Dwellings," *Quaternary International* 359 (2015): 47–57, https://www.sciencedirect.com/science/article/pii/S1040618214004595. Brian Handwerk, "A Mysterious 25,000-Year-Old Structure Built of the Bones of 60 Mammoths," *Smithsonian Magazine*, 16 March 2020, https://www.smithsonianmag.com/science-nature/60-mammoths-house-russia-180974426.

1341 To estimate the number of families, I'm counting the number of hearths per structure. And I'm assuming one family per hearth. Which may be an underestimate. See: François Djindjian, "Identifying the Hunter-Gatherer Systems Behind Associated Mammoth Bone Beds and Mammoth Bone Dwellings," *Quaternary International* 359–360 (2015): 47–57, https://doi.org/10.1016/j.quaint.2014.07.006, https://www.sciencedirect.com/science/article/pii/S1040618214004595, https://www.sciencedirect.com/science/article/pii/S1040618214004595 And look at the visually rich "Paleolithic Art," by Alena Buis in her Art And Visual Culture; Prehistory To Renaissance, https://pressbooks.bccampus.ca/cavestocathedrals/chapter/paleolithic. Also see the illustration of a Paleolithic dwelling in "The Paleolithic Period," Boundless Art History, College Sidekick, https://www.collegesidekick.com/study-guides/boundless-arthistory/the-paleolithic-period.

1342 Laurent Demay, Stéphane Péan, Marie Patou-Mathis, "Mammoths Used as Food and Building Resources by Neanderthals: Zooarchaeological Study Applied to Layer 4, Molodova I (Ukraine)," *Quaternary International* 276 (2012): 212–226, https://doi.org/10.1016/J.QUAINT.2011.11.019.

1343 Art Ramos, "Early Jericho," *World History Encyclopedia*, 19 September 2016, https://www.worldhistory.org/article/951/early-jericho.

1344 "The Neolithic Site of Catalhoyuk, a nomination document for the UNESCO World Heritage List," Republic of Turkey Ministry of Culture and Tourism, 2011, https://whc.unesco.org/uploads/nominations/1405.pdf .

1345 Howard Bloom, "Instant Evolution: The Influence of the City on Human Genes," *New Ideas In Psychology*, 13 September 2001, linkinghub.elsevier.com/retrieve/pii/S0732118X01000046.

1346 James Mellaart, Catal-Huyuk: A Neolithic Town in Anatolia (McGraw-Hill, 1967). Hans Helback, "First Impressions of the Catal Huyuk Plant Husbandry," *Anatolian Studies* 14 (1964): 121–123. Marija Gimbutas, "Wall Paintings of Catal Huyuk," *The Review of Archaeology* (Fall 1990).

1347 S. Ambrose, "Paleolithic Technology and Human Evolution," *Science* 291 (2001): 1748–1753, https://doi.org/10.1126/SCIENCE.1059487.

1348 J. M. Rabaey, "Homo Technologicus," 2018 International Symposium on VLSI Technology, Systems and Application (VLSI-TSA), 2018, 1-1, https://doi.org/10.1109/VLSI-TSA.2018.8403799.

1349 Isaac Asimov, "Reason," https://addsdonna.com/old-website/ADDS_DONNA/Science_Fiction_files/2_Asimov_Reason.pdf. Isaac Asimov, "The Truth Isn't Stranger Than Science Fiction—Just Slower," *New York Times*, 12 February 1984, https://archive.nytimes.com/www.nytimes.com/books/97/03/23/lifetimes/asi-v-truth.html.

1350 William Yardley, "Peter Glaser, Who Envisioned Space Solar Power, Dies at 90," *New York Times*, 5 June 2014, https://www.nytimes.com/2014/06/06/us/peter-glaser-who-envisioned-space-solar-power-dies-at-90.html. Peter E. Glaser, "Power from the Sun: Its Future," *Science* 162, no. 3856 (22 November 1968): 857–861, http://doi.org/10.1126/science.162.3856.857, https://www.science.org/doi/10.1126/science.162.3856.857.

1351 Gerard K. O'Neill, "Space Colonies and Energy Supply to the Earth: Manufacturing Facilities in High Orbit Could Be Used to Build Satellite Solar Power Stations from Lunar Materials," *Science* 190, no. 4218 (5 December 1975): 943–947, http://doi.org/10.1126/science.190.4218.943, https://www.science.org/doi/10.1126/science.190.4218.943.

1352 Gerard K. O'Neill, "The Colonization of Space – Gerard K. O'Neill, Physics Today, 1974," The National Space Society, https://space.nss.org/the-colonization-of-space-gerard-k-o-neill-physics-today-1974.

1353 Mail Today, "Abdul Kalam is India's 'most trusted' person," *India Today*, 28 February 2010, https://www.indiatoday.in/mail-today/story/abdul-kalam-is-indias-most-trusted-person-68396-2010-02-27.

1354 Muhammad Asif Hanif, Umer Rashid, "Solar Power Satellites: The SPS Cost Estimates Are Based On Point Design And Represent Forecasts Of Future Technology Development That Are Unlikely To Be Precise," *Advances in Energy Systems and Technology* 2 (1979), in *Science Direct*, https://www.sciencedirect.com/topics/earth-and-planetary-sciences/solar-power-satellites.

1355 European Space Agency, Advanced Concepts Team, Energy Systems, "Space-based Solar Power," 14 April 2013, https://www.esa.int/gsp/ACT/projects/sps.

1356 Howard Bloom, "Smart Tiles—an Energy Infrastructure for the Solar System," American Institute of Aeronautics and Astronautics, Space Conference, September 2016, http://doi.org/10.2514/6.2016-5328.

1357 Tereza Pultarova, "A Solar Power Plant in Space? The UK Wants to Build one by 2035," Space.com, 11 May 2022, https://www.space.com/space-based-solar-power-plant-2035.

1358 Peter J. Schubert, Sheylla Monteiro Pinto, Bruna C. Pires, Moises do Nascimento, Edward Barks, Jonathan Nderitu, Gabriel Oliveira Goncalves, Fatih Tokmo, "Analysis of a Novel SPS Configuration Enabled by Lunar ISRU," American Institute of Aeronautics and Astronautics, AIAA 2015-4648, AIAA Space 2015 Conference and Exposition, August 2015, https://arc.aiaa.org/doi/abs/10.2514/6.2015-4648.

1359 David L. Chandler, "Shining Brightly: Vast Amounts of Solar Energy Radiate to the Earth Constantly, but Tapping That Energy Cost-Effectively Remains a Challenge," *MIT News*, 26 October 2011, https://news.mit.edu/2011/energy-scale-part3-1026.

1360 Sabine Pongruber, "#SpaceWatchGL Opinion: What Is the Real Annually Generated Revenue of the Space Industry? PART 2," *SpaceWatch Europe*, https://spacewatch.global/2022/10/spacewatchgl-opinion-what-is-the-real-annually-generated-revenue-of-the-space-industry-part-2/.

1361 Isabelle Dicaire, Institute of Electrical and Electronics Engineers, https://ieeexplore.ieee.org/author/37085692754.

1362 Bjorn Carey, "Sahara Desert Was Once Lush and Populated," LiveScience, July 20, 2006, https://www.livescience.com/4180-sahara-desert-lush-populated.html.

1363 J. Russell "The Population of Medieval Egypt," *Journal of the American Research Center in Egypt* 5 (1966): 69–82, https://doi.org/10.2307/40000174.

1364 M. Krom, U. Stanley, R. Cliff, J. Woodward, "Nile River Sediment Fluctuations over the Past 7000 Years and Their Key Role in Sapropel Development,"

Geology 30 (2002): 71–74, https://pubs.geoscienceworld.org/gsa/geology/article/30/1/71/191804/Nile-River-sediment-fluctuations-over-the-past, https://doi.org/10.1130/0091-7613(2002)030<0071:NRSFOT>2.0.CO;2.

1365 John Romer, *A History of Ancient Egypt, From the First Farmers to the Great Pyramid* (New York: Thomas Dunne Books, 2013).

1366 Rosalie David, *The Pyramid Builders of Ancient Egypt: A Modern Investigation of Pharaoh's Workforce* (Routledge, 1997). Dr. Joyce Tyldesley, "The Private Lives of the Pyramid-Builders," Ancient History in Depth, BBC, 17 February 2011, https://www.bbc.co.uk/history/ancient/egyptians/pyramid_builders_01.shtml.

1367 The phrase normally credited to Herodotus is, "Egypt is the gift of the Nile." Herodotus 2.5, 2.10.

1368 James Cloyd Bowman, Pecos Bill, *The Greatest Cowboy of All Time* (Albert Whitman, 2017). Edward O'Reilly, "The Saga of Pecos Bill," *Century Magazine*, 1923.

1369 National Oceanic and Atmospheric Administration, "Layers of the Atmosphere," https://www.noaa.gov/jetstream/atmosphere/layers-of-atmosphere.

1370 Hosea 8:7.

1371 Pablo A. Tedesco et al., "Estimating How Many Undescribed Species Have Gone Extinct," Conservation Biology 28, no. 5 (2014): 1360–1370.

1372 Pablo A. Tedesco, "Macroecological Patterns & Processes - Spatial & Temporal Dynamics," Institut de Recherche pour le Développement, France http://tedesco1.free.fr.

1373 Colonies of nitrifying bacteria live in biofilms. See Armin Gieseke et al., "Community Structure and Activity Dynamics of Nitrifying Bacteria in a Phosphate-Removing Biofilm," *Applied and Environmental Microbiology* 67, no. 3 (2001): 1351–1362, https://journals.asm.org/doi/full/10.1128/aem.67.3.1351-1362.2001.Li-Hung Lin et al., "Long-Term Sustainability of a High-Energy, Low-Diversity Crustal Biome," *Science* 314, no. 5798 (2006): 479–482, https://www.science.org/doi/full/10.1126/science.1127376.

1374 Karen G. Lloyd et al., "Mysterious Microbes in Earth's Crust Might Help with the Climate Crisis," *Scientific American*, 29 March 2023, https://www.scientificamerican.com/video/mysterious-microbes-in-earths-crust-might-help-with-the-climate-crisis. Adam Hadhazy, "Life Might Thrive 12 Miles Beneath Earth's Surface," Space.com, 4 February 2015, https://www.space.com/28447-earth-life-extremophiles-underground.html.

1375 Stephanie A. Napieralski et al., "Microbial Chemolithotrophy Mediates Oxidative Weathering of Granitic Bedrock," *Proceedings of the National Academy of Sciences* 116, no. 52 (2019): 26394–26401. Jennifer Frazer, "Scientists Waited Two and a Half Years to See Whether Bacteria Can Eat Rock: The Mystery of Dirt's Origins Is a Thorny Experimental Problem," *Scientific American*, 1 May 2020, https://www.scientificamerican.com/blog/artful-amoeba/scientists-waited-two-and-a-half-years-to-see-whether-bacteria-can-eat-rock.

1376 Todd Stevens, "Lithoautotrophy in the Subsurface," *FEMS Microbiology Reviews* 20, no. 3-4 (1997): 327–337. Eric S. Boyd et al., "Chemolithotrophic Primary Production in a Subglacial Ecosystem," *Applied and Environmental Microbiology* 80, no. 19 (2014): 6146–6153.

1377 Peter M. Vitousek et al., "Human Appropriation of the Products of Photosynthesis," *BioScience* 36, no. 6 (1986): 368–373, https://doi.org/10.2307/1310258.

1378 Ankit Gupta, Rasna Gupta, Ram Lakhan Singh, "Microbes and Environment," *Principles and Applications of Environmental Biotechnology for a Sustainable Future* (2017): 43–84, https://www.ncbi.nlm.nih.gov/pmc/articles/PMC7189961.

1379 Helmut Haberl, Karl-Heinz Erb, Fridolin Krausmann, Institute of Social Ecology, Klagenfurt University, Vienna, Austria, "Human Appropriation of Net Primary Production (HANPP)," Entry prepared for the Internet Encyclopaedia of Ecological Economics, International Society for Ecological Economics, March 2007, https://www.academia.edu/7163154/International_Society_for_Ecological_Economics_Internet_Encyclopedia_of_Ecological_Economics_Human_appropriation_of_net_primary_production_HANPP. Helmut Haberl, Fridolin Krausmann, Karl-Heinz Erb, Niels B. Schulz, "Human Appropriation of Net Primary Production," *Science* 296 (2002): 1968–1969, http://doi.org/10.1126/science.296.5575.1968.

1380 Yinon M. Bar-On, Rob Phillips, Ron Milo, "The Biomass Distribution on Earth," *Proceedings of the National Academy of Sciences* 115, no. 25 (2018): 6506–6511, https://www.pnas.org/doi/10.1073/pnas.1711842115. Fridolin Krausmann et al., "Global Human Appropriation of Net Primary Production Doubled in the 20th Century," *Proceedings of the National Academy of Sciences* 110, no. 25 (2013): 10324–10329, https://www.pnas.org/doi/10.1073/pnas.1211349110.

1381 M.S. Savoca, M.F. Czapanskiy, S.R. Kahane-Rapport, et al., "Baleen Whale Prey Consumption Based on High-Resolution Foraging Measurements," *Nature* 599 (2021): 85–90, https://doi.org/10.1038/s41586-021-03991-5.

1382 David B. Cook and Stacy B. Robeson, "Varying Hare and Forest Succession," *Ecology* 26, no. 4 (1945): 406–410, https://doi.org/10.2307/1931662.

1383 Sun Ling Wang, Roberto Mosheim, Eric Njuki, and Richard Nehring, "US Agricultural Output Has Grown Slower in Response to Stagnant Productivity Growth," *Amber Waves: The Economics of Food, Farming, Natural Resources, and Rural America*, 2022, https://ageconsearch.umn.edu/record/338870/. Nathan Childs, Sharon Raszap Skorbiansky, and William D. McBride, "U.S. Rice Production Changed Significantly in the New Millennium, but Remained Profitable," Economic Research Service, U.S. Department of Agriculture, https://www.ers.usda.gov/amber-waves/2020/may/u-s-rice-production-changed-significantly-in-the-new-millennium-but-remained-profitable.

1384 Genesis, 26:4, King James Bible, https://www.kingjamesbibleonline.org/Genesis-26-4. For a comparison of eleven different translations of Genesis, 26:4, see: Youversion, A Digital Ministry of Lifechurch, https://www.bible.com/bible/compare/GEN.26.4.

1385 Brad Plumer, "These Maps Show Where All the World's Cattle, Chickens, and Pigs Are," Vox, 5 Feb 2015, https://www.vox.com/2014/6/20/5825826/these-maps-show-where-all-the-worlds-cattle-chickens-and-pigs-live.

1386 Adam J. George and Sarah L. Bolt, "Livestock Cognition: Stimulating the Minds of Farm Animals to Improve Welfare and Productivity," *Livestock* 26, no. 4 (2021): 202–206.

1387 David Grimm, "What Are Farm Animals Thinking," *Science*, 7 December 2023, https://www.science.org/content/article/not-dumb-creatures-livestock-surprise-scientists-their-complex-emotional-minds.

1388 P. Smýkal, M. Nelson, J. Berger, and E. Wettberg, "The Impact of Genetic Changes during Crop Domestication on Healthy Food Development," *Agronomy* 8, no 7 (2018): 119, https://doi.org/10.3390/AGRONOMY8030026 On the other hand, Khoshbakht and Hammer contend that we have domesticated 35,000 species of plants. In that total, they include domesticated ornamental flowers and ornamental trees. Khoshbakht, K., and K. Hammer, "How Many Plant Species Are Cultivated?," *Genetic Resources and Crop Evolution* 55 (2008): 925–928, https://doi.org/10.1007/s10722-008-9368-0. S. Pironon, I. Ondo, M. Diazgranados, et al., "The Global Distribution of Plants Used by Humans," Science 383, no. 6680 (2024): 293–297.Pironon et al. say that 35,687 is the number of species we have utilized for food, medicine, decoration, etc.

1389 Adam Purcell, "Molecules of Life," *Basic Biology*, August 31, 2020, https://basicbiology.net/biology-101/molecules-of-life.

1390 J. Rios-garaizar et al., "A Middle Palaeolithic Wooden Digging Stick from Aranbaltza III, Spain," *PLoS One* 13 (2018), https://doi.org/10.1371/journal.pone.0195044.

1391 G. Topp, B. Dow, M. Edwards, et al., "Oxygen Measurements in the Root Zone Facilitated by TDR," *Canadian Journal of Soil Science* 80 (2000): 33–41, https://doi.org/10.4141/S99-037. E. Anderson, P. Millner, and H. Kunishi, "Maize Root Length Density and Mycorrhizal Infection as Influenced by Tillage and Soil Phosphorus," *Journal of Plant Nutrition* 10 (2008): 1349–1356, https://doi.org/10.1080/01904168709363667. K. Schroeder and T. Paulitz, "Root Diseases of Wheat and Barley During the Transition from Conventional Tillage to Direct Seeding," *Plant Disease* 90, no. 9 (2006): 1247–1253, https://doi.org/10.1094/PD-90-1247.

1392 Kimberly Totten, "What a Load of Guano: 5 Facts You Didn't Know About Bird Poop," National Museum Of American History, Behring Center, 17 February 2016, https://americanhistory.si.edu/blog/what-load-guano-5-facts-you-didnt-know-about-bird-poop.

1393 "A Historical Overview of Fertilizer Use: Almost 8,000 years ago farmers recognized its value," Crop Watch, Institute Of Agriculture And Natural Resources, University Of Nebraska-Lincoln, 15 March 2015, https://cropwatch.unl.edu/fertilizer-history-p1.

1394 Eviatar Nevo, "Evolution of Wild Emmer Wheat and Crop Improvement," *Journal of Systematics and Evolution*, 18 August 2014, https://doi.org/10.1111/jse.12124.

1395 Chad Jorgensen et al., "A High-Density Genetic Map of Wild Emmer Wheat from the Karaca Dağ Region Provides New Evidence on the Structure and Evolution of Wheat Chromosomes," *Frontiers in Plant Science* 8 (2017), https://doi.org/10.3389/fpls.2017.01798.

1396 Alex Thornton, "Food Security: This Is How Many Animals We Eat Each Year," World Economic Forum, 8 February 2019, https://www.weforum.org/agenda/2019/02/chart-of-the-day-this-is-how-many-animals-we-eat-each-year.

1397 Shelef O., P.J. Weisberg, F.D. Provenza, "The Value of Native Plants and Local Production in an Era of Global Agriculture," *Frontiers in Plant Science* 8 (2017), http://doi.org/10.3389/fpls.2017.02069. Food and Agriculture Organization of the United Nations, "What Is Happening to Agrobiodiversity?," https://www.fao.org/3/y5609e/y5609e02.htm.

1398 Clark L. Erickson, "Raised Field Agriculture in the Lake Titicaca Basin," *Expedition Magazine* 30, no. 3, 1988, https://www.penn.museum/sites/expedition/raised-field-agriculture-in-the-lake-titicaca-basin.

1399 Clark L. Erickson, "Raised Field Agriculture in the Lake Titicaca Basin," *Expedition Magazine* 30, no. 3 (1988), Penn Museum, https://www.penn.museum/sites/expedition/raised-field-agriculture-in-the-lake-titicaca-basin. Clark L. Erickson, "The Lake Titicaca Basin: A Pre-Columbian Built Landscape," in *Imperfect Balance: Landscape Transformations in the Pre-Columbian Americas* (Columbia University Press, 2000): 311–356, https://core.ac.uk/download/pdf/76392515.pdf. Global Nature Fund, "Lake Titicaca – Bolivia and Peru," https://www.globalnature.org/en/lake-titicaca.

1400 Weather Atlas, "Climate and Monthly Weather Forecast, Lake Titicaca, Peru," https://www.weather-atlas.com/en/peru/lake-titicaca-climate.

1401 Victor M. Ponce, "Facts On Lake Titicaca, Peru And Bolivia," San Diego State University, https://ponce.sdsu.edu/facts_on_lake_titicaca.html.

1402 Clark L. Erickson, "Raised Field Agriculture in the Lake Titicaca Basin."

1403 Lydia Pyne, "In Search of Prehistoric Potatoes," *Archaeology* 73, no. 2 (2020): 57–64.

1404 Ian G. Warkentin and Corey J.A. Bradshaw, "A Tropical Perspective on Conserving the Boreal 'Lung of the Planet'," *Biological Conservation* 151, no. 1 (2012): 50–52.

1405 Carolina Levis et al., "Persistent Effects of Pre-Columbian Plant Domestication on Amazonian Forest Composition," *Science* 355, no. 6328 (2017): 925–931, https://www.science.org/doi/full/10.1126/science.aal0157. Ben Panko, "The Supposedly Pristine, Untouched Amazon Rainforest Was Actually Shaped by Humans," *Smithsonian*, 3 March 2017, https://www.smithsonianmag.com/science-nature/pristine-untouched-amazonian-rainforest-was-actually-shaped-humans-180962378.

1406 Charles C. Mann, *1491: New Revelations of the Americas before Columbus* (New York: Knopf, 2005). Charles C. Mann, "1491: Before It Became the New World, the Western Hemisphere Was Vastly More Populous and Sophisticated Than Has Been Thought," *The Atlantic*, March 2002, https://www.theatlantic.com/magazine/archive/2002/03/1491/302445. Robinson Meyer, "The Amazon Rainforest Was Profoundly Changed by Ancient Humans," *The Atlantic*, March 2017, https://www.theatlantic.com/science/archive/2017/03/its-now-clear-that-ancient-humans-helped-enrich-the-amazon/518439/.

1407 Thomas W. Lee and John H. Walker, "Forests and Farmers: GIS Analysis of Forest Islands and Large Raised Fields in the Bolivian Amazon," *Land* 11, no. 5 (2022): 678, https://doi.org/10.3390/land11050678. (This article belongs to the Special Issue Archaeological and Historical Landscapes of South America: From Past Changes to Current Landscape Configurations.)

1408 U.S. Geological Survey, "New Map Shows 4.62 Billion Acres of Cropland Globally," 8 December 2017, Neogen: Solutions for Food and Animal Safety, https://www.neogen.com/neocenter/blog/new-map-shows-4-62-billion-acres-of-cropland-globally.

1409 M. Frazier, "A Short History of Pest Management: The Development of the Field of Integrated Pest Management," Penn State Extension, 30 June 2022, https://extension.psu.edu/a-short-history-of-pest-management. Catherine M. Hill, "Crop-Raiding by Wild Vertebrates: The Farmer's Perspective in an Agricultural Community in Western Uganda," *International Journal of Pest Management* 43, no. 1 (1997): 77–84, http://doi.org/10.1080/096708797229022. Alexander Campbell Martin, Herbert Spencer Zim, and Arnold L. Nelson, *American Wildlife & Plants: A Guide to Wildlife Food Habits: The Use of Trees, Shrubs, Weeds, and Herbs* by Birds and Mammals of the United States, Courier Corporation, 2013.

1410 4.5 times is an extrapolation from figures in E.C. Ellis and N. Ramankutty, "Putting People in the Map: Anthropogenic Biomes of the World,"

Frontiers in Ecology and the Environment 6 (2008): 439-447, https://doi.org/10.1890/070062. According to Ellis and Ramankuty, 75% of the world's available land is managed by humans. 25% is managed by nature. In other words, 25% is wild. But the human-managed land, including the land of cities, currently produces 89% of the earth's biomass. Wild lands only produce 11%. If you do the arithmetic, you discover that human-managed lands outproduce wild lands 4.5 to one.

1411   Hannah Ritchie, "Yields vs. Land Use: How the Green Revolution Enabled Us to Feed a Growing Population," OurWorldInData.org. https://ourworldindata.org/yields-vs-land-use-how-has-the-world-produced-enough-food-for-a-growing-population.

1412   Daisy A. John, Giridhara R. Babu, "Lessons From the Aftermaths of Green Revolution on Food System and Health," *Frontiers in Sustainable Food Systems* 5 (22 February 2021), https://doi.org/10.3389/fsufs.2021.644559

1413   Food and Agriculture Organization of the United Nations, "Land use in agriculture by the numbers," *Sustainable Food and Agriculture*, 7 May 2020, https://www.fao.org/sustainability/news/detail/en/c/1274219.

1414   Yinon M. Bar-On, Rob Phillips, and Ron Milo, "The Biomass Distribution on Earth," *Proceedings of the National Academy of Sciences* 115, no. 25 (13 April 2018): 6506–6511, https://www.pnas.org/doi/full/10.1073/pnas.1711842115.

1415   Jessica Leigh Hester, "The Incredibly Tricky Task of Measuring All Life on Earth," *Atlas Obscura*, May 22, 2018, https://www.atlasobscura.com/articles/biomass-of-everything-on-earth.

1416   Seth Borenstein, "Humans Account for Little Next to Plants, Worms, Bugs," Associated Press, 21 May 2018, https://apnews.com/article/science-fungi-plants-worms-us-news-72a5dd80a5004804ba1c458b6a438ec3.

1417   Hafiz I. Ahmad, Muhammad J. Ahmad, Farwa Jabbir, Sunny Ahmar, Nisar Ahmad, Abdelmotaleb A. Elokil, Jinping Chen, "The Domestication Makeup: Evolution, Survival, and Challenges," *Section on Evolutionary and Population Genetics, Frontiers in Ecology and Evolution* 8 (8 May 2020), https://doi.org/10.3389/fevo.2020.00103.

1418   Yinon M. Bar-On, Rob Phillips, and Ron Milo, "The Biomass Distribution on Earth," https://doi.org/10.1073/pnas.1711842115.

1419   Ron Sender, Shai Fuchs, and Ron Milo, "Revised Estimates for the Number of Human and Bacteria Cells in the Body," *PLoS Biology* 14, no. 8 (2016), https://journals.plos.org/plosbiology/article?id=10.1371/journal.pbio.1002533&mod. Alison Abbott, "Scientists Bust Myth That Our Bodies Have More Bacteria Than Human Cells: Decades-Old Assumption About Microbiota Revisited," *Nature*, 8 January 2016, https://www.nature.com/articles/nature.2016.19136.pdf.

1420 Michaeleen Doucleff, "Thank Your Gut Bacteria For Making Chocolate Healthful," NPR, 18 March 2014, https://www.npr.org/sections/the-salt/2014/03/18/290922850/chocolate-turns-into-heart-helpers-by-gut-bacteria. Jo Napolitano, "Exploring the Role of Gut Bacteria in Digestion," Argonne National Laboratory, 19 August 2010, https://www.anl.gov/article/exploring-the-role-of-gut-bacteria-in-digestion.

1422 Michael J. Morowitz, Erica M. Carlisle, and John C. Alverdy, "Contributions of Intestinal Bacteria to Nutrition and Metabolism in the Critically Ill," *Surgical Clinics of North America* 91, no. 4 (2011): 771–785, https://www.ncbi.nlm.nih.gov/pmc/articles/PMC3144392.

1423 Jason Daley, "Humans Make Up Just 1/10,000 of Earth's Biomass," *Smithsonian Magazine*, 25 May 2018, https://www.smithsonianmag.com/smart-news/humans-make-110000th-earths-biomass-180969141.

1424 Vaclav Smil, "Detonator of the Population Explosion," *Nature* 400 (1999): 415, https://doi.org/10.1038/22672.

1425 P.H. Raven and G.B. Johnson, *Biology* (McGraw-Hill Education, 2014), 112.

1426 As of 1999, world nitrogen use efficiency for cereal production was 33%. It has increased since then but as of 2019, that increased efficiency had not been measured. Peter Omara, Lawrence Aula, Fikayo Oyebiyi, and William R. Raun, "World Cereal Nitrogen Use Efficiency Trends: Review and Current Knowledge," Agrosystems, Geosciences & Environment 2, no. 1 (2019): 1–8, https://acsess.onlinelibrary.wiley.com/doi/10.2134/age2018.10.0045. See also: Nitrogen Facts, "World Nitrogen Use Efficiency for Cereal Production is 33%," Nitrogen Use Efficiency, Oklahoma State University, https://www.nue.okstate.edu/Crop_Information/Nitrogen_Facts1.htm.

1427 N.A. Campbell and J.B. Reece, *Biology* (Benjamin Cummings, 2005), 134.

1428 Timothy Rooks, "How Humboldt Put South America on the Map," Deutsche Welle, July 12, 2019, https://www.dw.com/en/how-scientist-alexander-von-humboldt-put-spanish-south-america-on-the-global-map/a-46693502. Deutsches Historisches Museum, Alexander Von Humboldt and The Ascent Of Chimborazo, https://www.dhm.de/blog/2016/11/17/humboldt-and-the-ascent-of-chimborazo. David Jim Nemeth, "Humboldt In North Africa," https://www.researchgate.net/publication/281652759_Humboldt_in_North_Africa

1429 Timothy Rooks, "How Humboldt Put South America on the Map."

1430 Gerd Kohlhepp, "Scientific Findings of Alexander von Humboldt's Expedition into the Spanish-American Tropics (1799-1804) from a Geographical Point of View," *Annals of the Brazilian Academy of Sciences* 77, no. 2 (June 2005), https://www.scielo.br/j/aabc/a/sLpM4QJZWhWKbgtCmVPdQNv/?lang=en#.

1431 Patrick Wildermann, "At the Easternmost Point of His Life," Alexander Von Humboldt Foundation (Alexander Von Humboldt Stiftung), 10 May 2019, https://www.humboldt-foundation.de/en/explore/alexander-von-humboldt/at-the-easternmost-point-of-his-life.

1432 David Kidd, Kingston University London, "The Humboldt Map: Natural Features and Human Artefacts that commemorate the great explorer and scientist Alexander von Humboldt (1768-1859)," StoryMaps, ArcGIS StoryMaps, 10 February 2020, https://storymaps.arcgis.com/stories/79eeffa9f54d429687c-17fa8267d3ba2.

1433 4877 Humboldt Asteroid Facts, Universe Guide, https://www.universe-guide.com/asteroid/8103/humboldt. Mark Andrew Holmes, "Humboldt," Mark Andrew Holmes' Personal Webpage, http://markandrewholmes.com/humboldt.html.

1434 Stanley Finger, Marco Piccolino, and Frank W. Stahnisch, "Alexander Von Humboldt: Galvanism, Animal Electricity, and Self-Experimentation Part 1: Formative Years, Naturphilosophie, and Galvanism," *Journal of the History of the Neurosciences* 22, no. 3 (2013): 225–260.

1435 Ewald Schnug, Frank Jacobs and Kirsten Stöven, "Guano: The White Gold of the Seabirds," In Heimo Mikkola, editor, Seabirds, United Kingdom: IntechOpen (5 November 2018): 79-100, https://www.intechopen.com/chapters/62618.

1436 Christopher Wasmuth, "A Name to Conjure With," Alexander Von Humboldt Stiftung (Alexander Von Humboldt Foundation), 2019, https://www.humboldt-foundation.de/en/explore/alexander-von-humboldt/a-name-to-conjure-with.

1437 "Humboldt Big-Eared Brown Bat," Animalia, https://animalia.bio/humboldt-big-eared-brown-bat.

1438 The Humboldt Orchid Society, https://www.humboldtorchids.org.

1439 California Native Plant Society, "Humboldt's Lily," Calscape, https://calscape.org/Lilium-humboldtii-().

1440 Christopher Wasmuth, "A Name to Conjure With," https://www.humboldt-foundation.de/en/explore/alexander-von-humboldt/a-name-to-conjure-with.

1441 R. Rosa, et al., "Dosidicus Gigas, Humboldt Squid," British Antarctic Survey, in NERC Open Research Archive, Natural Environment Research Council, UK, https://nora.nerc.ac.uk/id/eprint/500367.

1442 Johan Östling, *Humboldt and the Modern German University: An Intellectual History* (Sweden: Lund University Press, 2018).

1443 Gordon Chancellor, "Humboldt's Personal Narrative and Its Influence on Darwin," *Darwin Online*, http://darwin-online.org.uk/EditorialIntroductions/Chancellor_Humboldt.html

1444 You can find all of Humboldt's travel diaries at Carmen Götz and Ulrike Leitner (eds.), Edition Humboldt Digital, Version 8, 11 May 2022, Berlin-Brandenburg Academy of Sciences, https://edition-humboldt.de/reisetagebuecher/index.xql

1445 Paul F. Johnston, "The Smithsonian and the 19th-Century Guano Trade: This Poop is Crap," National Museum of American History, Behring Center, 31 May 2017, https://americanhistory.si.edu/blog/smithsonian-and-guano.

1446 Jodie King, "The Usefulness of Penguin Poo," Penguins International, 9 February 2020, https://www.penguinsinternational.org/the-usefulness-of-penguin-poo.

1447 P. Rodrigues and J. Micael, "The Importance of Guano Birds to the Inca Empire and the First Conservation Measures Implemented by Humans," *Ibis* 163 (2021): 283–291, https://doi.org/10.1111/ibi.12867.

1448 Ewald Schnug, Frank Jacobs and Kirsten Stöven, "Guano: The White Gold of the Seabirds," https://www.intechopen.com/chapters/62618

1449 Cara Giaimo, "When The Western World Ran on Guano," *Atlas Obscura*, 14 October 2015, https://www.atlasobscura.com/articles/when-the-western-world-ran-on-guano.

1450 Schnug, E., et al. "Guano: The White Gold of the Seabirds," 2018, https://www.intechopen.com/chapters/62618.

1451 Vaclav Smil, *Enriching the Earth: Fritz Haber, Carl Bosch, and the Transformation of World Food Production* (MIT Press, 2001).

1452 Anthony S. Travis, "Dirty Business: What Happened Before Humans Could Produce Fertilizer from the Air Itself, Courtesy of the Haber-Bosch Process?" *Distillations Magazine*, Science History Institute, 16 April 2013, https://www.sciencehistory.org/distillations/dirty-business.

1453 Gregory T. Cushman, *Guano and the Opening of the Pacific World: A Global Ecological History* (Cambridge University Press, 2013), 31.

1454 Ibid.

1455 Schnug, E., et al. "Guano: The White Gold of the Seabirds," 2018, https://www.intechopen.com/chapters/62618.

1456 Anne M. Streich, Kim A. Todd, "Classification and Naming of Plants,' University of Nebraska, Lincoln, 2014, https://alec.unl.edu/documents/cde/2017/natural-resources/classification-and-naming-of-plants.pdf.

1457 "19th Century Steamships," Bureau of Ocean Energy Management, U.S. Department Of The Interior, https://www.boem.gov/environment/19th-century-steamships.

1458 Jonathan Watts, "Vaclav Smil: 'Growth Must End. Our Economist Friends Don't Seem to Realize That,'" *The Guardian*, 21 Sep. 2019, https://www. theguardian.com/books/2019/sep/21/vaclav-smil-interview-growth-must-end-economists.

1459 Smil, *Enriching the Earth*.

1460 George C. Braverman, *The Fertilizer Industry: Mergers, Acquisitions, and Future Trends* (Boca Raton: CRC Press, 2018). Gregory T. Cushman, *Guano and the Opening of the Pacific World: A Global Ecological History* (Cambridge University Press, 2013). Douglas W. Allen, *The Nature of the Farm: Contracts, Risk, and Organization in Agriculture* (Cambridge, MA: The MIT Press, 1991).

1461 Cara Giaimo, "When The Western World Ran on Guano," *Atlas Obscura*, 14 October 2015, https://www.atlasobscura.com/articles/when-the-western-world-ran-on-guano.

1462 Ibid.

1463 Thomas Hager, *The Alchemy of Air: A Jewish Genius, a Doomed Tycoon, and the Scientific Discovery That Fed the World but Fueled the Rise of Hitler* (Broadway Books, 2008).

1464 Ross D. Milton, et al., "Bioelectrochemical Haber–Bosch Process: An Ammonia-Producing H2/N2 Fuel Cell," *Angewandte Chemie International Edition* 56, no. 10 (2017): 2680–2683. Paul Gabrielsen, "Flipping the Switch on Ammonia Production," Department of Chemistry, University of Utah, https://chem. utah.edu/news/ammonia-production.php.

1465 Hager, *The Alchemy of Air*, 176.

1466 American Experience, "The Man Who Tried To Feed The World, The Green Revolution: Norman Borlaug and the Race to Fight Global Hunger," 3 April 2020, PBS, https://www.pbs.org/wgbh/americanexperience/features/green-revolution-norman-borlaug-race-to-fight-global-hunger.

1467 Vikram Singh Gaur, Giresh Channappa, Mridul Chakraborti, Tilak Raj Sharma, Tapan Kumar Mondal, "'Green Revolution' Dwarf Gene Sd1 of Rice Has Gigantic Impact," *Briefings in Functional Genomics* 19, no. 5–6 (Sep.–Nov. 2020): 390–409, https://doi.org/10.1093/bfgp/elaa019.

1468 Patrick Kilby, *The Green Revolution: Narratives of Politics, Technology and Gender* (Taylor & Francis, 2019).

1469 Prabhu L. Pingali, "Green Revolution: Impacts, Limits, and the Path Ahead," *Proceedings of the National Academy of Sciences* 109, no. 31 (31 July 2012): 12302–12308, https://www.pnas.org/doi/full/10.1073/pnas.0912953109.

1470 William L. Cavert, "The Technological Revolution in Agriculture, 1910–1955 (In Part with Special Reference to the North Central States)," *Agricultural History* 30, no. 1 (Jan. 1956): 18–27, https://www.jstor.org/stable/3739967.

1471 Christine Clark, "Green Revolution Saved Over 100 Million Infant Lives in Developing World, Yet Could Go Further," *UC San Diego Today*, 17 December 2020, https://today.ucsd.edu/story/green-revolution-saved-over-100-million-infant-lives-in-developing-world-yet-could-go-further.

1472 G. Federico, *Feeding the World: An Economic History of Agriculture, 1800-2000* (Princeton University Press, 2010), 1.

1473 Elizabeth Pennisi, "Reforestation means more than just planting trees: Scientists are figuring out the best strategies to regrow lost forests," *Science*, 22 November 2022, https://www.science.org/content/article/reforestation-means-just-planting-trees.

1474 Hannah Ritchie, "Deforestation and Forest Loss," 2021, OurWorldInData. org, https://ourworldindata.org/deforestation.

1475 "Trillion Tree Campaign," *Wikipedia*, https://en.wikipedia.org/wiki/Trillion_Tree_Campaign.

1476 UN Environment Programme, "Plant for the Planet: The Billion Tree Campaign," 10 September 2008, https://www.unep.org/resources/publication/plant-planet-billion-tree-campaign.

1477 Plant for the Planet, Trillion Tree Campaign, https://www.trilliontreecampaign.org.

1478 The Bonn Challenge, https://www.bonnchallenge.org.

1479 Sean DeWitt, Sarah Weber and Mamadou Diakhite, "African Countries Aim to Restore 100 Million Hectares of Degraded Land," World Resource Institute, 6 December 2015, https://www.wri.org/insights/african-countries-aim-restore-100-million-hectares-degraded-land.

1480 A. Krause, et al., "Quantifying the Impacts of Land Cover Change on Gross Primary Productivity Globally," *Scientific Reports* 12, no. 1 (Nov. 2022), https://doi.org/10.1038/s41598-022-23120-0.

1481 Vaclav Smil, "Harvesting the Biosphere: The Human Impact," *Population and Development Review* 37, no. 4 (December 2011): 613–36, http://www.jstor.org/stable/41762374. Other sources on biomass: E.C. Ellis and N. Ramankutty, N. "Putting People in the Map: Anthropogenic Biomes of the World," *Frontiers in Ecology and the Environment* 6, no. 8 (2008): 439–447, https://doi.org/10.1890/070062. Steffen, Will, Wendy Broadgate, Lisa Deutsch, Owen Gaffney, and Cornelia Ludwig, "The Trajectory of the Anthropocene: The Great Acceleration," *The Anthropocene Review* 2, no. 1 (2015): 81–98. Wil Roebroeks, Katharine MacDonald, Fulco Scherjon, Corrie Bakels, Lutz Kindler, Anastasia Nikulina, Eduard Pop, Sabine Gaudzinski-Windheuser, "Landscape Modification by Last Interglacial Neanderthals," *Science Advances* 7, no. 51 (2021). Thompson, Jessica C., David K. Wright, Sarah J. Ivory, Jeong-Heon Choi, Sheila Nightingale, Alex Mackay, Flora Schilt et al., "Early Human Impacts and Ecosystem Reorganization in Southern-Central Africa," *Science Advances* 7, no. 19 (5 May 2021), https://www.science.org/doi/full/10.1126/sciadv.abf9776. P.H. Raven and G.B. Johnson, *Biology* (McGraw-Hill Education, 2014), 112. N.A. Campbell, J.B. Reece, *Biology* (7th ed.) (Benjamin Cumming, 2005), 134.

1482 Michael J. Roberts, et al., "Measurement of Plant Biomass and Net Primary Production," in *Techniques in Bioproductivity and Photosynthesis* (Pergamon, 1985): 1–19. See also: Bernhard Sonnleitner, Georg Locher, and Armin Fiechter, "Biomass Determination," *Journal of Biotechnology* 25, no. 1–2 (1992): 5–22.

1483 K. Niklas and B. Enquist, "Invariant Scaling Relationships for Interspecific Plant Biomass Production Rates and Body Size," *Proceedings of the National Academy of Sciences of the United States of America* 98 (2001): 2922–2927, https://www.pnas.org/doi/full/10.1073/pnas.041590298, https://doi.org/10.1073/pnas.041590298.

1484 M. Sarrafzadeh, H. La, S. Seo, H. Asgharnejad, and H. Oh, "Evaluation of Various Techniques for Microalgal Biomass Quantification," *Journal of Biotechnology* 216 (2015): 90–97, https://www.sciencedirect.com/science/article/pii/S0168165615301589. K. Niklas and B. Enquist, "Invariant Scaling Relationships for Interspecific Plant Biomass Production Rates and Body Size," *Proceedings of the National Academy of Sciences of the United States of America* 98 (2001): 2922–2927, https://pubmed.ncbi.nlm.nih.gov/11226342/.

1485 Michael Yarus, "Getting Past the RNA World: The Initial Darwinian Ancestor," *Cold Spring Harbor Perspectives in Biology* 3, no. 4 (2011): https://doi.org/10.1101/cshperspect.a003590. John Hoober, "Evolutionary Aspects of Chloroplast Development," in *Chloroplasts* (Springer, 1984): 263–273, https://doi.org/10.1007/978-1-4613-2767-7_10. I.M. Naumov, J.E. Wilberger, and C.D. Keyes, "Beginning and End of Biological Life," in *New Harvest: Contemporary Issues in Biomedicine, Ethics, and Society*, ed. C.D. Keyes (Humana Press, Totowa, NJ, 1991): 31–56, https://doi.org/10.1007/978-1-4612-0489-3_3, https://rdcu.be/dBsHh. M. Schidlowski, "Application of Stable Carbon Isotopes to Early Biochemical Evolution on Earth," *Annual Review of Earth and Planetary Sciences* 15 (1987): 47–72, https://doi.org/10.1146/ANNUREV.EA.15.050187.000403. John Hoober, "Evolutionary Aspects of Chloroplast Development," in *Chloroplast* (Springer, 1984): 263–273, https://doi.org/10.1007/978-1-4613-2767-7_10. E.A. Bell, P. Boehnke, T.M. Harrison, and W.L. Mao, "Potentially Biogenic Carbon Preserved in a 4.1 Billion-Year-Old Zircon," *Proceedings of the National Academy of Sciences* 112, no. 47 (2015): 14518–14521, https://www.pnas.org/doi/abs/10.1073/pnas.1517557112.

1486 Donna Weaver, Ray Villard, Space Telescope Science Institute, "5,000 Light Year Long Jet of Superheated Gas Ejected From a Supermassive Black Hole," *SciTechDaily*, August 23, 2013, https://scitechdaily.com/5000-light-year-long-jet-of-superheated-gas-ejected-from-a-supermassive-black-hole.

1487 Pamela Lyon, Fred Keijzer, Detlev Arendt, and Michael Levin, "Reframing Cognition: Getting Down to Biological Basics," *Philosophical Transactions of the Royal Society B* 376, no. 1820 (2021), https://royalsocietypublishing.org/doi/10.1098/rstb.2019.0750.

1488 Rowan Jacobsen, "Brains Are Not Required When It Comes to Thinking and Solving Problems—Simple Cells Can Do It," *Scientific American*, February 1, 2024, https://www.scientificamerican.com/article/brains-are-not-required-when-it-comes-to-thinking-and-solving-problems-simple-cells-can-do-it.

1489 "Prof. Eshel Ben-Jacob Late, Physics," Tel Aviv University, https://english.tau.ac.il/profile/benjacob. "Eshel Ben-Jacob," *Wikipedia*, https://en.wikipedia.org/wiki/Eshel_Ben-Jacob.

1490 Israel Physical Society, https://www.israelphysicsociety.org

1491 Eshel B. Ben-Jacob and Herbert Levine, "The Artistry of Microorganisms: Colonies of Bacteria or Amoebas Form Complex Patterns That Blur the Boundary Between Life and Nonlife," *Scientific American*, 1 October 1998, https://www.scientificamerican.com/article/the-artistry-of-microorganisms/.

1492 Eshel B. Ben-Jacob, Israela Becker, Yoash Shapira, Herbert Levine, "Bacterial Linguistic Communication and Social Intelligence," *Trends in Microbiology* 12, no. 8 (1 August 2004): 366–372, https://doi.org/10.1016/j.tim.2004.06.006. Eshel B. Ben-Jacob, A. Tenenbaum, O. Shochet, I. Cohen, A. Czirók, and T. Vicsek, "Communication, Regulation, and Control During Complex Patterning of Bacterial Colonies," Fractals 2, no. 1 (1994): 14–44. Eshel B. Ben-Jacob, "Bacterial Wisdom, Gödel's Theorem, and Creative Genomic Webs," *Physica A* 248 (1998): 57–76.

1493 Microbes Mind Forum, "Learning from Bacteria about Social Networks," Undated talk by Eshel Ben-Jacob, Google Headquarters, Mountain View, CA, https://microbes-mind.net/ben-jacob.

1494 Ibid.

1495 Eshel Ben-Jacob, "Learning from Bacteria about Natural Information Processing," *Annals of the New York Academy of Sciences*, 15 October 2009, https://doi.org/10.1111/j.1749-6632.2009.05022.x.

1496 Inbal Hecht, Sari Natan, Assaf Zaritsky, Herbert Levine, Ilan Tsarfaty, & Eshel Ben-Jacob, "The Motility-Proliferation Metabolism Interplay during Metastatic Invasion," *Scientific Reports*, 4 September 2015, https://www.nature.com/articles/srep13538.

1497 Eshel B. Ben-Jacob, "Learning from Bacteria About Natural Information Processing," https://doi.org/10.1111/j.1749-6632.2009.05022.x. N. Steinberg and I. Kolodkin-Gal, "The Matrix Reloaded: How Sensing the Extracellular Matrix Synchronizes Bacterial Communities," *Journal of Bacteriology* 197, no. 13 (2015): 2092–2103, https://doi.org/10.1128/JB.02516-14.

1498 S. Sonea and L. Mathieu, "Evolution of the Genomic Systems of Prokaryotes and Its Momentous Consequences," *International Microbiology* 4 (2001): 67–71, https://doi.org/10.1007/S101230100015. H. Felbeck and G. Somero, "Primary Production in Deep-Sea Hydrothermal Vent Organisms: Roles of Sulfide-Oxidizing Bacteria," *Trends in Biochemical Sciences* 7 (1982): 201–204, https://doi.org/10.1016/0968-0004(82)90088-3. "From Soup to Cells: The Origin of Life," UC Museum of Paleontology, University of California, Berkeley, 15 May 2017, https://evolution.berkeley.edu/from-soup-to-cells-the-origin-of-life/where-did-life-originate. Rachel Brazil and Chemistry World, "Life's Origins by Land or Sea? Debate Gets Hot, Volcanic Springs and Deep-Ocean Vents Get New Evidence," *Scientific American*, 15 May 2017, https://www.scientificamerican.com/article/lifes-origins-by-land-or-sea-debate-gets-hot.

1499 Lawrence S. Barham, "Systematic Pigment Use in the Middle Pleistocene of South-Central Africa," *Current Anthropology* 43 (February 2002): 182–190. Rudolf Botha and Chris Knight, eds., *The Cradle of Language* (Oxford: Oxford University Press, 2009): 2. Thomas H. Maugh, "Early Humans Found to Use Makeup, Tools," *Los Angeles Times*, 20 October 2007, https://www.latimes.com/archives/la-xpm-2007-oct-20-sci-humans20-story.html.

1500 Chukwunyere Kamal, "The Ishango Bone: The World's First Known Mathematical Sieve and Table of the Small Prime Numbers," AfricArXiv Preprints, 28 February 2021, https://osf.io/preprints/africarxiv/6z2yr.

1501 UNESCO, "Global Number of Museums 2021, by UNESCO Regional Classification," Statista Research Department, 39 March 2023, https://www.statista.com/statistics/1201800/number-of-museums-worldwide-by-region.

1502 Jody Schmidt, "How Many Paintings Exist in the World," Wet Canvas, Artistsnetwork, August 20, 2013, https://www.wetcanvas.com/forums/topic/how-many-paintings-exist-in-the-world.

1503 30,000 symphonies is a rough estimate from Anthropic's ai Claude. Google's Gemini timidly states that there have been "thousands." According to Indiana State University Professor of Musicology A. Peter Brown "The First Golden Age of the Viennese Symphony includes over 170 symphonies by" just four composers, "Haydn, Mozart, Beethoven, and Schubert." A. Peter Brown, "The First Golden Age of the Viennese Symphony: Haydn, Mozart, Beethoven, and Schubert," in *The Symphonic Repertoire, Vol. 2* (Indiana University Press, 2002).

1504 Niall, "How Many Official Songs Are There in the World?," BigTimeMusicians.com, 30 November 2023, https://bigtimemusicians.com/how-many-official-songs-are-there-in-the-world.

1505 Extrapolated very roughly from the number of languages we know of. There have been a mere 3,814 cultures according to Price's Atlas of Ethnographic Societies, cited in R.A. Foley and M.M.Lahr, "The Evolution of the Diversity of

Cultures," *Philosophical Transactions of the Royal Society B: Biological Science* 366, no. 1567 (April 2011): 1080–1089, https://doi.org/10.1098/rstb.2010.0370.

1506 Richard Armstrong, "Language Death," *Engines of Our Ingenuity, No. 2723,* Cullen College Of Engineering, University of Houston, https://www.uh.edu/engines/epi2723.htm.

1507 In 2021 alone, "16.5 million patents were in force." Meaning at least 16.5 mllion inventions. New patents were being filed at the rate of 3.4 million per year. See: "World Intellectual Property Indicators 2022," World Intellectual Property Organization, Geneva, Switzerland, 2022, https://www.wipo.int/edocs/pubdocs/en/wipo-pub-941-2022-en-world-intellectual-property-indica-tors-2022.pdf.

1508 Pierre Teilhard de Chardin, *The Phenomenon of Man* (Paris: Éditions du Seui, 1955).

1509 Petroc Taylor, "Total Installed Base of Data Storage Capacity in Global Datasphere 2020-2025," Statista, September 2021, https://www.statista.com/statistics/1185900/worldwide-datasphere-storage-capacity-installed-base/

1510 David Reinsel, John Gantz, John Rydning, "Data Edge 2025, the Digitiza-tion of the World: From Edge to Core," November 2018, Seagate white paper, https://www.seagate.com/files/www-content/our-story/trends/files/idc-sea-gate-dataage-whitepaper.pdf.

1511 This calculation comes from Barak Shoshany, Assistant Professor of Physics at Brock University in Saint Catherines, Ontario. See: "If the Entire Internet Were Printed in a Single Book, How Many Pages Would It Be?" *Quora,* 2015, https://www.quora.com/If-the-entire-internet-were-printed-in-a-single-book-how-many-pages-would-it-be Shoshany is the author of physics publications like: Freidel, Laurent, Florian Girelli, and Barak Shoshany, "2+ 1 D loop quantum gravity on the edge," *Physical Review D* 99, no. 4 (2019).

1512 Isaac L. Auerbach, "A static magnetic memory system for the ENI-AC," Association for Computing Machinery, ACM '52: *Proceedings of the 1952 ACM national meeting (Pittsburgh)* May 1952) 213–222, https://doi.org/10.1145/609784.609813.

1513 Maddie Stone, "Human-Made Materials Now Equal Weight of All Life on Earth," *National Geographic,* 9 December 2020, https://www.nationalgeograph-ic.com/environment/article/human-made-materials-now-equal-weight-of-all-life-on-earth.

1514 Marcos Hassan, "Remembering Queen's Infamous and History-Making Tour of South America," *Remezcla,* 21 November 2018, https://remezcla.com/features/music/we-remember-queens-infamous-tour-of-latin-america.

1515 Robin Fox, *Participant Observer: A Memoir of a Transatlantic Life* (Rout-ledge, 2018).

1516 Richard Sandomir, "Gerry Stickells, Who Helped Make Rock Shows Big, Dies at 76," *New York Times*, 3 April 2019, https://www.nytimes.com/2019/04/03/obituaries/gerry-stickells-dead.html.

1517 "Queen on Tour: S. America Bites The Dust 1981," QueenConcerts, https://www.queenconcerts.com/live/queen/1981-southam.html.

1518 Jacky Smith and Jim Jenkins, "What Happened When Queen Conquered South America," *Louder*, 31 January 2022, https://www.loudersound.com/features/what-happened-when-queen-conquered-south-america.

1519 Emile Durkheim, *The Elementary Forms of Religious Life* (Routledge, 2016), 52–67. Originally published 1912.

1520 Maaike Goudriaan, Victor Hernando Morales, Marcel T.J. van der Meer, Anchelique Mets, Rachel T. Ndhlovu, Johan van Heerwaarden, Sina Simon, Verena B. Heuer, Kai-Uwe Hinrichs, Helge Niemann, "A Stable Isotope Assay with 13C-Labeled Polyethylene to Investigate Plastic Mineralization Mediated by Rhodococcus Ruber," *Marine Pollution Bulletin* 186 (2023): https://doi.org/10.1016/j.marpolbul.2022.114369. Sabrina Weiss, "The Harmful Side Effect of Cleaning Up the Ocean: Patches of Floating Plastic Are Teeming with Life, and Cleanup Companies Hauling Trash Out of the Water Risk Destroying a Marine Habitat," *Wired*, May 2023, https://www.wired.com/story/ocean-cleanup-habitat-destruction. Jaydev Misraa and Priyadarshini Mallick, "Managing Wastewater Using Plastic-Eating Bacteria–A Sustainable Solution for Sewage-Fed Fisheries," *Journal of the Indian Chemical Society* 97 (2020): 513–519. Christina Reed, "The Other Plastic Plague," *New Scientist* 239, no. 3186 (2018): 37–39. Sukla Ghosh, "Plastic and Environment," *Journal of Social Science and Welfare* 4 (2019): 15–24, https://www.womenscollegekolkata.ac.in/uploads/1689761244_jssw-2019.pdf#page=21. Carly Cassella, "Bacteria Can Use Plastic Waste as a Food Source, Which Isn't as Good as It Sounds," *Science Alert*, 25 January 2023, https://www.sciencealert.com/bacteria-can-use-plastic-waste-as-a-food-source-which-isnt-as-good-as-it-sounds.

INDEX